SETON HALL UNIVERSITY
QD400 .A16 1976
An introduction to the chemistry
MAIN
3 3073 00252746 1

D1116467

AN INTRODUCTION TO THE CHEMISTRY OF HETEROCYCLIC COMPOUNDS

Third Edition

An Introduction to the Chemistry of Heterocyclic Compounds

R. M. ACHESON

Fellow and Tutor in Chemistry
The Queen's College, Oxford

Third Edition

A Wiley–Interscience Publication
JOHN WILEY & SONS
New York • London • Sydney • Toronto

Library of Congress Cataloging in Publication Data:

Acheson, Richard Morrin.
An introduction to the chemistry of heterocyclic compounds.

"A Wiley-Interscience publication."
Includes bibliographical references and indexes.
1. Heterocyclic compounds. I. Title.
QD400.A16 1977 547'.59 76-21319
ISBN 0-471-00268-2

Printed in the United States of America

10 9 8 7 6 5 4 3 2

To
my B.T. and 3 L.T's

PREFACE TO THE FIRST EDITION

In recent years many specialised works on specific topics in heterocyclic chemistry have appeared, while at the same time the more general text-books on organic chemistry have not given this important part of the subject the attention it deserves. The specialised works, while valuable to research workers and occasionally to advanced students, contain so much detail that it is virtually impossible for the average undergraduate to separate the wheat from the chaff and to recognise even the major features of heterocyclic chemistry in the time available to him. This time is proportionately too little in most teaching schools when it is considered that heterocyclic chemistry at the moment accounts for about 26% of the organic chemistry papers published. An attempt has been made, in this work, to present to the student a concise account of the more important properties, and chemical reactions, of the basic heterocyclic systems with which he should have some acquaintance. This account, if it is to serve its purpose, cannot be complete. It is more difficult to decide on what to leave out than what to put in, and every writer has his own ideas on this matter. The present author is no exception, in consequence a great deal has been left out, including the sugars and alkaloids, which have been described at the appropriate level elsewhere. However, attempts have been made to include recent pertinent physical data, to use modern electronic and mechanistic concepts where possible, and to deal briefly with biochemical discoveries concerning the metabolism and biosynthesis of some important compounds. A general bibliography has been placed at the end of each chapter, references have been kept to a minimum, and review articles are quoted whenever possible.

I should be most grateful to know of errors, for which I accept sole reponsibility and which have inevitably crept into this book in spite of the vigilance of many of my friends and pupils. I thank Professor R. L. Huang, of the University of Malaya, for reading most of the manuscript, and Dr. M. J. T. Robinson and Dr. R. Brettle for reading the complete manuscript and offering most constructive criticism. I am also indebted to Dr. E. Schlittler and Dr. N. F. Taylor for reading the proofs and for making valuable suggestions. I thank Professor W. C. Gibson, of the University of British Columbia, for encouragement which was much appreciated during the early stages of writing, the staff of the Radcliffe Science Library for their help and co-operation, Mrs. M. Little for the typing, and the officers of Interscience Inc., who have made my task as easy as possible. I also thank my wife for her help at all stages of the book.

R. M. ACHESON

Department of Biochemistry
University of Oxford

PREFACE TO THE THIRD EDITION

In the last 10 years, the time since the second edition was prepared, many significant developments have taken place in organic chemistry, particularly in regard to mechanism. The investigation of mechanism is usually time consuming, is only possible in favourable cases, and has not been carried out for most of the chemical conversions described here. Mechanistic *speculation*, preferably on the basis of established close analogy, is of great value in debate, as a spur to further experiment, and as an aid to memory. It is very important, in my opinion, to distinguish between the speculation and mechanisms which have been reasonably established by experimental scrutiny, there being, of course, a large grey area in between. Established mechanisms are therefore included, while rather speculative mechanistic suggestions are generally avoided.

The flow of original papers and review articles in heterocyclic chemistry continues to increase yearly, and attempts have been made to include the more important work published up to mid-1975. It has proved increasingly difficult to keep up to date, and much material of the previous edition has been superseded and replaced. A small amount of quite new material, concerning eight- and nine-membered rings, has been included, but this has necessitated deletion of some less important subjects, as the book would lose its purpose if it grew too large. I will be very pleased, as in the past, if those who note errors or important omissions will kindly bring them to my attention, and I thank all those who have kindly corresponded about, or discussed, such matters with me in the last few years. In response to requests I have included a greater proportion of references to recent work.

I thank Professor C. A. Grob for the hospitality of the Institute of Organic Chemistry, The University of Basel, and Oregon State University for a Visiting Professorship, for during these periods much of the literature work was done. I am most grateful to Mr. P. J. Abbott and Miss M. B. Acheson for a great deal of help in many directions during the preparation of the book, and to Mrs. C. L. Lindsay-Scott for the typing. I also thank Mr. M. P. Acheson, Mr. J. P. Gibbs, Miss P. Lloyd, and Mr. G. Procter for assistance with the proofs and the index.

R. M. ACHESON

The Queen's College
and The Department of Biochemistry,
University of Oxford

CONTENTS

AN INTRODUCTION TO THE CHEMISTRY OF HETEROCYCLIC COMPOUNDS

Third Edition

INTRODUCTION AND NOMENCLATURE

A heterocyclic compound is one which possesses a cyclic structure with at least two different kinds of atoms in the ring. The most common types, discussed here, contain largely carbon atoms. Nitrogen, oxygen, and sulphur are the most common heteroatoms, but many other elements, including even bromine, can also serve. The heterocyclics containing the less common atoms have been subject to much investigation in recent years but are not considered in this book.

Heterocyclic compounds are very widely distributed in nature and are essential to life in various ways. Most of the sugars and their derivatives, including vitamin C, for instance, exist largely in the form of five-membered (furan) or six-membered (pyran) rings containing one oxygen atom. Most members of the vitamin B group possess heterocyclic rings containing nitrogen. One example is vitamin B_6 (pyridoxine), which is a derivative of pyridine essential in amino acid metabolism. Many other examples of the importance of heterocyclic compounds in biological systems can be given. Most of the alkaloids, which are nitrogenous bases occurring in plants, and many antibiotics, including penicillin, also contain heterocyclic ring systems. A large number of heterocyclic compounds, obtainable only by laboratory syntheses, have valuable properties as chemotherapeutic agents, drugs, dyestuffs, copolymers, etc.

Heterocyclic compounds can be aliphatic or aromatic in character, depending on their electronic constitution. In general, the aliphatic heterocyclics, where specific effects due to the constitution of the compound are excluded, are very similar chemically to their open-chain aliphatic analogues. For instance, tetrahydrofuran (**1**) has many properties characteristic of diethyl ether (**2**). In a

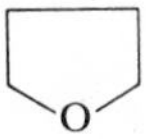

O — **1** | CH_3 CH_3 H_2C O CH_2 — **2** | N — **3**  | **4**

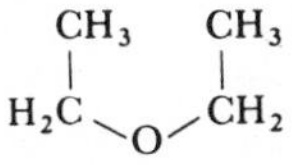

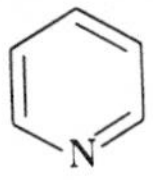

similar way the aromatic heterocyclic compounds have many properties resembling their aromatic carbocyclic analogues, a specific example being a comparison of pyridine (**3**) and benzene (**4**). In general, many well-developed syntheses of aromatic heterocyclic rings are available, in contrast to the benzene series, where very few practical syntheses of the aromatic ring itself are known. This is because, while the parent compound, benzene, is readily available and easy to substitute, and in consequence direct syntheses of the ring have attracted little attention, most unsubstituted heterocyclic systems are either difficult to obtain or are not susceptible to substitution.

There are several conventions for numbering the atoms and substituents in simple heterocyclic rings that are generally accepted and are used by *Chemical Abstracts*, the world's most important and useful chemical abstract journal and index. The atoms of a simple heterocyclic ring are numbered from the heteroatom, which is counted as 1. Substituents are given the lowest possible numbers and then are arranged in alphabetical order. For example, the chloromethylpyridine shown below could be numbered and described as 5-methyl-6-chloropyridine (**5**), but it is better numbered the other way round and called 2-chloro-3-methylpyridine (**6**). It should be found under the last name in most chemical indexes. In *Chemical Abstracts* it would be found under

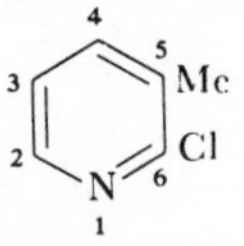

5 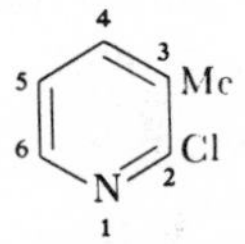 **6** 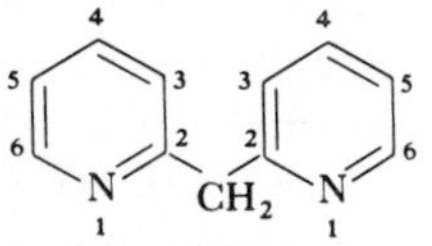 **7**

'pyridine, 2-chloro-3-methyl-,' as derivatives are indexed under what is arbitrarily considered the 'parent' part of the molecule. Compound **6** is therefore not usually known as a derivative of methane; it could be called 3-(2-chloropyridyl)-methane. However, in the case of **7**, when the pyridine ring might be considered as 'parent' and the compound named 2-(2-pyridylmethyl)pyridine, it is customary to treat the compound as a derivative of methane and to name it bis-2-pyridylmethane. Because of difficulties of this sort it is always advisable to look up the several alternative names, when ambiguity exists in the first instance, when conducting a literature search.

Where the heterocyclic ring contains more than one heteroatom the order of preference for position 1 is oxygen, sulphur, and then nitrogen. If there are two ways of numbering the ring the way which gives the second heteroatom the lowest possible number is chosen. Isoxazole (**8**) and thiazole (**9**) are therefore numbered as shown below.

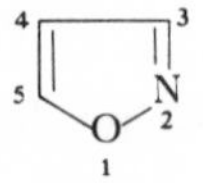

8 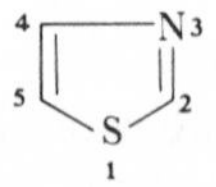 **9**

Nomenclature and numbering become more complicated for condensed or fused-ring systems, where a part of one ring is also a part of another. Generic terminology showing the genesis of the structures may be used. On this basis structure **11**, usually known by the trivial name of quinoline, may be called 2,3-benzopyridine, showing that a benzene ring has been fused onto the 2,3 side of the pyridine ring (**10**). Another system that is commonly used is to label the side of the heterocyclic system – *a, b, c,* . . . as shown for pyridine (**12**), starting from the atom number 1. Structure **11** can then be called benzo[*b*]pyridine. These types of systematic nomenclatures are often used for complex molecules, the last system being employed by *Chemical Abstracts*.

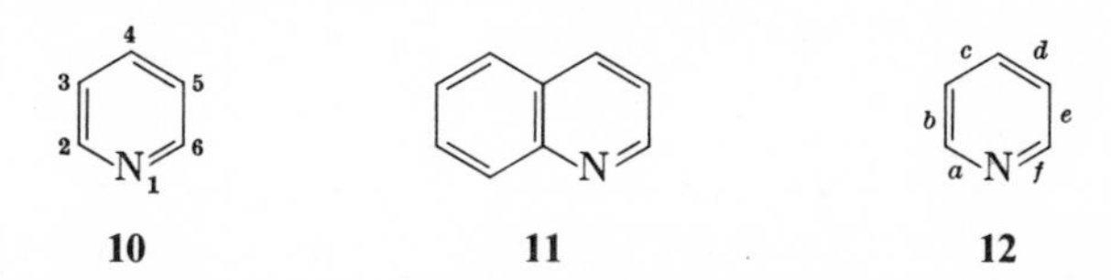

10 **11** **12**

Although general agreement has been reached for the naming and numbering of many heterocyclic compounds, agreement has not been reached with regard to others for which alternative names and numberings are used. This can lead to much difficulty in the use of original literature. It is therefore advisable to ascertain at an early stage which name is in vogue for the compound in the particular journal, when making a search, for the name can change from journal to journal, and occasionally different names are used for the same compound in the same journal. The *Ring Index* (Patterson, Capell, and Walker, American Chemical Society, New York, 1960) is a most useful compilation of mainly heterocyclic ring systems with their alternative names and numbers. It is often very useful to consult this reference work before making a search for less common heterocyclic compounds in the literature.

CHAPTER I

HETEROCYCLIC ANALOGUES OF CYCLOPROPANE

All three-membered rings have one major property in common – a strained ring that confers great reactivity on the compounds in comparison with their open-chain analogues. This strain is reflected in the compression of the normal bond angles and by a shortening of the bond lengths from normal, as is shown by modern physical methods. The presence of a double bond in the ring increases the strain, and the molecular dimensions of cyclopropane (**1**) and cyclopropene (**2**) are given here for comparison with those of the heterocyclic compounds described later in this chapter. The bonds between the atoms are

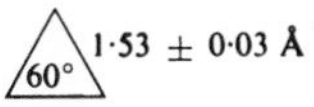

60° 1·53 ± 0·03 Å

1

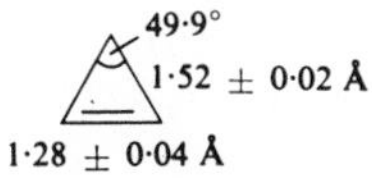

49·9° 1·52 ± 0·02 Å 1·28 ± 0·04 Å

2

$Me(CH_2)_7$ — $(CH_2)_7CO_2H$

3, Sterculic acid

drawn straight, although it is suspected that a bent line in the form of an arc would better represent the electronic distribution in these molecules.[1] A cyclopropene derivative, sterculic acid (**3**), occurs in fats used as foodstuffs.

Three-membered rings containing a nitrogen, oxygen, or sulphur atom as the heteroatom are called aziridine, oxirane, and thiirane, respectively. While the bonds between the three heteroatoms and the carbon atoms are a trifle longer than corresponding bonds in an unstrained molecule (e.g., dimethyl ether) the carbon–carbon bonds are all very similar in length but are substantially shorter than those of cyclopropane (**1**). The rings of these compounds are much more easily opened than that of cyclopropane, since the heteroatoms facilitate the attack of ionic and free-radical reagents. The synthesis of three-membered heterocyclic rings from appropriate three-atom intermediates is on the whole quite easy, for although the 'strain energy' of the molecule must be provided, the atoms that must combine are largely oriented so that cyclization is preferred to intermolecular reaction and polymerization.

Molecular orbital (MINDO) calculations[2] suggest that 2-azirine (**4**) (the position from which the double bond starts being numbered), and oxirene and thiirene where the CH_2 group of cyclopropene is replaced by an oxygen or sulphur atom, respectively, should be stable enough for isolation, even though the possession of four π-electrons could bring them into the 'antiaromatic' class of compounds, which includes cyclobutadiene. In the last 10 years or so this class of compound has attracted much research effort, culminating in the detection of derivatives of all three ring systems as reaction intermediates. 1-Azirines (**5**) have also been detected as intermediates in a number of thermal and photochemical reactions and are relatively stable compounds of considerable synthetic potential. The formal abstraction of a hydride ion from these heterocycles gives cations that have two π-electrons and are isoelectronic with the 'aromatic' cyclopropenium cation. Calculations[3] suggest that the azirinylium cation (**6**) might be stable enough to exist.

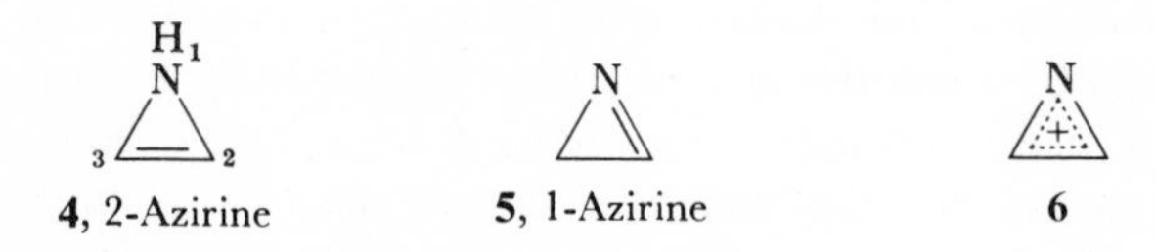

4, 2-Azirine **5**, 1-Azirine **6**

1. AZIRINES

A. 1-Azirines

1. *Chemical Properties*

The reactions of azirines are characterised by the ring strain, which is probably over 170 kJ (40·5 kcal)/mole and which leads to ring opening, the low basicity and relative stability to mineral acid, and the tendency of the double bond to undergo additional reactions similar to those of substituted imines. Their low basicity, which discourages reactions at physiological pH's, makes these compounds much more pleasant to handle than aziridines.

Electrophilic reagents attack 1-azirines at the nitrogen atom. 3-Methyl-2-phenyl-1-azirine (**7**) does not[4] dissolve in aqueous 2M hydrochloric acid but does in the 10M acid. The ring is relatively stable to acids as indicated by the fact that about half may be recovered after the solution is left standing for 5 minutes at room temperature and is made basic. This azirine with benzoyl chloride at 0° yields *cis*- and *trans*- **8** in roughly equal proportions.[4] Pyridinium perchlorate and 4-toluenesulphinic acid with 3,3-dimethyl-2-phenyl-1-azirine

H N + Ph Me ⇌ (H⁺ / base) Ph N Me (**7**) —PhCOCl→ COPh N Ph Me Cl (**8**)

(**10**) yield the aziridines **9** and **11**, respectively, acid-catalysed addition to the double bond having taken place.

H
N
Ph Me ⟵ N Ph Me Me ⟶ H N Ph Me
N Me MeC$_6$H$_4$SO$_2$ Me
+ ClO_4^-

9 **10** **11**

Neber in 1932 isolated an unstable intermediate in one of his syntheses of α-amino ketones from oxime 4-toluenesulphonates and considered that it must be the 1-azirine **13**. In 1953 Cram and Hatch[5] confirmed this conclusion and reawakened interest in unsaturated small-ring heterocycles. The oxime derivative, on successive treatment with pyridine and sodium carbonate, gave the 1-azirine (**13**), presumably via the aziridine **14**, which is similar to **9** mentioned above. This 1-azirine with acetic anhydride–acetic acid was converted to the acylamino ketone **15**, which can also be obtained from the oxime (**12**) by treatment with pyridine followed by acetic acid–acetic anhydride. The opening of 1-azirine rings in this way is a general method for making α-acetamido ketones, but the synthesis of azirines by Neber's route requires an acidic hydrogen atom in the oxime derivative **12**.

N–OTs
‖
RCH$_2$CMe

12

↓ $-H^+$

[RC̄HC(=N–OTs)Me] —TsO⁻⟶ 1-azirine (N 1, C 3, C 2) R, Me **13** —pyridinium TsO⁻⟶ / ⟵Na$_2$CO$_3$— H N R Me N+ (pyridinium) TsO⁻ **14**

↓ Ac$_2$O, AcOH

[Ac N R Me OAc] ⟶ NHAc R–CH–COMe **15**

R = 2,4-dinitrophenyl

It is interesting that the 1,3 bond of the azirine **13** was broken on hydrogenation over palladium on charcoal in acetic anhydride and the *cis* and *trans* isomers of **16** were obtained; on hydrolysis both gave 2,4-dinitrophenylacetone (**17**). Lithium aluminium hydride reduction of **13** gave a poor yield of the aziridine **18**, presumably with the substituents *cis*, as is usually observed[6] with this sort of reduction. It is noteworthy that in these reductions the double bond of the azirine ring is reduced in preference to the nitro groups. The double

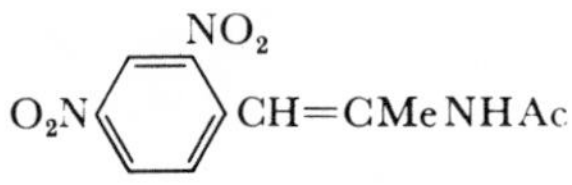

O_2N–$C_6H_3(NO_2)$–CH=CMeNHAc

16

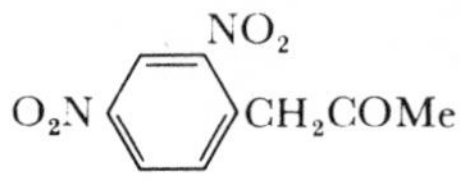

O_2N–$C_6H_3(NO_2)$–CH_2COMe

17

2,4-$(O_2N)_2C_6H_3$-aziridine (NH) with Me

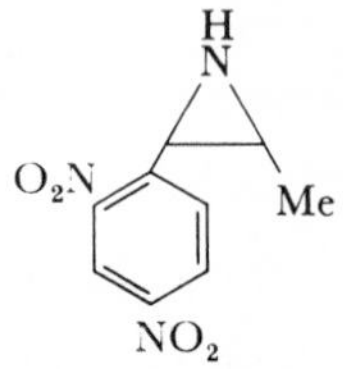

18

bond of the 1-azirine **7** can react[7] with dienes in normal Diels–Alder reactions and accepts[8] a methylene group from dimethylsulphonium methylide to give the first 1-aza[1.1.0]bicyclobutane known (**19**). Phenyl magnesium bromide adds[4] to the C=N of **7** in the same way as to a carbonyl group, giving **20**; this last

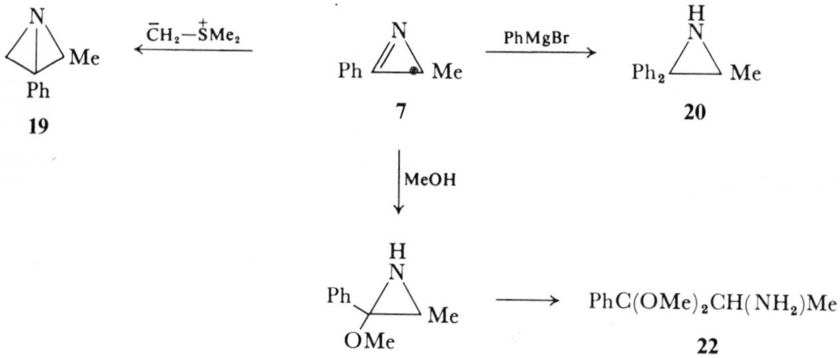

19 ← $\bar{C}H_2$–$\overset{+}{S}Me_2$ — **7** (Ph, Me azirine) — PhMgBr → **20** (Ph_2, Me aziridine, NH)

7 — MeOH ↓ → **21** (Ph, OMe, Me aziridine, NH) ⟶ $PhC(OMe)_2CH(NH_2)Me$ **22**

compound is also obtained from the chloroaziridine **8** with the same Grignard reagent. The azirine **7** adds methanol in the presence of a catalytic amount of sodium methoxide to give the alkoxyaziridine (**21**), and continuation of the reaction causes ring opening to the aminoketone acetal **22** (see p. 9).

The thermolysis of 2,3-diphenyl-1-azirine (**23**) gives[9] a 75% yield of phenylindole (**24**), possibly via a nitrene; phenyl cyanide is not an intermediate. In the presence of the thiazoledione (**25**), which decomposes as shown, a symmetry allowed [$\pi 4s + \pi 2s$] cycloaddition is thought[10] to occur, yielding **26**.

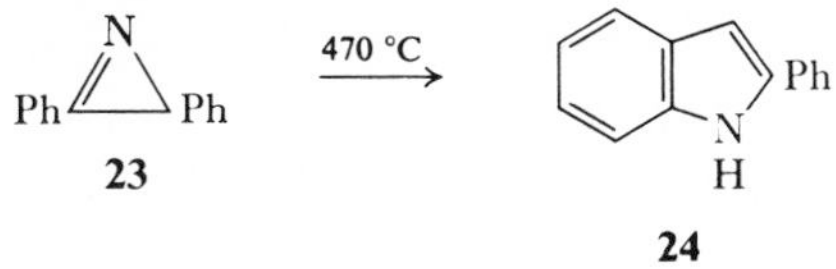

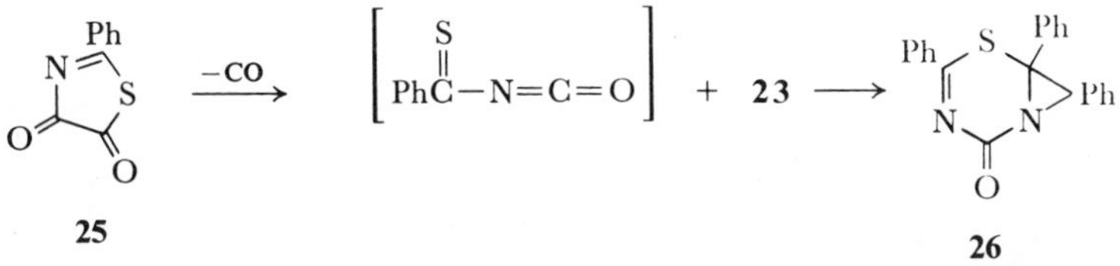

N
Ph Ph
23
470 °C →
Ph
N
H
24

Ph
N S
O O
25
−CO →
[S ‖ PhC—N=C=O] + 23 →
Ph S Ph
N N Ph
O
26

The photolysis of azirines breaks the ring and leads to charged intermediates[11] that are stable at low temperatures. For example,[12] the azirine **23** at −196°C with 250 nm radiation opens to the ylid **27**, which cyclises again on irradiation at 350 nm. Warming the ylid to −150°C causes reaction with unchanged azirine to form **28**. The ylid (**27**) undergoes normal cycloaddition reactions in the expected orientation with activated olefines and carbonyl compounds. It also undergoes a remarkable addition to the carbonyl group of methyl trifluoroacetate to give **29** with a small proportion of the geometrical isomer. Many heterocyclic photochemical rearrangements involve azirines (Chapter VII).

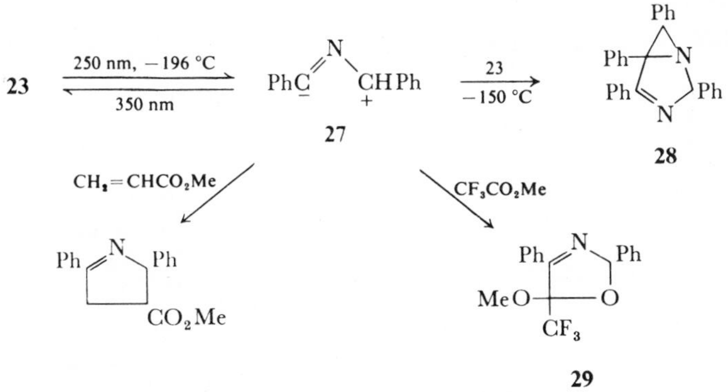

23 ⇌ (250 nm, −196 °C / 350 nm) $Ph\underline{C}$ =N— CHPh (+)
27
23, −150 °C →
Ph
Ph N
Ph N Ph
28

CH_2=$CHCO_2Me$ ↙
Ph N Ph
CO_2Me

CF_3CO_2Me ↘
Ph N Ph
MeO O
CF_3
29

2. *Synthetic methods*

(1) The most useful route[13] starts from vinyl azides, which on thermolysis in the gas phase or on photolysis[6] using a black light phosphor lamp are converted to 1-azirines in excellent yield. Many substituted derivatives have been obtained by the latter procedure. Photolysis in methanol[6] with a trace of sodium methoxide leads via the 1-azirine to the α-aminoketal (p. 8).

$PhCH{=}CH_2 \xrightarrow{Br_2} PhCHBrCH_2Br \xrightarrow[(b)\ NaOH/DMSO]{(a)\ NaN_3/DMSO} PhC(N_3){=}CH_2$

$PhC(N_3){=}CH_2 \xrightarrow{h\nu,\ MeOH,\ MeONa} PhC(OMe)_2CH_2NH_2$

$PhC(N_3){=}CH_2 \xrightarrow[\text{or } h\nu,\ \text{quartz, pentane}]{\Delta,\ \text{gas phase, 55–65\% overall}}$ 2-phenyl-2-azirine

(2) The Neber reaction (p. 7) is limited in its application, for in the absence of nitro groups, alkoxyaziridines (cf. **21**) are obtained. A similar cyclization of the dimethylhydrazone methiodide (**30**) with sodium isopropoxide rapidly gives[14] the azirine (**31**). Subsequent base-catalysed addition of isopropanol forms the aziridine (**32**).

$PhC(=N-\overset{+}{N}Me_3)CHMe_2\ I^-$ (**30**) ⟶ azirine (**31**, Ph, Me_2) $\underset{(b)}{\overset{(a)}{\rightleftharpoons}}$ aziridine (**32**, H, Ph, OiPr, Me_2)

(a) *i*PrOH and base, (b) azetropic distillation with PhMe

(3) The photolysis of isoxazoles[15] can give high conversions to azirines, but then further reactions may occur (p. 370).

3,5-diphenylisoxazole $\underset{>3000\ Å}{\overset{2537\ Å}{\rightleftharpoons}}$ PhCO–azirine–Ph

B. 2-Azirines

Theoretical calculations[16] suggest that 2-azirines are nonplanar with an unusually high [147 kJ (35 kcal)/mole] nitrogen inversion barrier, but no

33 (Ph, Me, Me-triazole) ⟶ Ph(MeN=)C–C̈Me (750 °C 0·01 torr) ⇌ **34** (1-Me-2-Ph-3-Me-1H-azirine) ⇌

35 (Me, Me, Ph-triazole) ⟶ Me(MeN=)C–C̈Ph ⟶ 3-methylisoquinoline and other compounds

compounds of this type have yet been isolated. However, the pyrolyses[17] of the isomeric triazoles **33** and **35** gave some 3-methylisoquinoline. This is probably formed via the 2-azirine **34**, and the results of a number of experiments lead to the conclusion that 2-azirines are very likely as reaction intermediates.[17]

2. AZIRIDINE

A. Introduction

Dihydroazirine (**36**), better known as aziridine or ethylenimine and occasionally as azacyclopropane, was first obtained by Gabriel in 1888, although its structure as a cyclic compound was first recognized by Markwald in 1900.

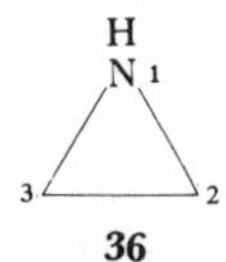

36

Interest in ethylenimine and its derivatives has increased greatly in recent years, and these compounds have much academic and industrial importance today.

B. Physical properties and structure

Aziridine is a colourless liquid with a strong ammoniacal smell. It is miscible with water and boils at 56°C. It is strongly caustic to the skin, and it should always be used under a hood (cf. p. 24). Inhalation of the vapour causes acute inflammation of the eyes, nose, and throat, and an individual may become sensitized to it. It is strongly basic (pK_a 7·98) but is much less so than dimethylamine (pK_a 10·87). The microwave spectrum[18] of the vapour has given the following structural data:

Aliphatic C—C 1·54 Å
Aliphatic C—N 1·47 Å

Me, Me ⟩N 111 ± 3°

H_2C; 1·481 Å; H 1·084 Å; 115·72° C; 59·87 Å; 60·25°; 109·31° H; 1·016 Å; N; 1·475 Å; 1·083 Å H

Distances ±0·001 Å
Angles ±0·01°

The molecule is slightly twisted because the H–C–H plane is at an angle of 87° to the ring plane. Although inversion of the nitrogen atom (like ammonia) is too fast for easy measurement a ΔG of 72·4 kJ (17·3 kcal)/mole has been estimated[19] from gas phase n.m.r. spectra. The dipole moment of aziridine is 1·89 D. The strain energy of the ring, due to the distortion of most of the bond angles and distances from normal, has been estimated[20] as approximately 96 kJ (23 kcal)/mole from combustion data. The strain in the ring is also reflected in the change of the C–H bending frequency in the infrared[21] from normal (1465 cm^{-1}) to 1475 cm^{-1}, and the N–H vibration frequency (1441 cm^{-1}) is lower than normally encountered in secondary amines (1460 cm^{-1}). These physical data completely confirm the much earlier deduction, made from the many facile ring-opening reactions, that the aziridine ring, like other three-membered rings, must be highly strained, but nevertheless it is more easily formed than the corresponding four-membered ring (p. 53).

C. Chemical properties

Aziridine is not a very stable compound as normally obtained, but in the absence of catalysts is stable at 150°C. It is best stored over sodium hydroxide in sealed bottles in a refrigerator. It can behave as a secondary amine, although the ring strain often makes other reactions of more consequence. With phenyl isocyanate and isothiocyanate it gives the ureas, it complexes with metals like ammonia, and with acid anhydrides or chlorides in the presence of alkali the *N*-acyl derivatives are formed. It is interesting that acetylation of aziridine under forcing conditions gives the *N*-protonated salt (**37**), which is stable at 0°C, but protonation of 1-acetylaziridine occurs on the oxygen atom.[22] As a secondary

H N (aziridine) $\xrightarrow[SO_2,\ -60\ °C]{Me\overset{+}{C}O\ SbF_6^-}$ H COMe N⁺ (**37**) $\xrightarrow{Na_2CO_3}$ MeC=O N (1-acetylaziridine) $\xrightarrow{FSO_3H,\ SbF_5\ |\ SO_2,\ -60\ °C}$ MeC(OH)=N⁺ (aziridine)

37

amine aziridine gives 1-chloro and 1-bromo derivatives with aqueous hypohalite[23] and adds to activated double bonds:

H N (aziridine) + $CH_2{=}CHCN$ ⟶ CH_2CH_2CN N (aziridine) and with methyllithium the *N*-lithium derivative is obtained. This last compound with alkyl or activated aryl halides gives the corresponding *N*-substituted aziridines, which are otherwise preferably obtained by direct synthesis and not from attempted *N*-substitution. Even 1,1′-biaziridine can be obtained from the lithium derivative and 1-chloroaziridine.[23] Aziridine with nitrosyl chloride at −60°C in ether yields an unstable *N*-nitroso derivative that decomposes at 0°C to nitrous oxide and ethylene. Although 2-chloroethyldimethylamine with silver perchlorate cyclizes to 1,1-dimethylaziridinium perchlorate (**38**) in 94% yield,[24] aziridine with methyl iodide gives only the ring-opened quaternary salt (**39**). Presumably the perchlorate anion, in contrast to the iodide anion, is too weakly nucleophilic to attack the ring. Steric factors are also important for when both carbon atoms have substituents, methyl iodide yields[25] stable quaternary iodides.

$Me_2NCH_2CH_2Cl \longrightarrow$ (1,1-dimethylaziridinium) ClO_4^- **38** $\quad Me_3\overset{+}{N}CH_2CH_2I \quad I^-$ **39**

Although pure dry aziridine is comparatively stable, it polymerizes in the presence of traces of water and rapidly, and occasionally explosively, in the presence of acids. Carbon dioxide is sufficiently acidic to promote polymerization. Polymerization is almost certainly not free radical in nature, as free-radical polymerization inhibitors do not affect the reaction. Aziridine is stable to bases, and it is thought that the polymerization proceeds through iminium intermediates as indicated. The polymer is linear.

Aziridine $\xrightarrow{H^+}$ [$H_2\overset{+}{N}$ aziridinium **40** $\overset{?}{\rightleftharpoons}$ $NH_2CH_2\overset{+}{C}H_2$ **41**] $\xrightarrow{\text{aziridine}}$ [H $CH_2CH_2NH_2$ $\overset{+}{N}$ aziridinium] $\xrightarrow{\text{aziridine}}$ Polymer

Kinetic and other evidence[26] shows that nucleophilic attack (here by an uncharged aziridine molecule) usually takes place on the cation (**40**) with simultaneous ring opening rather than on the corresponding carbonium ion (**41**).

Under proper conditions aqueous hydrochloric acid opens the aziridine ring to give 2-chloroethylamine in a way similar to its reaction with oxirane and thiirane. This ring opening is approximately second order, recyclisation with

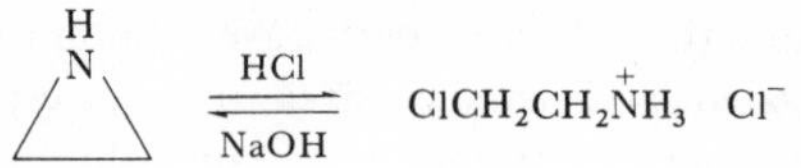

$$\text{aziridine (NH)} \underset{\text{NaOH}}{\overset{\text{HCl}}{\rightleftharpoons}} ClCH_2CH_2\overset{+}{N}H_3\ Cl^-$$

sodium hydroxide is first order and quantitative, and the rate constants for both types of reaction are greatly influenced by substitution. *N,N*-Dialkylaminoalkyl halides cyclise reversibly in dilute aqueous solution to aziridinium salts **(42)**, and these can react with a variety of anions to yield the corresponding products. The reaction with thiosulphate is very fast and is used for estimating aziridines.

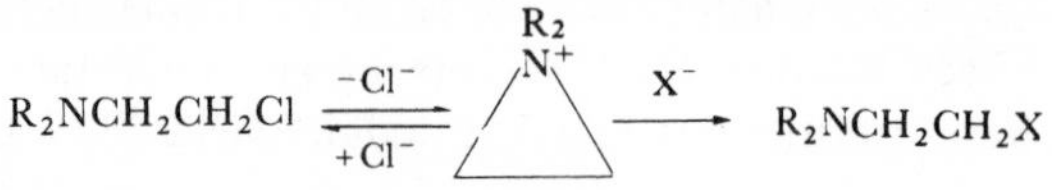

$$R_2NCH_2CH_2Cl \underset{+Cl^-}{\overset{-Cl^-}{\rightleftharpoons}} \mathbf{42}\ (R_2N^+\text{ aziridinium}) \xrightarrow{X^-} R_2NCH_2CH_2X$$

$X^- = HO^-, Cl^-, EtCO_2^-, PhCO_2^-, HCO_3^-$, and $S_2O_3^{2-}$

Dialkylaminoethyl chlorides are well known for dimerizing to the cyclic bisquaternary salts **(43)**, probably via reaction of the halide with a cation molecule **(42)**. This side reaction is of little consequence at the low pH used in the recent kinetic work. Structural evidence supports the partial formation of

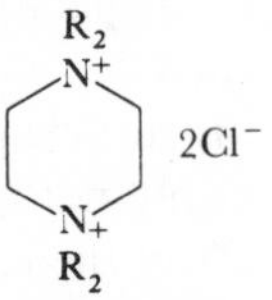

$$\mathbf{43}:\ R_2N^+\text{–(piperazinium ring)–}N^+R_2\quad 2Cl^-$$

aziridines as intermediates in some alkylation reactions involving 2-chloroethylamines. Diphenylacetonitrile **(44)** reacts[27] with both 2-chloro-1-dimethylaminopropane **(45)** and 1-chloro-2-dimethylaminopropane **(47)** in the presence

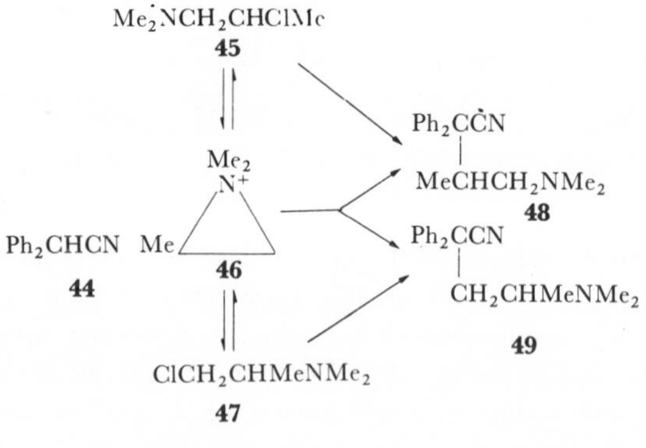

$Me_2NCH_2CHClMe$ **45** ⇌ **46** (Me_2N^+ aziridinium, Me-substituted) ⇌ $ClCH_2CHMeNMe_2$ **47**

Ph_2CHCN **44**

45, **46** → $Ph_2C(CN)$–$MeCHCH_2NMe_2$ **48**

46, **47** → $Ph_2C(CN)$–$CH_2CHMeNMe_2$ **49**

of potassium *t*-butoxide to give both **48** and **49**. The best explanation of this is that some cyclization to an aziridinium cation (**46**) must occur and that its ring can open in both ways.

Aziridine reacts with hydrogen sulphide to give 2,2′-diaminodiethyl sulphide (**50**). The reaction has not been halted at the 2-mercaptoethylamine stage, perhaps because aziridine reacts especially rapidly with thiols; with the essential amino acid cysteine, the product is **51**. Water in the presence of dilute nitric or sulphuric acid opens the ring to give ethanolamine (**52**), aqueous sulphur dioxide gives taurine (**53**), while phenol and thiophenol behave similarly, yielding **54** and **55**, respectively.

$(H_2NCH_2CH_2)_2S$ **50** $H_2NCH_2CH_2SCH_2CHNH_2CO_2H$ **51** $NH_2CH_2CH_2OH$ **52**

$NH_2CH_2CH_2SO_3H$ **53** $PhOCH_2CH_2NH_2$ **54** $PhSCH_2CH_2NH_2$ **55**

Aziridine reacts vigorously with carbon disulphide to give 2-thiathiazolidone (p. 20), and heating *N*-benzoylaziridine yields a polymer and some 2-phenyloxazoline.

Aziridine reacts with acetaldehyde, benzaldehyde, acetone, and some other aldehydes and ketones to give oxazolidines (**56**) at 5–10°C, while *N,N*-dimethylaziridinium perchlorates behave similarly, leading to salts such as **57**.[24]

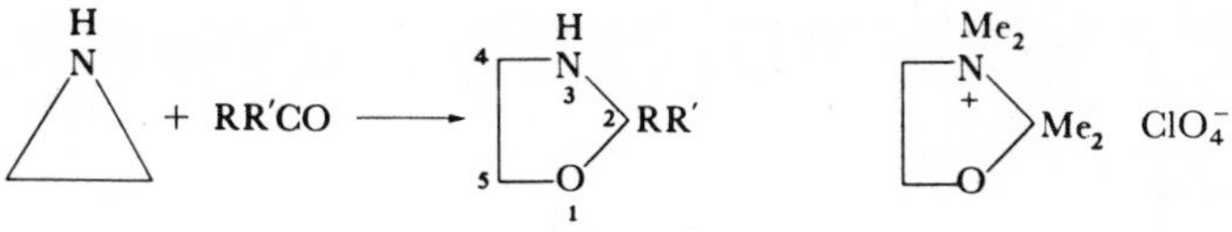

56 **57**

D. Derivatives of aziridine

1. *Nitrogen Inversion*

As the bonds from a trivalent nitrogen atom are tetrahedrally arranged, it would appear that aziridines of type **58** might be capable of resolution. Many

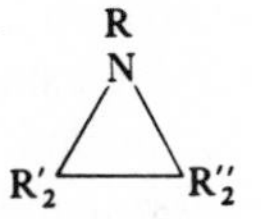

58

attempts to resolve a number of aziridines of this type have, however, been unsuccessful. It appears that the nitrogen atom inverts too easily to permit isolation of the optical enantiomorphs at ordinary temperatures. From nuclear magnetic resonance spectroscopic measurements it has been calculated that the rate of inversion for *N*-cyclohexylaziridine is about 100 times a second at 95°C, and *N*-phenylaziridine inverts[28] even more rapidly at –77°C. As *N*-(4-bromobenzoyl)aziridine is trihedral in the crystal but effectively planar at –155°C, as regards its n.m.r. spectrum,[29] conjugation between the substituent and nitrogen atom is of major importance.

When the substituent attached to the nitrogen atom is chlorine or methoxyl,[30] the inversion rate is enormously slowed, and the geometrical isomers **59** and **60** have recently been obtained from the corresponding aziridine with aqueous sodium hypochlorite[31]; *N*-chlorosuccinimide[32] and *t*-butyl hypochlorite[33] convert, respectively, 2-methyl- and 2-benzoyl-3-phenylaziridines to mixtures of corresponding geometrical isomers, which have also been separated by thin layer chromatography. The more strained *endo* compound (**60**) is isomerised to the *exo* isomer **59** (b.p. 57–58°C at 11 torr) on attempted distillation. Oxidation of 2,3-diphenylaziridine by (1*R*, 2*R*)-(–)isobornyl hypochlorite at –60°C gave optically active 1-chloro-2,2-diphenylaziridine (**61**) in 87% purity, but it racemised completely on standing for 4 days at 0°C.[34]

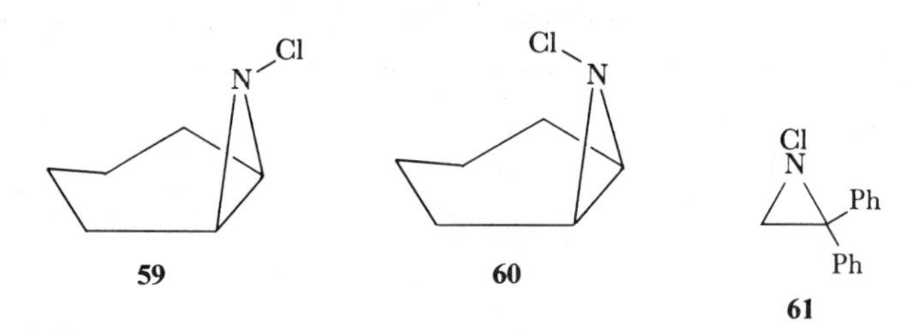

2. *Aziridinones*

The first genuine aziridinone (**64**) was obtained in 1962, and its structure follows from its nuclear magnetic resonance spectra and the reactions outlined below.[35] The ring opens in two ways with liquid ammonia.[36]

A complementary cyclisation (68% yield) of the bromo amide **65** to the aziridinone **66** has been carried out.[37] This type of synthesis usually gives better results and has been used in a number of cases. Hydrolysis of the dichloroaziridine **68**, obtained (5%) from dichlorocarbene and the imine **67**, gives 1% of **66**.[37] 1,3-Di-*t*-butylaziridinone (**66**), an unusually stable 2-aziridinone, can be distilled and even stands up to gas–liquid chromatography. With proton-containing nucleophiles the ring opens to give amides (cf. **63**), and with aprotic

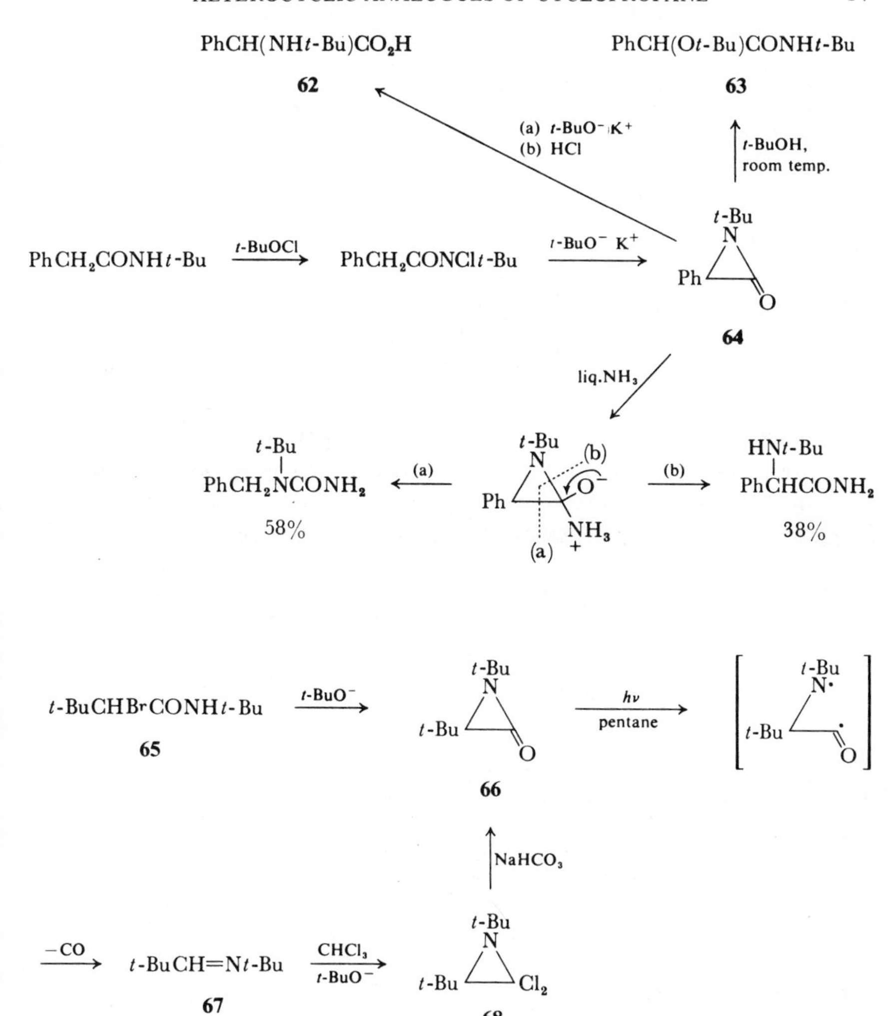

nucleophiles the ring opens to give amino acids (cf. **62**). Photolysis[38] opens the ring and splits out carbon monoxide to form **67**.

The aziridinone **69** on lithium aluminium hydride reduction gives none of the corresponding aziridine, for the ring is split leading to the amino aldehyde and amino alcohol. On standing, or more quickly in refluxing ether, the ring cleaves between positions 1 and 3, and the compound also undergoes rearrangement to an oxirane, which leads to the minor products formed.[37]

$$\text{69} \xrightarrow{LiAlH_4} Me_2C(NH\textit{t}\text{-Bu})-CHO + Me_2C(NH\textit{t}\text{-Bu})-CH_2OH$$

69 —(Et$_2$O, boil)→ $CH_2{=}CMeCONH\textit{t}\text{-Bu}$ (Main product)

69 —(Et$_2$O, boil)→ [2,2-dimethyl-3-(N-*t*-Bu)imino-oxirane] ⟶ Me_2CO + *t*-BuNC (Minor products)

Bond-length data (below) for the aziridinone ring have been obtained from an X-ray structure determination[39] for 1,3-diadamantylaziridinone. The nitrogen atom is pyramidal, but the short bond between this atom and the carbonyl group suggests that the bond has some double-bond character.

N(Ad)–C(Ad): 1·509 Å ± 0·015; N–C(O): 1·328 ± 0·013 Å; C–C: 1·446 ± 0·11 Å; C=O: 1·199 ± 0·008 Å; angles: 60·9° (at N), 53·4°, 65·7°

Ad = 1-adamantyl

3. *Ring openings of aziridines*

Ring-opening reactions of asymmetrically substituted aziridines can theoretically take place in three ways, but often one product is largely obtained. Most of the reactions examined involve breaking a carbon–nitrogen bond and have been acid catalysed, but as base-catalysed ring openings are essentially simpler, they are considered first.

Several substituted aziridines have been cleaved by aqueous ammonia and ethylamine. The reactions were bimolecular and proceeded through attack by the amine at one of the ring carbon atoms.[40] Walden inversions at the position of attack therefore occur, and the ring openings consequently take place in a *trans* manner. Where the aziridine is unsymmetrical the attacking amine usually reacts with the carbon atom possessing most hydrogen atoms, but sometimes the

NHEt NHEt | MeC—CMe | H H ←(a)— Et N, H, Me, Me, H ; Et N, Me, Me, H, H —(a)→ NHEt H | MeC—CMe | H NHEt + H NHEt | MeC—CMe | NHEt H

meso, optically inactive (*erythro* series) — *trans*, optically active — *cis*, non-resolvable — *dl*- or (±)-resolvable (*threo* series)

(a) $EtNH_2$ at 120° for 16 days

ring opens in both possible directions. These reactions are similar to those of oxiranes (pp. 28–31) and thiiranes (p. 43).

Acid-catalysed ring openings of aziridines give mixtures much more often than reactions carried out under basic conditions. The *cis* and *trans* isomers of 2,3-diphenylaziridine are taken as an example. On treatment with hydrogen chloride these compounds are first converted into the aziridinium chlorides (**70**). Now if the chloride ion were to attack the 2-position of the aziridinium ring, then by *trans* reaction the *cis* isomer should give only a *threo* product, while the *trans* should give an *erythro* compound. In fact, a mixture of both racemates is obtained from both the *cis* and the *trans* aziridines. This leads to the conclusion that in these cases the cyclic cations **70** with the one open-chain form (**71**), or a similar species permitting changes in stereochemistry, are in equilibrium. The presence of the phenyl groups, in contrast to the otherwise similar case described on page 13, facilitates the formation of carbonium ions. In some instances, however, only *trans* reaction takes place, and this implies combination of the nucleophile with the cyclic cation, or the aziridine.

H N, Ph, Ph → [H_2 ^+N, Ph, Ph] (**70**) ⇌ [NH_2 | $PhCH\overset{+}{C}HPh$] (**71**) —Cl^-→ $PhCH(NH_2)CHClPh$

The asymmetric 2,2-dimethylaziridine reacts with benzene and aluminium chloride to give **72**, and similar openings with methanol and with hydrochloric acid have been described. This aziridine opens in the reverse way, however, with hydrogen and Raney nickel, as *t*-butylamine is the sole product. 2-Phenylaziridine and its 1-*p*-toluenesulphonyl derivative open between positions 1 and 2 to give the expected products with hydrogen halides, and with hydroxylic

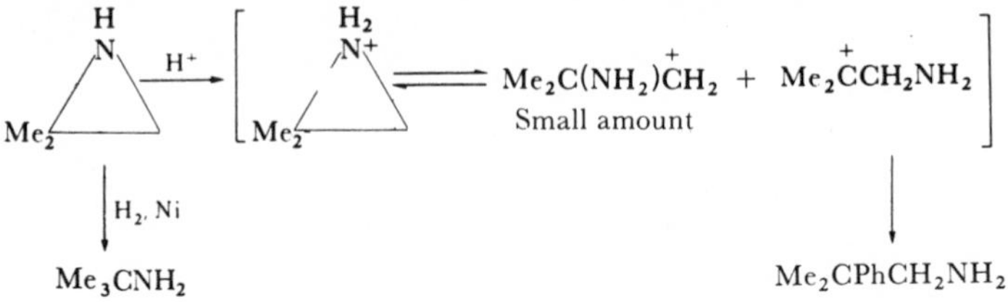

$$Me_2C(NH_2)\overset{+}{C}H_2 + Me_2\overset{+}{C}CH_2NH_2$$

Small amount

Me_3CNH_2

$Me_2CPhCH_2NH_2$

72

compounds in the presence of acids. 2-Phenylaziridine, like aziridine, combines with carbon disulphide, yielding 5-phenylthiazolid-2-thione.

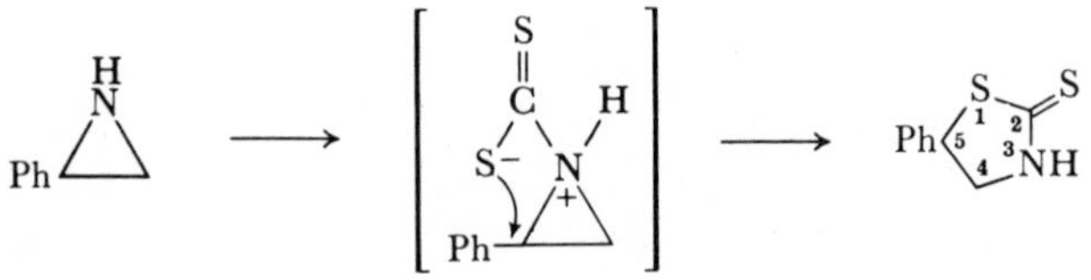

1-*n*-Butylaziridine gave a normal von Braun product ($BrCH_2CH_2NBuCN$) with cyanogen bromide, while asymmetrically substituted aziridines give mixtures.[41]

cis- and *trans*-2,3-Dimethylaziridine with nitrosyl chloride yield *cis*- and *trans*-but-2-ene, respectively, in over 99% purity, while in the second case working at low temperatures the *N*-nitroso derivative (**73**) was obtained as a

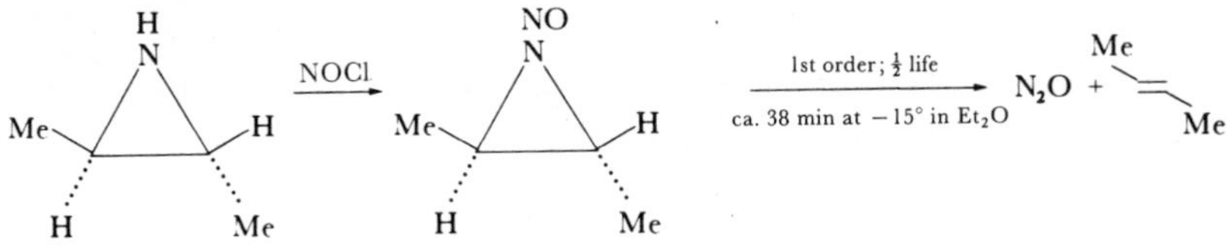

NOCl

1st order; ½ life ca. 38 min at −15° in Et_2O → N_2O +

73

yellow oil decomposing explosively at room temperature.[42] These ring openings appear to be concerted and may be examples of nonlinear chelotropic processes.

Oxidation of *cis*-1,2,3-triphenylaziridine (**74**) by 3-chloroperbenzoic acid gave a complex mixture, the *N*-oxide being a possible intermediate.[43] Heating this aziridine gives an equilibrium mixture with 22% of the *trans*-isomer at 150°C. Equilibrium is established via the formation of the ylids. The *cis*-aziridine, by a conrotatory process, gives the *trans*-ylid (**75**), which can be trapped as **76**, before rotation leading to the isomeric configuration can occur,[44] if dimethyl acetylenedicarboxylate is present. A number of reactions of this type are now known. The interesting ring opening[45] of both *cis*- and *trans*-**77** may proceed by a similar

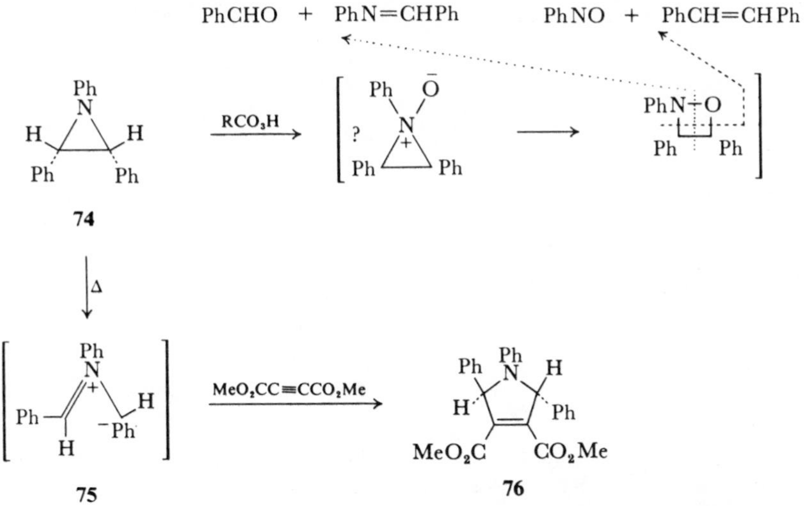

PhCHO + PhN=CHPh   PhNO + PhCH=CHPh

RCO₃H → ? → PhN–O / Ph Ph

74

Δ

75

$MeO_2CC{\equiv}CCO_2Me$

MeO_2C CO_2Me

76

intermediate, but the opening of the ring of **78** can be regarded as a reverse Michael reaction.[46]

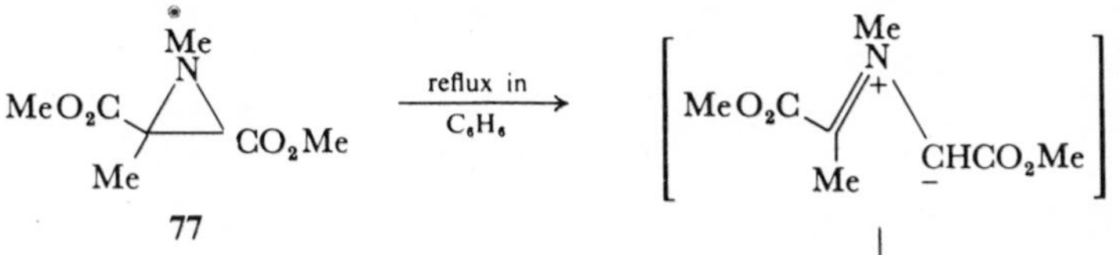

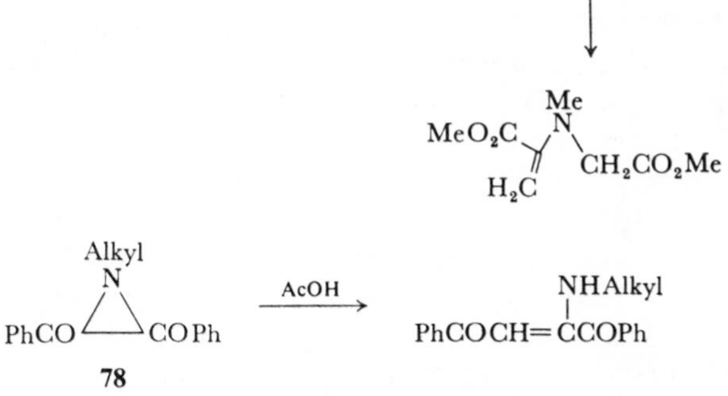

MeO_2C Me CO_2Me

77

reflux in C_6H_6

MeO_2C Me $CHCO_2Me$

MeO_2C H_2C CH_2CO_2Me

Alkyl

PhCO COPh

78

AcOH

NHAlkyl

PhCOCH=CCOPh

The solvolysis of 1-chloroaziridines (**79**) to carbonyl compounds is thought[47] to involve a concerted loss of halide and cleavage of the 2,3-bond.

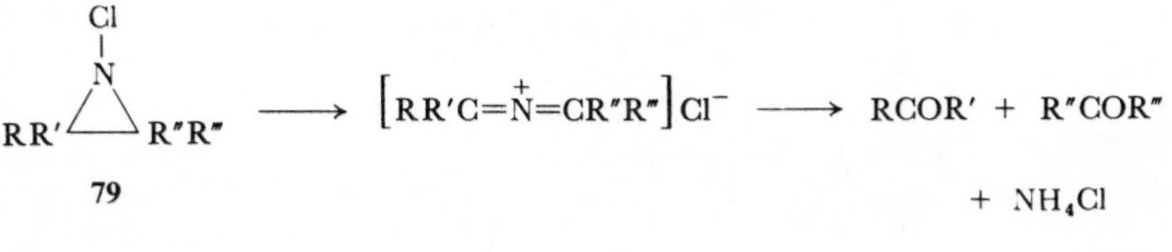

RR′ R″R‴

79

$[RR'C{=}\overset{+}{N}{=}CR''R''']Cl^-$ → RCOR′ + R″COR‴ + NH_4Cl

The aziridine ring can also be split by photolysis in suitable cases (e.g., **80**).[48]

80 $\xrightarrow[n \to \pi^*]{h\nu,\ \text{pentane}}$ [$\xrightarrow{1,3\text{-shift}}$ →] $\xrightarrow{H_2O}$ PhCO(NH*t*-Bu)C=CH(Ph)

E. Synthetic methods

Attempts to prepare aziridine from acetylene and ammonia, and by heating ethylenediamine, which gives some piperazine, have failed.

It should be noted that methods (1) and (2) shown below are often complementary, and when one gives bad results the other is satisfactory.

(1) The best preparative method[49] for aziridine, also useful for the synthesis

$$H_2NCH_2CH_2OH \xrightarrow{H_2SO_4} H_3\overset{+}{N}CH_2CH_2OSO_3^- \xrightarrow{NaOH} \text{aziridine } (34\text{–}37\%)$$

of both *N*- and *C*-alkyl and *N*- and *C*-aryl derivatives, is to heat 2-aminoethylsulphuric acid with concentrated aqueous sodium hydroxide when aziridine, b.p. 56°C, distils over. A small amount of the dimer, 1-(2-aminoethyl)aziridine, b.p. 127°C, is also formed.

(2) Another widely used method is the original one of Gabriel for aziridine, the cyclization of a 2-chloro- or 2-bromoethylamine by silver oxide or better by concentrated aqueous potassium hydroxide. Even 1-aza[1.1.0]bicyclobutanes can be made in this way.[50] The rate of reaction depends largely on the substituents, and the synthesis has also been used for the corresponding four-, five-, and six-membered rings; the five- and six-membered rings are by far the most easily formed. When substituents are present that make observation of the centre of inversion possible, it is found that a Walden inversion takes place at the

$$NH_2CH_2CH_2OH \xrightarrow{SOCl_2} H_3\overset{+}{N}CH_2\overset{*}{C}H_2Cl\ Cl^- \xrightarrow{^-OH} \text{aziridine}$$

$$BrCH_2-\underset{^+NH_3\ \ Br^-}{\overset{R}{C}}-CH_2Br \xrightarrow{^-OH} \text{1-azabicyclobutane} \qquad R = H, Me$$

carbon atom marked with an asterisk during the cyclization, and a second inversion occurs if the ring is opened during a subsequent reaction, as is also the case with the oxiranes (p. 36) and thiiranes (p. 44).

The reduction of α-chloronitriles by lithium aluminium hydride,[51] and of the chloroacetyl derivatives[52] of primary amines by aluminium hydride also gives aziridines, presumably via the β-chloroethylamines. A valuable synthesis from olefines through derived β-iodoazides[53] is outlined below.[54]

$$\text{cyclohexene} \xrightarrow[\text{MeCN}]{IN_3(ICl + NaN_3),} \text{I, } N_3 \text{ adduct} \xrightarrow[Et_2O,\ 0\,°C]{LiAlH_4,} \text{NH (aziridine)}$$

N-Alkyl, *N*-aryl, or benzenesulphonyl derivatives of 2-chloroethylamines also cyclize with alkali to the *N*-substituted aziridines.

(3) A remarkable synthesis of the aziridine ring occurs when alkyl- or arylmagnesium halides react with aryl alkyl ketoximes. The reaction proceeds via an azirine, and in the second stage the original reagent adds preferentially to the less hindered side.[55]

$$PhC(=N{-}OH)Me \xrightarrow{EtMgBr} \text{Ph-azirine (N)} \longrightarrow \text{Et, Ph-aziridine (H, N)}$$

(4) The decomposition[56] of ethyl azidoformate in cyclohexene yields an aziridine:

$$\text{cyclohexene} + N_3CO_2Et \xrightarrow[354\ m\mu]{h\nu} \text{NCO}_2\text{Et (aziridine)} + N_2$$

Similar nitrene additions are known, and one has provided a route to an unstable 1-aminoaziridine.[57]

$$PhthNNH_2 \xrightarrow{Pb(OAc)_4} [PhthN\ddot{N}] \xrightarrow{Ph_2C{=}CH_2} \text{Ph}_2\text{-aziridine-N-NPhth}$$

$$\xrightarrow{NH_2NH_2} \text{phthalhydrazide (O, NH, NH, O)} + \text{Ph}_2\text{-aziridine-N-NH}_2 \xrightarrow[\text{few hours}]{\text{room temp.}} N_2 + Ph_2CHMe + Ph_2C{=}CH_2$$

(5) Enamine perchlorates (e.g., **81**) readily add diazomethane to form aziridinium salts that are of special interest because on treatment with water or alcohols azepines (e.g., **82**) are formed.[24]

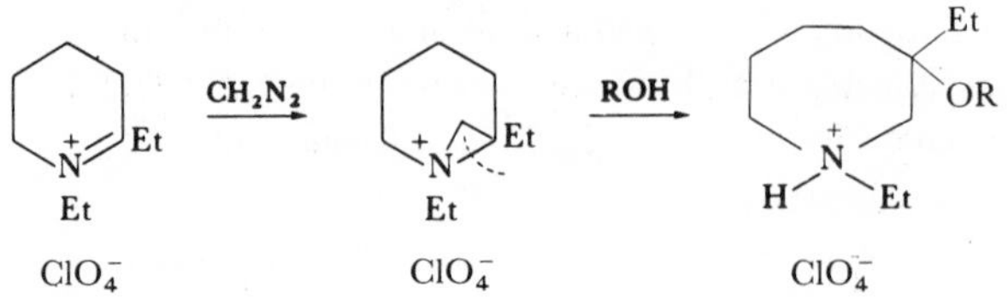

81 $\xrightarrow{CH_2N_2}$ (aziridinium salt, ClO_4^-) $\xrightarrow{ROH}$ **82**

F. Natural occurrence and compounds of special interest

Aziridine itself is prepared commercially on a large scale and is used industrially to alter the properties of hydroxylic polymers.

The unpleasant handling properties of aziridine are due to its great reactivity, and it is worth noting that it has some carcinogenic activity in rats, while its *N*-acetyl, and many other *N*-acyl, derivatives are more powerful in this respect. However, some of its derivatives, notably **83**, Tretamine (**84**), and Tetramin are

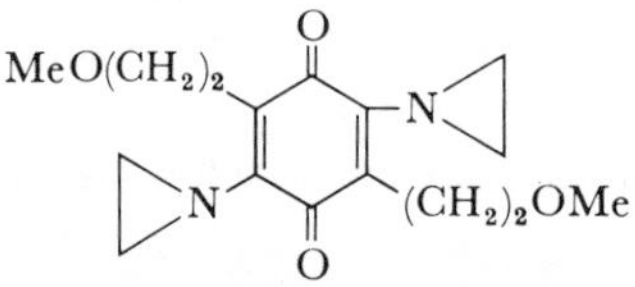

83 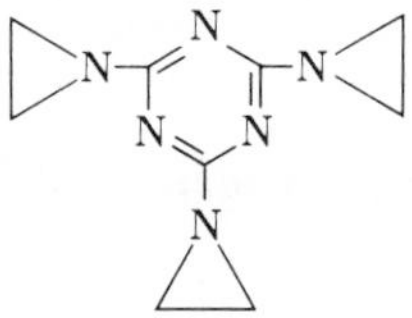 **84**

under clinical trial as agents against leucaemia and related cancers. Tetramin[58] is a 2 : 1 mixture of **85** and **86** obtained from 2-vinyloxirane with aziridine. One class of antibiotics, an example of which is Mitomycin A (**87**), possesses both the aziridine ring and anticancer activity.[59] Some of the toxic properties of the 'nitrogen mustards,' for example, $MeN(CH_2CH_2Cl)_2$, may be related to the fact that they cyclise to reactive aziridines under physiological conditions, and aziridines can even *S*-alkylate methionine $[MeS(CH_2)_2CH(NH_2)CO_2H]$.[60]

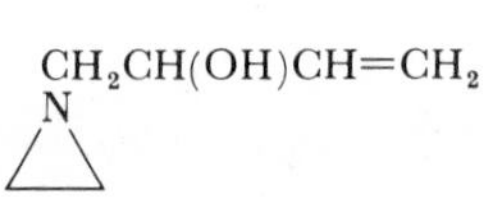

85 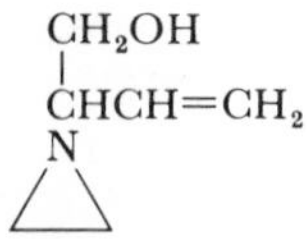 **86** 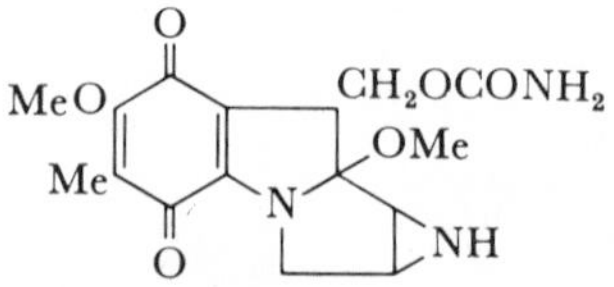 **87**

3. OXIRANE OR ETHYLENE OXIDE

A. Introduction

Oxirane or ethylene oxide (**1**) is also known as 1,2- or β-oxidoethane, as α,β-epoxyethane, and, rarely, as oxacyclopropane. It was first obtained by Wurtz in 1859 but attracted little attention until about 1925, when ethylene chlorohydrin became commercially available from ethylene. Treatment of the

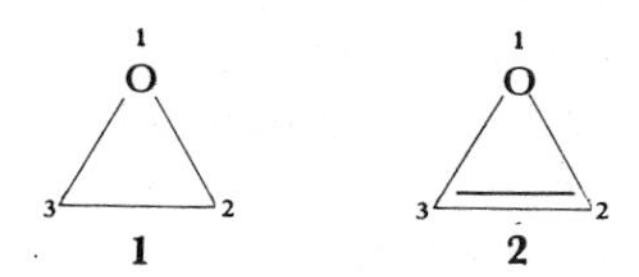

crude chlorohydrin with caustic soda gives up to 95% of oxirane. In 1931 a patent was taken out on the direct oxidation of ethylene to oxirane by oxygen, and the first plant utilizing this process began production in 1937. Since then many new plants have been opened, and next to ethanol, oxirane is the most important commercial derivative of ethylene produced in the United States. About 650 million pounds of oxirane were produced in 1954. Approximately half of this was used in the form of derivatives in the automobile industry, as antifreeze bases, brake-fluid components, resins, solvents for finishes, and so forth. The simplest oxygen-containing heterocycle, oxirene (**2**), has not yet been described, although there is evidence from ^{13}C scrambling[61] and oxygen migration studies[62] that oxirenes can be reaction intermediates, and they may be in rapid equilibrium with corresponding oxocarbenes.

$R{-}C({=}O){-}C(N_2)R^1 \xrightarrow{h\nu} R{-}{}^{13}C({=}O){-}\ddot{C}R^1$ (via various electronic states) $\longrightarrow$ $R{-}{}^{13}C{=}CR^1$ (oxirene) $\longrightarrow$ $RR^1C{=}C{=}O$ (Scrambled)

R = Pr, R^1 = Et; $h\nu$, Et_2O:

$PrC({=}O){-}\ddot{C}Et \longrightarrow PrCOC_3H_5$ (Several isomers)

$PrC({=}O){-}\ddot{C}Et \rightleftarrows PrC{=}CEt$ (oxirene) $\rightleftarrows Pr\ddot{C}{-}C({=}O)Et \longrightarrow C_4H_7COEt$ (Several isomers)

B. Physical properties and structure

Oxirane is a colourless liquid, b.p. 10·7°C (Me_2O, b.p. –23·6°C). The most accurate structural data for the molecule are derived from microwave spectrum measurements.[63] It is between the accepted C–C distances in ethane (1·54 Å) and ethylene (1·35 Å). The H–C–H angle is between the tetrahedral and the 120° in ethylene, and the dipole moment is 1·88 D. Dimethyl ether, for comparison, has a dipole moment of 1·30 D, the C–O distance is 1·416 ± 0·003 Å, and its C–O–C angle is 111·5°. In water the H–O–H angle is 104·5°, while for pure *p* orbitals it would be 90°. Oxirane has been assumed to have p^2 bonding, with 'bent' bonds lying along the arc, tangents to the C and O orbitals. The strain in the ring is also reflected by the C–H vibration frequency in the infrared,[21] which is at 1500 cm^{-1} instead of at the normal value of 1465 cm^{-1} for aliphatic compounds. The strain energy, from combustion data,

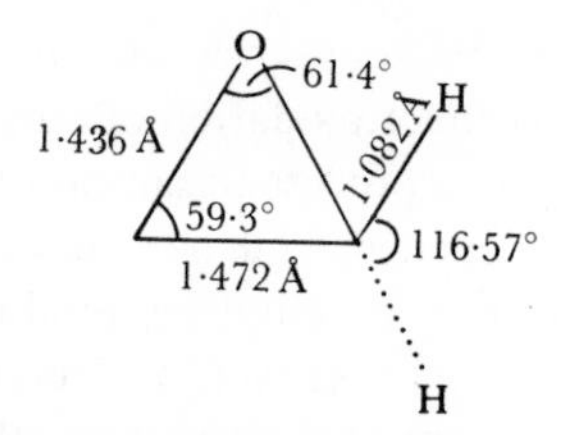

has been estimated[20] at about 117 kJ (28 kcal)/mole, a similar figure to those obtained for thiirane and aziridine. The ring is, of course, very easily opened. Oxirane, qualitatively, has a similar effect to that of a carbonyl (C=O) or alkenyl (C=C) group on the ultraviolet absorption spectrum of a conjugated molecule.

C. Chemical properties

Oxirane, or ethylene oxide, is a well known stable laboratory reagent, which, because of its low b.p. (10·7°C), is available in sealed glass tubes or metal cylinders. Care should be exercised in its manipulation, for application of even a 1% solution to human skin for 15 minutes gives bad blisters; it should always be used under a hood. Although its ring is highly strained and liable to open, oxirane does, like other ethers, form a 1 : 1 addition compound with boron trifluoride at –78·8°C. This compound decomposes reversibly on heating, unlike

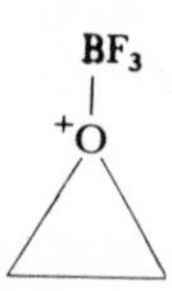

comparable addition compounds of the four-, five-, and six-membered ring cyclic ethers, which decompose irreversibly and give some hydrogen fluoride. Oxirane is a much poorer electron donor than these other cyclic ethers, as is shown by the heats of mixing with chloroform (Table 1).

TABLE 1. Heats[64] of mixing with chloroform at 13°

Compound	Heat of mixing [J (cal)/mole]
Oxirane	1530 (365)
Oxetane (trimethylene oxide)	3180 (760)
Tetrahydrofuran (tetramethylene oxide)	3120 (750)
Tetrahydropyran (pentamethylene oxide)	2680 (640)
Diethyl ether	2720 (650)

Oxirane polymerizes slowly on standing with sodium hydroxide or zinc chloride, while the reaction is faster with stannic chloride. Peroxides and ultraviolet light do not cause polymerization, which therefore proceeds by ionic mechanisms. The acid-catalysed polymerization is probably similar to that of

$$\text{oxirane} \xrightarrow{H^+} [\text{protonated oxirane } (O^+H)] \xrightarrow{\text{oxirane}} HOCH_2CH_2O^+(\text{oxirane ring}) \xrightarrow{\text{oxirane}} \text{Polymer}$$

aziridine (p. 13), but under the correct condition dioxane (**3**) can be prepared and is obtained industrially in this way. The alkaline polymerization, which has

3

industrial application mentioned later, probably proceeds by the primary addition of an hydroxyl ion:

$$\text{oxirane} \xrightarrow{OH^-} HOCH_2CH_2O^- \xrightarrow{\text{oxirane}} HO(CH_2)_2OCH_2CH_2O^- \xrightarrow{\text{oxirane}} \text{Polymer}$$

Water, preferably in the presence of catalytic amounts of acids or in the vapour phase, converts oxirane into ethylene glycol; the first formed glycol also combines with unused oxirane to give some diethylene glycol (**4**) as a

$$\text{oxirane} \xrightarrow{H^+} \left[\text{protonated oxirane } (\overset{+}{O}H)\right] \begin{cases} \xrightarrow[-H^+]{H_2O} HOCH_2CH_2OH \\ \xrightarrow[-H^+]{\text{glycol}} HOCH_2CH_2OCH_2CH_2OH \quad \mathbf{4} \end{cases}$$

by-product. The reaction with water has great commercial importance. The use of alcohols instead of water gives the corresponding glycol monoethers, which are useful industrial solvents. Hydrogen fluoride in ether, and the other hydrogen halides, give the corresponding halogen hydrins with oxirane, while treatment with even aqueous sodium bromide or nitrate gives appreciable quantities of ethylene bromohydrin and ethylene glycol mononitrate, respectively. Acetyl chloride (reacting as $CH_3CO^+Cl^-$) and a number of other acids,

$$\text{oxirane} \xrightarrow{H^+} \left[\text{protonated oxirane } (\overset{+}{O}H)\right] \xrightarrow{X^-} HOCH_2CH_2X$$

including carboxylic acids, sodium hydrogen sulphite, hydrogen sulphide, and hydrogen cyanide, behave similarly. The last reaction is very important commercially, as an easy dehydration of the resulting 3-hydroxypropionitrile (**5**) yields acrylonitrile (**6**), which is a very valuable polymer intermediate used in the manufacture of Acrilan and Orlon.

$$\underset{\mathbf{5}}{HOCH_2CH_2CN} \longrightarrow \underset{\mathbf{6}}{CH_2{=}CHCN}$$

Oxirane combines very easily with many aromatic and aliphatic amines, yielding 'ethanolamines.' With ammonia, for instance, 'monoethanolamine' (**7**), 'diethanolamine' (**8**), and 'triethanolamine' (**9**) are obtained according to the conditions. This type of reaction is very general.

$$\text{oxirane} + NH_3 \longrightarrow \begin{cases} H_2NCH_2CH_2OH & \mathbf{7} \\ HN(CH_2CH_2OH)_2 & \mathbf{8} \\ N(CH_2CH_2OH)_3 & \mathbf{9} \end{cases}$$

D. Nucleophilic openings of oxirane rings

Asymmetrically substituted oxirane rings are opened by nucleophilic (anionic) attack under alkaline conditions. The opening proceeds by an S_N2 mechanism, inversion at the atom being attacked, and is *trans*, as in the case of analogous aziridines (p. 18). This is illustrated by the case of *cis*-2,3-dimethyloxirane, which with ammonia and a trace of water yields a threo alcohol,[65] which can be resolved.

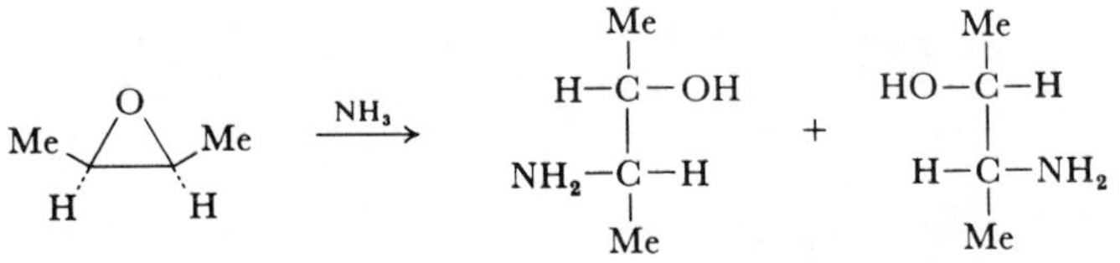

$$\text{Me}\overset{\text{O}}{\triangle}\text{Me (H, H)} \xrightarrow{NH_3} \begin{array}{c} \text{Me} \\ | \\ \text{H}-\text{C}-\text{OH} \\ | \\ \text{NH}_2-\text{C}-\text{H} \\ | \\ \text{Me} \end{array} + \begin{array}{c} \text{Me} \\ | \\ \text{HO}-\text{C}-\text{H} \\ | \\ \text{H}-\text{C}-\text{NH}_2 \\ | \\ \text{Me} \end{array}$$

(*threo* series)

Where two carbon atoms are otherwise equivalent, anionic attack usually occurs at the least hindered atom, the atom possessing most hydrogen atoms. Thus 2-methyloxirane (propylene oxide, **10**) with amines, and with sodium derivatives of phenols, alcohols, and so forth yields products exemplified by **11** and **12**. Sodium derivatives of compounds with activated methylene groups, such as dimethyl malonate, react similarly, but the end products are often lactones. For instance, **13**, which is formed from oxirane.

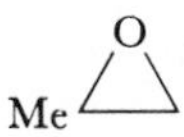

10

$MeCH(OH)CH_2NR_2$

11

$MeCH(OH)CH_2OR$

12

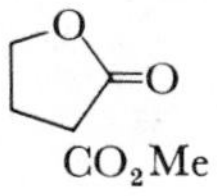

CO_2Me

13

Epichlorhydrin is a useful synthetic intermediate obtained from glycerol. Its reactions with secondary amines are of interest. First the oxide ring is opened normally. If more amine is present it acts as a base and re-forms an oxirane, while if excess is available the diamino alcohol is formed.

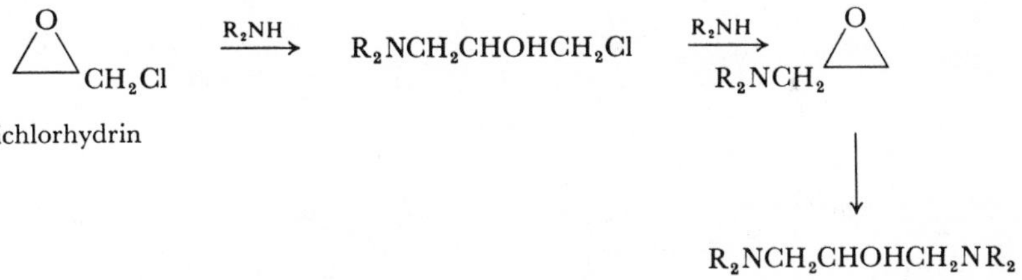

$$\overset{\text{O}}{\triangle}CH_2Cl \xrightarrow{R_2NH} R_2NCH_2CHOHCH_2Cl \xrightarrow{R_2NH} R_2NCH_2\overset{\text{O}}{\triangle} \longrightarrow R_2NCH_2CHOHCH_2NR_2$$

Epichlorhydrin

2-Ethyloxirane with lithium borohydride gives exclusively 2-butanol (EtCHOHMe) via a *trans* delivery of the hydride ion.[66] 2-Methyloxirane with lithium aluminium hydride similarly gives mainly isopropanol, while Raney nickel and hydrogen yield *n*-propanol. The production of secondary and primary alcohols by these reagents is quite general,[67] but the direction of ring opening by lithium aluminium hydride can be altered by addition of aluminium chloride, which may lead to intermediate carbonium ion formation (see next section).

The reaction between oxirane and Grignard reagents has been used widely to obtain primary alcohols.[68] In the case of ethylmagnesium bromide an addition compound is first formed, which upon distillation of the ether from the reaction mixture is thought to rearrange. Subsequent hydrolysis yields *n*-butyl alcohol,

EtMgBr

$\longrightarrow EtCH_2CH_2OMgBr \xrightarrow{H^+} n\text{-}BuOH$

which is anticipated as the sole product, but occasionally secondary alcohols are formed to some extent (10%). This side reaction can be avoided by the use of dialkylmagnesiums instead of the Grignard reagents. Magnesium bromide,

$$2\ RMgBr \rightleftharpoons MgR_2 + MgBr_2$$

formed from the Grignard reagent, isomerizes the oxirane to acetaldehyde (see also p. 35), which then reacts in the normal way with more Grignard reagent. The products obtained from oxiranes and Grignard reagents depend enormously on the conditions, as exemplified below.[69]

add to PhMgBr in Et_2O → Ph, OH — 84%

add PhMgBr slowly to oxirane → [CHO] → Ph OH — 73%

$CH_2{:}CHMgBr$

	Ph / OH	OH / Ph
Addition of Grignard, in Et_2O	80%	19%
Addition of oxirane, in THF	18%	80%

Lithium dimethyl cuprate ($LiCuMe_2$), which effectively provides a methyl carbanion, opens oxirane rings in an exclusively *trans*[70] mode.

cis-Cyclooctene (**14**) has been converted[71] to the *trans* isomer (**16**) in ca. 60% yield through peracid oxidation to the oxirane (**15**) followed by the reactions outlined below:

(a) Ph_2PLi
(b) $AcOH, H_2O_2$

NaH
DMSO

14 **15** **16**

$+ Ph_2PO_2^- Na^+$

Nucleophilic attack by a Wittig reagent on *cis*-1,2-dimethyloxirane (**17**) proceeds by inversion of both carbon atoms and leads[72] to the cyclopropane (**18**). This is

$(EtO)_2PO\bar{C}HCO_2Et$

17 **18**

a convenient way of converting an oxirane to a cyclopropane.

E. Acid-catalysed openings of oxirane rings

In acid-catalysed openings of oxirane rings the direction of opening is not always the same as in the simple nucleophilic reaction. It is clear that protonation of the oxygen atom occurs (e.g., **20**) and then direct attack on the protonated oxirane, or ring opening to a carbonium ion followed by nucleophilic attack at the positive centre, can occur. Either pathway can be followed, depending on the precise conditions and substituents, and of course substituents that stabilise carbonium ions favour the latter pathway.[73]

The rates of opening of the epichlorhydrin ring (**19**), when the ring is protonated, by Cl^- and OH^- have been examined and are about 400 times as fast as the comparable basic reactions. If the protonated ring can open to give a

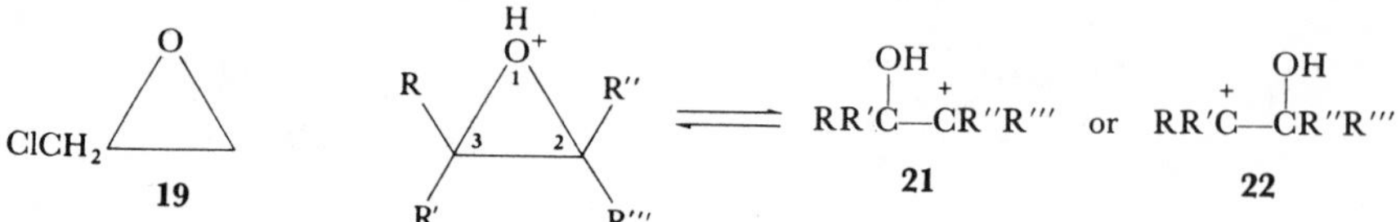

$ClCH_2$ **19** R R' R'' R''' H O^+ 1 2 3 **20** ⇌ OH $RR'C$—$\overset{+}{C}R''R'''$ **21** or OH $RR'\overset{+}{C}$—$CR''R'''$ **22**

carbonium ion, which can then by a fast reaction combine with the anion to give the product, then it would appear that the ring would open to give the more stabilized carbonium ion, which in general has the positive charge on the most highly substituted carbon atom. This leads to the racemization of the carbon atom bearing the charge and the opposite mode of ring opening to that obtained in the alkaline reaction. In the case of 2-methyloxirane (**23**), which would not be expected to give carbonium ions readily, acid-catalysed ring opening in the presence of chloride ions obeys second-order rate equations[74] between pH 7 and 2 when the proportion of the 'abnormal' product (**25**) increases from 14 to 39%. In this case it is clear that attack of the anion occurs on the protonated oxirane (**20**), as for aziridine (p. 13), rather than on a carbonium ion (**21** or **22**). 2-Methyloxirane (propylene oxide) with hydrogen chloride gives 80–90% of **24**

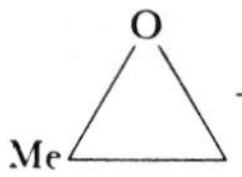

$$\text{Me-oxirane } (\mathbf{23}) \longrightarrow \underset{\mathbf{24}}{MeCHOHCH_2Cl} + \underset{\mathbf{25}}{MeCHClCH_2OH}$$

with 20–10% of **25**. In the acid-catalysed opening with alcohols both isomers are obtained in roughly equivalent proportions. Hydrogen chloride with 2,2-dimethyloxirane gives about 60% of 2-chloro-2-methylpropanol, while in the case of 2-phenyloxirane (styrene oxide) the product ($PhCHClCH_2OH$) of the more stabilised carbonium ion route is obtained virtually exclusively.

Oxirane with aluminium chloride alkylates benzene to give $PhCH_2CH_2OH$, while with benzoyl chloride in pyridine $PhCOCH_2CH_2OH$ is obtained.[75] *cis*- (**26**) and *trans*-2,3-dimethyloxirane with acetonitrile in concentrated sulphuric acid yield the *trans*- (**27**) and *cis*-oxazolines exclusively,[76] so nucleophilic attack on the intact ring must occur.

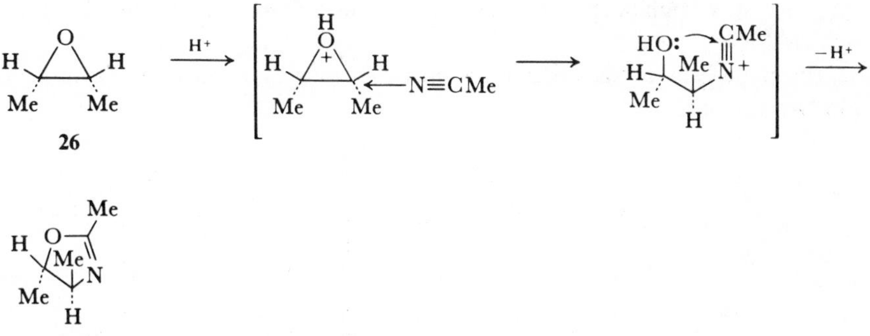

$$\mathbf{26} \xrightarrow{H^+} \left[\text{protonated epoxide} \leftarrow N{\equiv}CMe\right] \longrightarrow \left[HO\text{:}\ \ CMe{\equiv}N^+\right] \xrightarrow{-H^+} \mathbf{27}$$

Epichlorhydrin (**28**) is used in the preparation[77] of triethyloxonium tetrafluoroborate (**29**) a most powerful alkylating agent.

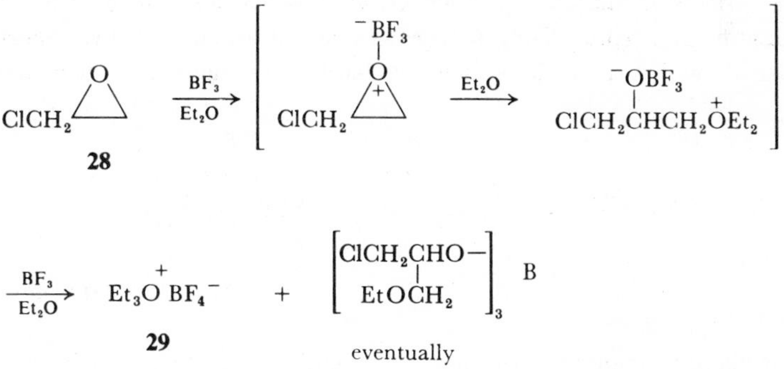

$$ClCH_2\text{-oxirane}\ (\mathbf{28}) \xrightarrow[Et_2O]{BF_3} \left[ClCH_2\text{-oxirane-}O^+\text{-}BF_3^- \xrightarrow{Et_2O} ClCH_2CH(O^-BF_3)CH_2\overset{+}{O}Et_2\right]$$

$$\xrightarrow[Et_2O]{BF_3} Et_3\overset{+}{O}\ BF_4^-\ (\mathbf{29}) + \left[\begin{matrix}ClCH_2CHO-\\ EtOCH_2\end{matrix}\right]_3 B$$

eventually

F. Other openings of oxirane rings

Tetracyanooxirane (**30**) reacts very easily with nucleophilic reagents, for example, boiling with water displaces a cyano group yielding **31**. With pyridine at 0°C the zwitterion (**32**) and mesoxalic nitrile (**33**) are obtained, while reaction with acetylene and ethylene gives the furans **34** and **35**, respectively. These are

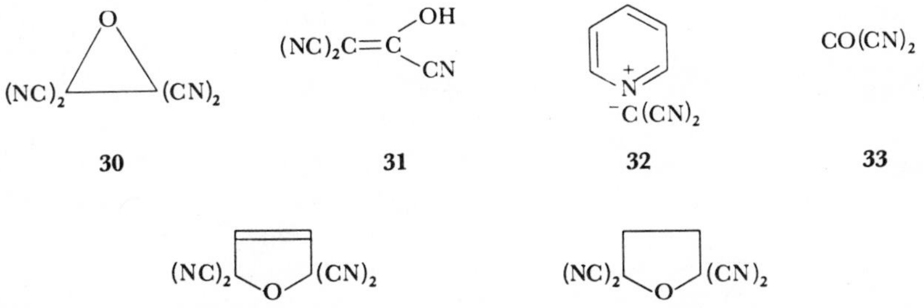

30: $(NC)_2C$—O—$C(CN)_2$ (oxirane)

31: $(NC)_2C{=}C(OH)CN$

32: $C_5H_5\overset{+}{N}{-}\overset{-}{C}(CN)_2$

33: $CO(CN)_2$

34: 2,2,5,5-tetracyano-2,5-dihydrofuran, $(NC)_2$ / $(CN)_2$

35: 2,2,5,5-tetracyanotetrahydrofuran, $(NC)_2$ / $(CN)_2$

the earliest examples of the scission of the carbon–carbon bond of an oxirane.[78]

The photolysis of the oxirane **36** opens the ring by a radical mechanism to give two products.[79]

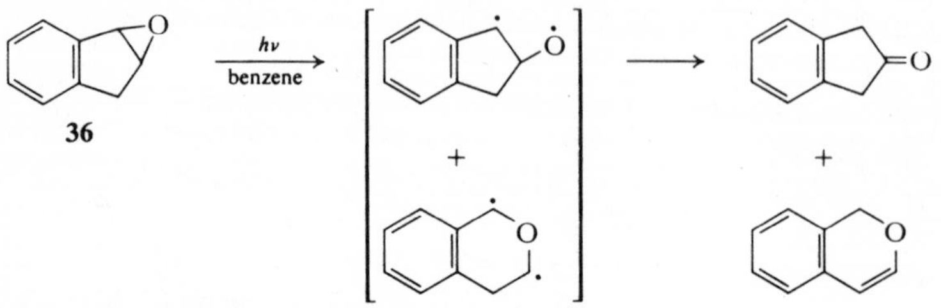

36

G. Functional derivatives of oxiranes

Some 2-methoxyoxiranes (**37**) have been obtained from the α-chloro ketones and sodium methoxide. They behave as reactive acetals and decompose very easily as shown.[80] A small number of 2-acetoxyoxiranes have been obtained

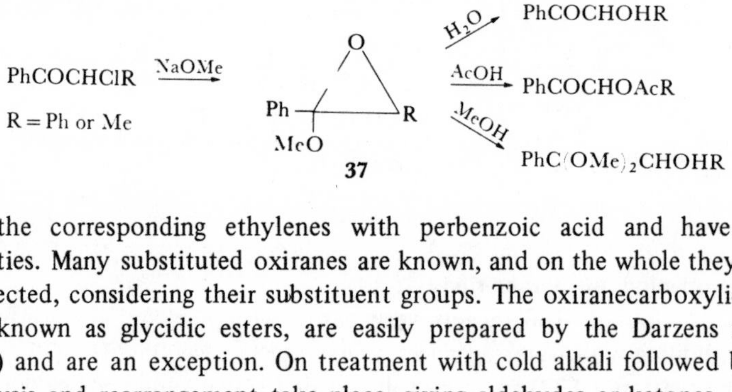

PhCOCHClR $\xrightarrow{\text{NaOMe}}$ **37** → (H_2O) PhCOCHOHR; (AcOH) PhCOCHOAcR; (MeOH) $PhC(OMe)_2CHOHR$

R = Ph or Me

from the corresponding ethylenes with perbenzoic acid and have similar properties. Many substituted oxiranes are known, and on the whole they behave as expected, considering their substituent groups. The oxiranecarboxylic esters, often known as glycidic esters, are easily prepared by the Darzens reaction (p. 37) and are an exception. On treatment with cold alkali followed by acid, hydrolysis and rearrangement take place, giving aldehydes or ketones, often in excellent yield.[81] The isomerization of oxiranes not possessing an ester group

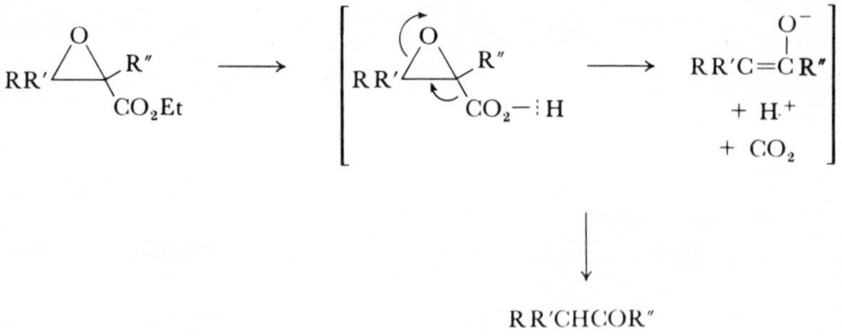

RR′, R″, CO_2Et → [RR′, R″, CO_2—H] → $RR'C{=}C(O^-)R''$ + H^+ + CO_2

↓

RR′CHCOR″

usually proceeds via a hydride shift giving mostly the aldehyde and is catalysed by magnesium bromide, by phosphoric acid, and by bases; ketones and allyl alcohols are occasionally obtained, however.

The only oxirane-2-one that has been obtained[82] at room temperature is the bistrifluoromethyl derivative (**38**), which is a gas with a half-life of 8 hours. It is a very reactive compound and the ring opens with ethanol. 2,2-Dibutyloxiranone, obtained[83] by a similar photolysis, is stable at $-196°C$, but at $-100°C$ it rapidly gives a polymeric ester. α-Bromopropionic acid, when treated with

$$(CF_3)_2C(COF)_2 \xrightarrow[MeSO_3H]{H_2O_2,} (CF_3)_2\text{(cyclic peroxide)} \xrightarrow[CCl_4,\ -15\ °C]{h\nu} CO_2 + (CF_3)_2\text{(oxiranone)}\ \mathbf{38}$$

$$\xrightarrow{EtOH} (CF_3)_2C(OH)CO_2Et + (CF_3)_2C(OEt)CO_2H$$

dilute aqueous alkali, yields lactic acid with the same configuration as the original bromo acid. This is best explained on the basis of the formation of an intermediate oxiran-2-one, inversions of configuration occurring both when it is formed and when it is hydrolysed.

H. Synthetic methods

Attempts to cyclize ethylene glycol to oxirane have not been successful. Vigorous dehydration initially gives the enol form of acetaldehyde, which then rearranges and polymerizes, and other conditions yield dioxane (p. 27).

(1) The direct oxidation of ethylene by air over a silver catalyst under the correct conditions gives about 50% oxirane. The reaction has much industrial value.

$$CH_2{=}CH_2 \xrightarrow{[O]} \text{oxirane} + H_2O + CO_2$$

(2) The oxidation of olefines with isolated double bonds by perbenzoic, the crystallizable 3-chloro derivative, or monoperphthalic acids[84] gives the corresponding oxirane. The reaction is carried out in organic solvents and gives excellent results. The rate of reaction depends greatly on substituents, and the reaction proceeds by *cis* addition to the double bond. Buffered two phase systems are good for sensitive compounds.[85] Even a spiro bisoxirane (**39**) can be made by treating[86] the appropriate allene with buffered peracetic acid, and optically active oxiranes can be made using optically active peracids.[87] The addition[88] of a radical inhibitor when oxidations are carried out hot with

unreactive olefines, such as cyclooctene and methyl methacrylate, gives excellent yields of oxiranes and prevents competitive polymerisations. If the double bond

$$RR'C{=}CR''R''' \xrightarrow{PhCO_3H} \text{oxirane}$$

39

is conjugated with a carbonyl group the reaction rate is extremely slow. However, in such cases hydrogen peroxide and sodium hydroxide, a combination that does not attack isolated double bonds, usually yields the oxirane. Reaction may proceed via a type of Michael addition as indicated:

$$RR'C{=}CHCOMe \xrightarrow{HOO^-} \left[RR'C(OOH)-\bar{C}HCOMe \right] \xrightarrow{-HO^-} RR'\text{(oxirane)}COMe$$

(3) The dehydrohalogenation of ethylene halohydrins by alkali at room temperature is a widely used general method. There is an inversion of configuration (Walden inversion) at the carbon atom losing the halogen and a *trans* cyclization, which has been amply demonstrated in the formation of substituted oxiranes. Many examples of the completely analogous cyclization of glycol monotoluenesulphonyl derivatives to epoxides with the elimination of the toluenesulphonyl group are well known in the sugar series, and an oxirane can be formed only if the OH and *O*-tosyl groups are initially *trans*.

$$HO-CH_2-CH_2-X \xrightarrow[-H^+]{base} {}^-O-CH_2-CH_2-X \longrightarrow \text{oxirane} + X^-$$

Similar types of synthesis are the Hofmann decomposition of amino alcohols

$$HO-CH_2-CH_2-{}^+NMe_3 \xrightarrow{-Me_3N} \left[CH_2-CH_2\ (\overset{+}{O}H) \right] \xrightarrow{-H^+} \text{oxirane}$$

and the action of nitrous acid on certain amino alcohols.

$$\text{2-aminocyclohexanol} \xrightarrow{HNO_2} [\text{cation}] \xrightarrow{-H^+} \text{epoxide}$$

(4) Condensation of an aldehyde or ketone with an α-chloro ester in the presence of sodium ethoxide or amide, the Darzens reaction,[81] gives an 'epoxy' ester. The reaction can be used with both aromatic and aliphatic compounds and is mainly of interest as a route to aldehydes and ketones by decomposition of the acid formed on hydrolysis (see p. 34).

$$RR'CO + ClCH_2CO_2Et \xrightarrow[-H^+]{base} \left[RR'C(O^-)CHClCO_2Et \right] \xrightarrow{-Cl^-} RR'C(O)CHCO_2Et \text{ (oxirane)}$$

(5) Diazomethane reacts with aldehydes and ketones to yield mixtures of higher aldehydes, ketones, and oxiranes.[89] Substitution has a very great effect on the course of the reaction, and in an unknown case it is difficult to predict the major product.

$$RR'CO \xrightarrow{CH_2N_2} \left[RR'C(O^-)-CH_2\overset{+}{N}_2 \right] \xrightarrow{-N_2} RR'C(O)CH_2 \text{ (oxirane)} + RCOCH_2R' + R'COCH_2R$$

(6) Oxiranes are obtained in high yield[90] from aldehydes and ketones, lauryl (dodecyl) dimethylsulphonium salts, and 15M NaOH in benzene.

$$C_{12}H_{25}\overset{+}{S}(O^-)(Me)=CH_2 \xrightarrow{PhCHO} Ph\text{-oxirane}$$

I. Natural occurrence and compounds of special interest

The first oxirane found in nature is probably 'auraptin,' a coumarin derivative (**40**). Naturally occurring oxiranes have been reviewed.[91] Benzene oxides (**41**) are intermediates[92] in the hydroxylation of aromatic compounds in plants and animals and are in equilibrium with the monocyclic oxepins (**42**) (p. 450).

MeO, CH_2, O

40

41 ⇌ **42**

A recently discovered[93] antibiotic (**44**) is particularly interesting, as it is the first compound discovered containing a 'benzene dioxide' system, and it can be reversibly converted to a 1,4-dioxocin (**43**).

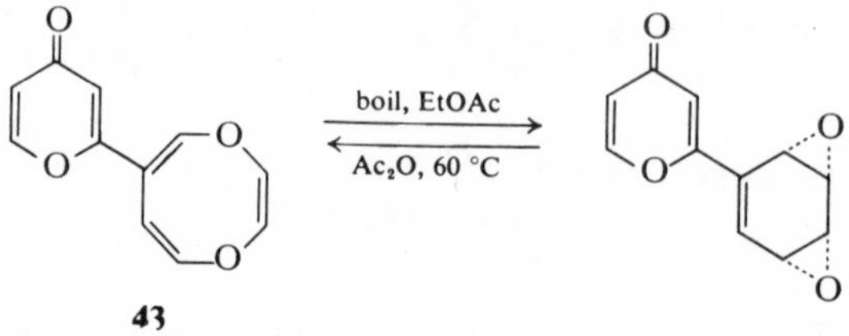

boil, EtOAc

Ac_2O, 60 °C

43

44

Oxirane reacts with proteins at physiological pH and it also gives useful commercial products with nylon, a 'synthetic protein.' 2,2-Bisoxirane (**45**) is active against certain cancer tumours in rats, and the antibiotic fumagillin (**46**) also contains two oxirane rings.[94]

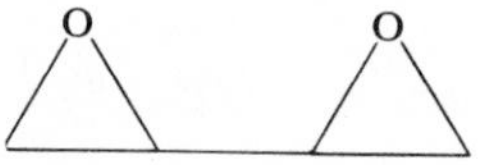

45

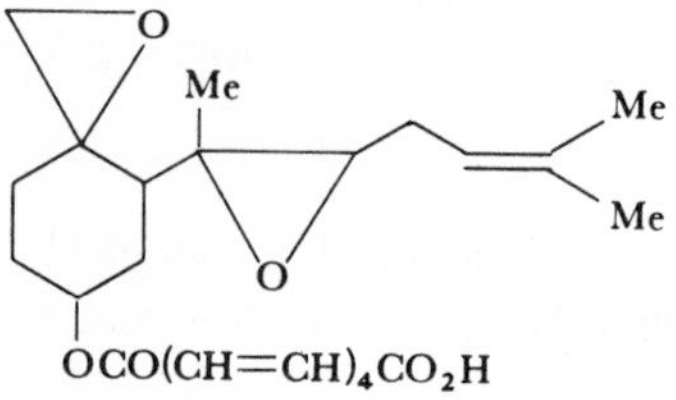

Me

Me

Me

$OCO(CH{=}CH)_4CO_2H$

46

Detergents must contain both fat-soluble and water-soluble groups in the same molecule; the fat-soluble part is largely hydrocarbon and the water-soluble group is usually ionic. However, by treating crude highly alkylated phenols ('largely hydrocarbon'), obtained from phenol, olefines, and boron trifluoride, with oxirane and a trace of sodium hydroxide, both addition and polymerization occur. When $n = 6$–12 in the formula below, the oxygenated side chain has

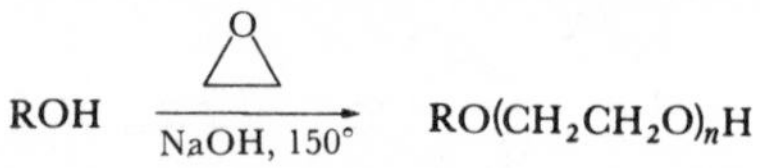

$$ROH \xrightarrow[\text{NaOH, 150°}]{\text{oxirane}} RO(CH_2CH_2O)_nH$$

sufficient hydrophilic properties that are not dependent on a charged ion to make the compound an excellent detergent.[95]

4. THIIRANE

A. Introduction

Thiirene (**1**), the analogue of cyclopropene, has not yet been prepared, but a few derivatives are known. Its dihydro derivative, thiirane (**2**), was first prepared in 1920 by Delépine.[96] It is commonly known as ethylene sulphide and occasionally as thiacyclopropane; many thiiranes are most commonly referred to

as the olefine sulphides. Thiirane attracted little attention until 1948, when its potentialities as a reactive analogue of oxirane were appreciated. No naturally occurring thiiranes are yet known. The hybridisation of the electrons round the sulphur atom in compounds such as **4**, **5**, **17**, and **19** is a subject of speculation.

B. Thiirenes

Thiirenium salts (e.g., **4**) have been detected as intermediates in the solvolysis of vinyl sulphonates (**3**) and in the reaction of phenylsulphenyl chloride with acetylenes,[97] as the eventual products are formed in virtually the same ratio.

Ph O_3SAr / PhS Tolyl (**3**) $\xrightarrow{-ArSO_3^-}$ Ph Tolyl / + / S / Ph (**4**) $\xleftarrow{-Cl^-}$ PhC≡CTolyl + PhSCl

Ar = 2,4,6-$(NO_2)_3C_6H_2$—
Tolyl = 4-MeC_6H_4—

4 ⟶ Ph Cl / PhS Tolyl + Cl Tolyl / Ph SPh

Thiirene 1,1-dioxides (**5** and **6**) can be obtained in two ways[98,99] and readily lose sulphur dioxide to give acetylenes.

$MeCHBrSO_2Cl \xrightarrow{Et_3N} [MeCBr{=}SO_2] \xrightarrow[7\,°C]{CH_2N_2}$ Me, Br, O_2, S $\xrightarrow{Et_3N}$ O_2, S, Me **5**

SO_2, PhCHBr CHBrPh $\xrightarrow{Et_3N}$ O_2, S, Ph Ph **6** $\xrightarrow{\Delta}$ SO_2 + PhC≡CPh

6 $\xrightarrow{NH_2OH}$ $PhCH_2C(Ph){=}NOH$

6 $\xrightarrow{PhMgBr}$ $PhSO_2MgBr$ + PhC≡CPh

6 $\xrightarrow{NaOH}$ Ph Ph, H $SO_3^-\,Na^+$

C. Physical properties of structure and thiirane

Thiirane is a colourless liquid very sparingly soluble in water but easily soluble in organic solvents. Its b.p. is 55–56°C, somewhat higher than that of methyl sulphide, whose b.p. is 38°C. The structural constants for the molecule, shown below, have been derived from microwave spectrum measurements.[63] Although the C–S bonds have almost the same length as in dimethyl sulphide (1·81 Å), the C–C bond is intermediate in length between that of ethane

S 48·4° H 1·819 Å 1·078 Å 65·8° 116·0° 1·492 Å H

(1·54 Å) and ethylene (1·35 Å), and the H–C–H angle is between the tetrahedral and the 120° in ethylene. It may be noted that the H–S–H angle in hydrogen sulphide is 92·1°. The dipole moment (1·66 ± 0·03 D) is higher than that of methyl sulphide (1·40 D), the bond angle of which is 104°. These indications that the thiirane ring is highly strained are reinforced by the small change of the C–H vibration frequency (1475 cm^{-1}) in the infrared[21] from that of a normal C–H in a CH_2 group (1465 cm^{-1}). The strain energy of the ring calculated from the heat of combustion gives[20] a value of about 78 kJ (18·6 kcal)/mole, similar to the values obtained for oxirane and aziridine. The ring is likewise easily opened.

D. Chemical properties

Thiirane is much less stable than oxirane, and polymerizes quite easily, even on standing in the dark. It is stabilized by small amounts of mercaptans or hydrogen sulphide, while acetic acid, mineral acids, or strong alkali greatly catalyse the polymerization. The mechanism of the polymerization has not yet been examined. Both ionic (cf. aziridine) and free-radical propagation mechanisms are possible. Substituted thiiranes are usually more stable than the parent compound.

S
Me Me
7
MeI
$-I^-$
Me
S
+
Me Me
8
MeI
\+ $Me_3S^+I^-$
9
BuLi
Bu
S^-
Me Me
10
Li^+
11
\+ BuSLi

cis-2,3-Dimethylthiirane **7** with methyl iodide gives[100] the crystalline salt (**9**) and *cis*-butene. Retention of the initial stereochemistry is usual for this sort of reaction, which proceeds via the thiiranium salt **8**. Salts such as **8** can be reaction intermediates, for example, in the conversion[101] of **12** and **13** to the same alkyl chloride.

$EtSCHMeCH_2OH$
12
$EtSCH_2CHMeOH$
13
HCl
$SOCl_2$
Et
S
+
Me
$EtSCH_2CHMeCl$

Both *cis*- and *trans*-cyclooctene sulphides with trimethyloxonium 2,4,6-trinitrobenzenesulphonate (a very nonnucleophilic cation) give[102] the corresponding thiiranium salts (e.g., **14**), which with iodide give dimethyldisulphide, iodine, and the corresponding cyclooctene stereospecifically. The *cis* salt (**14**) with tetraphenylarsonium chloride yields[103] first the tetravalent sulphur derivative

15, which is thought to have the square pyramidal arrangement shown, and which isomerises to the cyclooctane **16**.

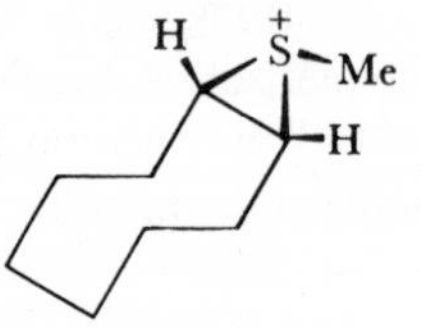

14 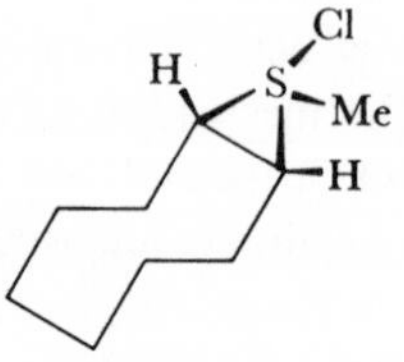 **15** 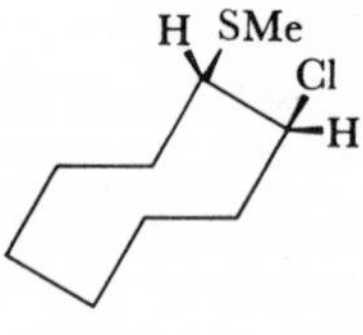 **16**

Thiiranes such as **7** are stereospecifically converted to the corresponding butenes, in this case to *cis*-butene (**11**), by iodine in benzene,[100] by triphenylphosphine, and by butyllithium[104] when the reaction probably proceeds via the anion **10**.

In contrast to thietane, the ring of thiirane is opened by hydrogen peroxide to form 2-hydroxyethylsulphuric acid. However, with sodium periodate thiirane gives the 1-oxide (**17**), which is a liquid, b.p. 46–48°C at 2 torr, with an A_2B_2 proton n.m.r. spectrum proving that the oxygen atom lies significantly away from the plane of the ring.[105] Pyrolysis gives ethylene and sulphur monoxide, which can be trapped as a Diels–Alder adduct (**18**),[106] and with acids the 1-oxide (**17**) decomposes.

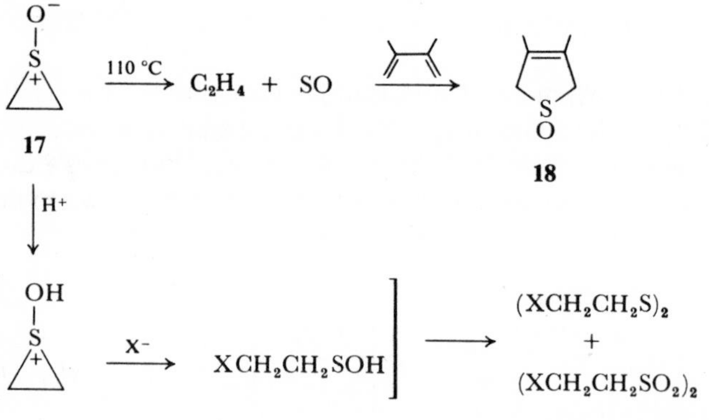

$$\mathbf{17} \xrightarrow{110\ ^\circ\mathrm{C}} C_2H_4 + SO \longrightarrow \mathbf{18}$$

$$\mathbf{17} \xrightarrow{H^+} \left[\text{protonated 1-oxide} \xrightarrow{X^-} XCH_2CH_2SOH \right] \longrightarrow (XCH_2CH_2S)_2 + (XCH_2CH_2SO_2)_2$$

A few highly substituted thiirane 1,1-dioxides (sulphones), for example, **19**, have

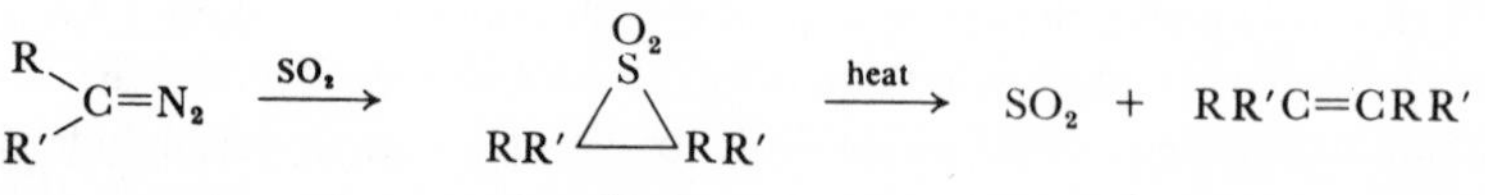

$$RR'C{=}N_2 \xrightarrow{SO_2} \mathbf{19} \xrightarrow{\text{heat}} SO_2 + RR'C{=}CRR'$$

been prepared[107] by the oxidation of the thiiranes by hydrogen peroxide, or as shown below, and on heating decompose to olefines and sulphur dioxide.

Thiirane has many similar ring-opening reactions to its oxygen and nitrogen analogues. The conditions for most of these reactions are very much more critical for good results than those of these analogues, owing to the much greater tendency to polymerization. Some polymer is almost invariably obtained even under the most favoured conditions. Nucleophilic ring opening, for example, of 2-methylthiirane (**21**) by lithium aluminium hydride, gives exclusively[108] the secondary mercaptan (**20**) by the usual S_N2 reaction, and all reactions with nucleophiles are of this type. In contrast to earlier claims, it now appears that acidic reagents usually give[108] mixtures because of the protonation of the sulphur atom and carbonium ion formation, as in the oxirane (p. 31) and aziridine series (p. 19).

$MeCH(SH)CH_3$ (**20**) $\xleftarrow{LiAlH_4}$ 2-methylthiirane (**21**) $\xrightarrow{HCl}$ $MeCH(SH)CH_2Cl$ 76% + $MeCHClCH_2SH$ 24%

No functional derivative of the thiirane ring, such as 2-oxothiirane, which would be a hypothetical dehydration product of 2-mercaptoacetic acid, has yet been authenticated.

E. Synthetic methods

(1) The best synthesis[109] of thiirane is from 2-mercaptoethanol, which with phosgene in ethyl acetate and pyridine yields monothiolethylene carbonate (**22**). This can be decarboxylated to thiirane in 80–88% yield.

CH_2SH–CH_2OH ⟶ **22** $\xrightarrow[\text{trace } Na_2CO_3]{200^\circ}$ thiirane

(2) Treatment of the appropriate oxirane, including oxirane itself, with aqueous or aqueous–ethanolic potassium thiocyanate,[110,111] or less frequently thiourea,[112] at about –7°C has been the most common synthesis of thiiranes. The mechanism is thought to involve two Walden inversions and a rearrangement. In the case of cyclohexene oxide, where the oxide ring always opens in a *trans* fashion (p. 28), the first product must be **23**.

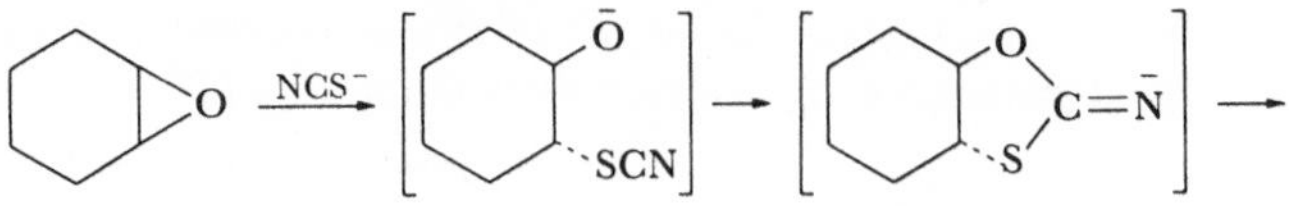

23 24

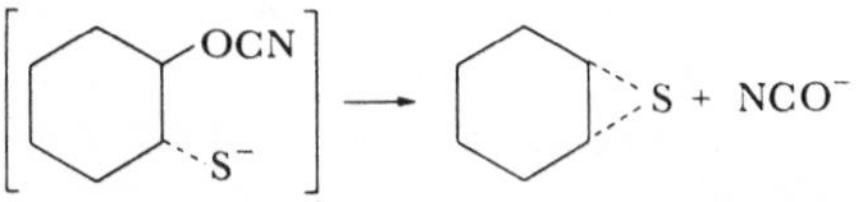

Rearrangement via the hypothetical intermediate **24** gives the *trans* ion, which is correctly oriented for cyclization to the sulphide. The protonated form of the intermediate **24** has been obtained crystalline, and the free base spontaneously loses cyanic acid in ether solution giving the sulphide. The synthesis fails, in support of the mechanism, with cyclopentene oxide, as the requisite hypothetical intermediate would involve two fused five-membered rings and be in a state of considerable strain. Confirmation of the above mechanism has been obtained from a quantitative study of the cyclization of both the *O*- and *S*-monoacetyl derivatives of 2-mercaptoethanol during their conversion by sodium hydroxide into thiirane. From estimations of free thiol and acetyl groups it is clear that the *O*-acetate is converted into the *S*-acetate before cyclization occurs. The mechanism also explains the formation of L-(–)-2,3-dimethylthiirane from D-(+)-2,3-dimethyloxirane.

(3) A small number of 2-chlorothiols obtained from the hydroxythiols by hydrogen chloride have been cyclized to the thiiranes by sodium bicarbonate. This synthesis is complementary to the first in that it can be used for cyclopentene sulphide. The chlorothiol must be *trans* in order to cyclize, and in this case it was obtained from the *trans*-2-hydroxy compound with hydrochloric acid. A Walden inversion to a *cis* product would be expected unless a three-membered sulphonium ring can be formed as an intermediate.

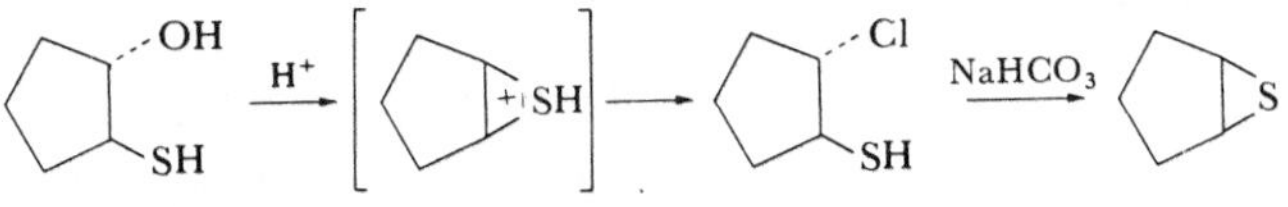

5. DIAZIRENE AND DIAZIRIDINE

Diazirene (**1**), first obtained in 1961,[113] is a colourless stable gas, b.p. –14°C, with a dipole moment of 1·59 ± 0·06 D. Its molecular dimensions, calculated from the microwave spectrum,[114] show that the molecule is highly strained and eliminate other possible structures. In contrast to 'ordinary' diazomethane, for which the diazirene structure was once considered, diazirene is decomposed

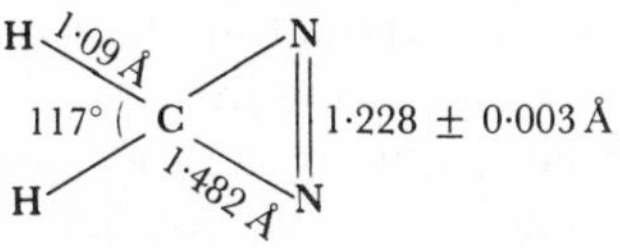

1 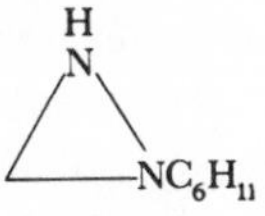 2

comparatively slowly by sulphuric acid with evolution of nitrogen. It is nonbasic and is stable to potassium dissolved in *t*-butyl alcohol, but it is reduced by sodium amalgam to ammonia and methylamine. Cyclohexylmagnesium bromide adds to yield the diaziridine **2**, an example of a general reaction.

Both the thermal decomposition and the photolysis of dimethyldiazirine yield nitrogen and propene.[115]

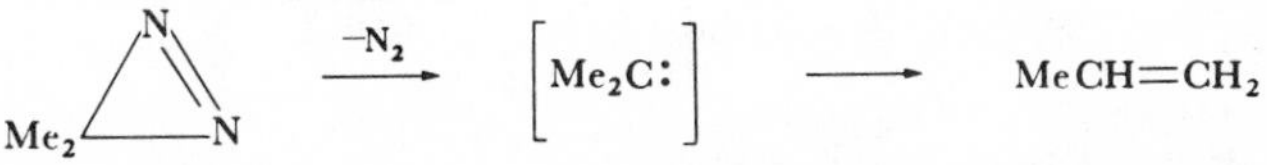

$$Me_2C(N{=}N) \xrightarrow{-N_2} [Me_2C:] \longrightarrow MeCH{=}CH_2$$

Diazirenes, including **1**, are usually obtained from the appropriate aldehyde or ketone, ammonia, and chloramine or *O*-hydroxylaminesulphonic acid, followed by *in situ* oxidation of the resulting diaziridine by chromium trioxide, or better

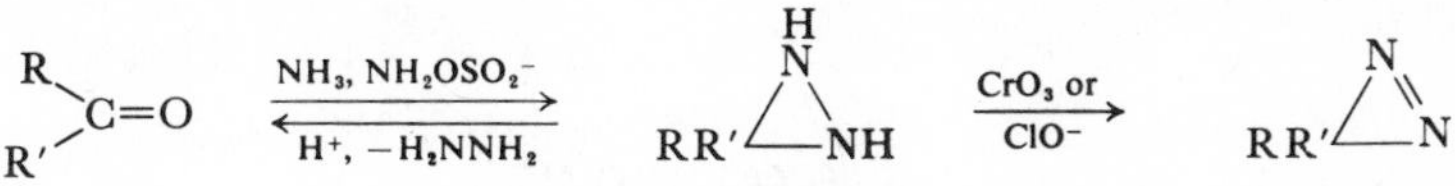

$$RR'C{=}O \underset{H^+,\ -H_2NNH_2}{\overset{NH_3,\ NH_2OSO_2^-}{\rightleftarrows}} RR'C(NH)NH \xrightarrow[ClO^-]{CrO_3\ \text{or}} RR'C(N{=}N)$$

3

iodine and triethylamine,[116] by simple variants of this synthesis,[117] and by the oxidation of amidines by hypohalites.[117] The intermediate diaziridines (**3**), which can be isolated, are hydrolysed by acid to the original carbonyl compound and hydrazine.

Electron diffraction studies[118] have given the size of the diaziridine ring. The length of the N–N bond is between that of hydrazine (1·449 Å) and tetrafluorohydrazine (1·492 Å) and the C–N bond is longer than that of trimethylamine (1·455 Å) and almost the same length as in diazirine (**1**), aziridine (p. 11) and azetidine (p. 55).

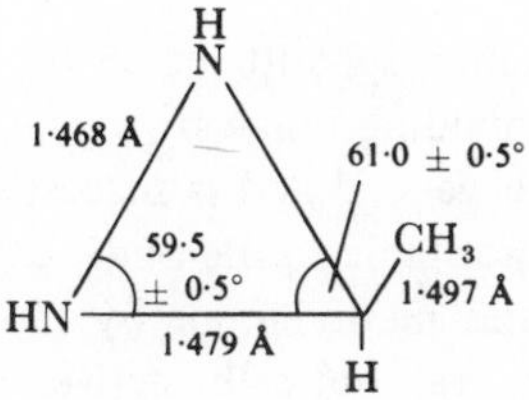

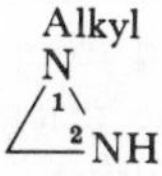

4

Nuclear magnetic resonance studies have shown that for diaziridines such as **4** where the 2-substituent is hydrogen, or different from the alkyl group at position 1, two sets of resonances can be seen up to ca. 100°C, indicating that the nitrogen atom inverts slowly at lower temperatures, and the activation energy for inversion is about 88 kJ (21 kcal)/mole.[119] Diaziridines (e.g., **3**) can be di-*N*-benzoylated under normal conditions and recovered on mild hydrolysis.[120]

Di-*t*-butyldiaziridinone (**5**), prepared as indicated,[121] has high thermal stability and is only slowly decomposed by potassium in *t*-butanol. The N–N bond is broken by many reducing agents[122] to give *N,N′*-di-*t*-butyl urea (**6**).

O=C(NH*t*-Bu)(NCl*t*-Bu) —*t*-BuOK→ **5** (*t*-Bu, N, N*t*-Bu, O) ⇌ [(a) dry HCl, (b) H_2O / *t*-BuOK] ClCON*t*-BuNH*t*-Bu

5 —H_2, Pd/C→ *t*-BuNHCONH*t*-Bu (**6**)

5 —$NaBH_4$→ *t*-BuNH*t*-Bu NCHO

6. OXAZIRIDINE

Structures involving the oxaziridine ring (**1**) have been suggested, without experimental justification, for a number of compounds. Subsequently such structures fell into disrepute, but in 1956 derivatives of oxaziridine were prepared and their ring structures were confirmed chemically beyond all doubt.[123] Oxazirene (**2**) and derivatives are not yet known.

1 (1 O, 2 NH, 3) **2** (1 O, 2 N, 3)

Most authentic oxaziridines contain a substituent on the nitrogen atom and have been prepared from the corresponding imine by oxidation with anhydrous peracetic acid. The reaction is quite general and is successful when the parent imine and the resulting oxaziridine are stable to the acidic conditions. Oxidation of the imine from benzophenone and methylamine by (1*S*-)peroxycamphoric acid gave the oxaziridine **3**, which was optically active, indicating that the nitrogen atom in this ring system inverts slowly; this is the earliest example of asymmetric induction involving trivalent nitrogen.[124]

$$RR'C{=}NR'' \xrightarrow{MeCO_3H} \text{RR'}\overset{O}{\triangle}\text{NR''} \qquad \text{Ph}_2\overset{O}{\triangle}\text{NMe}\ (\mathbf{3})$$

Oxaziridines have also been obtained[125] by the irradiation of nitrones, for example, of **5** yielding **4**, and by treatment of aldehydes or ketones with hydroxylamine *O*-sulphonic acid.[126] The simple oxaziridines are colourless liquids, and 2-*t*-butyloxaziridine has a b.p. of 52–54°C at 75 torr.

$$RR'C{=}O + NH_2OSO_3H \longrightarrow \text{RR'}\overset{O}{\triangle}\text{NH}$$

The ring structure was rigorously proved in the case of 2-*t*-butyl-3-phenyloxaziridine by the degradation sequences shown. This oxaziridine (**4**) can be considered an 'electronic tautomer' of the nitrone (**5**), but their separate identities are beyond all question. The ultraviolet absorption spectrum of the oxaziridine (**4**) is that expected of a compound possessing an isolated phenyl ring, while that of the nitrone shows a strong additional band. Oxaziridines, in contrast to the nitrones, oxidize both iodide and chloride ions to the corresponding halogens, and the iodide oxidation has been used quantitatively.

$$PhCH{=}NBu\text{-}t \underset{(a)}{\overset{MeCO_3H}{\rightleftharpoons}} \text{Ph}\overset{O}{\triangle}\text{NBu-}t\ (\mathbf{4}) \xrightarrow{H_2SO_4} t\text{-BuNHOH} + PhCHO$$

$$\mathbf{4} \underset{\text{MeCN, boil}}{\overset{\text{u.v. light}}{\rightleftharpoons}} PhCH{=}\overset{+}{N}(O^-)Bu\text{-}t\ (\mathbf{5}) \xrightarrow{(a)} PhCH_2N(OH)Bu\text{-}t$$

$$\mathbf{4} \longrightarrow PhCONHBu\text{-}t\ (\mathbf{6})$$

(a) $LiAlH_4$, same conditions

A further differentiation has proved possible in the case of the oxaziridine (**7**), which has been obtained partially resolved. It contains an asymmetric carbon atom, while the corresponding nitrone (**8**) does not. Ferrous ions initiate a free-radical reaction with **4** leading to **6**.

Me, i-Pr, O, NPr-n
7

$Me(i\text{-}Pr)C{=}\overset{+}{N}Pr\text{-}n$, O^-
8

Oxaziridine rings are normally opened by acids in the presence of 2,4-dinitrophenylhydrazine, which traps the carbonyl compounds liberated. If the *N*-substituent is tertiary a migration of one group onto the nitrogen atom can take place to allow an otherwise similar decomposition; an alternative scission of the oxaziridine ring by acid is, however, mentioned in the previous paragraph. Oxaziridines are stable to sodium methoxide if the substituent on the

$$\text{(O, NCMe}_3\text{ oxaziridine)} \xrightarrow{H^+} \left[HOCH_2\overset{+}{N}CMe_3\right] \longrightarrow \left[HOCH_2N\overset{+}{C}Me_2 \atop Me\right] \longrightarrow CH_2O + Me_2CO + MeNH_2\ (67\%)$$

nitrogen atom is tertiary. If it is secondary, the molecule decomposes to the same products as those obtained with acids. If there is no substituent *N*-acylation is effected by acyl halides.[127]

The oxadiaziridine **9**, a heterocyclic compound with no ring carbon atoms, is much less stable than structurally similar oxaziridines and has been prepared[128] as shown below.

$$t\text{-}Bu\overset{+}{N}(O^-){=}Nt\text{-}Bu \underset{\Delta}{\overset{h\nu}{\rightleftharpoons}} t\text{-}BuN\text{—}Nt\text{-}Bu\ (\text{O bridged})$$

9

GENERAL BIBLIOGRAPHY

ALL RINGS

Heterocyclic Compounds with Three and Four Membered Rings. ed. A. Weissberger, Part I. *Ethylene Oxides*, A. Rosowsky; *Aziridines*, P. E. Fanta; *Ethylene Sulphides*, D. D. Reynolds and D. L. Fields; *Oxaziranes*, W. O. Emmons. *Chemistry of Heterocyclic Compounds*, Vol. 19, Interscience, Wiley, New York, 1964.

AZIRIDINE

Ethylenimine and Other Aziridines, O. C. Dermer and G. E. Ham, Academic Press, New York, 1969.

AZIRINE

F. W. Fowler, *Adv. Heterocycl. Chem.*, **13**, 45 (1972).

OXAZIRIDINE

J. F. Dupin, *Bull. Soc. Chim. Fr.*, 3085 (1967).

OXIRANE

S. Winstein and R. B. Henderson, in *Heterocyclic Chemistry*, ed. R. C. Elderfield, Vol. I, Wiley, New York (1950); R. E. Parker and N. S. Isaacs, *Chem. Rev.*, **59**, 737 (1959).

THIIRANE

D. S. Tarbell and D. P. Harnish, *Chem. Rev.*, **49**, 1 (1951); M. Sanders, *Chem. Rev.*, **66**, 297 (1966).

REFERENCES

1. C. A. Coulson and T. H. Goodwin, *J. Chem. Soc.*, 2851 (1962) and earlier papers; cf. D. T. Clark, *Theor. Chem. Acta* (Berl.), **10**, 111 (1968).
2. M. J. S. Dewar and C. A. Ramsden, *Chem. Comm.*, 688 (1973).
3. C. U. Pittman, A. Kress, T. B. Patterson, P. Walton and L. D. Kispert, *J. Org. Chem.*, **39**, 373 (1974).
4. A. Hassner and F. W. Fowler, *J. Amer. Chem. Soc.*, **90**, 2875 (1968).
5. D. J. Cram and M. J. Hatch, *J. Amer. Chem. Soc.*, **75**, 33, 38 (1953).
6. A. Hassner and F. W. Fowler, *J. Amer. Chem. Soc.*, **90**, 2869 (1968).
7. A. Hassner and D. J. Anderson, *J. Org. Chem.*, **39**, 3070, (1974); V. Nair, *ibid.*, **37**, 2508 (1972).
8. A. G. Hartmann and D. A. Robertson, *J. Amer. Chem. Soc.*, **89**, 5974 (1967).
9. J. H. Bowie and B. Nussey, *J.C.S. Perkin I*, 1693 (1973).
10. V. Nair and K. H. Kim, *J. Org. Chem.*, **39**, 3763 (1974).
11. A. Padwa, M. Dharan, J. Smolanoff, and S. I. Wetmore, *J. Amer. Chem. Soc.*, **95**, 1954 (1973).
12. W. Sieber, P. Gilgen, S. Chaloupka, H-J. Hansen and H. Schmid, *Helv. Chim. Acta*, **56**, 1679 (1973).
13. A. G. Hartmann, D. A. Robertson, and B. K. Gillard, *J. Org. Chem.*, **37**, 322 (1972); G. R. Harvey and K. W. Ratts, *ibid.*, **31**, 3907 (1966).
14. R. F. Parcell, *Chem. Ind. (Lond.)*, 1396 (1963).
15. E. F. Ullman and B. Singh, *J. Amer. Chem. Soc.*, **88**, 1844 (1966).
16. D. T. Clark, *Theor. Chem. Acta*, **15**, 225 (1969).
17. T. L. Gilchrist, G. E. Gymer, and C. W. Rees, *J.C.S. Perkin I*, 1 (1975) and earlier papers.
18. B. Bak and S. Skaarup, *J. Mol. Structure*, **10**, 385 (1971).
19. R. E. Carter and T. Drakenberg, *Chem. Comm.*, 582 (1972).
20. J. D. Cox, *Tetrahedron*, **19**, 1175 (1963).
21. L. J. Bellamy, *The Infra-Red Spectra of Complex Molecules*, Methuen, London, and Wiley, New York, 1954.
22. G. A. Olah and P. J. Szilagyi, *J. Amer. Chem. Soc.*, **91**, 2949 (1969).
23. A. F. Graefe and R. E. Meyer, *J. Amer. Chem. Soc.*, **80**, 3939 (1958).
24. D. R. Crist and N. J. Leonard, *Angew. Chem. Int. Ed.*, **8**, 962 (1969).
25. A. T. Bottini and R. L. Van Etten, *J. Org. Chem.*, **30**, 575 (1965).
26. J. E. Early, C. E. O'Rourke, L. B. Clapp, J. O. Edwards, and B. C. Lawes, *J. Amer. Chem. Soc.*, **80**, 3458 (1958).
27. E. M. Schultz and J. M. Sprague, *J. Amer. Chem. Soc.*, **70**, 48 (1948).
28. T. G. Traylor, *Chem. Ind. (Lond.)*, 649 (1963).
29. G. R. Boggs and J. T. Gerig, *J. Org. Chem.*, **34**, 1484 (1969).

30. F. A. Carey and C. J. Hayes, *J. Org. Chem.*, **38**, 3107, (1973); S. J. Brois, *J. Amer. Chem. Soc.*, **92**, 1079 (1970).
31. D. Felix and A. Eschenmoser, *Angew. Chem. Int. Ed.*, **7**, 224 (1968).
32. S. J. Brois, *J. Amer. Chem. Soc.*, **90**, 506, 508 (1968).
33. A. Padwa and A. Battisti, *J. Org. Chem.*, **36**, 230 (1971).
34. R. Annunziata, R. Fornasies, and F. Montanari, *Chem. Comm.*, 1133 (1972).
35. H. E. Baumgarten, J. R. Fuerholzer, R. D. Clark and R. D. Thompson, *J. Amer. Chem. Soc.*, **85**, 3303 (1963).
36. H. E. Baumgarten, R. D. Clark, L. S. Endres, L. D. Hagemeier, and V. J. Elia, *Tetrahedron Lett.*, 5033 (1967).
37. I. Lengel and J. C. Sheehan, *Angew. Chem. Int. Ed.*, **7**, 25 (1968).
38. J. C. Sheehan and M. Mehdi Nafissi-V., *J. Amer. Chem. Soc.*, **91**, 1176 (1969).
39. I. C. Paul, A. H-J. Wang, E. R. Talaty and A. E. Dupuy, *Chem. Comm.*, 43 (1972).
40. R. Ghirardelli and H. J. Lucas, *J. Amer. Chem. Soc.*, **79**, 734 (1957).
41. H. A. Hageman, *Org. React.*, **7**, 198 (1953).
42. R. D. Clark and G. K. Helmkamp, *J. Org. Chem.*, **29**, 1316 (1964).
43. A. Padwa and L. Hamilton, *J. Org. Chem.*, **31**, 1995 (1966).
44. J. H. Hall, R. Huisgen, C. H. Ross and W. S. Hear, *Chem. Comm.*, 1188 (1971).
45. P. Dowd and K. Kang, *Chem. Comm.*, 258 (1974).
46. A. B. Turner, H. W. Heine, J. Irving and J. B. Bush, *J. Amer. Chem. Soc.*, **87**, 1050 (1965).
47. P. G. Gassman and D. K. Dygos, *J. Amer. Chem. Soc.*, **91**, 1543 (1969).
48. A. Padwa and W. Eisenardt, *J. Amer. Chem. Soc.*, **90**, 2442 (1968).
49. C. F. H. Allen, F. W. Spangler, and E. R. Webster, *Org. Synth.*, **30**, 38 (1950).
50. W. Funke, *Angew. Chem. Int. Ed.*, **8**, 70 (1969).
51. K. Ichimura and M. Ohta, *Bull. Chem. Soc. Jap.*, **40**, 432 (1967).
52. Y. Langlois, H. P. Husson, and P. Potier, *Tetrahedron Lett.* 2085 (1969).
53. A. Hassner, G. J. Matthews, and F. W. Fowler, *J. Amer. Chem. Soc.*, **91**, 5046 (1969).
54. L. A. Paquette, D. E. Kuhle, J. H. Barrett, and R. J. Haluska, *J. Org. Chem.*, **34**, 2866 (1969).
55. A. Laurent and A. Muller, *Tetrahedron Lett.*, 759 (1969).
56. W. Lwowski and T. W. Mattingly, *Tetrahedron Lett.*, 277 (1962).
57. R. Annunziata, R. Fornasier, and F. Montanari, *J. Org. Chem.*, **39**, 3195 (1974).
58. A. T. Bottini and V. Dev, *J. Org. Chem.*, **27**, 968 (1962).
59. J. B. Patrick, R. P. Williams, W. E. Meyer, W. Fulmor, D. B. Cosulich, R. W. Broschard, and J. S. Webb. *J. Amer. Chem. Soc.*, **86**, 1889 (1964).
60. P. A. Capps and A. R. Jones, *Chem. Comm.*, 321 (1974).
61. J. Fenwick, G. Frater, K. Ogi, and O. P. Strausz, *J. Amer. Chem. Soc.*, **95**, 124 (1973).
62. S. A. Matlin and P. G. Sammes, *J.C.S. Perkin I*, 2623 (1972).
63. G. L. Cunningham, A. W. Boyd, R. J. Meyers, W. D. Gwinn, and W. I. le Van, *J. Chem. Phys.*, **19**, 676 (1951).
64. S. Searles, M. Tamres, and E. R. Lippincott, *J. Amer. Chem. Soc.*, **75**, 2775 (1953).
65. S. P. McManus, C. A. Larson, and R. A. Hearn, *Synth. Comm.*, **3**, 177 (1973).
66. R. H. Cornforth, *J. Chem. Soc.* (*C*), 928 (1970).
67. W. G. Brown, *Org. React.*, **6**, 469 (1951).
68. N. G. Gaylord and E. I. Becker, *Chem. Rev.*, **49**, 413 (1951).
69. C. B. Rose and S. K. Taylor, *J. Org. Chem.*, **39**, 578 (1974).
70. G. H. Posner, *Org. React.*, **19**, 1 (1972).
71. A. J. Bridges and G. H. Whitham, *Chem. Comm.*, 142 (1974).

72. R. A. Izydore and R. G. Ghirardelli, *J. Org. Chem.*, **38**, 1790 (1973).
73. J. G. Pritchard and I. A. Siddiqui, *J.C.S. Perkin II*, 452 (1973); G. Berti, B. Macchia and F. Macchia, *Tetrahedron Letters*, 3421 (1965).
74. J. K. Addy and R. E. Parker, *J. Chem. Soc.*, 915 (1963).
75. D. P. N. Satchwell and V. F. Shvets, *J.C.S. Perkin II*, 995 (1973).
76. R. A. Wohl and J. Cannie, *J. Org. Chem.*, **38**, 1787 (1973).
77. H. Meerwein, *Org. Synth.*, **46**, 113 (1966).
78. W. J. Linn, O. W. Webster, and R. E. Benson, *J. Amer. Chem. Soc.*, 87, 3651 (1965).
79. H. Kristinsson, R. A. Mateer, and G. W. Griffin, *Chem. Comm.*, 415 (1966).
80. C. L. Stevens, M. L. Weiner, and R. C. Freeman, *J. Amer. Chem. Soc.*, **75**, 3977 (1953); C. L. Stevens and E. Farkas, *ibid.*, **74**, 618 (1952).
81. M. S. Newman and B. J. Magerlein, *Org. Reactions*, **5**, 413 (1949); M. Ballester, *Chem. Rev.*, **55**, 283 (1955).
82. W. Adam, J-C. Liu, and O. Rodriguez, *J. Org. Chem.*, **38**, 2269 (1973).
83. W. Adam and R. Rucktäschel, *J. Org. Chem.*, **37**, 4128 (1972).
84. D. Swern, *Chem. Rev.*, **45**, 1 (1949); *Org. Reactions*, **7**, 378 (1953).
85. W. K. Anderson and T. Veysoglu, *J. Org. Chem.*, **38**, 2267 (1973).
86. J. K. Crandall, W. H. Machleder, and M. J. Thomas, *J. Amer. Chem. Soc.*, **90**, 7346 (1968).
87. D. R. Boyd and R. Graham, *J. Chem. Soc. (C)*, 2648 (1969).
88. Y. Kishi, M. Aratani, H. Tanino, T. Fukuyaman, and T. Goto, *Chem. Comm.*, 64 (1972).
89. C. D. Gutsche, *Org. React.*, **8**, 364 (1954).
90. Y. Yano, T. Okonogi, M. Sunaga, and W. Tagaki, *Chem. Comm.*, 527 (1973).
91. A. D. Cross, *Quart. Rev. (Lond.)*, **14**, 317 (1960).
92. J. W. Daly, D. M. Jerina, and B. Witkop, *Experientia*, **28**, 1129 (1972).
93. D. B. Borders and J. E. Lancaster, *J. Org. Chem.*, **39**, 435 (1974).
94. S. T. Young, J. R. Turner, and D. S. Tarbell, *J. Org. Chem.*, **28**, 928 (1963).
95. S. F. Birch, *J. Inst. Petr.*, **38**, 69 (1952).
96. M. Delépine, *Bull. Soc. Chim. Fr.*, **27**, 740 (1920); **29**, 136 (1921); **33**, 703 (1923).
97. G. Modena, G. Scorrano, and U. Tonellato, *J.C.S. Perkin II*, 493 (1973).
98. L. A. Carpino, L. V. McAdams, R. H. Rynbrandt, and J. W. Spiewak, *J. Amer. Chem. Soc.*, **93**, 476 (1971), and earlier papers.
99. L. A. Paquette and L. S. Wittenbrook, *Chem. Comm.*, 471 (1966).
100. G. K. Helmkamp and D. J. Petit, *J. Org. Chem.*, **27**, 2942 (1962), and earlier papers.
101. W. H. Mueller, *Angew. Chem.*, 8, 462 (1969).
102. D. C. Owsley, G. K. Helmkamp, and S. N. Spurlock, *J. Amer. Chem. Soc.*, **91**, 3606 (1969).
103. D. C. Owsley, G. K. Helmkamp, and M. F. Rettig, *J. Amer. Chem. Soc.*, **91**, 5239 (1969).
104. B. M. Trost and S. D. Ziman, *J. Org. Chem.*, **38**, 932 (1973).
105. G. E. Hartzell and J. N. Paige, *J. Org. Chem.*, **32**, 459 (1967); *J. Amer. Chem. Soc.*, 88, 2616 (1966) and earlier papers.
106. R. M. Dodson and R. F. Sauers, *Chem. Comm.*, 1189 (1967).
107. D. C. Dittmer and G. C. Levy, *J. Org. Chem.*, **30**, 636 (1965).
108. N. V. Schwartz, *J. Org. Chem.*, **33**, 2895 (1968).
109. D. D. Reynolds, *J. Amer. Chem. Soc.*, **79**, 4951 (1957).
110. H. R. Snyder, J. M. Stewart, and J. B. Ziegler, *J. Amer. Chem. Soc.*, **69**, 2672 (1947).
111. C. C. Price and P. F. Kirk, *J. Amer. Chem. Soc.*, **75**, 2396 (1953).

112. C. C. J. Culvenor, W. Davies, and N. S. Heath, *J. Chem. Soc.*, 282 (1949).
113. E. Schmitz, *Adv. Heterocycl. Chem.*, **2**, 83 (1963); *Angew. Chem. Int. Ed.*, **3**, 333 (1964).
114. L. Pierce and V. Dobyns, *J. Amer. Chem. Soc.*, **84**, 2651 (1962).
115. H. M. Frey and I. D. R. Stevens, *J. Chem. Soc.*, 3514 (1963).
116. R. F. R. Church and M. J. Weiss, *J. Org. Chem.*, **35**, 2465 (1970).
117. W. H. Graham, *J. Org. Chem.*, **30**, 2108 (1965); *J. Amer. Chem. Soc.*, 87, 4396 (1965).
118. V. S. Mastryukow, O. V. Dorfeeva, and L. L. Vilkov, *Chem. Comm.*, 397 (1974); cf. A. Nabeya, Y. Tamura, T. Kodama, and Y. Iwakura, *J. Org. Chem.*, **38**, 3758 (1973).
119. A. Mannschreck and W. Seitz, *Angew. Chem. Int. Ed.*, **8**, 212 (1969).
120. E. Schmitz, D. H. Habisch, and C. Gründemann, *Chem. Ber.*, **100**, 142 (1967).
121. F. D. Greene and J. C. Stowell, *J. Amer. Chem. Soc.*, **86**, 3569 (1964); F. D. Greene, W. R. Bergmark, and J. G. Pacifici, *J. Org. Chem.*, **34**, 2263 (1969).
122. A. J. Fry, W. E. Britton, R. Wilson, F. D. Greene, and J. G. Pacifici, *J. Org. Chem.*, **38**, 2620 (1973).
123. W. D. Emmons, *J. Amer. Chem. Soc.*, **78**, 6208 (1956); **79**, 5739 (1957).
124. F. Montanari, I. Moretti, and G. Torre, *Chem. Comm.*, 1694 (1968).
125. J. S. Splitter and M. Calvin, *J. Org. Chem.*, **23**, 651 (1958).
126. E. Schmitz and S. Schramm, *Chem. Ber.*, **100**, 2593 (1967); E. Schmitz, R. Ohme, and S. Schramm, *Chem. Ber.*, **100**, 2600 (1967).
127. E. Schmitz, R. Ohme, and S. Schramm, *Tetrahedron Lett.*, 1857 (1965).
128. S. S. Hecht and F. D. Greene, *J. Amer. Chem. Soc.*, **89**, 6761 (1967).

CHAPTER II

HETEROCYCLIC ANALOGUES OF CYCLOBUTANE

Four-membered rings are not as highly strained as the corresponding three-membered rings, but are more difficult to prepare by the direct cyclization of straight-chain intermediates. This is partly because the atoms that must combine, unlike those that join to give a three-membered ring, can alter their relative positions considerably under the influence of thermal motion. It is only when the cyclizing atoms happen to be suitably oriented and the appropriate stimulus applied that the ring can form. The rate of such cyclizations should therefore be much less than those of the corresponding three-membered rings. Studies of the rate of cyclization of 2-bromoethylamine (**1**) and 3-bromopropylamine (**2**), and

$BrCH_2CH_2NH_2$ (**1**) $BrCH_2CH_2CH_2NH_2$ (**2**)

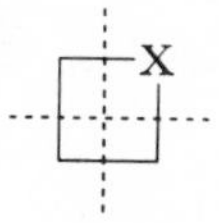

X

3

the alkaline cyclization of **4**, which gives the aziridine **5**, and not the less strained azetidine **6**, support this contention. Four-membered heterocyclic rings are, in general, fairly easily opened, but they can also be split into fragments as shown in **3**. They are, however, more stable than their three-membered analogues to ring-opening reagents, and although similar reactions ensue, more vigorous conditions are usually required. This is perhaps largely because the normal bond angles and distances are less distorted in four-membered than three-membered rings, and there is therefore less 'strain energy' associated with the larger ring.

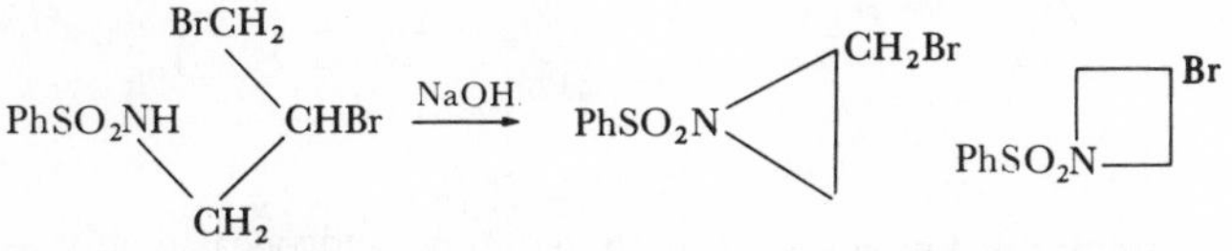

$BrCH_2$, $PhSO_2NH$, $CHBr$, CH_2 (**4**) $\xrightarrow{NaOH}$ $PhSO_2N$ CH_2Br (**5**) $PhSO_2N$ Br (**6**)

Owing to their comparative inaccessibility, the four-membered heterocyclics attracted little attention until about 1943, when the discovery of an azetidin-2-one ring in penicillin gave new interest to the compounds described in this chapter. The first derivative of a four membered heterocyclic ring containing one double bond was clearly demonstrated in 1961. It contains a sulphur atom (p. 77), and the ring is doubtless strained like that of cyclobutene. The preparation of compounds analogous to cyclobutadiene, such as 2,3,4-tris(dimethylamino)azete[1] and the benzo[*b*]azete **10**, has been achieved recently.[2]

1. AZETIDINE

A. Introduction

The unsaturated compounds azete (**7**) and azetine (**8**) have not yet been prepared, but derivatives of both have been obtained in the last few years. The pyrolysis of the benzotriazoline (**9**) yields[2] (64%) 2-phenylbenzo[*b*]azete (**10**),

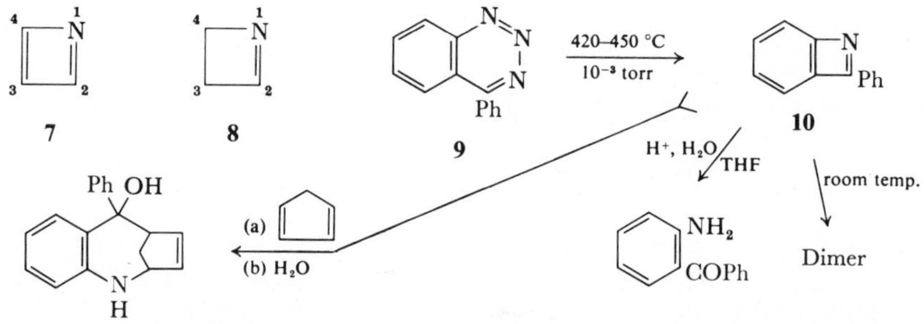

which is isoelectronic with the corresponding cyclobutadiene. It was collected on a cold finger at –78°C, is stable up to ca –40°C, and undergoes Diels–Alder reactions. The four-membered ring is easily opened. A few substituted azetines are known (p. 61), and 2-phenyl-1-azetine has been obtained[2a] by pyrolysis of 1-phenylcyclopropyl-1-azide.

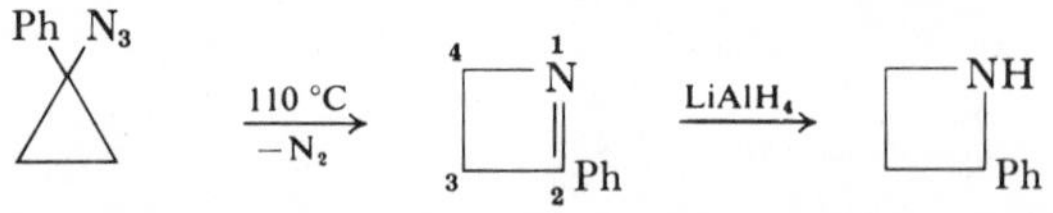

Azetidine, commonly known as trimethylenimine and occasionally as azacyclobutane, was first obtained pure in 1899.

B. Physical properties and structure

Azetidine is a colourless liquid, b.p. 61°C, which smells like ammonia and fumes in air. It is miscible with water or ethanol, and is more basic (pK_a 11.29) than aziridine. An electron diffraction study[3] has shown that the ring is not flat. The C–C–C and C–N–C planes subtend an angle of 37° to each other and, reflecting the ring strain, the bond lengths are a little longer than those of related noncyclic molecules. The ring puckering depends greatly on the substituents, for it is 11 and <1° for the zwitterionic 2- (**22**) and 3-carboxylic acids,[4] respectively. The ring inversion barrier[3] for azetidine is only 5.3 kJ/mole.

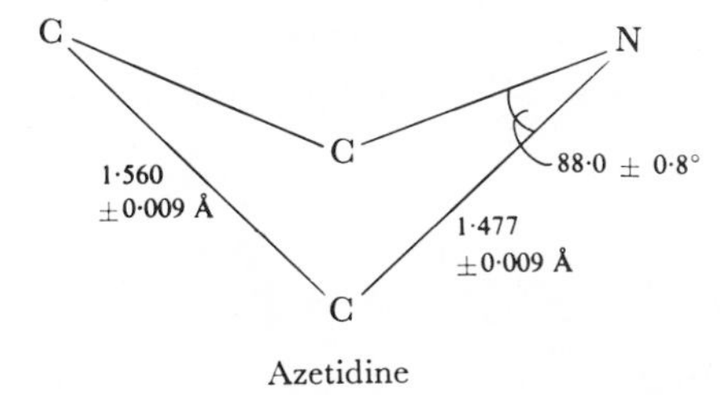

Azetidine

Azetidine appears to form a much more stable complex with trimethylboron than aziridine. It has been suggested that the ring strain of aziridine increases when the nitrogen becomes tetracovalent, but the heats of formation of the other trimethylboron complexes (**11**) given in Table 1 are not in simple agreement with this postulate.

$$(CH_2)_n \quad H\overset{+}{N}-\bar{B}Me_3$$

11

TABLE 1. Trimethylboron complexes (**11**)[5]

Amine	Number of atoms in ring (n + 1)	Heat of formation kJ (kcal)/mole	Extrapolated b.p. at 760 mm
Aziridine	3	73·9 (17·59)	159·5
Azetidine	4	94·5 (22·48)	201·4
Pyrrolidine	5	85·8 (20·43)	191·3
Piperidine	6	80·8 (19·26)	176·5

C. Chemical Properties of Azetidines

Azetidine itself is comparatively stable and is largely unchanged when passed over alumina at 360°C. The azetidine ring is less reactive than that of aziridine. Nevertheless it can be opened quite easily, but much polymerization often takes place. With hydrogen chloride 3-chloropropylamine hydrochloride is formed, while hydrogen peroxide causes decomposition to acrolein and ammonia. The

$$\text{azetidine (NH)} \xrightarrow{2HCl} ClCH_2CH_2CH_2NH_3^+Cl^-$$

acid conditions required to hydrolyse the 4-toluenesulphonyl derivative of azetidine are such that the azetidine formed is instantly decomposed.

Azetidines have many properties typical of secondary aliphatic amines, examples being *N*-benzoylation and the formation of asymmetric ureas with potassium cyanate and thiocyanate. Azetidine with formaldehyde gives the methylolamine (**12**), which has not been obtained pure, as further reaction to form **13** takes place very readily. Azetidine with nitrous acid gives an oily yellow

NCH_2OH (azetidine) **12** $N-CH_2-N$ (bis-azetidine) **13**

N-nitroso compound, b.p. 197°C, and the corresponding *cis*-1,3-dimethyl derivative (**14**) is reduced by lithium aluminium hydride to the substituted hydrazine **15**. This compound is interesting as oxidation to the nitrene (**16**) is followed by expulsion of nitrogen and a disrotatory closure to give mainly (84.5%) the *trans*-cyclopropane (**17**).[6]

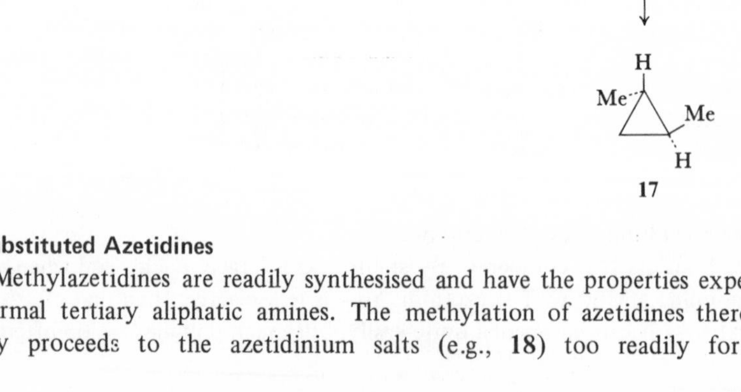

14 (Me, H, N–NO, Me, H) $\xrightarrow{LiAlH_4}$ **15** (Me, H, N–NH_2, Me, H) $\xrightarrow{HgO}$ [**16** (Me, H, N–N:, Me, H)] → **17** (H, Me, Me, H)

D. Substituted Azetidines

N-Methylazetidines are readily synthesised and have the properties expected of normal tertiary aliphatic amines. The methylation of azetidines therefore usually proceeds to the azetidinium salts (e.g., **18**) too readily for the

$$\text{(18)}\ \xrightarrow{Br^-} Me_2N(CH_2)_3\overset{+}{N}Me_2(CH_2)_3Br \longrightarrow \text{Trimer} \longrightarrow \text{Polymer}$$

18

intermediate tertiary amine to be isolated. These salts are crystalline solids, soluble in water but not in nonpolar solvents. The lower members are hygroscopic or deliquescent, and some tend to polymerize. For example, 1,1-dimethylazetidinium bromide is converted into a linear polymer slowly on standing, or rapidly at 200°C, but the 1,1-diethyl compound does not polymerize under these conditions. These salts, in which the C–N bond is relatively long, in contrast to the corresponding aziridines do not usually readily solvolyse, ring open, or undergo 1,3-dipolar additions.[7] The pyrolysis of some azetidinium chlorides merely yields the corresponding dialkylaminopropyl chlorides; this reaction is reversible. The azetidinium hydroxides, prepared with silver oxide and water, are similarly decomposed to 3-hydroxypropylamines. The von Braun decomposition of 1-*n*-butylazetidine with cyanogen bromide[8] is normal and yields the bromocyanamide.

$$\text{NBu (azetidine)} \xrightarrow{BrCN} Br(CH_2)_3NBuCN$$

The 3-chloroazetidine **19** behaves normally with potassium cyanide, but on heating alone gives the aziridine **21**. This is the first reported[9] conversion of an azetidine to an aziridine and probably proceeds through the intermediate formation of the [1.1.0]azabicyclobutane **20**.

NC–(azetidine)–N*t*-Bu ← KCN — Cl–(azetidine)–N*t*-Bu (**19**) —Δ→ [azabicyclobutanium $\overset{+}{N}$*t*-Bu Cl^-] (**20**) → *t*-Bu–N (aziridine)–CH_2Cl (**21**)

An interesting photolysis of an azetidine gives a pyrrole.[10]

Ph–C(=O)–(azetidine, H, Ph)–N*t*-Bu —$h\nu$→ [Ph–C(OH)·, ·Ph, N*t*-Bu → Ph OH bicyclic, Ph, N*t*-Bu] —$-H_2O$→ 2,3-diphenyl-1-*t*-Bu-pyrrole (N, *t*-Bu, Ph, Ph)

E. Synthetic methods

(1) Azetidine can be prepared by treating 1,3-dibromopropane with 4-toluenesulphonamide and alkali. The product cannot be hydrolysed to azetidine, but the *N*-substituent can be reduced off by sodium and amyl alcohol[11] or by sodium bis(2-methoxyethoxy)aluminium hydride.[12]

$$Br(CH_2)_3Br \longrightarrow \text{azetidine-}NSO_2C_6H_4Me \longrightarrow \text{azetidine-}NH + PhMe, \text{ etc.}$$

An interesting variation[13] gives a new route to azetidine-2-carboxylic acid (see also next section).

$$\gamma\text{-butyrolactone} \xrightarrow[\text{(b) } PhCH_2OH,\ HCl]{\text{(a) } Br_2,\ P} (CH_2)_2Br\text{–}CHBrCO_2CH_2Ph \xrightarrow{Ph_2CHNH_2} \text{azetidine-}NCHPh_2\ (2\text{-}CO_2CH_2Ph)$$

$$\xrightarrow{H_2,\ Pd/C} \text{azetidine-}NH\ (2\text{-}CO_2H) + PhMe + Ph_2CH_2$$

(2) The cyclisation of 3-aminopropyl halides or sulphonates gives very poor yields, in contrast to the aziridine synthesis (p. 22), except when substituents are present. Azetidinium halides are obtained from dialkylaminopropyl halides.

$$XCH_2CH_2CH_2NHR \xrightarrow[\text{base}]{-HX} \text{azetidine-}NR$$

X = halogen or OSO_3H $\qquad$ R = H, alkyl, or aryl

A recent modification starting with 3-aminopropanols gives very good results and the best synthesis of azetidine.[14] Michael addition to ethyl acrylate followed by chlorination and cyclisation yields the *N*-substituted azetidine. This

$$CH_2{=}CHCO_2Et + NH_2(CH_2)_3OH \longrightarrow CH_2NH(CH_2)_2CO_2Et\text{–}(CH_2)_2OH$$

$$\xrightarrow[\text{(b) } Na_2CO_3]{\text{(a) } SOCl_2/CH_2Cl_2} \text{azetidine-}N(CH_2)_2CO_2Et \xrightarrow[250\ °C]{KOH} \text{azetidine-}NH$$

is a β-amino ester and undergoes a typical elimination reaction to give azetidine in 54% overall yield from the aminopropanol.

(3) A photochemical synthesis[15] of a 3-hydroxyazetidine, proceeding via a Norrish type II process, has been achieved.

$$\text{H–CH(Me)–NEt–CH}_2\text{–C(=O)Ph} \xrightarrow[\text{Et}_2\text{O}]{h\nu} \left[\cdot\text{CH(Me)–NEt–CH}_2\text{–}\dot{\text{C}}\text{(OH)Ph}\right] \longrightarrow \text{1-ethyl-2-methyl-3-phenyl-3-hydroxyazetidine}$$

F. Natural occurrence and compounds of special interest

Azetidine-2-carboxylic acid (**22**) has been isolated from the leaves of *Convallaria majalis* Lin. (Lily of the Valley), and its structure has been determined by degradation and synthesis.

$$CH_2NH_2\text{–}CH_2CHBrCO_2H \xrightarrow{Ba(OH)_2} \text{azetidine-2-}CO_2H\ (\mathbf{22}) \qquad CH_2NH_2\text{–}CH_2CH(NH_2)CO_2H\ (\mathbf{23})$$

It is formed in the plant from methionine.[16] Both azetidine-2-carboxylic acid and the related amino acid (**23**) occur in *Polygonatum.*

2. AZETIDINONES

A. Introduction

Before 1912, Staudinger[17] initiated work on the chemistry of the azetidin-2-ones or β-lactams. Interest in these compounds was largely lost until 1943, when it was suggested that the penicillins might contain azetidinone rings. Since then a great deal of work has been done on these compounds, including Sheehan's[18] synthesis of penicillin V. This synthesis is still one of the most outstanding of complex syntheses of recent years, especially considering the very intensive unsuccessful efforts made by many groups of workers during World War II. It led the way to other methods of building penicillins.

B. Physical properties and structure

Azetidin-2-one (**1**) was first obtained[19] in 1949. It is a colourless solid, m.p. 73–74°C, b.p. 106°C at 15 mm, very soluble in ethanol and chloroform and moderately so in ether and benzene. X-Ray studies[20] show that the ring is flat and strained.

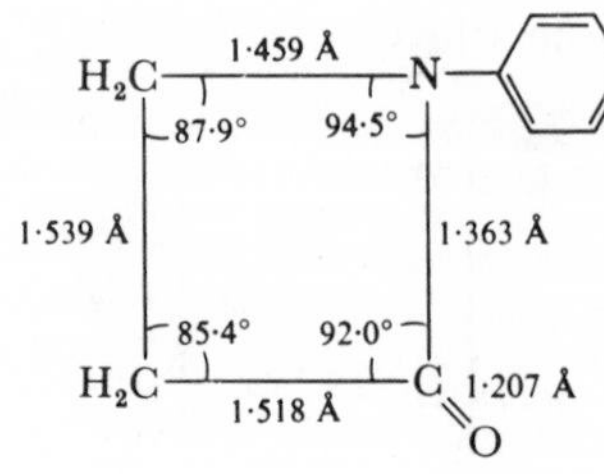

H_2C—(1·459 Å)—N—C_6H_4—Br; angles 87·9°, 94·5°, 85·4°, 92·0°; H_2C—H_2C 1·539 Å; N—C 1·363 Å; H_2C—C 1·518 Å; C=O 1·207 Å

C. Chemical properties

Azetidin-2-one (**1**) is not nearly as reactive as oxetan-2-one (p. 74), although it is much more reactive than normal amides. Both alkaline and acid hydrolyses

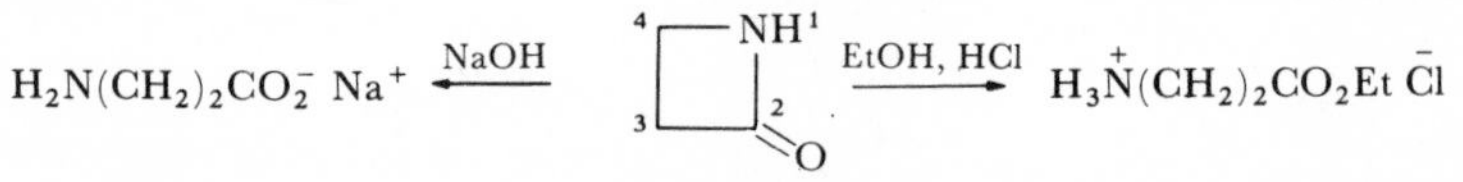

$$H_2N(CH_2)_2CO_2^- \, Na^+ \xleftarrow{\text{NaOH}} \mathbf{1} \xrightarrow{\text{EtOH, HCl}} H_3\overset{+}{N}(CH_2)_2CO_2Et \, \overset{-}{Cl}$$

open the ring, but no reaction occurs with saturated aqueous sodium chloride (contrast oxetan-2-one, p. 74). Alkaline hydrolysis opens the ring very much more slowly than in the case of benzylpenicillin.

D. Substituted derivatives, excluding penicillin

The most noteworthy property of alkyl- and, more particularly, of aryl-substituted azetidin-2-ones is the difficulty with which they are hydrolysed by alkali. This contrasts very markedly with the easy opening of the fused azetidinone ring in the penicillins. In some cases the ring has been opened with

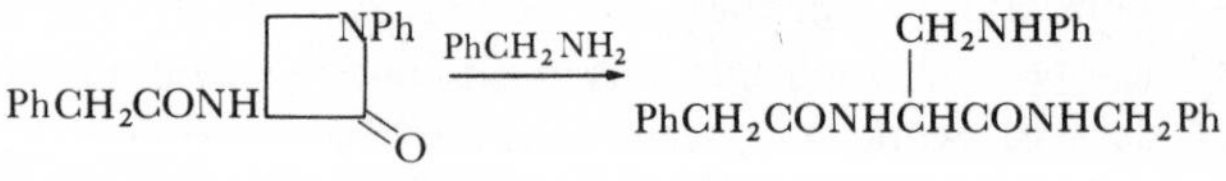

$$\text{PhCH}_2\text{CONH-(azetidinone, NPh)} \xrightarrow{\text{PhCH}_2\text{NH}_2} \text{PhCH}_2\text{CONHCH(CH}_2\text{NHPh)CONHCH}_2\text{Ph}$$

amines, as is shown above. Hydrogenation also sometimes opens an azetidinone ring, when aromatic groups are present; lithium aluminium hydride can also

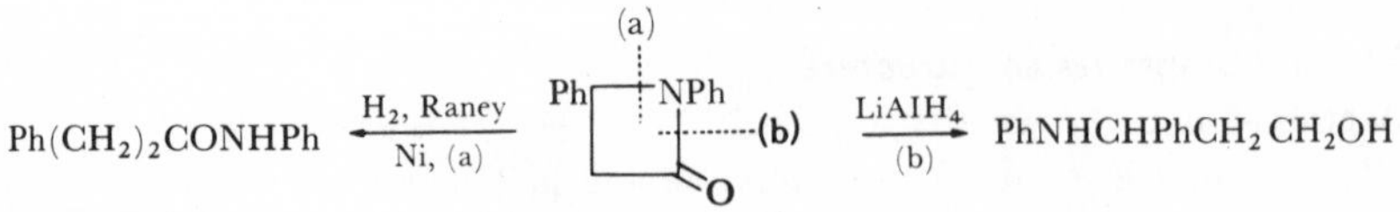

$$\text{Ph(CH}_2)_2\text{CONHPh} \xleftarrow{\text{H}_2\text{, Raney Ni, (a)}} \text{(4-Ph-1-Ph-azetidin-2-one)} \xrightarrow{\text{LiAlH}_4\text{ (b)}} \text{PhNHCHPhCH}_2\text{CH}_2\text{OH}$$

cause both reduction to the azetidine[21] and reductive scission. Photolysis, and thermolysis, can split azetidin-2-ones in alternative ways.[22]

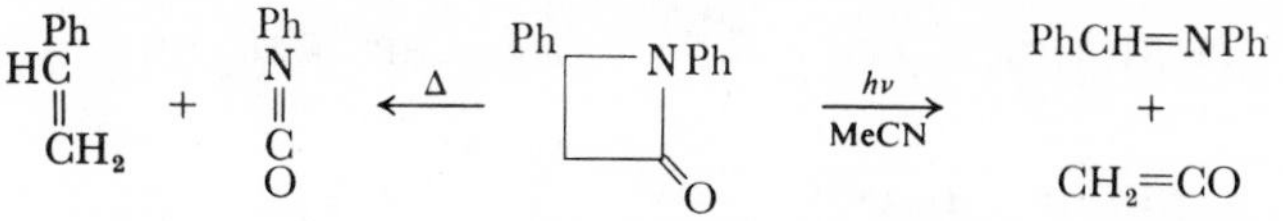

The first azetinone synthesised (2) was obtained[23] by an elimination reaction, and on hydrogenation gave the known 1,4-diphenylazetidin-2-one.

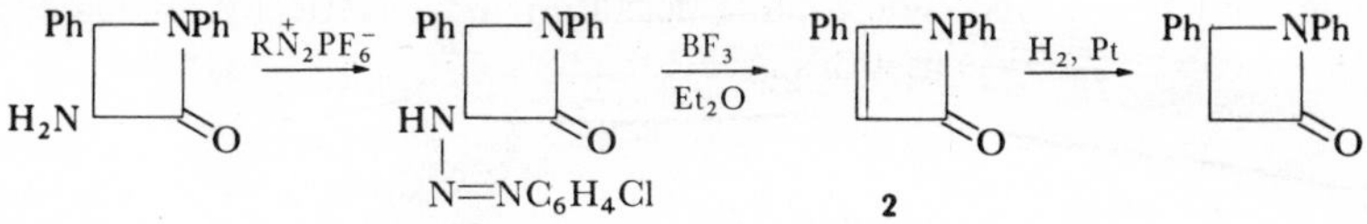

2

The azetidin-2-one **3** is *N*-methylated under the usual conditions with dimethyl sulphate and alkali, but in the absence of alkali the methoxyazetinium salt **6**, is obtained.[24] This is a remarkable transformation, for such *O*-alkylations usually need triethyloxonium tetrafluoroborate (p. 33) or similar reagents. The free base (**5**) is a reactive iminoether and undergoes the expected easy hydrolysis to the amino ester (**4**).

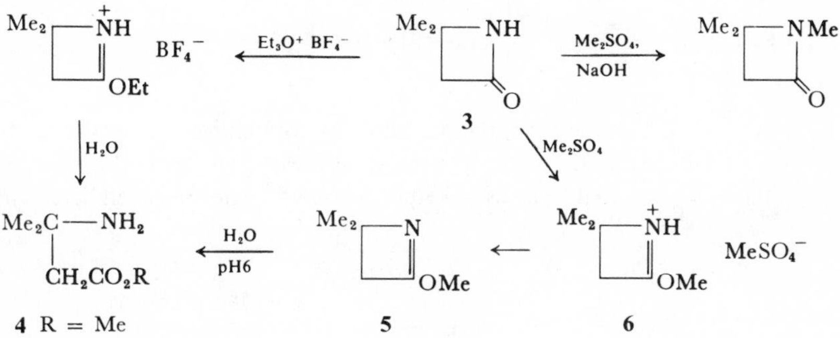

4 R = Me **5** **6**

E. Synthetic methods, excluding penicillin

A large number of routes to azetidin-2-ones have been developed recently because of a new research impetus in the penicillin area. Only the most interesting or important are outlined here.

(1) Thermal decomposition of β-amino acids can proceed in two ways, but no azetidin-2-one is formed. However, cyclization in some instances has been carried out with acetyl chloride or phosphorus trichloride.

$$PhNH(CH_2)_2CO_2H \xrightarrow{heat} PhNH_2 + CH_2{=}CHCO_2H$$

$$PhNHCHPhCPh_2CO_2H \xrightarrow{heat} PhN{=}CHPh + Ph_2CHCO_2H$$

$$PhNHCHPhCH_2CO_2H \xrightarrow{PCl_3}$$ (4-phenyl-1-phenylazetidin-2-one: Ph, NPh, O)

(2) Ethyl 3-aminopropionate is cyclised to azetidin-2-one in very poor yield (0.76%) by ethyl magnesium bromide, but 2,4,6-trimethylphenyl magnesium bromide, which is a poor nucleophile because it is so sterically hindered, gives 50–75% yields in such cyclisations.

$$H_2N(CH_2)_2CO_2Et \longrightarrow$$ (azetidin-2-one: NH, O)

(3) Certain chloroacetamides have been cyclised by triethylamine,[25] or an Amberlite (alkaline) ion-exchange resin[26] to azetidin-2-ones, a carbon–carbon bond being established at cyclisation.

$$(EtO_2C)_2CH{-}NPh(ClCH_2{-}CO) \xrightarrow{-H^+} [(EtO_2C)_2\underline{C}{-}NPh(Cl{-}CH_2{-}CO)] \xrightarrow{-Cl^-}$$ ($(EtO_2C)_2$, NPh, O azetidinone)

(4) One of the earliest syntheses, used by Staudinger,[17] is the direct combination of an imine with a ketene, or with an acid chloride and triethylamine, which could act as a ketene source.[25] The four-membered ring

$$PhCH{=}NPh + R_2C{=}CO \longrightarrow$$ (Ph, NPh, R_2, O azetidinone) R = H or Me

can also be built up in the latter way from an olefine, the stereochemistry of which is retained.[27] As hydrolysis of azetidin-2-ones yields β-amino acids, this could be a useful stereospecific synthesis from olefines.

(Ph/Me olefine) + $NSO_2Cl{=}CO$ $\longrightarrow$ (Ph, Me, NSO_2Cl, O azetidinone) $\xrightarrow{PhSH}$ (Ph, Me, NH, O azetidinone)

(5) Cyclisation of an unsaturated amide, under the conditions of the Michael reaction, unexpectedly gave an azetidin-2-one[28] instead of the less strained alternative five-membered ring.

$(EtO_2C)_2CH$—NPh, $EtO_2CCH{=}CH$—CO $\xrightarrow{\text{piperidine}}$ $(EtO_2C)_2$, NPh, EtO_2CCH_2, O

(6) A ring contraction process gives a number of azetidin-2-one-3-carboxylic acids.[29]

NMe, O, O $\xrightarrow[\text{24 hours}]{IO_4^-,\ pH7,}$ NMe, HO_2C, O

30% yield

(7) A ring expansion process[30] leads sterospecifically to 3-chloroazetidin-2-ones in up to 79% yield.

t-Bu, N, HO_2C, Me, H, H $\xrightarrow[\text{THF}]{SOCl_2}$ [*t*-Bu, H, N, Me, H, C, OSOCl, O ⟶ H, *t*-Bu, N^+, Me, Cl^-, H, O] ⟶ H, Me, N*t*-Bu, Cl, H, O

(8) Photolysis of an unsaturated amide gives[31] an azetidin-2-one, the original stereochemistry being largely retained, and photolysis of 1-methyl-2-pyridone yields a fused azetidin-2-one (p. 264).

Ph, NHPh, CO, Ph $\xrightarrow[C_6H_6]{h\nu}$ Ph, NPh, Ph, O (37%) + Ph, NPh, Ph, O (2·3%)

(9) Pyrrolidine-2,4-diones can be obtained from α-amino acids. Photolysis of the corresponding diazo compounds causes ring contraction to the azetidin-2-ones and this may be compared with synthesis (6) above.[32]

$$\xrightarrow[Et_3N]{MeSO_2N_3,} N_2 \quad \xrightarrow[t\text{-}BuO_2CNHNH_2]{h\nu,} t\text{-}BuO_2CNHNHCO$$

F. The penicillins and cephalosporins

Although 'penicillin' was first reported by Fleming in 1929, he did not investigate it as a potential chemotherapeutic agent. It attracted comparatively little attention until 1940, when Florey, Chain, and their collaborators isolated the material in an inhomogeneous form from a culture of the mould *Penicillium notatum* and showed its enormously powerful effect *in vivo* against various pathogenic organisms. A very great deal of work was done over the next few years towards the commercial production and synthesis of the antibiotic, and in spite of very intensive efforts, a chemical synthesis confirming the structure and giving a reasonable yield was first carried out only in 1957. Penicillin is prepared commercially from the fluid obtained by growing selected strains of *Penicillium* species in suspension in a culture medium. Cold extraction with water-immiscible organic solvents is usually the first stage towards isolation. The antibiotic is marketed as the calcium and sodium salts and as the salts of several organic bases.

Penicillins, and the related cephalosporins, work as antibacterial agents by preventing the normal development of the bacterial cell wall. At present a large research effort is being made on both the chemistry and biochemistry of these antibiotics because many bacteria have developed resistant strains that produce penicillinase. This is an enzyme that opens the four-membered ring before this ring can itself open while acylating and thereby blocking the enzyme responsible for a cross-linking process essential for the bacterial cell wall biosynthesis.

There is a whole family of penicillins, consisting of the basic structure 7 with various substituent groups, which is produced by different strains of *Penicillium notatum.* The most common substituent groups (R) are benzyl, 4-hydroxybenzyl, 2-pentenyl, *n*-amyl, *n*-heptyl, and phenoxymethyl. Penicillin N, an antibiotic obtained from *Cephalosporium sp.,* also contains the same fused-ring system (7), but the side chain ($R = HO_2CCH(NH_2)CH_2CH_2CH_2-$) is derived from D-α-aminoadipic acid. Degradation studies have been carried out on all these penicillins, and in almost every instance the products could have come from more than one parent structure. The rings in the molecule are very easily opened, like thiazolidines (p. 380), but unlike typical azetidine-2-ones. The accepted structure was first proposed on the basis of degradative work, but it remained for Mrs. D. Crowfoot Hodgkin to provide real proof from X-ray crystallographic data. Two of the many important degradation sequences for

Degradation of sodium benzylpenicillin (R = $PhCH_2$)

RCONH– [penicillin, 7] –Me_2, CO_2H —(pH 10–11, H_2O)→ $PhCH_2CONHC(H)(HO_2C)$– [thiazolidine] –Me_2, H, CO_2H

7 Benzylpenicilloic acid

Raney Ni on Na^+ salt ↓ pH 2–5, 27 °C ↓

$PhCH_2CONH$– [azetidinone] –N–CH($CHMe_2$)(CO_2H) + other products

$PhCH_2CONHCH_2$– [thiazolidine] –Me_2, H, CO_2H

8 Benzylpenilloic acid, **9**

$PhCH_2NH_2$ ↓ (a) $HgCl_2$ (b) H_2S ↓

$PhCH_2CONHCH(PhCH_2NHCO)–CH_2$ $NHCH(CHMe_2)(CO_2H)$

$PhCH_2CONHCH_2CHO$ + $HS–CMe_2–C(H)(H_2N)(CO_2H)$

10 **11** D-Penicillamine, **12**

sodium benzylpenicillin are outlined above. The penicillin molecule undergoes a remarkable number of rearrangements and novel reactions; new ones are still being found.[33]

The very easy opening of the fused thiazolidine–azetidinone ring by alkali is in contrast to the behaviour of most azetidin-2-ones, which are quite stable under such conditions. This can be associated with the fact that the ring nitrogen atom in penicillin is pyramidal and the molecule is strongly folded, which reduces resonance stabilisation, of the type that occurs in amides, from that possible in, for example, 1-methylazetidin-2-one where there are no steric constraints. In the desulphurisation step (**7** → **8**) it is very interesting that the configuration at position 5 is completely retained, the entering hydrogen exactly replacing the sulphur atom.[34] The benzylamine opening of **8** to **10** is normal.

The opening of the thiazolidine ring of benzylpenilloic acid (**9**) is also expected. The mercuric ion forms a complex with the penicillamine (**12**), while the aldehyde (**11**) is left in solution. It should be noted that the penicillamine is of the D-configuration, as is the case with a few other amino acids obtained from the degradation of other antibiotics. This is remarkable, as amino acids obtained from other natural sources usually have the L-configuration.

Sheehan and Henery-Logan's[18] synthesis of penicillin V is shown below. The most notable points are (*a*) the use of anhydrous hydrochloric acid to split a tertiary butyl ester to the acid and butylene and/or tertiary butyl chloride, and (*b*) the use of dicyclohexylcarbodiimide (**13**), which is hydrated to the urea (**14**) during the reaction (p. 278), as the cyclodehydration reagent.

Phthalimido: $C_6H_4(CO)_2NCH(CO_2Bu\text{-}t)CHO$ + $HSCMe_2CH(NH_2)CO_2H$ ⟶ $C_6H_4(CO)_2NCH(CO_2Bu\text{-}t)$–thiazolidine($Me_2$, CO_2H) $\xrightarrow{(a)\ H_2NNH_2;\ (b)\ HCl}$

$H_3\overset{+}{N}CH(CO_2Bu\text{-}t)\ \bar{C}l$–thiazolidine($Me_2$, CO_2H) $\xrightarrow[Et_3N]{PhOCH_2COCl}$ $PhOCH_2CONHCH(CO_2Bu\text{-}t)$–thiazolidine($Me_2$, CO_2H) $\xrightarrow{\text{anhydrous HCl}}$

$PhOCH_2CONHCH(CO_2H)$–thiazolidine(Me_2, CO_2H) $\xrightarrow[C_6H_{11}N{=}C{=}NC_6H_{11}\ (\mathbf{13})]{KOH\ (1\ mole)}$ $PhOCH_2CONH$–β-lactam–thiazolidine(Me_2, CO_2H)

Penicillin V

+

$C_6H_{11}NHCONHC_6H_{11}$

14

The biosynthesis of benzylpenicillin by the mould has been studied by the addition of labelled possible intermediates to the nutrient media followed by isolation of the penicillin. The penicillin was then degraded by known routes and the positions of the labelled atoms (C, H, N, and/or S) in the products determined. Labelled phenylacetic acid is incorporated into the side chain, probably before the bicyclic system is formed (below). The locations of two of the hydrogen atoms of added labelled cysteine are shown by asterisks. Using L-valine, both uniformly labelled and with individual atoms labelled, it has been shown that the whole of the carbon skeleton is incorporated into the penicillamine part of the molecule. When each of the methyl groups of the valine is specifically labelled with ^{13}C, the label is incorporated into the appropriate methyl of the penicillin molecule as shown below. If the methyl groups are fully deuteriated no deuterium is lost in the biosynthesis.

The source of the ring nitrogen atom may be valine, but the $^{15}N/^{14}C$ ratio incorporated into the penicillin from doubly labelled valine is much lower than the ratio in the initial valine. This may, or may not, be due to exchange of ^{15}N for ^{14}N by transamination (p. 280).

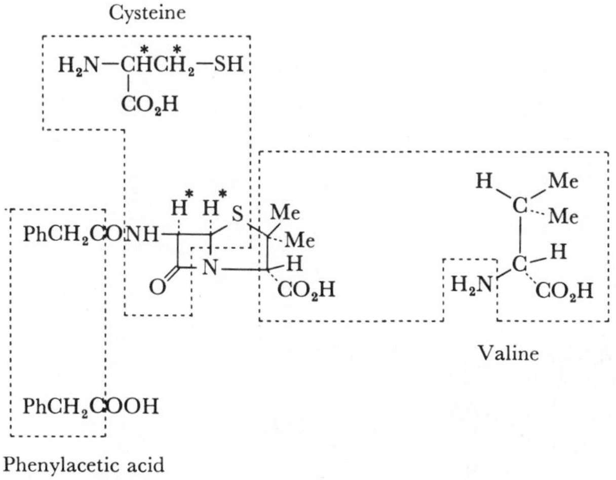

Cysteine

$H_2N-\overset{*}{C}H\overset{*}{C}H_2-SH$ / CO_2H

$PhCH_2CONH$ … H^* H^* S Me Me H CO_2H

H Me C Me H C H_2N CO_2H

Valine

$PhCH_2COOH$

Phenylacetic acid

When the side-chain precursor is omitted from the culture medium in the case of one *Penicillium* species, 6-aminopenicillanic acid (**15**) is obtained instead of penicillin. This discovery has been of great interest, as this acid can be acylated on the amino group by acids that are not incorporated by the mould to give new penicillins that have advantages over the natural ones. Although the isolation of

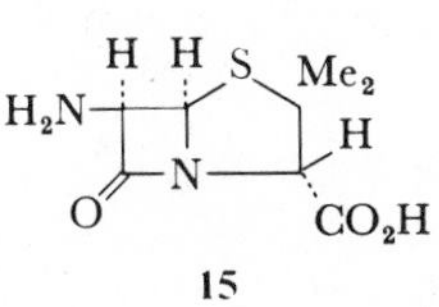

H H S Me_2 H_2N H N O CO_2H

15

this compound could suggest that the moulds build up the 6-aminopenicillanic acid structure and then acylate it to give the various penicillins, it appears that it is not the case. Recent experiments[35] suggest that the tripeptide L-α-aminoadipyl-L-cystenyl-D-valine, synthesised by the mould from L-valine, is the key intermediate. The synthetic tripeptide, with a deuterium atom replacing the α-hydrogen of the valine moiety, is also converted by the mould into penicillin N (7, R = $HO_2CH(NH_2)CH_2\ CH_2CH_2-$) without loss of deuterium. It is then likely that this penicillin is deacylated to 6-aminopenicillanic acid, which is then reacylated to give the 'penicillin' actually isolated from the mould species employed.

A number of interesting schemes for the biosynthesis of the penicillin ring system have been suggested,[36] for example, that shown below, but none have as yet received any biochemical support.

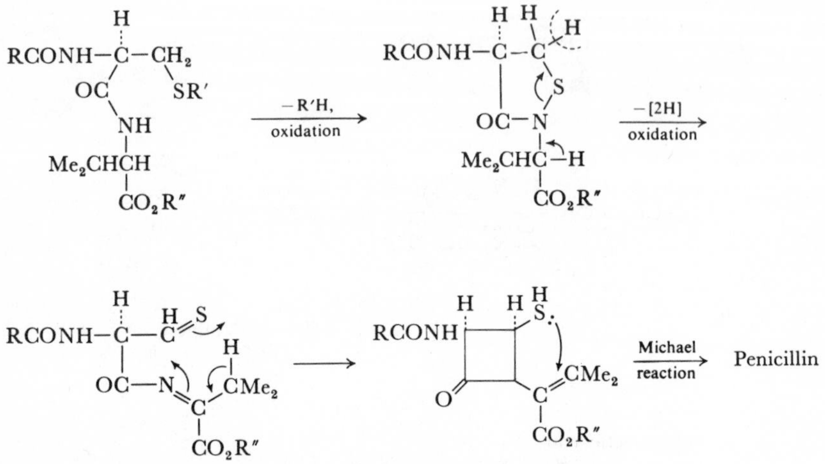

The *Cephalosporium sp.* that yields penicillin N also produces, by an analogous route[37], the closely related antibiotic cephalosporin C (**16**), which has been synthesised chemically.[38] Treatment with nitrosyl chloride removes the

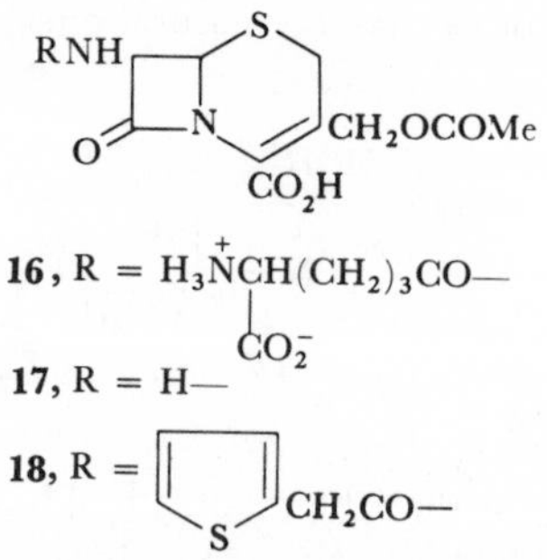

16, R = $H_3\overset{+}{N}CH(CH_2)_3CO$— with CO_2^- on the CH

17, R = H—

18, R = 2-thienyl-CH_2CO—

side chain without opening the four-membered ring and gives the amine (**17**), from which many new cephalosporins, for example, the clinically useful **18**, have been obtained by acylation. The acetoxy group of **16**, is remarkably easily displaced by nucleophilic reagents, even by pyridine, when a pyridinium salt is obtained.

3. OXETENE AND OXETANE

A. Introduction

Although a few reports of oxetenes are scattered throughout the literature, the first derivatives of oxetene (**1**) clearly possessing this ring system were described in 1965. Oxetane (**2**), often known as 1,3-trimethylene oxide or just as trimethylene oxide, was first obtained in 1878 but received little attention until recently. It has a somewhat less strained ring than oxirane, which it resembles in many respects, although it is more sluggish in its reactions.

B. Oxetenes

The first compound (**3**) definitely possessing the oxetene ring was obtained from trifluoroacetone and ethoxyacetylene at -78°C. It isomerises to the unsaturated ester (**4**) at room temperature.[39]. Photolysis of mesityl oxide (**5**)

$(CF_3)_2C{=}O$ + $HC{\equiv}COEt$ $\xrightarrow{-78\ ^{\circ}C}$ $(CF_3)_2$-oxetene-OEt (**3**) $\longrightarrow$ $(CF_3)_2C{=}CHCO_2Et$ (**4**)

under the correct conditions causes a $2\pi + 2\pi$ (concerted?) cyclisation to the oxetene (**6**), which has a half-life of about 12 hours in boiling pentane and can be hydrogenated to the corresponding oxetane.[40]

$Me_2C{=}CH{-}C({=}O)Me$ (**5**) $\xrightarrow[\text{pentane}]{h\nu}$ Me_2-oxetene-Me (**6**) $\xrightarrow[\text{Pd/CaCO}_3]{H_2,}$ Me_2-oxetane-Me

C. Physical properties and structure of oxetane

Oxetane is a colourless liquid, b.p. 47–48°C, which is miscible with water and most organic solvents. It can donate electrons more easily than oxirane, as it gives out more heat on mixing with chloroform (p. 27). Microwave studies[41] have given molecular dimensions and show that the molecule is effectively planar in contrast to the puckered cyclobutane. Its dipole moment is 1.93 D.

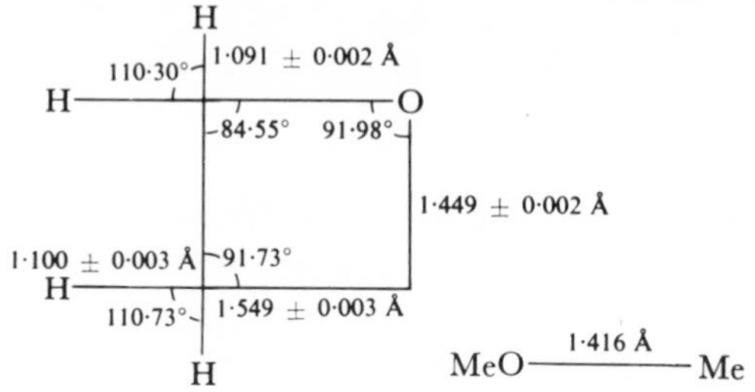

H
110·30° 1·091 ± 0·002 Å
H — O
84·55° 91·98°
1·449 ± 0·002 Å
1·100 ± 0·003 Å 91·73°
H
110·73° 1·549 ± 0·003 Å
H

MeO —1·416 Å— Me

D. Chemical properties of oxetane

Most of the reactions that have been examined are ring-opening reactions analogous to those of oxirane, but they require much more vigorous conditions. In many cases yields are good, but some polymerization occurs. Thermal decomposition over alumina at 250°C yields some propionaldehyde among other products.

Oxetane with hydrogen bromide gives 1,3-dibromopropane, presumably via the bromohydrin, and 1,3-dichloropropane is obtained with phosphorus pentachloride. Reaction also takes place with sodium sulphite, presumably yielding 3-hydroxypropylsulphonic acid, although with potassium bisulphite in the presence of excess sulphur dioxide the bisulphite addition compound of propionaldehyde is formed. With alkyl- or aryllithiums and Grignard reagents, the ring opens as anticipated.

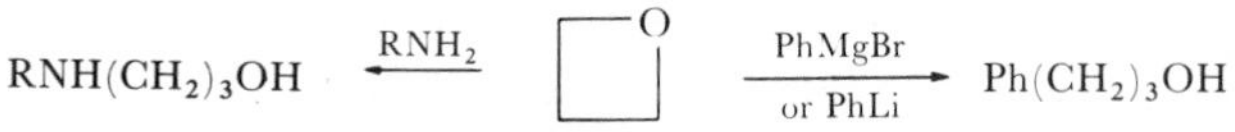

$$RNH(CH_2)_3OH \xleftarrow{RNH_2} \text{oxetane} \xrightarrow[\text{or PhLi}]{\text{PhMgBr}} Ph(CH_2)_3OH$$

Grignard reagents, particularly those from secondary and tertiary bomides, also give 3-bromopropanol, which is formed from the oxetane and the magnesium bromide present in the Grignard reagent (p. 30). The ring opens similarly with alkali-metal mercaptides and phenoxides, with alcohols in the presence of acid catalysts, with amines and with benzene in the presence of aluminium chloride 3-phenylpropanol is formed. Lithium aluminium hydride cleaves the ring, forming *n*-propyl alcohol.

E. Derivatives of oxetane

The few known alkyl- and aryloxetanes behave like the parent compound. Asymmetrically substituted oxetanes are reported to give a single alcohol with

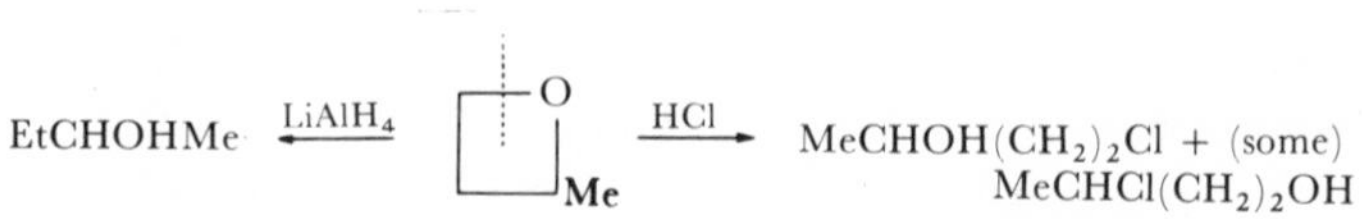

$$EtCHOHMe \xleftarrow{LiAlH_4} \text{2-methyloxetane} \xrightarrow{HCl} MeCHOH(CH_2)_2Cl + \text{(some)}\ MeCHCl(CH_2)_2OH$$

lithium aluminium hydride, while hydrogen chloride and bromide yield mixtures of the isomeric halogen hydrins. The ring usually opens in the same way as that of the analogous oxirane, but variations do occur.[42]

$$Ph(CH_2)_3OH \xleftarrow{LiAlH_4} \text{2-phenyloxetane} \xrightarrow{HCl} PhCHCl(CH_2)_2OH + \text{(some) } PhCHOH(CH_2)_2Cl$$

Pyrolysis[43] of *trans*-2,3-diphenyloxetane yields *trans*-stilbene and paraformaldehyde quantitatively, presumably by a concerted pathway. The *cis* isomer decomposes more slowly and also gives largely *trans*-stilbene, presumably because steric interactions in the transition state make a radical route energetically preferable.

$$\text{trans-2,3-diphenyloxetane} \xrightarrow{200\ °C} CH_2{=}O + \text{trans-PhCH=CHPh};\quad CH_2{=}O \longrightarrow (CH_2O)_n$$

trans-2,3-Dimethyloxetane also ring opens in a stereoselective manner.[44]

$$\text{trans-2,3-dimethyloxetane} \xrightarrow[{[Rh(CO)_2Cl]_2}]{\Delta,\ AgBF_4 \text{ or}} \text{MeCH=C(Me)CH}_2\text{OH (96\%)} + \text{MeCH=C(Me)CH}_2\text{OH (4\%)}$$

F. Synthetic methods

(1) Oxetane can be prepared from 3-chloropropanol, or preferably its acetate (7), and potassium hydroxide. Heating propane-1,3-diol with sulphuric acid, when the disulphonic ester (**8**) is probably obtained, followed by treatment with sodium hydroxide, is a convenient laboratory synthesis.[45]

$$\underset{\mathbf{7}}{Cl(CH_2)_3OAc} \longrightarrow \text{oxetane} \longleftarrow \underset{\mathbf{8}}{HSO_3O(CH_2)_3OSO_3H}$$

Ac = MeCO—

(2) The photolysis of mixtures of aldehydes or ketones with olefines gives a remarkably general synthesis of oxetanes. It can be a stereospecific reaction,[46]

which is expected for a $2\pi + 2\pi$ concerted reaction under the Woodward–Hoffmann rules.

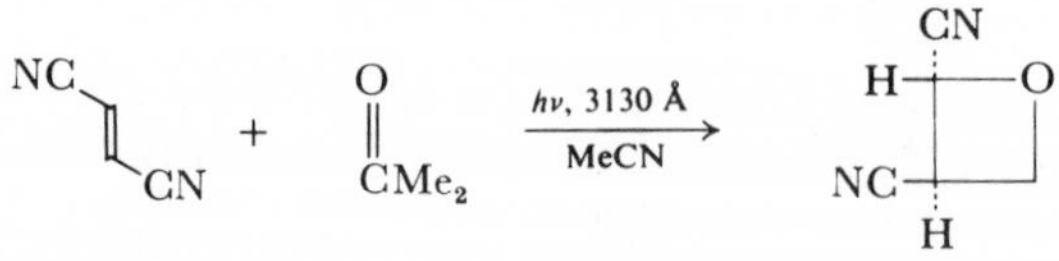

NC–CH=CH–CN + O=CMe₂ —(hν, 3130 Å, MeCN)→ oxetane

In practice nonstereospecificity is sometimes observed, because the olefine employed is partially photoisomerised before addition to the carbonyl compound occurs, and the isomerised olefine may be isolable. Even quinones react to give spirooxetanes (e.g., **9**),[47] the 2,3-bond of furan can act as the olefine (p. 129), and methyl benzoate[48] and diethyl oxalate[49] react with the appropriate olefines to give **10** and **11**, respectively.

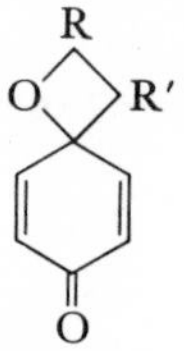

9 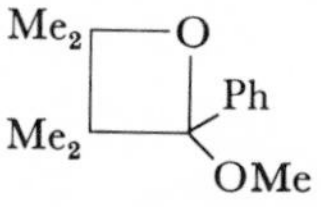 **10** 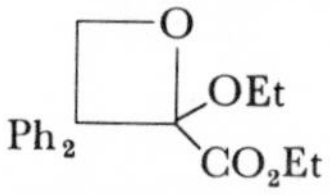 **11**

(3) The photolysis of certain ketones leads to 3-hydroxyoxetanes[50] via a Norrish II process.

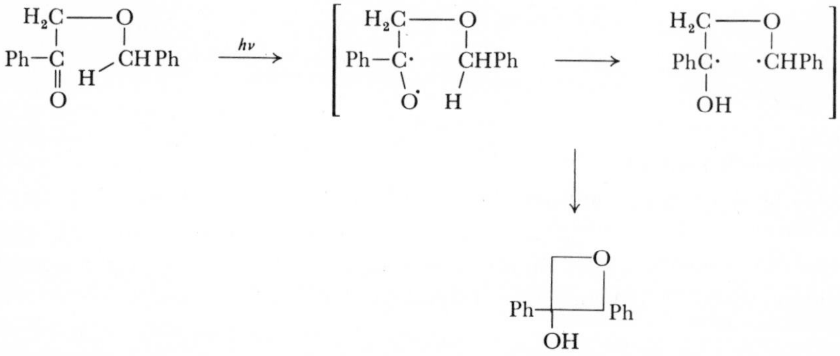

(4) Ring expansion[51] of the oxirane **12** to the oxetane **14** probably occurs through the [1.1.0] oxabicyclobutane **13**.

12 —(tosyl chloride, pyridine)→ [→ (−TsO⁻) → **13**] → **14**

4. OXETANONES

A. Introduction

The most important functional derivatives of oxetane are the oxetan-2-ones[52] or β-lactones (**1**), while a few oxetan-3-ones (**2**) have also been prepared. The strained ring of the oxetan-2-ones makes these compounds much more reactive

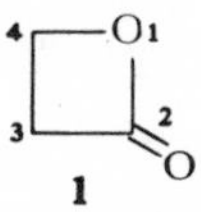

1 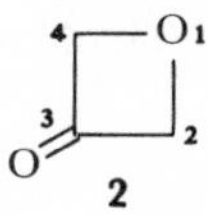 **2**

and difficult to prepare than the γ- and δ-lactones. β-Propiolactone or oxetan-2-one (**1**) is commercially available and may become an important synthetic intermediate. It is, however, a carcinogen which may be related to the fact[54] that it can alkylate guanine (p. 421). The technical material is both a lachrymator and a vesicant.

B. Physical properties and structure

Oxetan-2-one is a colourless liquid, b. p. 51°C at 10 mm, which is miscible with water. It has a dipole moment of 3·10 D, and its structure as found by electron-diffraction measurements is shown.[55]

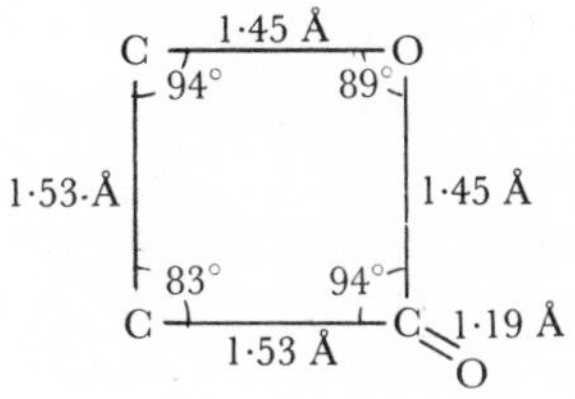

C–O 1·45 Å; O–C 1·45 Å; C–C 1·53 Å; C–C 1·53 Å; C=O 1·19 Å; angles 94°, 89°, 94°, 83°

The molecule is planar.[56]

C. Chemical properties

Oxetan-2-one polymerises slowly on standing. The polymerisation is catalysed by acids, bases, and salts and can be explosive. Alkaline hydrolysis is second order[57] and gives salts of 3-hydroxypropionic acid. Pyrolysis yields acrylic acid. Oxetan-2-one can react with a reagent RX in two ways:

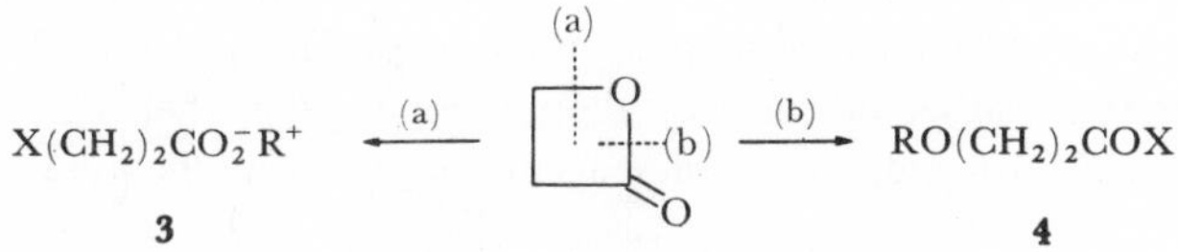

$$\underset{\mathbf{3}}{X(CH_2)_2CO_2^-R^+} \xleftarrow{(a)} \text{oxetan-2-one} \xrightarrow{(b)} \underset{\mathbf{4}}{RO(CH_2)_2COX}$$

Ionic reagents, such as aqueous sodium salts (halides, acetate, phenate, etc.), break the bond labelled (a), yielding products of type **3** and polymer; sodium hydrogen sulphide gives a good yield of 3-mercaptopropionic acid. Alcohols, in the absence of a catalyst, slowly give 3-alkoxypropionic acids. In strongly basic or strongly acidic solution the ring opens in the alternative position (b), leading to 3-hydroxypropionic esters (**4**) and their dehydration products (acrylic esters), respectively.

Oxetan-2-one reacts easily with amines by both possible routes, giving 3-aminopropionic acids and 3-hydroxypropionamides. Aromatic amines usually

$H_2N(CH_2)_2CO_2H$ ← (a) NH_3, MeCN — oxetan-2-one — (b) NH_3, H_2O → $HO(CH_2)_2CONH_2$

$Me_2\overset{+}{S}(CH_2)_2CO_2^-$ ← (a) Me_2S — oxetan-2-one — (a) R_3N → $R_3\overset{+}{N}(CH_2)_2CO_2^-$

give amino acids, but the course of reaction with aliphatic amines depends, in order of importance, on the amine used, the solvent, and the order of mixing the reactants. Water and acetonitrile are the best solvents for hydroxyamide and amino acid production, respectively. With tertiary amines (including pyridine) and dimethyl sulphide, betaines are formed.

Grignard reagents give complex mixtures with oxetan-2-one, while alkyllithiums yield 1,1-dialkylpropane-1,3-diols.[58]

D. Derivatives of oxetanone

Ketene dimer (**5**), the structure of which has been confirmed by X-ray and nuclear magnetic resonance measurements, is of special importance, as with ethanol it yields ethyl acetoacetate. This is an industrial synthesis of the ester.

5 $\xrightarrow{EtOH}$ $MeCOCH_2CO_2Et$

6 $\xrightarrow{80\ °C}$ $Me(Et)C{=}CH_2$ + CO_2

Heating oxetan-2-ones (e.g., **6**) can cause elimination of carbon dioxide to give an olefine.[59]

Oxetan-3-one, b.p. 106°C, has been obtained[60] from the corresponding alcohol by oxidation with chromium trioxide in pyridine. Oxetan-3-ones are of special interest because, in spite of their strained ring, their carbonyl group behaves normally towards Grignard reagents, hydroxylamine, and 2,4-dinitrophenylhydrazine. Tetramethyloxetan-3-one (7) is obtained from tetramethylacetone by oxidation with lead tetraacetate. Photolysis in polar solvents gave acetone and dimethyl ketene almost exclusively, but in nonpolar solvents carbon monoxide and tetramethyloxirane were also formed.[61]

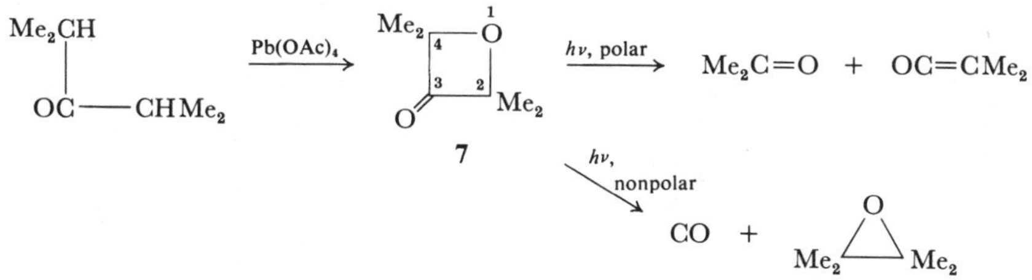

Me_2CH—OC—$CHMe_2$ $\xrightarrow{Pb(OAc)_4}$ **7** $\xrightarrow{h\nu,\ polar}$ $Me_2C{=}O + OC{=}CMe_2$

7 $\xrightarrow{h\nu,\ nonpolar}$ CO + tetramethyloxirane

The few known naturally occurring oxetanes include trichothecin (**8**), an antifungal compound, and anhydroscymnol, which is a steroid.[62] The oxetane **9** is 25 times as active as an insecticide as is DDT.[63]

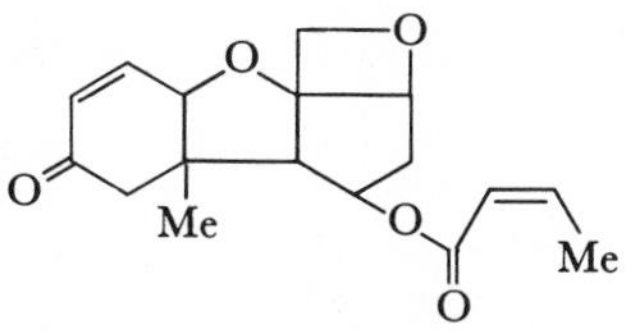

8

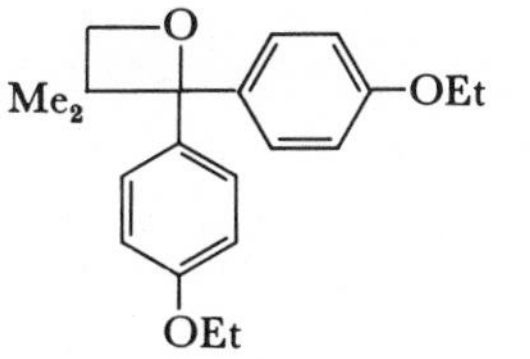

9

E. Synthetic methods

(1) Oxetan-2-ones cannot be obtained by the thermal dehydration of 3-hydroxy acids; acrylic acids and polymerization products are formed. However, treatment of a 3-bromo, or occasionally of a 3-iodo, acid with one equivalent of sodium carbonate, bicarbonate, or silver oxide effects cyclization. Excess base causes hydrolysis of the lactone, and ethylenes are also formed as by-products. A Walden inversion at the carbon atom bearing the halogen takes place, and if steric factors render this impossible, a hydroxy acid, rather than the cyclic lactone, is the main product.

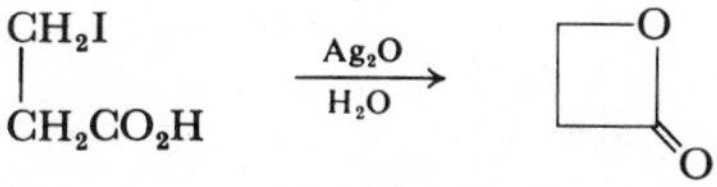

CH_2I—CH_2CO_2H $\xrightarrow[H_2O]{Ag_2O}$ oxetan-2-one

A new synthesis from certain 3-hydroxy acids appears promising,[64] and the photolysis of 2-pyrones yields fused oxetanones (p. 285).

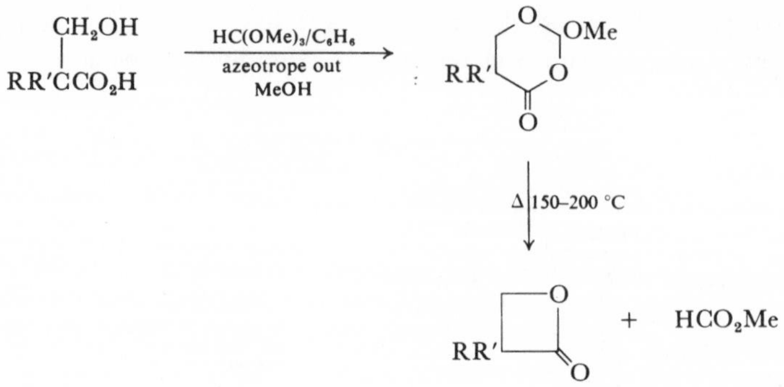

$$RR'C(CH_2OH)CO_2H \xrightarrow[\text{azeotrope out MeOH}]{HC(OMe)_3/C_6H_6} \text{RR' (cyclic orthoester, OMe)} \xrightarrow{\Delta\ 150\text{–}200\ ^\circ C} \text{RR' (oxetan-2-one)} + HCO_2Me$$

(2) Ketenes react with carbonyl compounds in the presence of acid catalysts. Oxetan-2-one is prepared commercially in this way, and the synthesis has wide application. The self-condensation of ketene is the industrial synthesis of ketene dimer (**5**), mentioned earlier (p. 74).

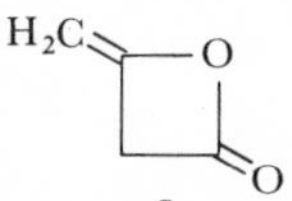

$$CH_2{=}O + CH_2{=}CO \xrightarrow{0^\circ - 20^\circ} \text{(oxetan-2-one)}$$

$H_2C{=}$ (4-methyleneoxetan-2-one) **5**

5. THIETE AND THIETANE

A. Introduction

Thiete (**1**), also called thietene and thiacyclobutene, was synthesised in 1967. It is of great theoretical interest, as loss of a proton could lead to a six-π system (**3**) in a four-membered ring, and sulphur atoms stabilise carbanions (e.g., thiamine, p. 381). Thietane (**2**), known also as trimethylene sulphide or 1,3-propylene sulphide, was first obtained in 1916. There is much current interest in this group of compounds.

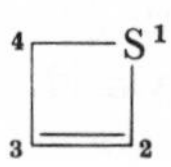

1 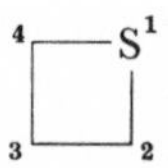 **2** 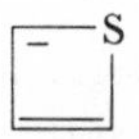 **3**

B. Thiete

Thiete (1) has been prepared[65] by the synthesis outlined below and is a liquid, b.p. 30°C at 20 torr. It is usable for about an hour after preparation if kept at room temperature, but it soon polymerises. In the mass spectrometer it shows the molecular ion and a base peak corresponding to **4**, and it can be converted to 3-mercaptopropionaldehyde 2,4-dinitrophenylhydrazone (5). The

ring is opened by *n*-butyllithium to give a mixture of allyl *n*-butyl and *n*-butyl propenyl sulphides. Treatment with sodium dissolved in deuteriated methanol in the hope of obtaining the anion 3, which could possess six π electrons, exchanged all the hydrogen atoms of the ring, but no thiete could be recovered afterwards.

Thiete 1,1-dioxide (6) has been obtained [66] from the alcohol (20) by successive treatment with thionyl chloride and triethylamine. Its double bond is reactive and readily adds nucleophiles. Thermolysis or photolysis expands the

ring to give the oxathiolene 1-oxide 8. The species 7 is an intermediate, as it is trapped by phenol.[67] Thiete 1,1-dioxide (6) behaves as a normal dienophile in many Diels–Alder reactions, for example, with furan and cyclopentadiene.[68]

C. Physical properties and structure of thietane

Thietane is a colourless liquid, m.p. –73°C, b.p. 95°C at 760 torr. Its dipole moment is 1·85 D and its molecular dimensions have been calculated from electron-diffraction data.[69] The ring is bent, the C–C–C and C–S–C planes

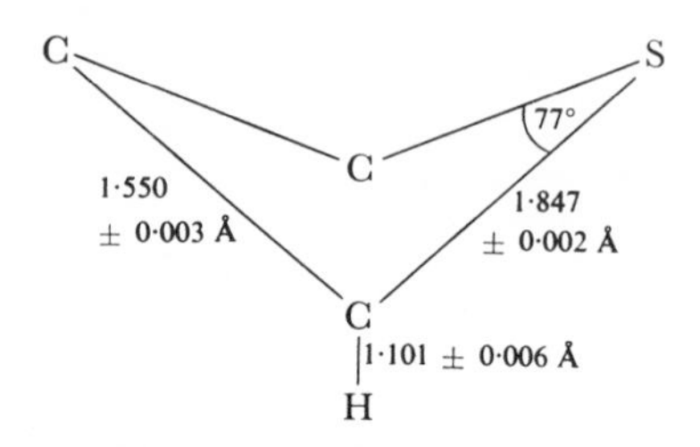

meeting at 26 ± 4°. The strain energy, the difference between the measured and calculated heats of formation, is 79 kJ (18·9 kcal)/mole.[70]

D. Chemical properties

Unlike thiirane, thietane can be oxidized by hydrogen peroxide successively to a stable 1-oxide,[71] and to the sulphone. The sulphone is reduced back to thietane by lithium aluminium hydride,[72] a procedure that does not affect

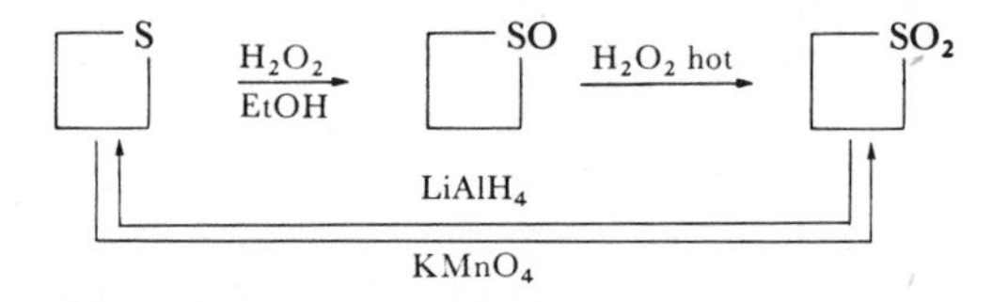

open-chain aliphatic sulphones. Thietane is also reported to polymerize with mineral acids under conditions that leave tetrahydrothiophene unchanged.

The ring of thietane, like that of thiirane and its heterocyclic analogues, is strained and ruptures quite easily. With chlorine in chloroform[73] it gives **9** and **10**, while with butyllithium some **11** is formed.[74] Tetrahydrothiophene is not

$Cl(CH_2)_3SCl$	$Cl(CH_2)_3SS(CH_2)_3Cl$	*n*-BuSPr-*n*
9	**10**	**11**

attacked by this last reagent. With aqueous ammonia at 200°C thietane is stated to give 3-aminopropyl mercaptan.[75]

Thietanes are desulphurized[76] in the expected way by Raney nickel (p. 160); for example, **17** gives isopropyl alcohol. Like thiirane, thietane is cleaved[75] by methyl iodide, because of nucleophilic attack by the iodide anion on the intermediate cation **12**.

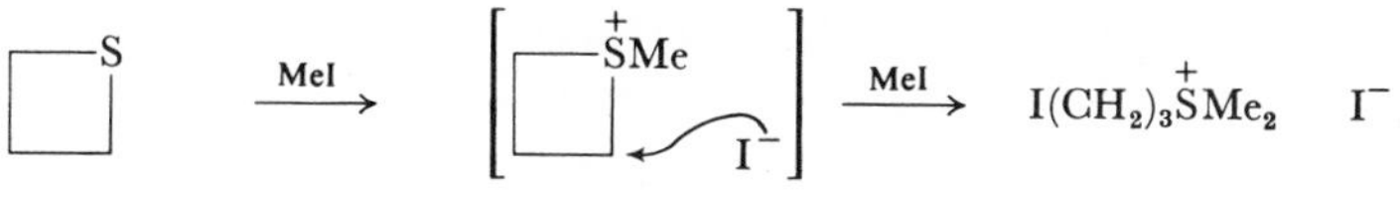

However, using trimethyloxonium tetrafluoroborate stable salts (e.g., **13**) can be obtained. Treatment of these with butyl lithium causes a simultaneous breaking of both carbon–sulphur bonds and a *con*-rotatory cyclisation to give the appropriate cyclopropane (e.g., *cis*-**14**) in high stereochemical purity.[77]

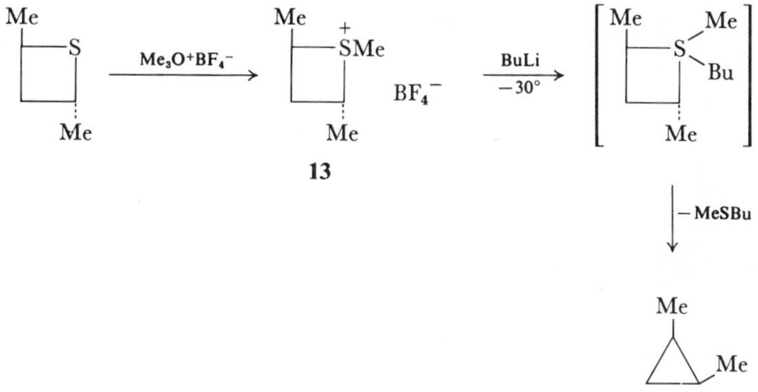

Thietan-3-one[78] (**16**), m.p. 63°C, is readily obtained by hydrolysis of the acetal (**15**). It has the usual reactions of a ketone in forming common derivatives, and with sodium borohydride gives the alcohol (**17**). Secondary amines open the ring, giving a mixture of mercaptoacetone ($MeCOCH_2SH$) and the amide ($MeCOCSNR_2$). The enamine[79] (**18**) and the related sulphone[80] (**21**)

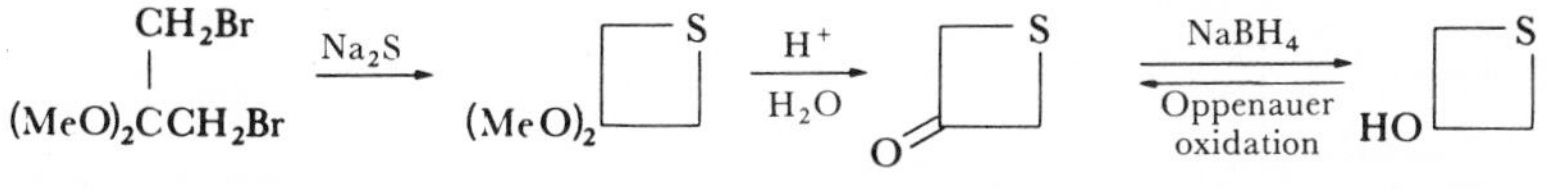

are easily hydrolysed to thietan-3-one 1,1-dioxide (**19**), which is acidic enough (pK_a 4·1) to liberate carbon dioxide from aqueous sodium bicarbonate. No evidence for the presence of an enol form of this ketone was found in its infrared or nuclear magnetic resonance spectra. The acetal (**21**), perhaps surprisingly, is decomposed by sodium hydroxide to the acid (**22**).

$$\mathbf{18} \xrightarrow[\text{resin}]{\text{acidic ion-exchange}} \mathbf{19} \xrightarrow[\text{dioxane}]{B_2H_6} \mathbf{20}$$

$$\mathbf{21} \xrightarrow{\text{conc. HCl}} \mathbf{19}$$

$$\mathbf{21} \xrightarrow{5\%\ \text{NaOH aq.}} [\text{EtO-thiete } SO_2\ ?] \longrightarrow MeSO_2CH_2CO_2H\ (\mathbf{22})$$

E. Synthetic methods

(1) The best synthesis[72] for thietane itself starts with 3-bromopropyl chloride. Treatment with thiourea gives the thiouronium salt, which with potassium hydroxide is converted into thietane in 53·5% yield.

$$Cl(CH_2)_3Br \longrightarrow Cl(CH_2)_3SC(=\overset{+}{N}H_2)NH_2\ \bar{Br} \xrightarrow{KOH} \text{thietane}$$

(2) The cyclization of 1,3-dibromo- or 1,3-dichloropropanes with sodium sulphide is quite a general method for substituted thietanes, and related intermediates, such as epichlorohydrin (**23**), can also be used. Thietan-2-one has been made in this way from 3-chloropropionyl chloride.

$$\mathbf{23} \xrightarrow{H_2S,\ Ba(OH)_2} [CH_2Cl{-}CHOH{-}CH_2SH] \longrightarrow \text{3-hydroxythietane} \xleftarrow{Na_2S} CH_2Cl{-}CHOH{-}CH_2Cl$$

(3) Methylsulphonyl chloride and triethylamine, perhaps acting as a source of sulphene ($CH_2{=}SO_2$), reacts with ketene diethyl acetal[80] and similar compounds[79] to give thietane or thietene 1,1-dioxides; the type of product is determined by the ketene derivative used.

$$CH_3SO_2Cl \xrightarrow[-HCl]{Et_3N} [CH_2{=}SO_2] \xrightarrow{CH_2=C(OEt)_2} (EtO)_2\text{-thietane } SO_2$$

$$[CH_2{=}SO_2] \xrightarrow{CH_2=CR_2} R\text{-thiete } SO_2 \qquad R = \text{—N(morpholino)O}$$

6. FOUR-MEMBERED RINGS WITH TWO HETEROATOMS

A. Introduction

As three-membered rings, even those possessing a double bond, are readily prepared, it is perhaps surprising that interest in the synthesis and properties of heterocycles of this type is only just awakening.[81] It has received much impetus by the realisation that many chemiluminescent reactions produce their light by the opening of dioxetane rings (p. 83).

B. Nitrogen-containing rings

A number of diazetine and diazetidine derivatives have been obtained in the last few years. The product (**1**) from butadiene and dimethyl azodicarboxylate can be converted to the diene **2**. Photolysis[82] causes cyclisation to the diazetidine **3**, which has also been obtained[83] by the alternative route shown, and which gives the normal Diels–Alder adduct **4**. Irradiation of this presumably leads to **8**, which has not been detected but undergoes a retro Diels–Alder reaction to give dimethyl Δ^3-1,2-diazetine-1,2,dicarboxylate (**7**). This unstable

NCO_2Me / NCO_2Me **1** —(a) NBS; (b) 2, 6-Me_2 pyridine→ NCO_2Me / NCO_2Me **2**

$Fe(CO)_3$ + NCO_2Me ‖ NCO_2Me —Ce^{IV}→ NCO_2Me / NCO_2Me **3** —(Ph, Ph, Me, Me cyclopentadienone, =O)→ Me, Ph, Ph, CO, Me, NCO_2Me / NCO_2Me **4**

4 —$-CO$ $h\nu$→ [Ph, Ph, Me, Me, NCO_2Me / NCO_2Me] **8** —(− Ph, Ph, Me, Me benzene)→ NCO_2Me / NCO_2Me **7**

7 —H_2, Pd/C→ NCO_2Me / NCO_2Me **5**

7 —Δ→ MeO_2CN=CH–CH=NCO_2Me **6**

compound has a half-life of 6·9 hours at 20°C, decomposing to **6**. It does not show significant aromatic character. Hydrogenation gives the diazetidine **5**, and compounds of this last type can sometimes be obtained directly from dimethyl azoformate and the ethylene.[84] Azo compounds also combine with ketenes[85] to give 1,2-diazetidinones, for example, **9**, which possesses a slowly inverting nitrogen atom. The ring is easily opened by nucleophiles or bases.

$PhCOCHN_2$ —($h\nu$, $-N_2$)→ [PhCH=C=O] + PhN=NPh → **9** —(MeO⁻, $-H^+$)→ [PhC=NPh / C(–O)=NPh] —(H^+)→ PhC=NPh / CONHPh

9 —(MeOH)→ [Ph, HO, OMe / NPh, NPh] → PhCH—NPhNHPh / CO_2Me

In contrast, the 1,3-diazetidine[86] **10** is stable to lithium aluminium hydride and is unchanged by long refluxing with concentrated aqueous alkali or 20% aqueous sulphuric acid. It forms salts with acids and is recovered on basification. This is quite unexpected for the nitrogen analogue of an acetal.

PhCH=NR ⇌ (PhCOMe, EtOH, $h\nu$, 25 °C; reverse: 235°C, sealed tube) Ph, NR, RN, Ph

10, 34%

Oxazetidines with an *N*-substituent can be obtained from perfluoroethylene and nitroso compounds, and recently an example (**11**) not possessing an *N*-substituent has been obtained. It boils at 10°C and is stable in the gas phase at 0°C for short periods.[87] The 2-chloro derivative (**12**), obtained by direct

$CF_2{=}CF_2$ + O=$NCF_2CFClOEt$ → F_2, O, F_2, $NCF_2CFClOEt$ —(H_2SO_4)→ [F_2, O, F_2, NCF_2CO_2Et] → [F_2, O, F_2, $NCOCO_2Et$] → F_2, O, F_2, NH (**11**) —(Cl_2)→ F_2, O, F_2, NCl (**12**)

chlorination, gives temperature-dependent n.m.r. spectra indicating that the nitrogen atom inverts slowly, as in the case of 1-chloroaziridines (p. 16).

Reaction of methoxycarbonylsulphamyl chloride (**13**) with sodium hydride gives the electrophilic species **14**, which with styrene forms the thiazetidine **15** and oxathiazine **16** in a 1 : 1 ratio.[88] The formation of the four-membered ring is stereospecific and may be a $\pi 2s + \pi 2a$ concerted thermal reaction.

$ClSO_2NHCO_2Me$ (**13**) $\xrightarrow{NaH, MeCN, 4°C}$ $SO_2=NCO_2Me$ (**14**)

$CH_2=CHPh$ + $SO_2=NCO_2Me$ (**14**) ⟶ **15** + **16**

15: four-membered ring S^1O_2–N^2CO_2Me–C^3(Ph)–C^4

16: six-membered ring S^1O_2, N^2, C(OMe), O^4, CH(Ph), CH_2

C. Dioxetane

1,2-Dioxetanes, which are cyclic peroxides, can be made by irradiating olefines at low temperatures in the presence of oxygen and a sensitiser (e.g., tetraphenylporphrin). Singlet oxygen adds stereospecifically to the double bond.[89,90] The dioxetane **17** (R = Et) is remarkably stable with a half-life of 102 minutes at 56°C.

EtOCH=CHOR $\xrightarrow[CFCl_3]{O_2}$ **17** (3-EtO-4-RO-1,2-dioxetane) $\xrightarrow[R = Ph]{\Delta, 50\ °C}$ EtOCHO + HCOPh

cis ⟶ *cis*

trans ⟶ *trans*

R = Et or Ph

The 1,2-dioxetan-3-one system is of great interest, as decomposition in a concerted manner[91] gives carbon dioxide and an excited carbonyl group that is responsible for the light emitted in many photochemical reactions (p. 385). The first compound of this type isolated[92] was **18**. It decomposes in a few minutes at room temperature, and on warming in the dark some chemiluminescence is observed.

t-$BuCHLiCO_2Li$ ⟶ t-$BuCH=C(OSiMe_3)_2$ $\xrightarrow[photooxidation]{^1O_2}$ t-$BuCH(OOSiMe_3)COOSiMe_3$

$\xrightarrow{MeOH}$ t-$BuCH(OOH)-CO_2H$ $\xrightarrow[-10\ °C, CCl_4]{C_6H_{11}N=C=NC_6H_{11}}$ **18** (t-Bu-1,2-dioxetan-3-one) $\xrightarrow[room\ temp.]{5\text{–}8\ minutes}$ CO_2 + t-BuCHO

18

D. Dithietanes

Four-membered rings with two sulphur atoms seem to be easily formed. Hexafluorobut-2-yne (**19**) with boiling sulphur gives the 1,2-dithietene (**20**) in 80% yield. It is a yellow liquid, b.p. 95°C, which with iodine and a trace of triethylamine gives the tetrathiaoctadiene (**21**), and with tetramethylethylene the 1,4-dithiin[93] (**22**). The stability of this dithietene is in accord with the results of molecular-orbital calculations, which suggest that the parent ring system should have considerable delocalization energy.[94]

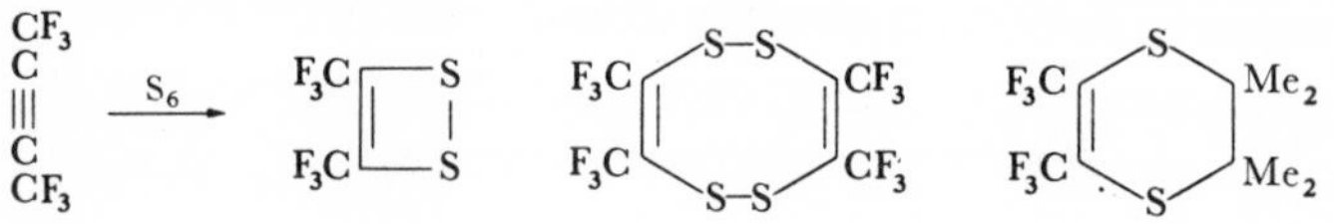

$CF_3C{\equiv}CCF_3$ (**19**) $\xrightarrow{S_6}$ **20** ; **21** ; **22**

Photolysis of **23** gives the dithioketone **24**, which is in equilibrium with the dithietene **25**.[95]

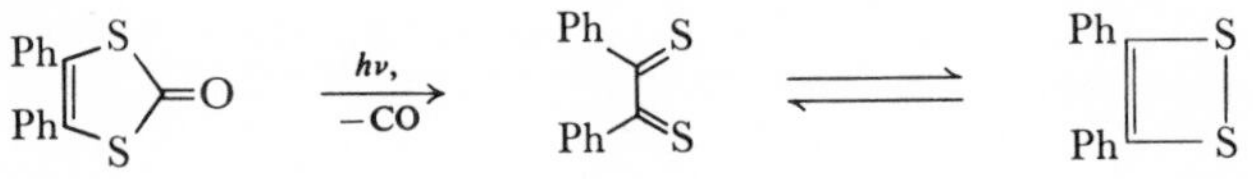

23 $\xrightarrow[-CO]{h\nu,}$ **24** $\rightleftharpoons$ **25**

A number of 1,3-dithietane derivatives have been made, such as the 1,1,3,3-tetraoxide **26**.[96] The desaurins (e.g., **27**), first made and given correct

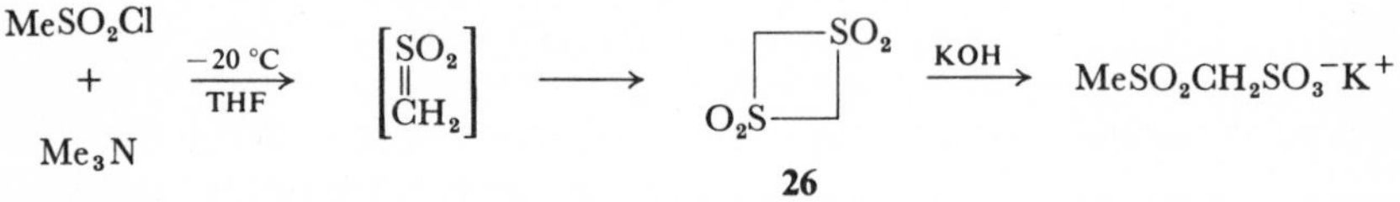

$MeSO_2Cl + Me_3N \xrightarrow[THF]{-20\,°C} [SO_2{=}CH_2] \longrightarrow$ **26** $\xrightarrow{KOH} MeSO_2CH_2SO_3^-K^+$

structures in 1888, are remarkably stable. Compound **27** is unchanged by concentrated sulphuric acid or zinc and acetic acid at room temperature, and is little affected by boiling concentrated hydrochloric acid.[97]

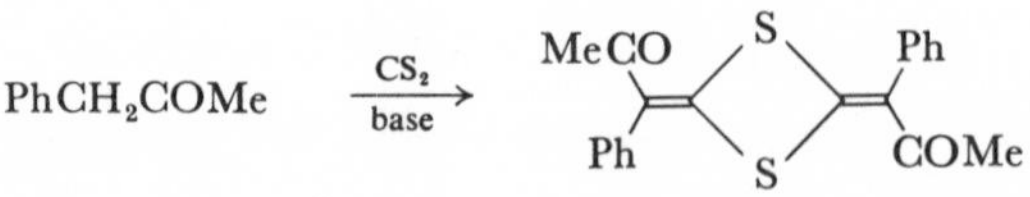

$PhCH_2COMe \xrightarrow[base]{CS_2}$ **27**

GENERAL BIBLIOGRAPHY

A. Weissberger (ed.), *Heterocyclic Compounds with Three and Four Membered Rings*, Part 2. *Thietane*, Y. Etienne, R. Soulas and H. Lambroso; *β-Lactones*, Y. Etienne and N. Fischer; *Trimethylenimines*, J. A. Moore; *Four-Membered Rings containing Two Heteroatoms,* W. D. Emmons; *Oxetanes*, S. Searles; *Chemistry of Heterocyclic Compounds*, Vol. 19, Interscience, Wiley, New York, 1964.

L. E. Muller and J. Hamer, *1,2-Cycloaddition Reactions, The Formation of Three and Four Membered Heterocycles,* Interscience, Wiley, New York, 1967.

AZETIDINE

S. A. Ballard and D. S. Melstrom, in *Heterocyclic Compounds*, Vol. I, ed. R. C. Elderfield, Wiley, New York, 1950.

AZETIDIN-2-ONES, PENICILLIN, AND CEPHALOSPORIN

H. T. Clarke, J. R. Johnson, and R. Robinson, *The Chemistry of Penicillin*, Princeton University Press, New Jersey, 1949.

E. P. Abraham and G. G. F. Newton, *Endeavour,* **20**, 92 (1961).

E. P. Abraham, *Q. Rev.,* **21**, 231 (1967).

E. H. Flynn, ed. *Cephalosporins and Penicillins*, Academic Press, New York, 1972.

R. J. Stoodley, 'Recent Penicillin Chemistry,' *Progr. Org. Chem.,* **8**, 102 (1973).

M. S. Mankas and A. K. Bose, *β-Lactones*, Wiley-Interscience, New York, 1971.

R. D. G. Cooper, L. D. Hatfield and D. O. Spry, 'Chemical Interconversion of the β-Lactone Antibiotics,' *Acc. Chem. Res.,* **6**, 32 (1973).

OXETANE

S. Winstein and R. B. Henderson, in *Heterocyclic Compounds* Vol. 1, ed. R. C. Elderfield, Wiley, New York, 1950.

G. Pruckmayr, in *Cyclic Monomers*, High Polymers, Vol. 26, ed. K. C. Frisch, Interscience, New York, 1972, p. 54.

THIETANE

M. Sander, *Chem. Rev.,* **66**, 341 (1966).

F. Lautenschlaeger and R. T. Woodham, p. 358 in *Cyclic Polymers*, High Polymers, Vol. 26, ed. K. C. Frisch, Interscience, New York, 1972.

REFERENCES

1. M. D. Cook and C. D. Johnson, *Ann. Rep. Chem. Soc.,* **70**, 479 (1973).
2. B. M. Adger, C. W. Rees and R. C. Storr, *J.C.S. Perkin I,* 45 (1975); R. M. Adger, M. Keating, C. W. Rees, and R. C. Storr, *Chem. Comm.*, 19 (1973).

2a. G. Szeimies, U. Sifken and R. Rinck, *Angew. Chem. Int. Ed.,* **12**, 161 (1973).

3. O. V. Dorofeeva, V. S. Mastryukov and L. V. Vilkov, *Chem. Comm.,* 772 (1973).
4. A. G. Anderson and R. Lok, *J. Org. Chem.,* **37**, 3953 (1972) and papers cited.
5. H. C. Brown and M. Gerstein, *J. Amer. Chem. Soc.,* **72**, 2926 (1950).
6. J. P. Freeman, D. G. Pucci, and G. Birsch, *J. Org. Chem.,* **37**, 1894 (1972).
7. D. R. Crist and N. J. Leonard, *Angew. Chem. Int. Ed.,* **8**, 962 (1969).
8. H. A. Hageman, *Org. React.,* **7**, 198 (1953).
9. V. R. Gaertner, *Tetrahedron Lett.,* 5919 (1968).
10. A. Padwa, R. Gruber, and L. Hamilton, *J. Amer. Chem. Soc.,* **89**, 3077 (1967).
11. W. R. Vaughan, R. S. Klonowski, R. S. McElkinney, and B. B. Milward, *J. Org. Chem.,* **26**, 138 (1961).

12. E. H. Gold and E. Babad, *J. Org. Chem.*, **37**, 2208 (1972).
13. R. M. Rodebaugh and N. H. Cromwell, *J. Heterocycl. Chem.*, **6**, 435 (1969).
14. D. H. Wadsworth, *J. Org. Chem.*, **32**, 1184 (1967).
15. R. A. Classen and S. Searles, Jun., *Chem. Comm.*, 289 (1966).
16. E. Leete, *J. Amer. Chem. Soc.*, **86**, 3162 (1964).
17. H. Staudinger, *Die Ketene*, Enke, Stuttgart, 1912.
18. J. C. Sheehan and K. R. Henery-Logan, *J. Amer. Chem. Soc.*, **81**, 3089 (1959).
19. R. W. Holley and A. D. Holley, *J. Amer. Chem. Soc.*, **71**, 2129 (1949).
20. J. A. Kartha and G. Ambady, *J.C.S. Perkin* II, 2042 (1973); cf. S. Abrahamson, D. C. Hodgkin and E. N. Masten, *Biochem. J.*, **86**, 514 (1963).
21. E. Testa, L. Fontanella, and V. Aresi, *Annalen*, **656**, 114 (1962).
22. M. Fischer, *Chem. Ber.*, **101**, 2669 (1968).
23. K. R. Henery-Logan and J. V. Rodricks, *J. Amer. Chem. Soc.*, **85**, 3524 (1963).
24. D. Bormann, *Annalen*, **725**, 124 (1969).
25. J. C. Sheehan and A. K. Bose, *J. Amer. Chem. Soc.*, **73**, 1761 (1951) and earlier papers.
26. B. G. Chatterjee, V. Venkateswara Rao, and B. N. G. Mazumdar, *J. Org. Chem.*, **30**, 4010 (1965).
27. T. Durst and M. J. O'Sullivan, *J. Org. Chem.*, **35**, 2043 (1970).
28. A. K. Bose, M. S. Manhas, and R. M. Ramer, *Tetrahedron*, **21**, 449 (1965).
29. D. R. Bender, L. F. Bjeldanes, D. R. Knapp, D. R. McKean, and H. Rapoport, *J. Org. Chem.*, **38**, 3439 (1973).
30. J. A. Deyrup and S. C. Clough, *J. Amer. Chem. Soc.*, **91**, 4590 (1969).
31. O. L. Chapman and W. R. Adams, *J. Amer. Chem. Soc.*, **89**, 4243 (1967).
32. J. R. Hlubucek and G. Lowe, *Chem. Comm.*, 419 (1974) and earlier papers.
33. R. J. Stoodley and N. R. Whitehouse, *Chem. Comm.*, 477 (1973).
34. S. Wolfe and S. K. Hasan, *Chem. Comm.*, 833 (1970).
35. H. Kluender, Fu-C. Huang, A. Fritzberg, H. Schnoes, C. J. Sih, P. Fawcett, and E. P. Abraham, *J. Amer. Chem. Soc.*, **96**, 4054 (1974) and earlier papers.
36. J. E. Baldwin, H. B. Huber, and J. Kitchin, *Chem. Comm.* 790 (1973).
37. E. P. Abraham and G. G. F. Newton, *Endeavour*, **20**, 92 (1961); E. P. Abraham, *Pharmacol. Rev.*, **14**, 473 (1962).
38. R. B. Woodward, K. Heusler, J. Gosteli, P. Naegeli, W. Oppolzer, R. Ramage, S. Ranganathan, and H. Vorbrüggen, *J. Amer. Chem. Soc.*, **88**, 852 (1966).
39. W. J. Middleton, *J. Org. Chem.*, **30**, 1307 (1965).
40. L. E. Friedrich and G. B. Schuster, *J. Amer. Chem. Soc.*, **91**, 7204 (1969).
41. S. I. Chen, J. Zinn and W. P. Gwinn, *J. Chem. Phys.*, **34**, 1319 (1969).
42. S. Searles, K. A. Pollart, and E. F. Lutz, *J. Amer. Chem. Soc.*, **79**, 948 (1957); S. Searles, K. A. Pollart, and F. Block, *ibid.*, **79**, 952 (1957).
43. N. Shimizu and S. Nishida, *Chem. Comm.*, 734 (1974).
44. H. A. J. Carless, *Chem. Comm.*, 982 (1974).
45. L. F. Schmoyer and L. C. Case, *Nature*, **183**, 389 (1959); C. R. Noller, *Org. Synth.*, **29**, 92 (1949).
46. J. C. Dalton, P. A. Wriede, and N. J. Turro, *J. Amer. Chem. Soc.*, **92**, 1318 (1970) and earlier papers.
47. D. Bryce-Smith, A. Gilbert and M. G. Johnson, *J. Chem. Soc.* (*C*), 383 (1967).
48. T. S. Cantrell, *Chem. Comm.*, 468 (1973).
49. T. Tominaga, Y. Odaira, and S. Tsutsumi, *Bull. Chem. Soc. Jap.*, **40**, 2451 (1967).
50. P. Yates and A. G. Szabo, *Tetrahedron Lett.*, 485 (1965).

51. H. G. Richey and D. V. Kinsman, *Tetrahedron Lett.*, 2505 (1969).
52. H. E. Zaugg, *Org. React.*, 8, 305 (1954): G. Machell, *Ind. Chem.*, **36**, 13 (1960).
53. A. Haddow, *Ann. Rev. Biochem.*, **24**, 689 (1955).
54. N. H. Colburn and R. K. Boutwell, *Cancer Res.*, **28**, 642, 653 (1968).
55. J. Bregman and S. H. Bauer, *J. Amer. Chem. Soc.*, **77**, 1955 (1955).
56. N. Kwak, J. H. Goldstein, and J. W. Simmons, *J. Chem. Phys.*, **25**, 1203 (1956).
57. M. G. Blackburn and H. C. H. Dodds, *J.C.S. Perkin* II, 377 (1974).
58. C. G. Stuckwisch and J. V. Bailey, *J. Org. Chem.*, **28**, 2362 (1963).
59. W. T. Brady and A. D. Patel, *J. Org. Chem.*, **37**, 3536 (1972).
60. J. A. Wojtowicz and R. J. Polak, *J. Org. Chem.*, **38**, 2061 (1973).
61. P. J. Wagner, C. A. Stout, S. Searles, and G. S. Hammond, *J. Amer. Chem. Soc.*, **88**, 1242 (1966).
62. A. D. Cross, *J. Chem. Soc.*, 2817 (1961).
63. G. Holan, C. Kowala, and J. A. Wunderlich, *Chem. Comm.*, 34 (1973).
64. R. C. Blume, *Tetrahedron Lett.*, 1047 (1969).
65. D. C. Pitman, P. L. Chang, F. A. Davis, I. K. Stamos, and K. Takahashi, *J. Org. Chem.*, **37**, 1116 (1972), and earlier papers.
66. D. C. Dittmer and M. E. Christy, *J. Amer. Chem. Soc.*, **84**, 399 (1962).
67. J. F. King, K. Piers, D. J. H. Smith, C. L. McIntosh, and P. de Mayo. *Chem. Comm.*, 31 (1969).
68. J. E. McCaskie, T. R. Nelson, and D. C. Dittmer, *J. Org. Chem.*, **38**, 3048 (1973).
69. K. Karakida, K. Kuchitsu, and R. K. Bohn, *Chem. Lett.*, 159 (1974); per *Chem. Abstr.*, **80**, 95626x (1974).
70. J. D. Cox, *Tetrahedron*, **19**, 1175 (1963).
71. M. Tamres and S. Searles, *J. Amer. Chem. Soc.*, **81**, 2100 (1959).
72. F. G. Bordwell and W. H. McKellin, *J. Amer. Chem. Soc.*, **73**, 2251 (1951).
73. J. M. Stewart and C. H. Burnside, *J. Amer. Chem. Soc.*, **75**, 243 (1953).
74. F. G. Bordwell, H. M. Andersen, and B. M. Pitt, *J. Amer. Chem. Soc.*, **76**, 1082 (1954).
75. D. S. Tarbell and D. P. Harnish, *Chem. Rev.* (*Lond.*), **49**, 1, (1951).
76. S. Searles, H. R. Hays, and E. F. Lutz, *J. Org. Chem.*, **27**, 2828 (1962).
77. B. M. Trost, W. C. Schinski, and J. B. Mantz, *J. Amer. Chem. Soc.*, **93**, 676 (1971) and earlier papers.
78. K. F. Funk and R. Mayer, *J. Prakt. Chem.*, **21**, 65 (1963).
79. R. H. Hasek, P. G. Gott, R. H. Meen, and J. C. Martin, *J. Org. Chem.*, **28**, 2496 (1963).
80. W. E. Truce and J. R. Norell, *J. Amer. Chem. Soc.*, **85**, 3231, 3236 (1963).
81. R. Livingstone, in *Chemistry of Carbon Compounds*, ed. E. H. Rodd, 2nd ed., Part IVA, Elsevier, London, 1973, pp. 72–82.
82. L. J. Altman, M. F. Semmelhack, R. B. Hornby, and J. C. Vederas, *Chem. Comm.*, 686 (1968).
83. E. E. Nunn and R. N. Warrener, *Chem. Comm.*, 818 (1972).
84. J. Firl and S. Sommer, *Tetrahedron Lett.*, 1133 (1969).
85. E. Fahr and W. Fischer, *Tetrahedron Lett.*, 3291 (1967), and earlier papers.
86. R. O. Kan and R. L. Furey, *J. Amer. Chem. Soc.*, **90**, 1666 (1968).
87. J. D. Readio and R. A. Falk, *J. Org. Chem.*, **35**, 927, 1607 (1970) and earlier papers.
88. G. M. Atkins and E. M. Burgess, *J. Amer. Chem. Soc.*, **94**, 4386 (1972) and earlier papers.
89. P. D. Bartlett and A. P. Schaaf, *J. Amer. Chem. Soc.*, **92**, 3223 (1970); S. Mazur and C. S. Foote, *ibid.*, **92**, 3225 (1970).

90. A. P. Schapp and N. Tontapanist, *Chem. Comm.,* 490 (1972).
91. F. McCapra, *Chem. Comm.,* 155 (1968), M. M. Rauhut, *Acc. Chem. Res.,* **2**, 80 (1969); see also D. R. Roberts, *Chem. Comm.*, 683 (1974).
92. W. Adam and J. C. Liu, *J. Amer. Chem. Soc.,* **94**, 2894 (1972).
93. C. G. Krespan, B. C. McKusick, and T. L. Cairns, *J. Amer. Chem. Soc.,* **82**, 1515 (1960).
94. G. Bergson, *Arkiv Kemi.,* **19**, 181, 265 (1962).
95. W. Küsters and P. de Mayo, *J. Amer. Chem. Soc.,* **95**, 2383, (1973).
96. G. Opitz and H. R. Mohl, *Angew. Chem. Int. Ed.,* 8, 73, (1969).
97. P. Yates and D. R. Moore, *J. Amer. Chem. Soc.,* **80**, 5577, (1958); J. A. Kapecki, J. E. Baldwin, and I. C. Paul, *J. Amer. Chem. Soc.,* **90**, 5800 (1968).

CHAPTER III

HETEROCYCLIC ANALOGUES OF CYCLOPENTADIENE WITH ONE HETEROATOM

Pyrrole, furan, and thiophene are all related to cyclopentadiene, from which they can be theoretically formed through replacement of the methylene group (CH_2) by the appropriate heteroatoms. In 1924 Robinson pointed out that these molecules all possess six π-electrons, which he associated for the first time with the concept of aromaticity. The aromatic and symmetrical cyclopentadienyl anion also has six π-electrons. Although superficially furan, pyrrole, and

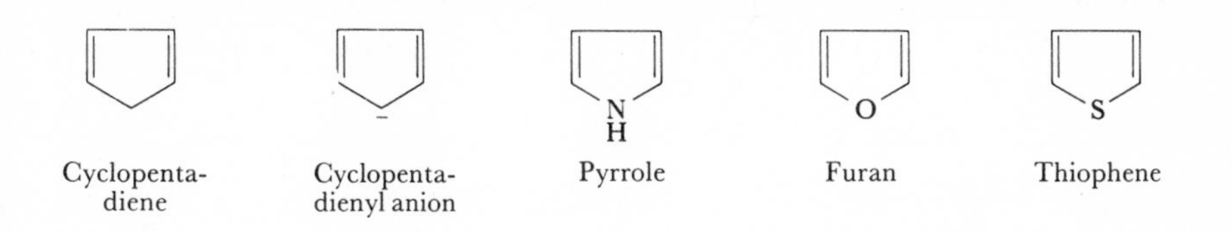

thiophene resemble each other in formula and in some properties, there are major fundamental differences, which affect the properties of the molecules to such an extent that a prediction of the behaviour of, for example, a thiophene, based on the known behaviour of the analogous pyrrole or furan could often be wrong or misleading. The three ring systems are therefore considered separately, and a summary comparison of their aromatic properties is given in section 4 of this chapter (p. 174).

1. PYRROLE

A. Introduction

In 1834 Runge found that the distillation of coal tar, bone oil, and other products derived from proteins gave an unknown substance in the ammonia fraction. It turned a pine splint, previously dipped in hydrochloric acid, a fiery red. He called the substance pyrrole, and in 1857 Anderson obtained the compound in a pure condition from a bone-oil distillate. Three years later it was synthesized by the pyrolysis of ammonium mucate. This is still a useful laboratory method, in spite of the availability of commercial pyrrole.

The pyrrole ring system soon became of great interest, as it was found in many compounds widely distributed in nature. It had been recognized in the dyestuff indigo, in haemin obtained from blood, and in chlorophyll before 1880. Two amino acids derived from pyrrole, proline (pyrrolidine-2-carboxylic acid) and 4-hydroxyproline, are constituents of many proteins. Many alkaloids also possess a pyrrole ring.

The positions of the substituents around the pyrrole ring may be shown by the use of numbers (**1**) or Greek letters (**2**). The first system is currently preferred and is obligatory when several substituents are present. Two

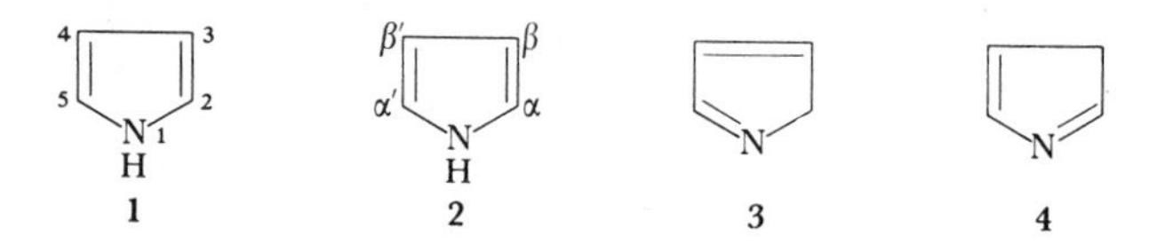

tautomeric formulations (**3** and **4**) can be written for pyrrole, and derivatives of both are known. They are named 2*H*-pyrrolenine or α-pyrrolenine (**3**) and 3*H*-pyrrolenine or β-pyrrolenine (**4**). Three dihydropyrroles or pyrrolines are

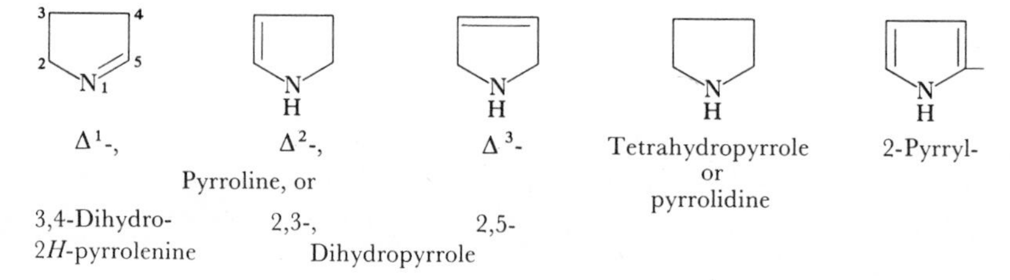

theoretically possible and tetrahydropyrrole is called pyrrolidine. The pyrrole ring, when considered as a substituent in another structure, is called pyrryl.

B. Physical properties and structure

Pyrrole is a colourless liquid, b.p. 129°C at 760 mm, f.p. –24°C, d_4^{20} 0·968, with an odour resembling that of chloroform. It turns brown on standing in air and is miscible with most organic solvents. It is somewhat soluble (6%) in water and dissolves 3% of its weight of water at 25°C. Substantial intermolecular hydrogen bonding takes place and is shown both by the lower boiling point of 1-methylpyrrole (114–115°C) and by the infrared absorption spectrum of pyrrole itself. Pyrrole is both a very weak acid (pK_a 17·5) and a very weak base (pK_a –3·8).

The structure of pyrrole, as usually written (**1**), was first suggested by Baeyer in 1870. This follows from the transformations outlined above.

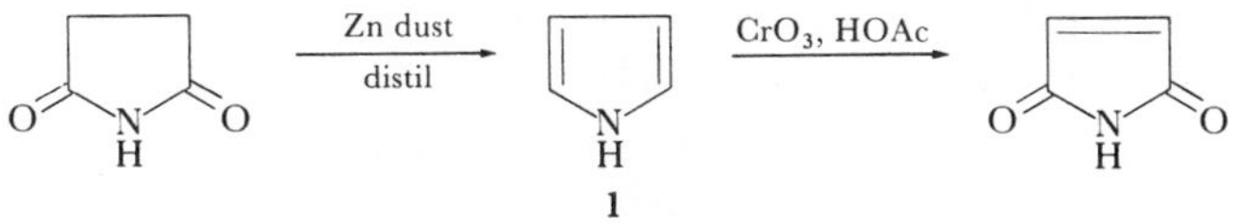

Zn dust, distil → 1 → CrO_3, HOAc

There has been no recent determination of the heat of combustion or heat of hydrogenation of pyrrole, and so the resonance energy of the molecule cannot be estimated with any certainty. Calculations based on the combustion data of Berthelot and André, obtained in 1899 with pyrrole of doubtful purity, suggest a value of 88–100 kJ (21–24 kcal)/mole. Molecular orbital calculations give similar values.

Microwave studies have shown that the pyrrole molecule is completely planar and have given molecular dimensions.[1]

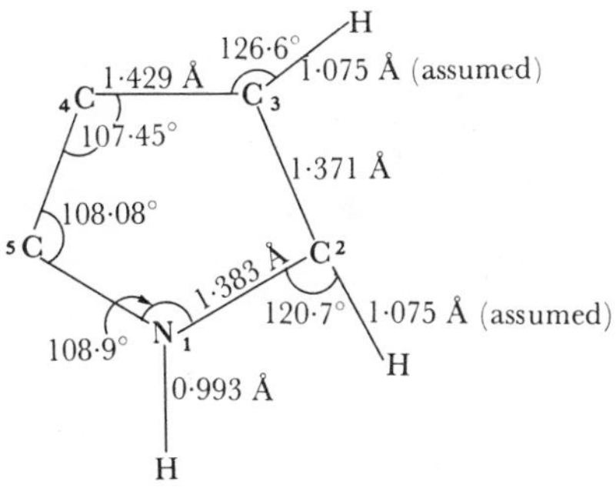

As the distances (1·37 Å) between carbon atoms 2 and 3, and 4 and 5 are greater than that between the carbon atoms of ethylene (1·34 Å), as the distance between carbon atoms 3 and 4 (1·43 Å) is less than that between the carbon atoms of ethane (1·54 Å) and approaches the carbon–carbon distance in benzene (1·40 Å), and as the carbon–nitrogen distances (1·38 Å) are less than those in trimethylamine (1·47 Å), the molecule cannot be adequately represented by the classical structure (**1**). It must be treated as a resonance hybrid of which the more important contributing structures are shown below; structure **1** is the major contributor. This conclusion is supported by the comparatively large resonance energy of the molecule and by dipole-moment studies.[2] These studies have shown decisively that the dipole moments of pyrrole (1·80 D) and 1-methylpyrrole (1·92 D) in benzene solution are directed so that the nitrogen

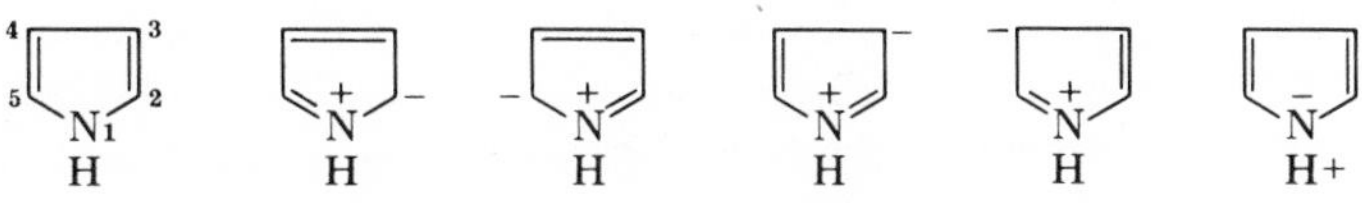

1 **1a** **1b** **1c** **1d** **1e**

atom is at the positive end of the dipole. Structure **1e** is therefore probably an insignificant contributor in pyrrole itself, although the anion of **1e** is the major contributor in metallic derivatives, such as potassium pyrrole. As the dipole of pyrrolidine (tetrahydropyrrole, 1·57 D) is in the opposite direction, having its negative end at the nitrogen atom, experimental proof of the importance of ionic structures in the resonance of pyrrole has been provided. Electron-density

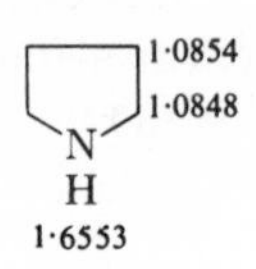

π-Electron densities

calculations by many procedures are not in good agreement, but recent calculations[3] by a self-consistent field method, including all valence electrons, suggest that the π-electron density is slightly higher at positions 3 and 4 than at positions 2 and 5.

The localization energies for electrophilic, nucleophilic, and free-radical attack are all lower for position 2 than position 3.[4] This is in agreement with the very facile electrophilic substitution of pyrrole by cationoid reagents that takes place preferentially at positions 2 and 5. The pyrrolenine tautomers (**3** and **4**),

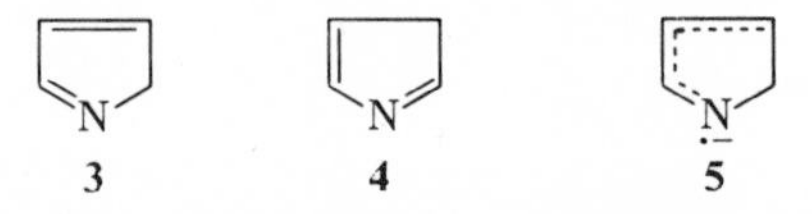

neither of which can possess much resonance energy, cannot be detected in pyrrole by examination of its infrared or Raman spectra.[5] However, the electron spin resonance spectrum of the anion radical derived from pyrrole by treatment with sodium, at 4°K in an argon matrix, shows that it must be the 2*H*-pyrrolenine **5**.[6]

C. Chemical properties

Pyrrole behaves mainly as a very reactive aromatic compound towards electrophilic reagents and has been compared to phenol in this respect. It also shows weakly acidic and basic properties, and it can behave as an enamine, and also as a 1,3-diene towards some reactive reagents. Many pyrroles, including 2,3,4,5-tetramethylpyrrole, which has only fully substituted carbon atoms, give positive Ehrlich reactions (red to violet colours with 4-dimethylaminobenzaldehyde and concentrated hydrochloric acid, see p. 194), and most form coloured azo compounds with diazonium salts.

1. *Opening of the pyrrole ring*

The pyrrole ring is not readily opened by acids or alkalis, but boiling with alcoholic hydroxylamine hydrochloride causes rupture, with the formation of

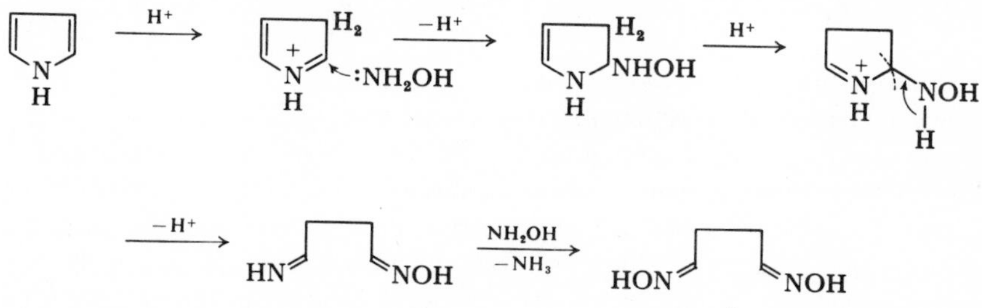

HON= ... =NOH

6

succindialdehyde dioxime (**6**). The reaction probably follows the type of course indicated (see also next section) and is almost the reverse of the synthesis of pyrrole from succindialdehyde (p. 114). The hydroxylamine may function by trapping the very reactive dialdehyde, thereby preventing polymerisations and reformation of the pyrrole. This opening is facilitated by alkyl groups and hindered by carbonyl and phenyl groups, which respectively assist and discourage the initial protonation.

The ozonolysis of pyrrole and derivatives at –60°C in chloroform breaks the ring. The production of both glyoxal and methylglyoxal (**7**) from 2,5-dimethyl-

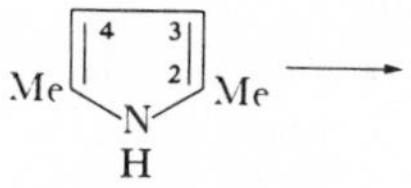

$$\longrightarrow \text{CHOCHO} + \text{MeCOCHO} + \text{MeCO}_2\text{H} + \text{NH}_3$$

7

pyrrole suggests that the bond between carbon atoms 2 and 3 is partly single, in confirmation of the presence of ionic structures in the resonance hybrid.

Pyrrole in aqueous silver nitrate is broken down by ultrasonic vibration into acetylene and cyanide ions.[7]

2. *Addition reactions*

In order to have aromatic properties the pyrrole ring must possess six π electrons, and as only four can be provided by the two double bonds of the uncharged formulation, the lone pair on the nitrogen atom must be involved. This is another way of illustrating the importance of ionic structures with a positive charge on the nitrogen in the resonance hybrid. The effect of this charge is to discourage an approaching proton, and in fact pyrrole is a very weak base ($pK_a = -3{\cdot}8$).[8]

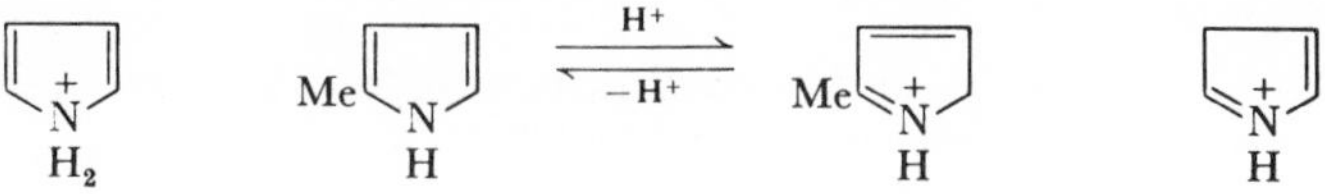

In spite of this, and the fact that it is an extremely weak acid (p. 96), nuclear magnetic resonance experiments have shown that the 1-*H* atom of pyrrole rapidly exchanges for deuterium under alkaline catalysis. It also exchanges at low acidities where, as under alkaline conditions, no detectable exchange at the 2 and 3 positions occurs. A very rapid *N*-protonation therefore gives the non-resonance stabilised cation **8**. This is present only in extremely low concentration, but cations of this sort have been detected in the solid phase by i.r. spectrum measurements.[9] N.m.r. spectra show that pyrrole and both its *N*- and *C*-alkyl derivatives in aqueous solution accept protons at the 2-position to give stable cations (e.g., **9**). The position and rate of protonation, often measured by deuterium exchange, are concentration dependent.[10] In some cases, particularly in more concentrated sulphuric acid, up to 30% of competitive protonation yields cations of type **10**. This is a typical enamine type of protonation[11] and is greatly facilitated by alkyl groups; 2,3,4,5-tetramethylpyrrole has a pK_a of 3·7. These cations (**9** and **10**, cf. furan, p. 127) have no π-electron sextets and no aromatic stability. They are very reactive and can readily polymerise by ionic[12] and/or radical mechanisms.

Treatment of pyrrole with 5·5 M hydrochloric acid at 0°C for 30 seconds yields[12] a crystalline trimer (**12**). This is presumably formed through electrophilic attack of the cation **10** on a neutral pyrrole molecule leading to **11**, followed by protonation of **11** at position 4 and subsequent attack on another

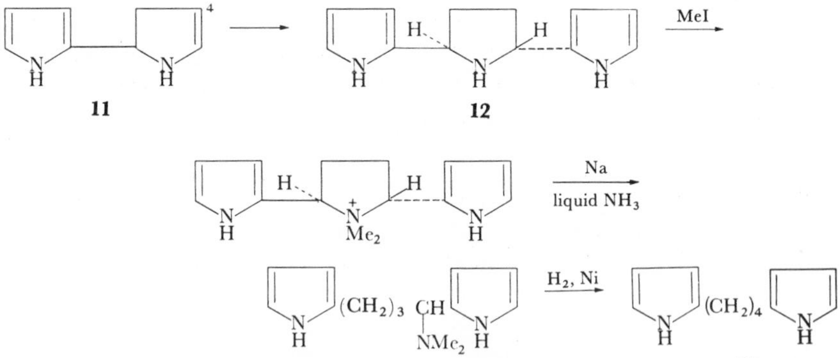

pyrrole. The trimer is basic, has the absorption spectrum required by two unreduced pyrrole rings, and forms a neutral monoacetyl derivative with an ultraviolet absorption spectrum identical to that of the parent base. The trimer was degraded as shown to 1,4-bis-2-pyrrylbutane (**13**), the structure of which was confirmed by synthesis. Pyrolysis of the trimer yields indole, pyrrole, and ammonia.

There are a few examples of pyrroles behaving as 1,3-dienes. Both pyrrole and its 1-phenyl derivative are reduced by zinc and acetic acid (1,4-addition of hydrogen occurring to the 1,3-diene system), yielding the Δ^3-pyrrolines, and the addition of triphenylmethyl radicals is similar. Singlet oxygen ($^1O_2$), from light, oxygen, and eosin, with pyrrole gives[13] **14**, presumably via an intermediate 2,5-peroxide. Oxidation of compound **14** with manganese dioxide gives a convenient synthesis of maleic imide. The only recorded Diels–Alder reaction of pyrrole itself is with an exceptional dieneophile and is shown below.[13a] A minor product (see also p. 106) from[14] 1-benzylpyrrole and acetylenedicarboxylic acid is **15**. 2-Fluorophenyllithium, probably reacting as benzyne (**16**), yields **17** with 1-methylpyrrole.[15] The structure of **17** was confirmed by successive treatment with methyl iodide and silver oxide when 1-dimethylamino-naphthalene was formed. Pyrrole itself with benzyne, from 2-bromofluoro-benzene and magnesium, gave 26% of 2-phenylpyrrole, no product corresponding to **17** being detected.[15]

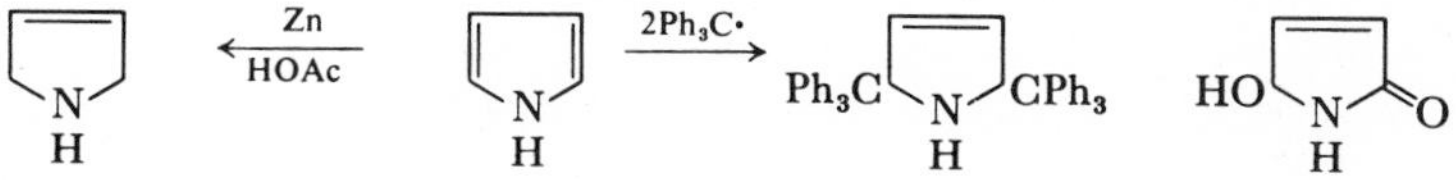

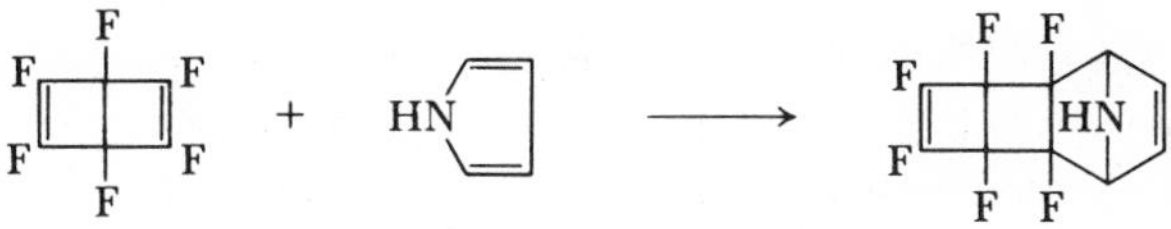

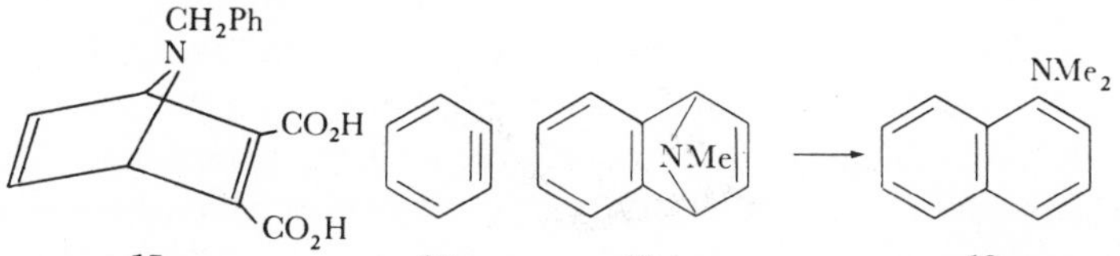

The high-pressure catalytic hydrogenation of pyrrole to pyrrolidine can be achieved over Raney nickel at 180°C; some Δ^1-pyrroline is formed under special conditions (p. 101). Pyrrole is much more easily hydrogenated over platinum in acid solution than under neutral conditions, where reaction proceeds very slowly, if at all. Protonation, either at nitrogen (**8**) or carbon (e.g., **10**) diminishes the aromaticity of the ring (p. 94) by removing the lone pair of the nitrogen atom. Introduction of a 1-ethoxycarbonyl group (**18**, for synthesis see below) has a similar effect, for hydrogenation over Raney nickel then proceeds at 70°C and Diels–Alder additions across the 2,5-positions take place readily with a number of dienophiles, especially when catalysed by aluminium chloride.[16] The products (e.g., **19**) can be aromatised (see also p. 447).

N
CO_2Et
18
$MeO_2CC{\equiv}CCO_2Me$
$AlCl_3$, 0 °C
30 min
CO_2Me
CO_2Me
NCO_2Et
19
$h\nu$
CO_2Me
CO_2Me
NCO_2Et
$AlCl_3$ or BF_3
CO_2Me
CO_2Me
NHAc

1-Benzoylpyrrole behaves exactly as furan towards photolysis with benzophenone (p. 129).[16a] The electron-withdrawing effect of the 1-ethoxycarbonyl group on the pyrrole ring is well illustrated by the fact that pyrrole couples with benzenediazonium chloride at position 2, even in weakly acid solution, while **18** is unaffected even by 2,4,6-trinitrobenzenediazonium salts.

3. *Substitution reactions of pyrrole*

a. At the nitrogen atom, and related substitutions at other positions. Pyrrole is an extremely weak acid. The anion (**20a**) may be compared with the much more highly symmetrical cyclopentadienyl anion (p. 89) and can be expected

N
H
$-H^+$
$+H^+$
N
20 a
N
20 b
N
20 c

to be somewhat more aromatic than pyrrole itself. Metallic potassium, but apparently not sodium, reacts with pyrrole, with the liberation of hydrogen. Other pyrroles bearing electron-attracting substituents which facilitate the loss of the proton do form crystalline *N*-sodium derivatives with sodium ethoxide, and pyrroles in general form sodium derivatives with sodamide in liquid ammonia. Potassium pyrrole is hydrolysed instantly by water and combines with methyl iodide, which does not react with pyrrole itself, acyl and arylsulphonyl chlorides,[17] and ethyl chloroformate, giving the corresponding 1-substituted pyrroles. This appears to be the general pattern, but the solvent and metal are important. Allyl bromide with potassium and dimethyl sulphoxide give 92% of 1-alkylation, while with sodium and tetrahydrofuran 80% of 2-allylpyrrole is formed. A Claisen type of rearrangement of 1-allyl- to 2-allylpyrrole has been excluded, and it is concluded that dissociation of the pyrrole–metal ion pair favours *N*-alkylation.[18] Sodium pyrrole in liquid ammonia with methyl iodide yields 95% of 1-methylpyrrole in agreement with this concept.

N
$CH{=}CH_2$
21

N
CH_2CH_2CN
22

N
CH_2OH
23

$HOCH_2$ N CH_2OH
H
24

Pyrrole, in the presence of alkaline catalysts, reacts with acetylene and other unsaturated compounds, such as acrylonitrile, to give **21** and **22** respectively. With formaldehyde and potassium carbonate at 40–55°C it gives **23**, while 2,5-dihydroxymethylpyrrole (**24**) is formed at 75–90°C. Pyrrole with methyl magnesium bromide, and other Grignard reagents, in ether gives pyrryl-magnesium bromide and the appropriate hydrocarbon. The pyrryl derivative usually reacts with electrophilic reagents at position 2, to a lesser extent at position 3, and very little at position 1. The magnesium must therefore be closely associated with the nitrogen atom, blocking attack, so the compound behaves as a resonance hybrid of the covalent **25** and ionic structures involving the pyrrole anion (**20a**, **20b**, and **20c**) under these conditions. Treatment of **25** with alkyl halides[19] and acyl halides, methyl chloroformate,[20] or ethyl formate at 150–200°C yields mainly the 2-alkyl- or 2-acylpyrrole, or pyrrole-2-carboxylic ester or pyrrole-2-carbaldehyde, respectively. Substitution at position 2 is direct in the case of the optically active 2-bromobutanes, for **25** gives both 2- and 3-*sec*-butylpyrroles with complete inversion of configuration.[21] At low

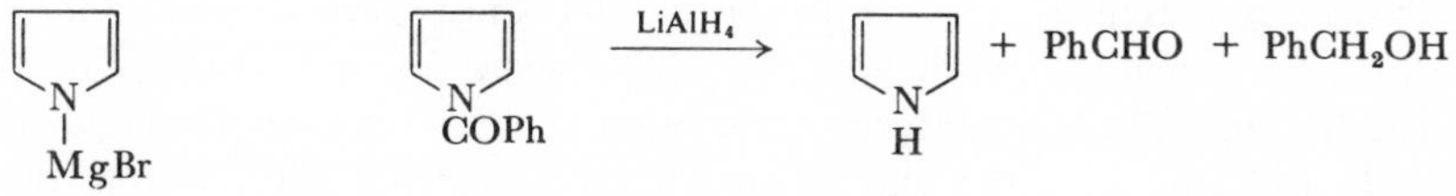

N
MgBr
25

N
COPh
$\xrightarrow{LiAlH_4}$
N
H
+ PhCHO + $PhCH_2OH$

temperatures pyrrylmagnesium bromide can sometimes give 1-substituted pyrroles, and with methyl iodide in hexamethylphosphoramide, an ionising solvent, 1-methylpyrrole is formed in high yield.

1-Acetylpyrrole is easily obtained by refluxing pyrrole with 1-acetylimidazole,[22] and 1-acylpyrroles on reduction with lithium aluminium hydride yield the parent pyrrole and an alcohol or aldehyde.

b. At carbon atoms. Pyrroles are attacked by electrophilic reagents *very* rapidly, and mainly at positions 2 and 5; electrophilic proton exchange is considered on p. 94. It is probable, in contrast to furan (p. 130) that these substitutions are direct and do not proceed through the formation of a Δ^3-pyrroline by addition to the 2- and 5-positions followed by an elimination.

Pyrrole cannot be sulphonated under ordinary conditions because polymerisation occurs (p. 94), but with pyridine–sulphur trioxide complex (p. 234) pyridinium 2-pyrrolesulphonate is formed, and with hydrochloric acid this gives 2-pyrrolesulphonic acid in 90% yield. The free acid is a very hygroscopic crystalline solid which forms the sulphonyl chloride with phosphorus pentachloride in carbon tetrachloride.

Pyrrole reacts extremely rapidly with halogens, and with iodine in aqueous potassium iodide, or bromine in methanol, the tetrahalogenated pyrroles are obtained. However, with 1 mole of sulphuryl chloride in ether at 0°C, 2-chloropyrrole is formed,[23] and further substitution occurs with more reagent leading to tetrachloropyrrole. 2-Chloropyrrole decomposes under nitrogen at 0°C in a few hours.

It is difficult to isolate pure compounds from the direct nitration of pyrrole owing to the formation of tar, but using nitric acid and acetic anhydride under controlled conditions which give acetyl nitrate 2- (**26**) and 3-nitropyrrole can be obtained in a 4 : 1 ratio and in satisfactory yield.[24] Pyrrole, ethyl nitrate, and sodium ethoxide do not give 3-nitropyrrole as commonly stated, but a 1% yield of 2-nitropyrrole (**26**)[24]; the product from the similar reaction involving ethyl nitrite is almost certainly sodium 2-nitrosopyrrole (**27**).

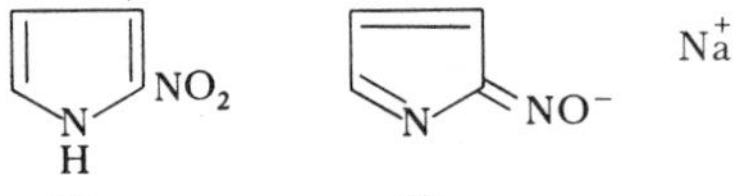

NO_2 N H **26** N NO^- $\overset{+}{Na}$ **27**

Pyrrole is extremely reactive in the Friedel–Crafts reaction and combines with acetic anhydride on heating, in the absence of any catalyst, to give 2-acetylpyrrole. The related Gattermann reaction, involving anhydrous hydrogen cyanide (or zinc cyanide) and hydrogen chloride, fails with pyrrole itself,

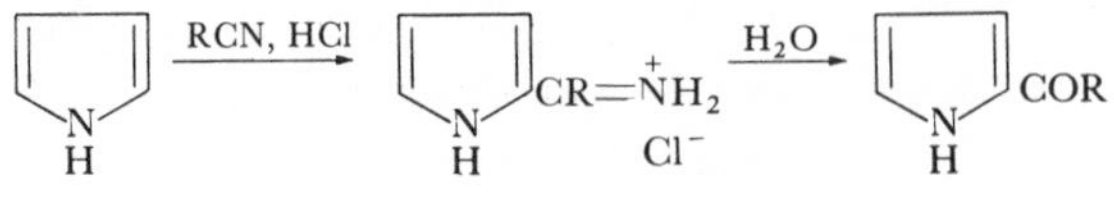

N H $\xrightarrow{RCN,\ HCl}$ N H $CR{=}\overset{+}{N}H_2$ Cl^- $\xrightarrow{H_2O}$ N H COR

although the reaction and the Houben–Hoesch modification, in which the hydrogen cyanide is replaced by an aliphatic or aromatic nitrile, proceed well with deactivated pyrroles. Pyrrole-2-carbaldehyde is prepared (90% yield) by the Vilsmeier reaction. This consists of treating pyrrole with phosphorus oxychloride and dimethylformamide and then boiling the product (**28**), which can

Me_2NCHO + $POCl_3$ $\xrightarrow{-Cl^-}$ $Me_2\overset{+}{N}{=}CH{-}OPOCl_2$ + pyrrole $\xrightarrow{-\bar{O}POCl_2}$ 2-($CH{=}\overset{+}{N}Me_2$)pyrrole Cl^- (**28**) $\longrightarrow$ 2-CHO-pyrrole

$Me_2\overset{+}{N}{=}CH{-}OPOCl_2$ + pyrrole $\xrightarrow{XNCCO_2Et}$ 2-($CXNHCO_2Et$)pyrrole (**29**)

$Me_2\overset{+}{N}{=}CH{-}OPOCl_2$ + pyrrole $\xrightarrow{Cl_3CCOCl,\ K_2CO_3}$ 2-($COCCl_3$)pyrrole $\xrightarrow[Et_3N]{MeOH}$ 2-(CO_2Me)pyrrole

be isolated, with aqueous sodium acetate.[25] Pyrrole and alkylpyrroles give[26] the corresponding phenyl 2- (or 5-) pyrryl ketones with benzoyl chloride and aqueous sodium hydroxide at 0°C. Pyrrole even combines with ethoxycarbonyl isocyanate or isothiocyanate at 0°C to give[27] **29** (X = O or S). A useful reaction for introducing the ester function is also shown.[28]

Pyrrole undergoes the Mannich reaction with formaldehyde and dimethylamine. Conversion of the product to the methiodide by methyl iodide, followed

2-(CH_2NMe_2)pyrrole $\xrightarrow{MeI}$ 2-($CH_2\overset{+}{N}Me_3$)pyrrole I^- $\longrightarrow$ 2-(CH_2OH)pyrrole

by treatment with alkali, gives pyrrole-2-methanol, while the methiodide with potassium cyanide yields pyrryl-2-acetonitrile.

Pyrrole, and many derivatives, couple preferentially at position 2 with aromatic diazonium compounds under weakly acid conditions to give highly coloured azo compounds; most phenols couple with benzene diazonium salts only in the presence of alkali.

Pyrrole with 30% hydrogen peroxide under controlled conditions[29] gives a mixture of the tautomeric pyrrolinones **30** and **31** (9 : 1 ratio) in 67% yield. In the presence of acid, protonation to **32** occurs, and if pyrrole (or another

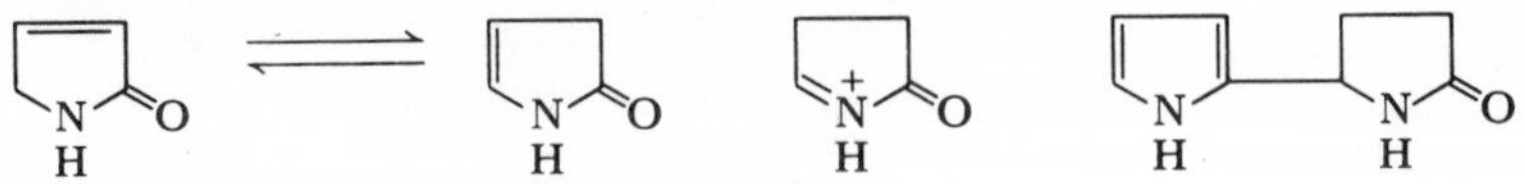

30 **31** **32** **33**

activated heterocycle) is present electrophilic attack occurs to give products such as **33**.

Photolysis of pyrrole in excess benzene gives the adduct **35**, which isomerises to the conjugated diene **36** with base. The reaction probably proceeds through the triplet diradical **34**, which also abstracts the 1-hydrogen atom, as 1-methylpyrrole is inert to the conditions employed.[30]

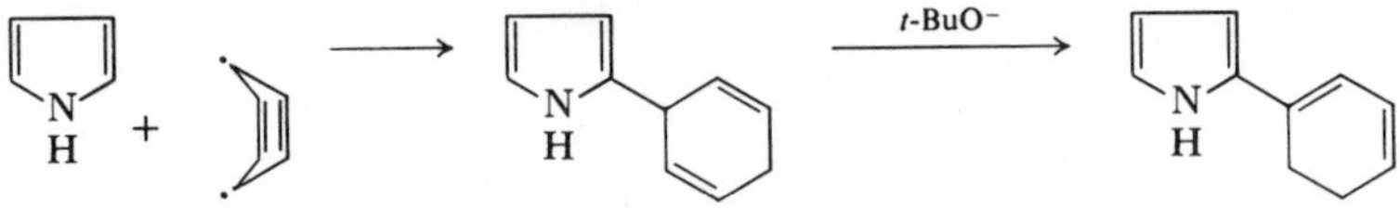

t-BuO⁻

34 **35** **36**

The Reimer–Tiemann reaction (boiling with chloroform and alcoholic potash) with pyrrole gives a mixture of the 2-aldehyde and 3-chloropyridine. The chloroform is first dehalogenated to dichlorocarbene, which then reacts with the pyrrole anion, or the neutral molecule,[31] as shown to give the respective

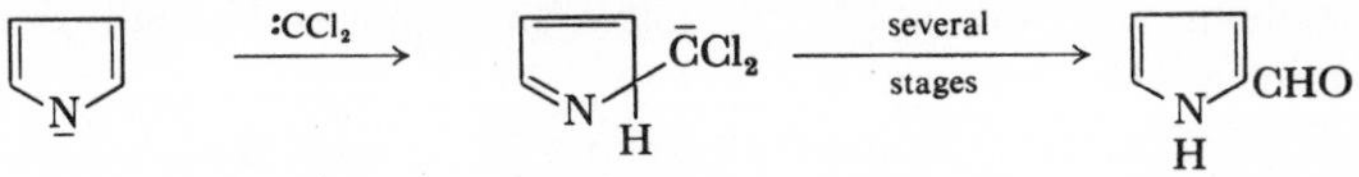

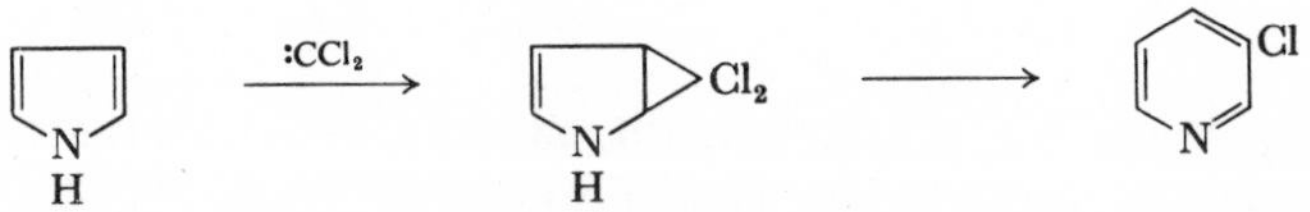

$:CCl_2$ several stages CHO

$:CCl_2$ Cl_2 Cl

37

products; the cyclopropane **37** is not a common intermediate as was once thought. Pyrrole can similarly be converted to pyridine (32%) by treatment with dichloromethane and methyllithium.[32] If the chloroform of the Reimer–Tiemann reaction with pyrrole is replaced by carbon tetrachloride, some 2-pyrrolecarboxylic acid is formed, but the mechanism of this conversion is not clear. Maleic anhydride does not undergo a Diels–Alder reaction with pyrrole, but addition takes place with the formation of **38**, or **39** if the anhydride is in excess. The reaction may be, in essence, a Michael addition of the more

$CH(CO_2H)CH_2CO_2H$ (N–H pyrrole, 2-substituted)
38

$HO_2CCH_2(HO_2C)CH$–(N–H pyrrole)–$CH(CO_2H)CH_2CO_2H$
39

reactive positions of pyrrole to the activated double bond. When the 2- and 5-positions are occupied, as with ethyl 3,5-dimethylpyrrole-2-carboxylate, maleic anhydride reacts at position 3 or 4.

D. Derivatives of pyrrole

The presence of electron-attracting substituents in a pyrrole ring greatly stabilizes the system to acids and reduces the avidity with which electrophilic reagents substitute; the comments made on p. 138 in connexion with furan are also applicable to pyrrole. It is therefore a practical proposition to monohalogenate, or nitrate, pyrrolecarboxylic acids or esters. At the same time the displacement of existing substituents by entering groups is particularly noticeable in the pyrrole series. For example, nitration of ethyl 5-acetyl-2,4-dimethylpyrrole-3-carboxylate gives ethyl 2,4-dimethyl-5-nitropyrrole-3-carboxylate, and bromination of 2,4-diethoxycarbonyl-3-methylpyrrole-5-carboxylic acid gives diethyl 5-bromo-3-methylpyrrole-2,4-dicarboxylate. The product of nitration of diethyl 2,4-dimethylpyrrole-3,5-dicarboxylate is particularly interesting, as both methyl substituents are apparently replaced by nitro groups. Presumably they are first oxidised to carboxylic acid groups, which are then displaced.

1. *Some reduced pyrroles*

Both the hydrogenation of pyrrole over a rhodium–alumina catalyst and the dehydrogenation of pyrrolidine give some Δ^1-pyrroline, which can be more readily prepared from pyrrolidine acetate as indicated. It shows no N–H

$\overset{+}{N}H_2$ $\bar{O}Ac$ —NaOCl→ N–Cl —alcoholic KOH→ Δ^1-pyrroline —pyrrole→ 2-(pyrrolidin-2-yl)pyrrole (N–H, N–H)

absorption band in the infrared, it reacts with pyrrole as shown and with hydrogen cyanide, giving the nitrile of proline (p. 116), and it also self-condenses to a trimer.

Although the zinc–hydrochloric acid reduction of pyrrole gives Δ^3-pyrroline (p. 95) containing 15% of pyrrolidine,[33] similar reduction of 2,5-dimethylpyrrole, presumably as the protonated pyrrole (see also p. 94), gives **40** and **41** in a 4 : 1 ratio. The Δ^3-pyrroline (**40**) has typical properties of an aliphatic secondary amine in that it forms a monobenzoyl derivative and reacts with 2 moles of methyl iodide. It also has the properties of an olefine in that it instantly decolourizes bromine water and acidified potassium permanganate. Ozone splits the double bond, giving **42**, and hydrogenation over platinum gives *trans*-2,5-dimethylpyrrolidine. The isomeric pyrroline (**41**), in contrast to **40**, shows no N–H vibration band in the infrared. It reacts with only 1 mole of methyl iodide and very slowly with bromine water or permanganate, and it resists ozone. Benzoylation opens the ring, yielding **43**, and hydrogenation gives *cis*-2,5-dimethylpyrrolidine. These reactions exclude the possible alternative

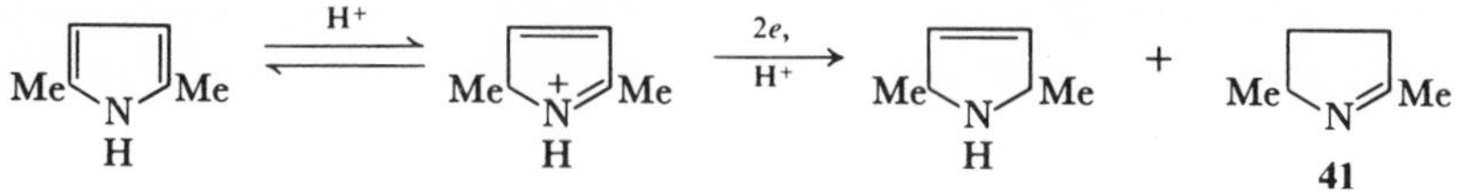

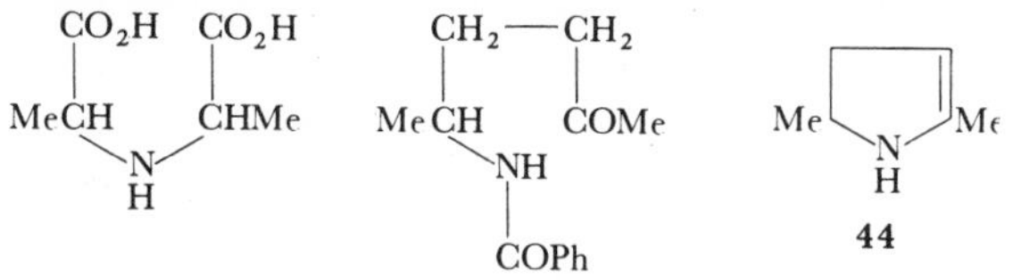

formulation (**44**) for the compound. The Δ^3-pyrroline (**40**) is isomerized to the Δ^1-compound (**41**) by boiling with Raney nickel in xylene. So far the authenticity of no Δ^2-pyrroline unsubstituted on the nitrogen atom has been confirmed; it appears that the tautomeric Δ^1-compounds are the more stable. Pyrroles are not reduced by lithium aluminium hydride, nor by sodium in liquid ammonia, but the latter reagent reduces an *N*-aryl substituent. Hydrolysis of the product, which probably has the structure shown but could tautomerise to an enamine, removes the aryl group from the pyrrole.[34]

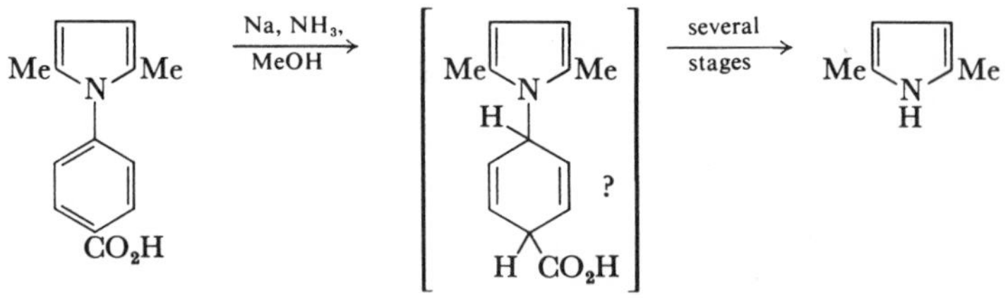

Tetrahydropyrrole, or pyrrolidine, b.p. 88°C, is an almost flat molecule which has the properties of a typical secondary aliphatic amine. It is much used in the preparation of enamines (e.g., **45**) from ketones; enamines have enormous synthetic value.[11]

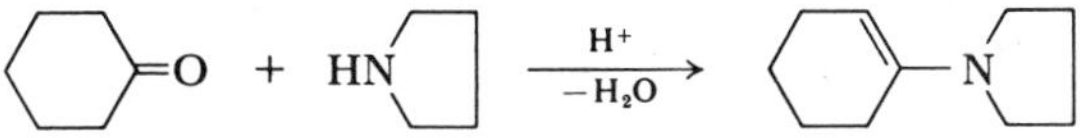

45

Pyrrolidine can be synthesized by many methods, examples being the reduction of pyrrole and the cyclization of butane-1,4-diol by ammonia over alumina at 325°C (68% yield). Pyrrolidines have been dehydrogenated to pyrroles by heating with platinized asbestos. Pyrrolidine rings can be opened by the Hofmann (p. 244) and von Braun (p. 245) methods.

2. *Alkyl- and arylpyrroles*

Many alkylpyrroles have been synthesized in connexion with work on the porphyrin pigments, and the Wolff–Kishner reduction of acylpyrroles obtained by Knorr's synthesis (p. 113) is a very common method. The direct alkylation of some pyrroles can be effected in a remarkable manner by heating to about 200°C with a sodium alkoxide. The mechanism of the reaction has not been investigated, but it is known that position 2 is attacked more easily than position 3. It is interesting that this is the same as for electrophilic attack, but it would be very remarkable if such were the case under such strongly alkaline conditions; an uncharged molecule, such as an aldehyde, is perhaps involved.[35] In this case the

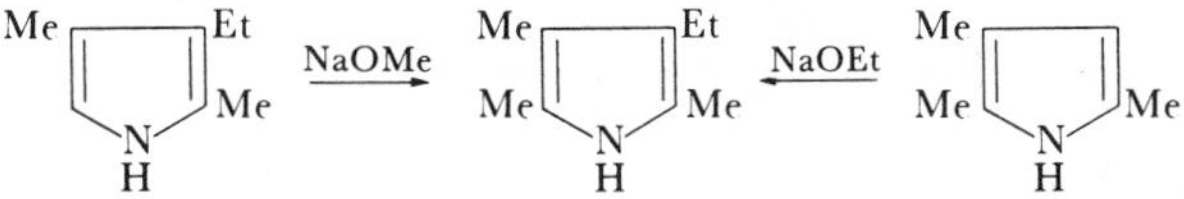

resulting hydroxyalkylpyrrole with sodium alkoxide could regenerate more aldehyde and give the corresponding alkylpyrrole, and an analogy is the conversion of 2-naphthol into 1-methyl-2-naphthol (**46**) by successive reaction with formaldehyde and sodium methoxide.

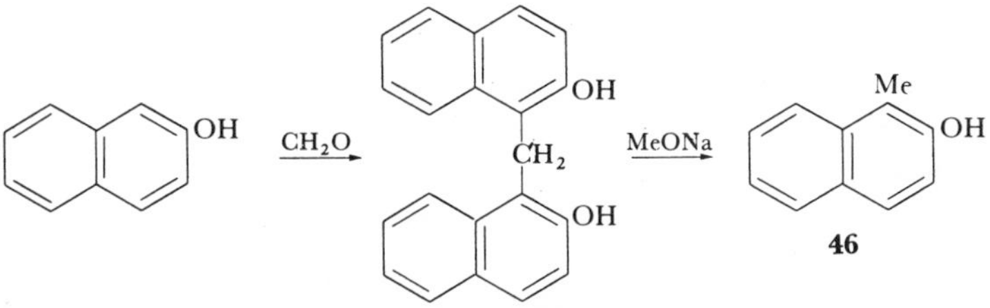

46

Alkylpyrroles are in general more reactive, more basic, and more sensitive to acids than pyrrole itself, as can be anticipated from the electron-donating properties of alkyl groups. Methyl groups at position 2 are much more easily attacked than those at position 3.

1-Alkyl- and 1-arylpyrroles rearrange to the 2-substituted pyrroles, along with some of the 3-isomers, at about 600°C. This reaction was employed in the first synthesis of the alkaloid nicotine. It is unimolecular in the gas phase, a high activation energy is involved, and in the case of (+)-*N*-(1-phenylethyl)pyrrole, the 2-substituted pyrrole formed retains (77%) the original configurations.[36] The idea of a radical pair as an intermediate accounts well for the results. The photolysis of 1-benzylpyrrole gives a mixture of the 2- and 3-benzylpyrroles,[37] but the mechanism by which this takes place is not clear.[38]

1-Methylpyrrole forms a hydrochloride. It also forms an *addition* compound with ethylmagnesium bromide, no ethane being liberated, and the addition compound gives no pyrrolecarboxylic acid with carbon dioxide. With acetyl chloride the addition compound surprisingly does give 2-acetyl-1-methylpyrrole. An explanation is that magnesium bromide is present in equilibrium in the

$$2EtMgBr \rightleftharpoons MgBr_2 + MgEt_2$$

mixture, and alone can catalyse the condensation of the pyrrole with the acetyl chloride. 1-Methylpyrrole reacts with 1 mole or an excess of butyllithium, yielding the 2-lithium or 2,4-dilithium derivative, respectively, as is shown by carbonation (Dry Ice) when the 2-carboxylic acid and the 2,4-dicarboxylic acid are formed.[39]

3,4-Dimethylpyrrole, in contrast to pyrrole (p. 94), undergoes a reversible reaction to form a dimer (**47**); this is the first example[40] of 2-protonation

6NHCl aq.

AcOH room temp.

AcOH

Zn(OAc)$_2$

47

48

initiating polymerisation in the pyrrole series. Under other conditions[41] this pyrrole gives the indolizine (**48**), presumably by ring opening to the aldehyde, which then combines with more of the pyrrole.

The nitration of 1-methylpyrrole in acetic anhydride gives[24] a mixture of the 2- and 3-nitro derivatives in a 2 : 1 ratio. These nitro compounds are of special interest, as they are among the very few nitro compounds known to undergo the Friedel–Crafts reaction. This gives an indication of the very great reactivity of the pyrrole ring, which is sufficient to overcome the deactivating properties of the nitro group.

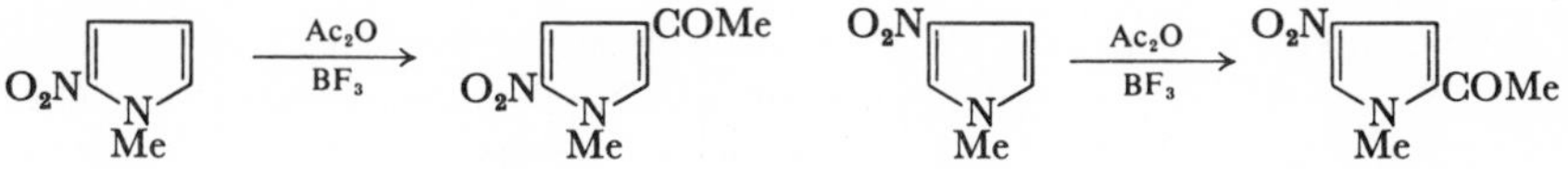

O_2N N Me $\xrightarrow[BF_3]{Ac_2O}$ O_2N N Me COMe O_2N N Me $\xrightarrow[BF_3]{Ac_2O}$ O_2N N Me COMe

The halogenation of alkylpyrroles has been widely studied, and even if the most favourable conditions for side-chain substitution are chosen, for example, using sulphuryl chloride, all unsubstituted carbon positions on the ring are chlorinated before attack occurs on a 2-methyl group. All three hydrogen atoms of the methyl group are then successively replaced. Bromination can be more easily controlled so that substitution ceases at the monobromomethyl stage. Careful hydrolysis of the halogenated methylpyrroles (cf. $PhCH_2Cl$, $PhCHCl_2$, and $PhCCl_3$) can yield the corresponding alcohols, aldehydes, and acids, but the self-condensation of the halogenated pyrroles may prevent their isolation. For example, the bromination of **49** goes through the sequence indicated, yielding the dipyrrylmethene **50**. Asymmetric dipyrrylmethenes can be obtained by using appropriate intermediates in the last stage of the synthesis.

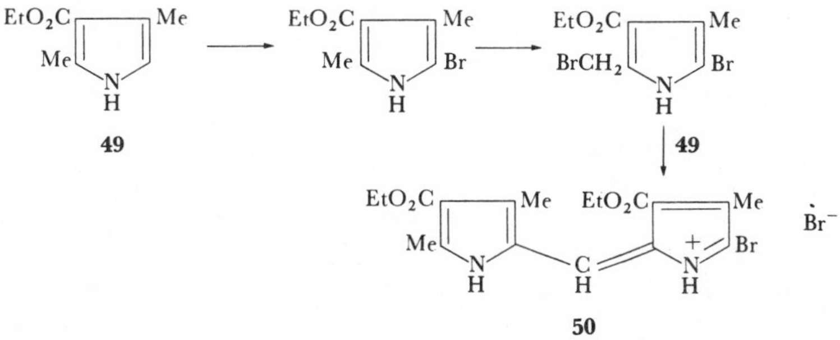

EtO_2C Me Me N H **49** ⟶ EtO_2C Me Me Br N H ⟶ EtO_2C Me $BrCH_2$ Br N H $\xrightarrow{49}$

EtO_2C Me Me N H C H EtO_2C Me Br $\overset{+}{N}$ H Br^-

50

1-Methylpyrrole on electrochemical (anodic) oxidation in methanol behaves like furan (p. 129), and the product (**51**) may be hydrolysed in two stages to *N*-methylmaleimide.[42]

1-Benzylpyrrole combines with acetylenedicarboxylic acid[14] giving the yellow **52** and **53**, with a small amount of colourless compound (see p. 95). On complete hydrogenation both yellow compounds give the same product.

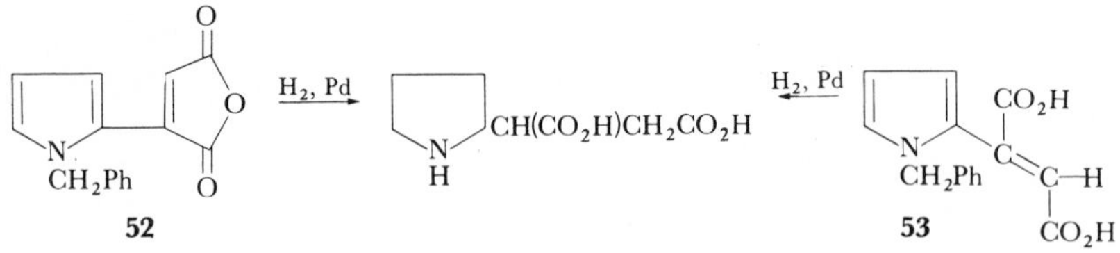

Tetraphenylpyrrole can be obtained by heating deoxybenzoin azine (from the ketone and hydrazine) with hydrochloric acid at 180°C. The reaction is reminiscent of Fischer's indole synthesis (p. 205). Tetraphenylpyrrole undergoes an unusual ring opening with nitrous acid.

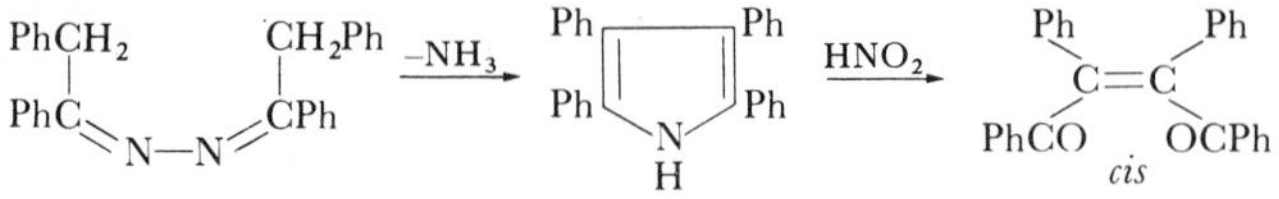

3. *Amino- and diazopyrroles*

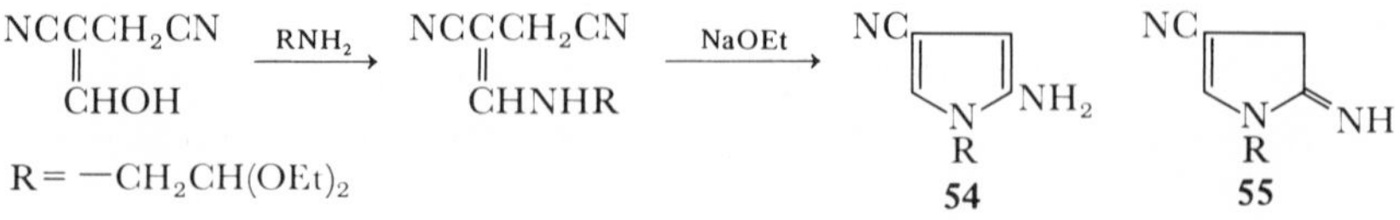

Comparatively few simple aminopyrroles are known. They can be prepared by direct synthesis as shown or by reduction of nitroso- or nitropyrroles, or of benzeneazopyrroles obtained by diazonium coupling. Infrared examination[43] of **54** has shown the absence of bands characteristic of the imine group. The compound is therefore not the pyrrolenine **55**. It gives positive Ehrlich's and pine splint tests, and no ammonia is evolved on boiling with 2N sodium hydroxide. Little is known about the diazotization and coupling of simple aminopyrroles. The diazonium compound **56** is obtained from ethyl 3,5-

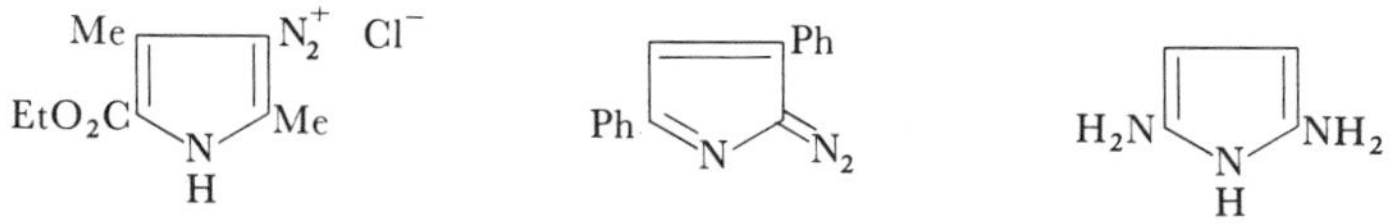

56 **57** **58**

dimethylpyrrole-2-carboxylate by successive nitrosation, reduction, and diazotization. It is comparatively stable, perhaps because of the two electron-attracting substituents, and couples in aqueous sodium bicarbonate[44] with phenols. Buffered nitrous acid with 2,4-diphenylpyrrole gives first 2-nitroso-3,5-diphenylpyrrole, but on prolonged treatment, the 2-diazopyrrole (**57**) is formed.[45] 2-Diazopyrroles are more reactive and less stable than their 3-isomers, but both couple with 2-naphthol. 2,5-Diaminopyrrole (**58**) has been obtained from succinonitrile and ammonia. Its exact structure is uncertain, although the amino–imino tautomer is excluded by the ultraviolet spectrum of the compound.

4. *Hydroxypyrroles*

As the ultraviolet absorption spectra of **59** and **62** differ widely, and as **62** can only have the structure given, **59** must exist largely as the 2-pyrrolone tautomer, as is shown. Hydrolysis and decarboxylation gave '2-hydroxypyrrole' as a colourless oil, b.p. 75°C at 0·05 torr, which resinified on standing for 24 hours.[46] N.m.r. studies have shown that this pyrrole consists mainly of the 3-pyrrolin-2-one tautomer (**60**) with a small proportion of 4-pyrrolin-2-one (**61**); the hydroxy tautomer could not be detected.

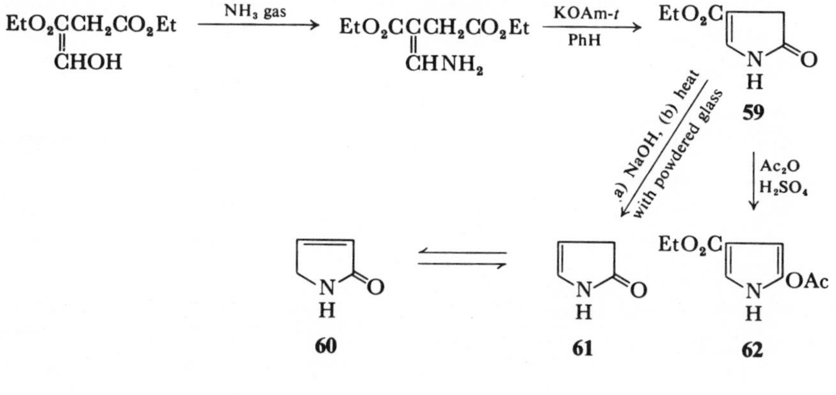

$EtO_2CC(=CHOH)CH_2CO_2Et$ —(NH_3 gas)→ $EtO_2CC(=CHNH_2)CH_2CO_2Et$ —(KOAm-*t*, PhH)→ **59**

59 —(a) NaOH, (b) heat with powdered glass→ **61** ⇌ **60**

59 —(Ac_2O, H_2SO_4)→ **62**

60 **61** **62**

3-Hydroxypyrrole has not yet been prepared, although its 5-carboxylic acid (**64**) has been obtained[47] by a long synthesis. This acid (**64**) gives no colour with ferrous salts (contrast pyridine-2-carboxylic acid), it gives a deep violet ferric chloride colour, and it is very easily oxidized to a blue dye. Although the ultraviolet absorption spectrum of the hydroxy acid (**64**) broadly resembles that of the ethoxy acid (**63**), it is likely that the former compound should be represented as the tautomer **65**. Spectral studies[48] have proved conclusively that the ester shown possesses structure **66**.

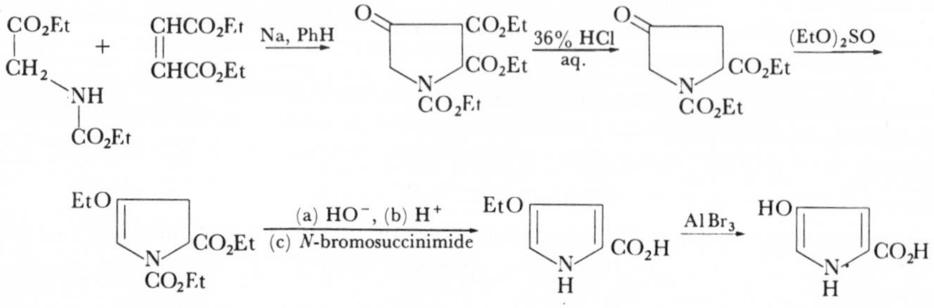

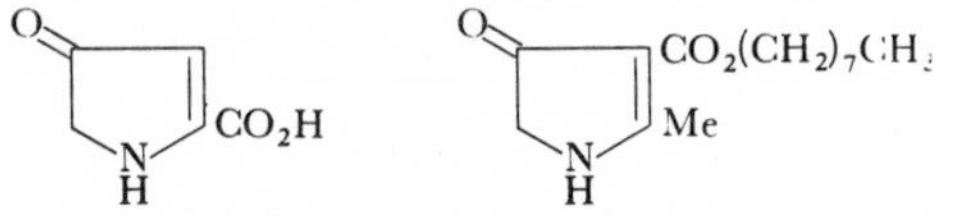

5. *Halogenated pyrroles*

2-Chloropyrrole is a very unstable substance (p. 98), in contrast to most halogenated pyrroles, but on sulphonation gives the 5-sulphonic acid.[23] Many pyrroles have been chlorinated by sulphuryl chloride, via a radical mechanism, substitution occurring at position 2 when possible. 3,4-Dichloropyrroles are unchanged by lithium aluminium hydride and some other nucleophiles,[49] unless a powerful electron-attracting group is present. The most interesting compound of this class is pyrrolnitrin (p. 117).

6. *Pyrrole alcohols*

Pyrrole-2-methanol (**68**) may be obtained by reducing pyrrole-2-carbaldehyde with sodium borohydride. Lithium aluminium hydride also gives 2-methylpyrrole[50] by the route indicated, as corresponding 1-alkylpyrroles are stable to this reagent. Several substituted pyrrole-2-methanols have been obtained from pyrroles, formaldehyde and a trace of alkali (p. 97; see also p. 111), but usually reaction proceeds further. Alcohols of this type are extremely reactive in the presence of either alkali (cf. **67**) or acid, when the resonance-stabilised cation **69** is formed. In consequence, the first compound usually isolated from pyrrole, formaldehyde, and acid is the dipyrrylmethane **70**, which is easily oxidised to the dipyrrylmethene **71**. A low yield of porphin (**72**) may be obtained by heating pyrrole, formaldehyde, and pyridine. Pyrrole-2-methanols, such as **73**, on heating lose formaldehyde to give the corresponding dipyrrylmethane (**74**).

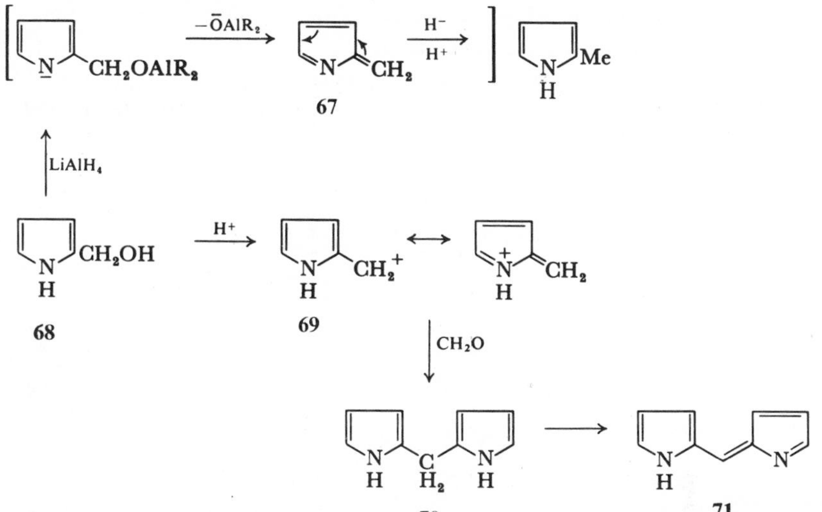

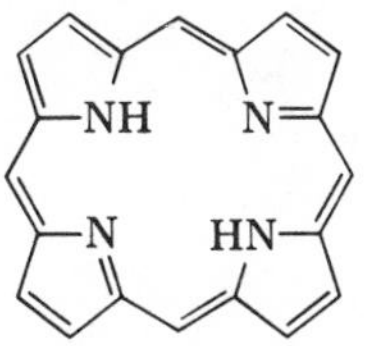

7. *Pyrrole aldehydes and ketones*

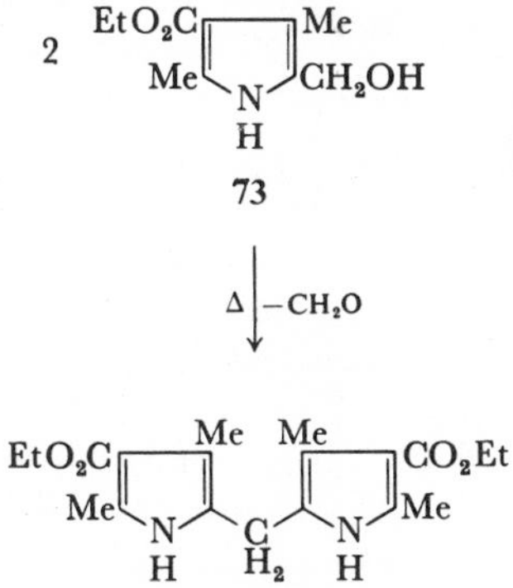

Pyrrole-2-carbaldehyde, and aldehydes of this type, are most easily obtained by the Vilsmeier reaction from the corresponding pyrrole, phosphorus oxychloride, and dimethylformamide (p. 99), although in earlier times the Gattermann reaction (p. 98) was often employed. Pyrrole-3-carbaldehyde can be synthesised as shown, the reaction conditions being important,[51] and the 1-methyl derivative is a minor product (6%) in the Vilsmeier formylation of 1-methylpyrrole, which gives mainly the 2-aldehyde.[51] The reduction of the ester **75** with hydrazine is an example of a useful new route to 3-alkylpyrroles.[52]

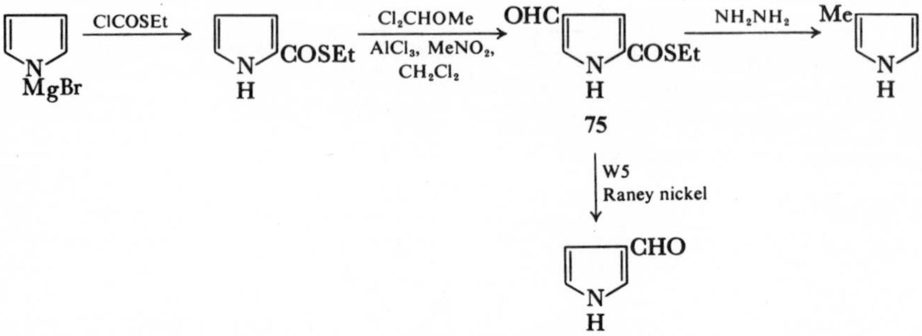

ClCOSEt; Cl_2CHOMe, $AlCl_3$, $MeNO_2$, CH_2Cl_2; NH_2NH_2; W5 Raney nickel

75

Pyrrole-2- and pyrrole-3-carbaldehydes have some of the properties of aromatic aldehydes, but the 1-hydrogen atom is relatively acidic because of the electron-attracting character of the aldehyde group. These aldehydes give no Schiff's or Fehling's reaction but do reduce Tollen's reagent. They do not undergo the Cannizzarro, benzoin, or (normal) Perkin reactions and cannot be converted to acetals.[51] These failures can be associated with the recognition of these aldehydes as vinylogous amides, the carbonyl groups being highly deactivated by resonance (e.g., **76** ↔ **77**). The carbonyl absorptions for the 2- and

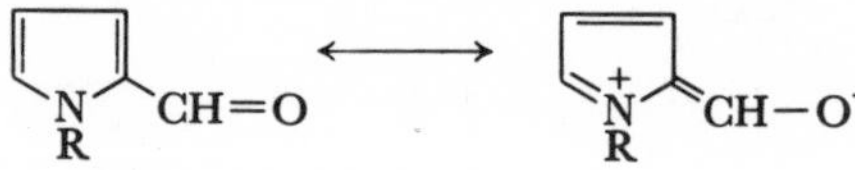

76 **77**

3-aldehydes are at 1667 and 1683 cm^{-1}, respectively (PhCHO, 1708 cm^{-1}), and are little altered by *N*-methylation. Pyrrole-2- and pyrrole-3-carbaldehydes undergo the Wolff–Kishner reduction to the corresponding methylpyrroles in good yield, and this is an important synthetic process. The 2-aldehyde undergoes the Wittig reaction.[51,53]

The combination of pyrrole-2-carbaldehydes with pyrroles in the presence of hydrogen bromide, to give dipyrrylmethenes for porphyrin syntheses, has proved most useful. The course of the reaction is not certain, and the route suggested is the formation of a tripyrrylmethane (**78**), which can split in two ways. However, the desired unsymmetrical products have often been obtained, and it is not impossible that they are formed directly. Bromination of pyrrole-2-carbaldehyde causes mostly 4-substitution.[54]

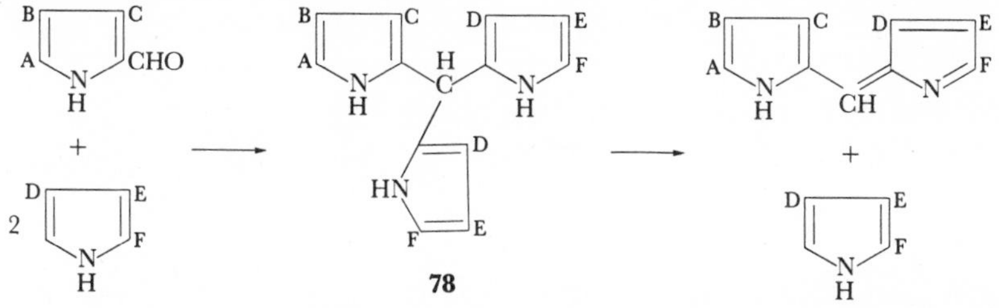

Pyrrole ketones may be obtained by a modification of Knorr's synthesis (p. 113) or by acylating a pyrrole directly with acetic anhydride (p. 98); a carboxyl group can be displaced in this second reaction. Pyrrole ketones are stable and undergo standard reactions.

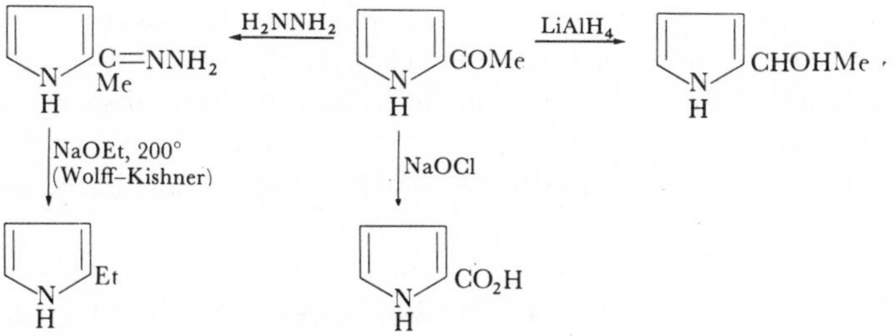

8. *Pyrrolecarboxylic acids and esters*

Pyrrolecarboxylic acids (see also pp. 100 and 104) are commonly obtained directly by synthesis involving ring closure (pp. 113–115), and less often by the oxidation of alkylpyrroles by fused potassium hydroxide, by carboxylation of metallic derivatives of pyrrole, and from ketones (flow sheet above). Pyrrole-2-carboxylic acid is best obtained from the aldehyde by oxidation with alkaline silver oxide[55] but has also been obtained from pyrrole itself by direct carboxylation with ammonium carbonate at 120°C. This last reaction has been little investigated but is similar to Kolbe's synthesis of phenolic acids from phenols.

Pyrrolecarboxylic acids, like phenolic acids, are usually decarboxylated readily on heating. This has been of great value in the preparation of alkylpyrroles via Knorr's synthesis. Alkyl groups facilitate and electron-attracting groups hinder the decarboxylation, which usually takes place preferentially at position 2. Pyrrole-2-carboxylic acid on heating with acetic anhydride gives the cyclic compound **79**.

Carboxyl groups can often be displaced by incoming substituents. Bromination, for instance, followed by catalytic hydrogenation can be a useful way of removing a carboxyl group when direct decarboxylation fails (e.g., with **80**).

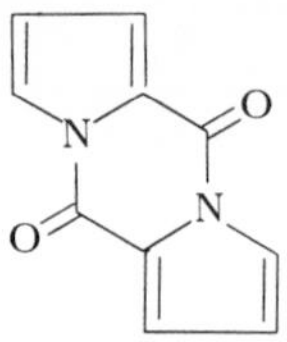

79

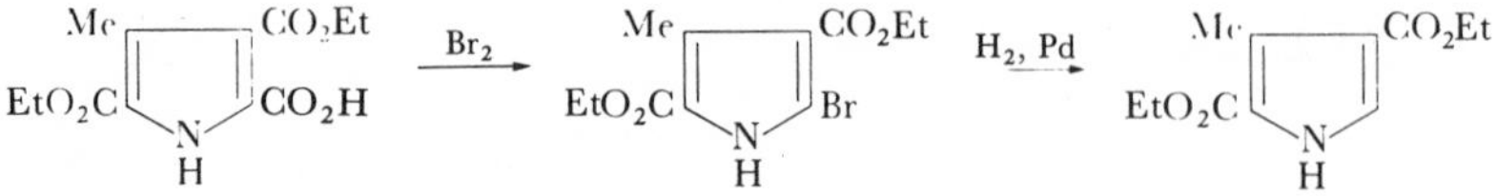

80

The halogenation of methyl pyrrole-2-carboxylate has been carefully investigated.[55] The highest yield (56%) of 4-substitution was obtained using bromine in acetic acid, while 80% of the 5-chloro compound was isolated after treatment with *t*-butyl hypochlorite in carbon tetrachloride. While it is tempting to assume that the last substitution is radical in mechanism, it has been shown that ethyl pyrrole-2-carboxylate gives mainly the 5-aldehyde in the Vilsmeier reaction (p. 99).[56] A 3-ester group directs electrophilic attack almost exclusively to the 5-position.

The difference in reactivity at positions 2 and 3 of the pyrrole ring is again shown in the hydrolysis of esters. The 2-carboxylic esters are the more rapidly hydrolysed by alkali,[57] and this is illustrated in the synthesis of ethyl pyrrole-3-carboxylate. Many 3-carboxylic esters are more rapidly attacked by

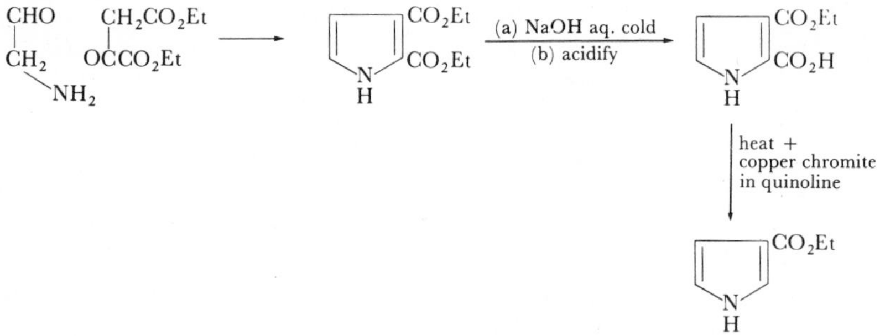

sulphuric acid at 40–60°C than the 2-isomers. This may be due to steric effects, as the pyrroles involved usually bear many substituents, but ethyl pyrrole-3-carboxylate, where there is no steric hindrance, is hydrolysed by hot alcoholic potassium hydroxide.

It is noteworthy that while *C*-ethoxycarbonyl groups inhibit or prevent the hydrogenation of the pyrrole ring, the presence of an *N*-ethoxycarbonyl group greatly facilitates it (p. 96). This is particularly useful, as the *N*-carboxylic acid, obtained on hydrolysis, immediately decarboxylates.

E. Synthetic methods

(1) The most general pyrrole synthesis is that of Knorr, who condensed an α-amino ketone with a ketone possessing a reactive methylene group in the presence of acetic acid. The α-amino ketone is usually prepared *in situ* by reducing the corresponding oximino compound with zinc. If the methylene

$$\left[\begin{matrix} MeCO \\ | \\ EtO_2CCH_2 \end{matrix} \longrightarrow \begin{matrix} MeCO \\ | \\ EtO_2CC{=}NOH \end{matrix} \longrightarrow \right] \begin{matrix} MeCO \\ | \\ YCHNH_2 \end{matrix} + \begin{matrix} {}^{*}CH_2X \\ | \\ OCMe \end{matrix} \longrightarrow$$

Me, X, Y, Me, N, H (pyrrole ring)

group marked with an asterisk is not sufficiently reactive the amino ketone self-condenses to a dihydropyrazine (p. 408). The synthesis usually fails or gives very poor results when X = alkyl. Y can be hydrogen, but better results are obtained when it is a COR group, while the methyl groups of the reactants may be replaced by aryl or RCO groups. The synthesis is therefore of very wide application. Recent improvements in the synthesis are (*a*) use of sodium dithionite as the reducing agent for the isonitroso compound, (*b*) use of benzyl acetoacetate, and (*c*) use of *t*-butyl acetoacetate. The advantages obtained respectively are:

(a) better yields,

(*b*) subsequent catalytic hydrogenolysis of the pyrrole gives toluene and a

$$RCO_2CH_2Ph \xrightarrow{H_2,\ Pd/C} RCO_2H + PhMe$$

free carboxylic acid without affecting other ester groups in the molecule, and

(*c*) subsequent catalytic hydrolysis of the pyrrole with a trace of 4-toluene-sulphonic acid at 200°C yields butylene and a pyrrole-3- or pyrrole-4-carboxylic acid, while concurrent decarboxylation occurs if the *t*-butyl

$$RCO_2Bu\text{-}t \longrightarrow RCO_2H + Me_2C{=}CH_2$$
$$RCO_2H \searrow RH + CO_2$$

ester was at position 2 or 5, leading to a pyrrole unsubstituted at this position. As in (*b*) ethoxycarbonyl groups in the pyrrole are not affected by the procedure.

Treatment of an α-amino ketone with an activated acetylene can give a dihydropyrrole (**81**) which, with acid, dehydrates to the pyrrole.[58] This new synthesis is very similar to that of Knorr, and the dihydropyrrole (**81**) may be a common intermediate.

$$RCO\text{–}R'CHNH_2 + MeO_2C\text{–}C{\equiv}C\text{–}CO_2Me \xrightarrow{NaOAc} \mathbf{81} \xrightarrow{\text{trace } H^+} \text{pyrrole (R, R', } CO_2Me, CO_2Me\text{)}$$

81

(2) The Paal–Knorr synthesis of pyrroles from 1,4-diketones by treating with ammonia or primary amines is an excellent method that works for sterically hindered ketones[59] but is limited by the availability of the diketones. However, new methods of preparing these diketones from furans have been discovered (p. 130).

$$\mathbf{82} \xrightarrow{\text{HCl aq.}} \underset{CHO\ \ CHO}{H_2C\text{—}CH_2} \xrightarrow{RNH_2} \text{1-R-pyrrole} \xleftarrow[\text{mechanism}]{\text{uncertain}} {}^{+}NH_3R\ {}^{-}O_2C(CHOH)_4CO_2^{-}\ {}^{+}NH_3R$$

82

Succindialdehyde and ammonia give pyrrole itself, in poor yield; the succindialdehyde, which need not be isolated, can be obtained from the acetal (**82**, p. 130) with dilute hydrochloric acid. Primary amines, however, give much better results, and by using appropriate amines and diketones it is possible to prepare pyrroles with substituents at any position. The first synthesis of pyrrole, the pyrolysis of ammonium mucate, is still the most convenient laboratory preparation for this compound, and if mucates derived from primary amines are used, then 1-alkylpyrroles are obtained.

A related synthesis starts from alkyl halides.[60] The crude product obtained from electrophilic addition of an acyl halide to the allyl chloride is heated with ammonia or a primary amine, giving the pyrrole.

$$RCOCl + CH_2{=}CR'CH_2Cl \longrightarrow RCO\text{–}H_2C\text{—}CClR'\text{–}CH_2Cl \longrightarrow [RC(O)\text{–}CH{=}CR'\text{–}CH_2Cl\ ?] \xrightarrow{NH_3} \text{2-R-4-R'-pyrrole (NH)}$$

(3) Another valuable synthesis[61] starts from vinyl ketones. Base-catalysed addition of the glycine derivative and dehydration give the dihydropyrrole. Treatment with sodium ethoxide, at room temperature in some cases, eliminates sodium 4-toluenesulphinate, yielding the pyrrole.

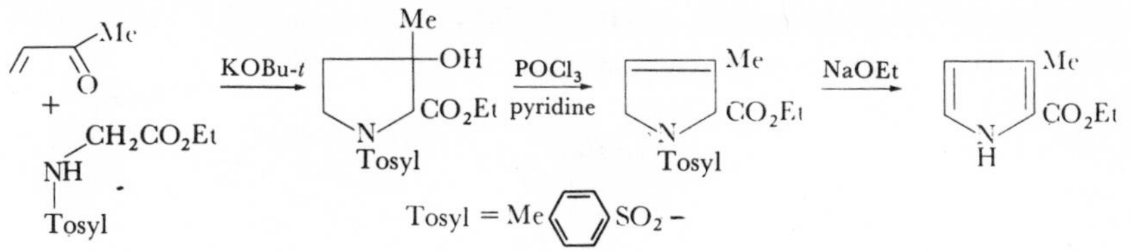

Tosyl = Me-C$_6$H$_4$-SO$_2$–

(4) The Hantzsch synthesis is much less generally applicable than that of Knorr, and requires the combination of a β-keto ester with an α-chloro ketone in the presence of ammonia or a primary aliphatic amine. Ethyl 2-aminocrotonate is almost certainly an intermediate in the example given. A big difficulty is that

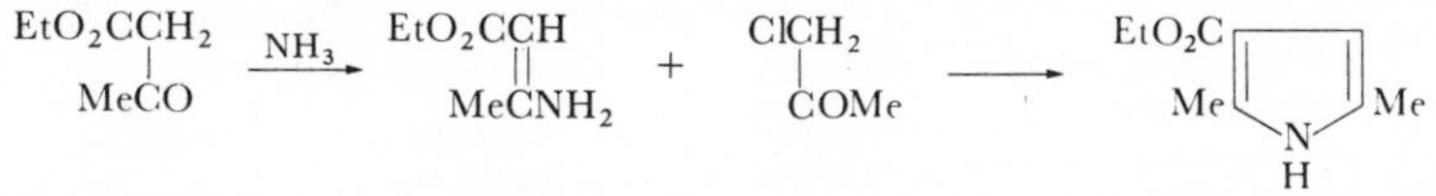

the chloro ketone can also react in another manner with the β-keto ester to give a furan (see p. 150). In this case the product would be ethyl 2,4-dimethylfuran-3-carboxylate.

(5) A potentially useful synthesis of 1-methylpyrroles involves a Vilsmeier formylation of a ketone. Examples reported include R = H, Me, and Ph and give yields of 38–74%.[62]

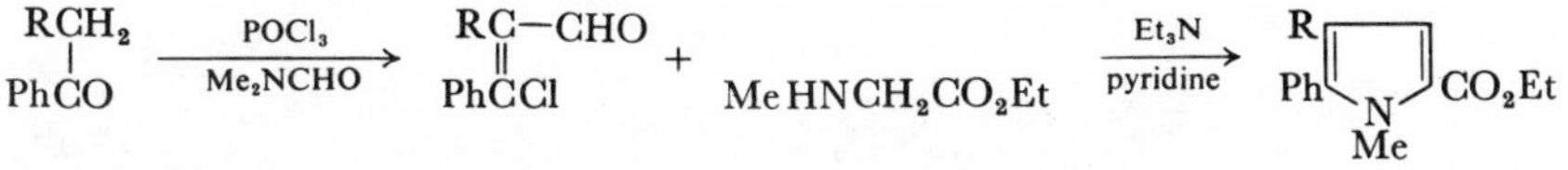

(6) An interesting synthesis involves a Diels–Alder reaction.

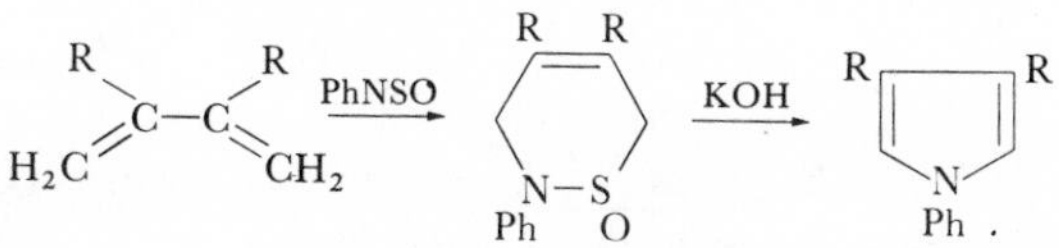

(7) Diacetylene, ammonia, and cuprous chloride at 140–160°C give pyrrole, and a number of derivatives have been similarly prepared.[63]

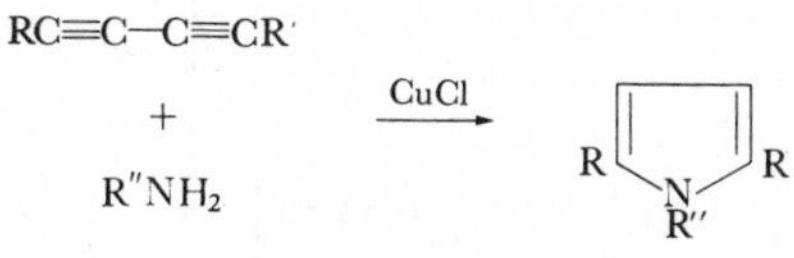

(8) Pyrrole is prepared commercially by passing furan, ammonia, and steam

over a heated aluminium oxide catalyst. This type of synthesis has been used for synthesizing isotopically labelled[64] and 1-substituted pyrroles.

RNH_2, Al_2O_3
steam

N
R

(9) Heating the *cis*-but-2-ene-1,4-diol with primary amines under dehydrogenation conditions can give excellent yields of pyrroles.[65]

HOH_2C CH_2OH + RNH_2 Pd 120 °C

N
R

46–95%

F. Natural occurrence and compounds of special interest

1. *Simple pyrroles*

Pyrrole has not been found in nature but does occur in tobacco smoke. Pyrrolidine itself occurs in carrot green, and two derivatives, proline (**83**) and 4-hydroxyproline (**84**), are amino acids of wide distribution. Indicaxanthin, one of the yellow cacti pigments related to betanidin (p. 211), is an *N*-substituted proline. Representative synthesies of these amino acids are outlined below.

N H (a) EtMgBr (b) $ClCO_2Et$ N CO_2Et CO_2Et H_2, Ni N CO_2Et CO_2Et NaOH N H CO_2H

83

HI

$ClCH_2$ O O CO_2Et (a) Cl_2 (b) hydrol. $ClCH_2$ O O Cl NH_3 HO N H CO_2H

84

– EtO⁻

$CH_2(CO_2Et)_2$
+
O
$ClCH_2$

NaOEt

[$ClCH_2CHCH_2CHCO_2Et$ O⁻ CO EtO]

[HOCH—CH_2 CH_2Cl $CHClCO_2H$]

A polymer obtained from l-vinylpyrrolidone (**85**) with aqueous hydrogen peroxide has been used as a blood-plasma substitute, but dextran sulphate is definitely superior.

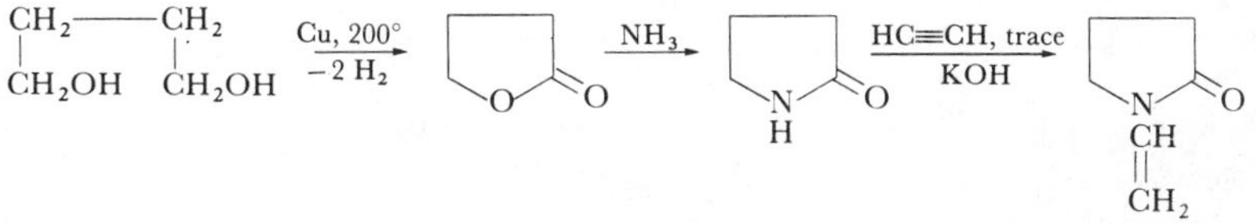

CH_2—CH_2 / CH_2OH CH_2OH —(Cu, 200°, $-2\,H_2$)→ [γ-butyrolactone] —(NH_3)→ [2-pyrrolidone, N–H] —(HC≡CH, trace KOH)→ [1-vinyl-2-pyrrolidone, N–CH=CH_2]

85

The antifungal antibiotic pyrrolnitrin (**86**) is synthesised by *Pseudomonas aurofaciens* from tryptophan.[66] It has been synthesised by a long route.[67] X-Ray studies show that the rings subtend an angle of 52° to each other,[68] and pyrrolnitrin is of special interest as it possesses a 3-chlorine atom. A dibromopyrrole (**87**) occurs in a marine sponge,[69] and an ant trail marker[70] is

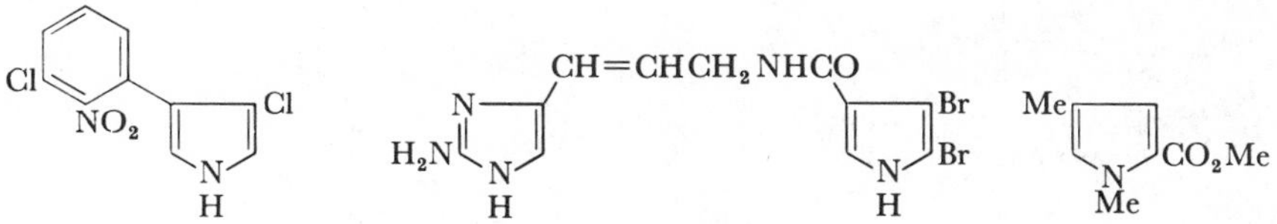

Cl, NO_2, Cl, N, H — **86**

H_2N, N, N, H, CH=CHCH_2NHCO, Br, Br, N, H — **87**

Me, CO_2Me, N, Me — **88**

the pyrrole ester **88**. One of the most toxic substances known, the Colombian arrow poison obtained from the frog *Phyllobates aurotaenia*, is an ester of 2,4-dimethylpyrrole-3-carboxylic acid.[71]

2. *The porphyrins*

Porphin, the simplest porphyrin,[71a] consists of four pyrrole rings linked together by methine bridges and is mentioned briefly on p. 109. A methine bridge is the unsaturated carbon–hydrogen system between each five-membered ring in, for example, protoporphyrin IX (**89**). The rings and substituents on the methine bridges should be labelled as indicated. The porphin system is one of great stability, the ring system is flat,[72] and n.m.r. studies show that the molecule can sustain an induced ring current. It is, however, not clear exactly where this ring current flows.

Haemoglobin and chlorophyll are very noteworthy derivatives of porphin, although there are others that are not considered here.

Haemoglobin is a conjugated protein made up of haem, the porphyrin moiety containing a ferrous atom, and the protein globin. Oxidation of the ferrous atom in haem to the ferric state occurs readily. The product is haemin (**90**), which was

Protoporphyrin IX, **89**

Haemin, **90**

Mesoporphyrin

$R,R' = H$ or Me; all 8 isomers obtained

Succinic acid

first crystallized in 1853 and was synthesized by Fischer and Zeile in 1929. Direct removal of the iron gives protoporphyrin IX (**89**).

The degradation studies outlined above left a number of possibilities for the structure of haemin, and as these could not be resolved by the degradation methods available, Fischer decided to synthesize all the possible structures. His successful sequence, discovered at a comparatively early stage, is shown on the opposite page.

A great deal of biochemical work, involving the biosynthesis of protoporphyrin from simple intermediates 'labelled' with heavy or radioactive atoms, has been done. It is clear that the biosynthesis of porphobilinogen (**91**), which is excreted in human urine under certain pathological conditions, proceeds as indicated. Four molecules of porphobilinogen are involved in essentially similar biosyntheses of chlorophylls in plants and of other porphyrins and corrins in bacteria and animals. Much effort has been devoted to the study of these biosyntheses, but a great deal still remains to be clearly understood. An outline of one of the pathways of porphyrin biosynthesis is given below, and the labelling (atoms marked with asterisks) observed when CH_2-labelled glycine is

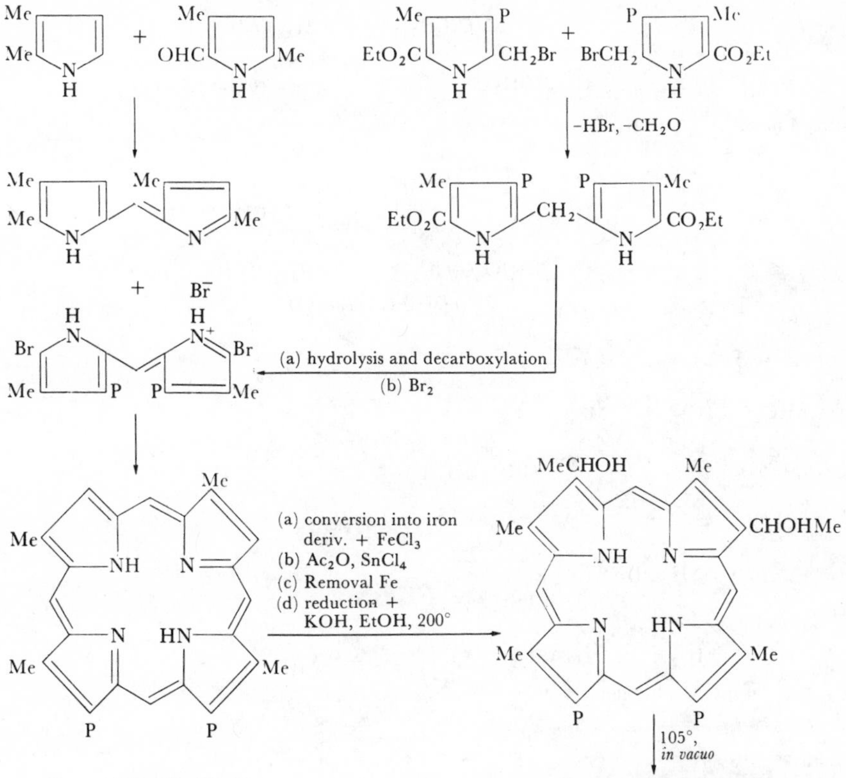

(a) hydrolysis and decarboxylation
(b) Br_2

(a) conversion into iron deriv. + $FeCl_3$
(b) Ac_2O, $SnCl_4$
(c) Removal Fe
(d) reduction + KOH, EtOH, 200°

105°, *in vacuo*

$P = —CH_2CH_2CO_2H$

Protoporphyrin IX, **89**

used for the biosynthesis is shown for protoporphyrin IX (**89**). One of the most interesting features is that one of the porphobilinogen (**91**) units is 'turned round' with respect to the orientation of the others. A mechanistic scheme accounting for this, and also for the established formation of the corrin system (p. 122) from porphobilinogen, has been put forward.[73]

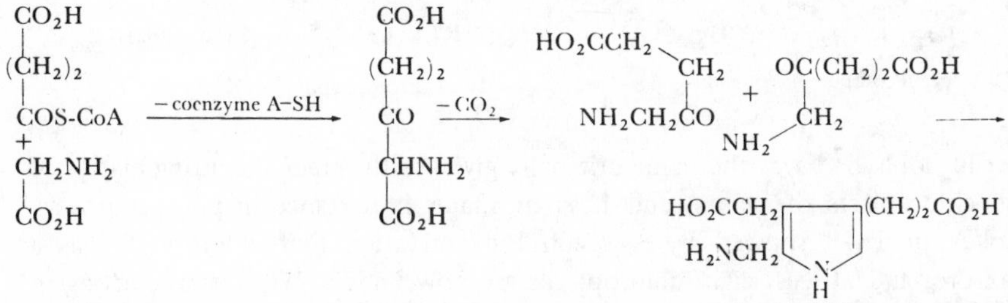

$$CO_2H(CH_2)_2COS\text{-}CoA + CH_2NH_2CO_2H \xrightarrow{-\text{coenzyme A–SH}} CO_2H(CH_2)_2CO\,CHNH_2CO_2H \xrightarrow{-CO_2} HO_2CCH_2CH_2\,NH_2CH_2CO + OC(CH_2)_2CO_2H\,CH_2NH_2 \longrightarrow$$

Porphobilinogen
91

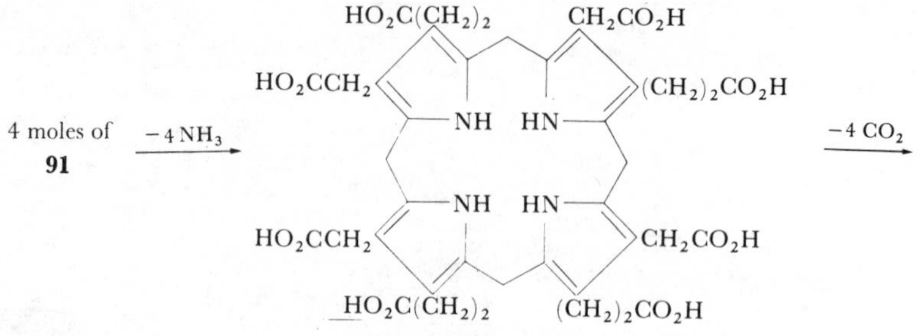

4 moles of **91** $\xrightarrow{-4\,NH_3}$

$HO_2C(CH_2)_2$ CH_2CO_2H
HO_2CCH_2 $(CH_2)_2CO_2H$
NH HN
NH HN
HO_2CCH_2 CH_2CO_2H
$HO_2C(CH_2)_2$ $(CH_2)_2CO_2H$

Uroporphyrinogen III

$\xrightarrow{-4\,CO_2}$

$HO_2C(CH_2)_2$ Me
Me $(CH_2)_2CO_2H$
NH HN
NH HN
Me Me
$HO_2C(CH_2)_2$ $(CH_2)_2CO_2H$

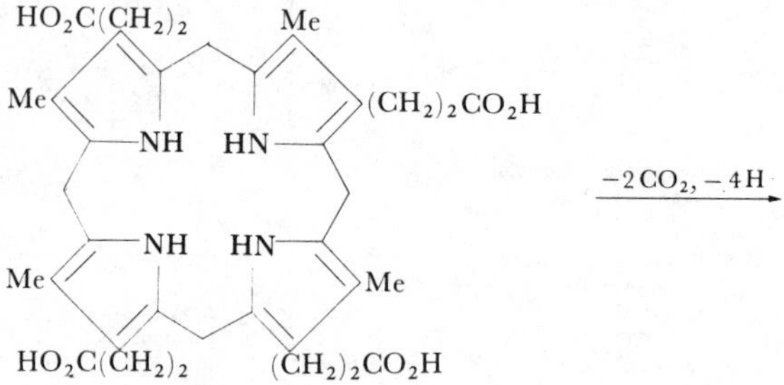

Coproporphyrinogen III

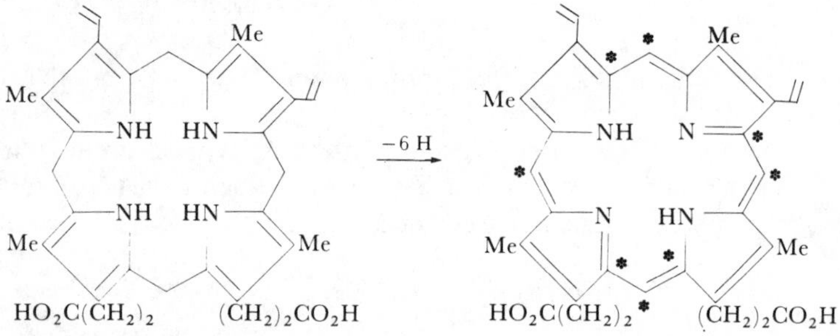

$\xrightarrow{-2\,CO_2,\,-4\,H}$

Me
Me
NH HN
NH HN
Me Me
$HO_2C(CH_2)_2$ $(CH_2)_2CO_2H$

Protoporphyrinogen

$\xrightarrow{-6\,H}$

Me
Me
NH N
N HN
Me Me
$HO_2C(CH_2)_2$ $(CH_2)_2CO_2H$

Protoporphyrin IX, **89**

Chlorophyll[74] was the name originally given to the green colouring matter of the chloroplasts of plants, and it is of major importance in photosynthesis. Stokes in 1818 showed by a spectral examination that chlorophyll was a mixture, and at least eight chlorophylls are now known. Willstätter, starting in 1906, found that all leaves contained two green pigments, which he named

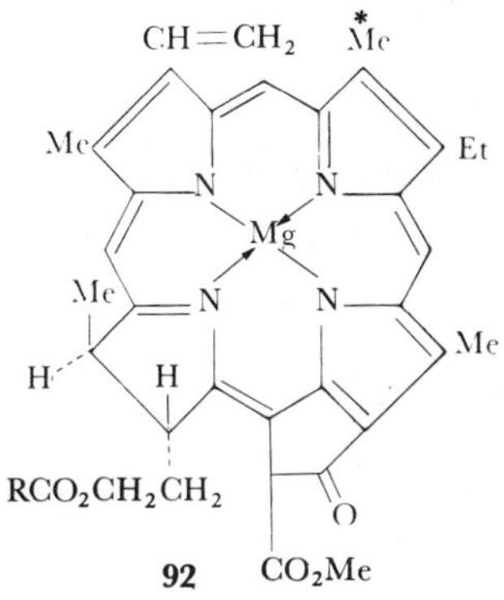

CH=CH$_2$ Me*
Me Et
N N
Mg
Me N N
Me
H H
RCO$_2$CH$_2$CH$_2$ O
92 CO$_2$Me

R = Phytyl =
$Me_2CH(CH_2)_3CHMe(CH_2)_3CHMe(CH_2)_3CMe{=}CHCH_2{-}$

chlorophyll *a* and *b*, that the molecules contained magnesium, and that the empirical formula of chlorophyll *a* was $C_{55}H_{72}N_4O_5Mg$. The structure of chlorophyll *a* (**92**) has been fairly well established by degradation, and it is a reduced porphyrin. The hydrogen atoms of the reduced pyrrole ring appear to be *trans*, and the absolute configuration has been established.[75] Chlorophyll *a* has been synthesised,[76] and protoporphyrin IX (**89**) can be a biosynthetic precursor of chlorophyll.[77]

Chlorophyll *b* has the methyl group marked with an asterisk replaced by a carbaldehyde group.

3. *Vitamin B_{12}*

Cyanocobalamin, or vitamin B_{12},[78] the antipernicious anaemia factor, was first crystallized in 1948, and although much intensive chemical work was carried out in order to determine its structure, comparatively little initial progress was made. The molecule was shown to have the composition $C_{61-64}H_{84-92}O_{13-14}N_{14}PCo$, and on hydrolysis yielded cyanide ion, 1-aminopropan-2-ol, 5–6 moles of ammonia, 5,6-dimethylbenzimidazole-1-α-D-ribofuranoside-3-phosphate, and a deep red gum from which a hexabasic acid of unknown constitution was obtained. Although only about 25% of the atoms of this large asymmetric molecule had been assigned to chemically identified degradation products, Professor D. Crowfoot Hodgkin and her collaborators, by a brilliant X-ray investigation of the vitamin and its derivatives, succeeded in determining its molecular structure.[79] The molecule of vitamin B_{12} contains a large planar group involving four reduced pyrrole rings that resembles a porphyrin. It should be specially noted that it differs from the porphyrin system, as it lacks one methine bridge. Recently a number of vitamin B_{12}-like compounds have been isolated in which the 5,6-dimethylbenzimidazole part of the vitamin molecule has been replaced by adenine, 2-methyladenine, or

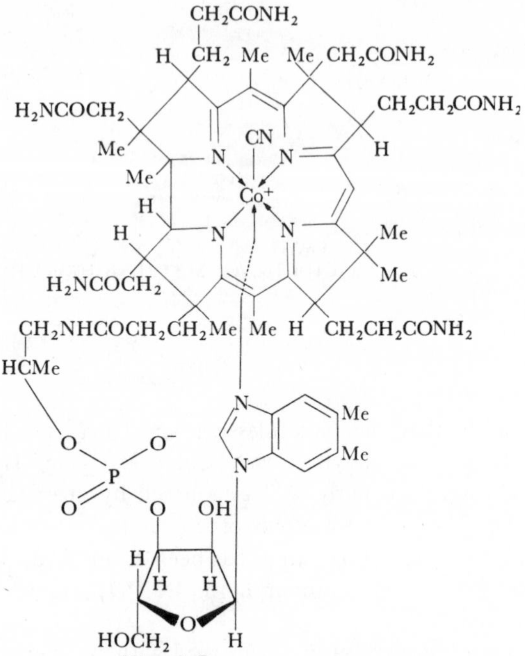

Vitamin B_{12}

2-methylhypoxanthine. Corrin, numbered as shown, is the name now given to the ring system from which vitamin B_{12} is derived.

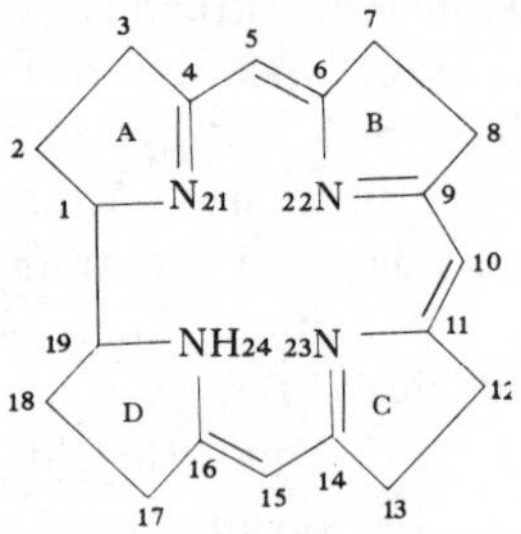

Corrin

Vitamin B_{12} has been synthesised,[80] and it is an important coenzyme for a number of transformations. One of the most interesting of these is the reversible conversion of succinyl-coenzyme A (**93**) to methylmalonyl-coenzyme A (**94**). Only one of the optical isomers of **94** undergoes the ester shift, and the ester and

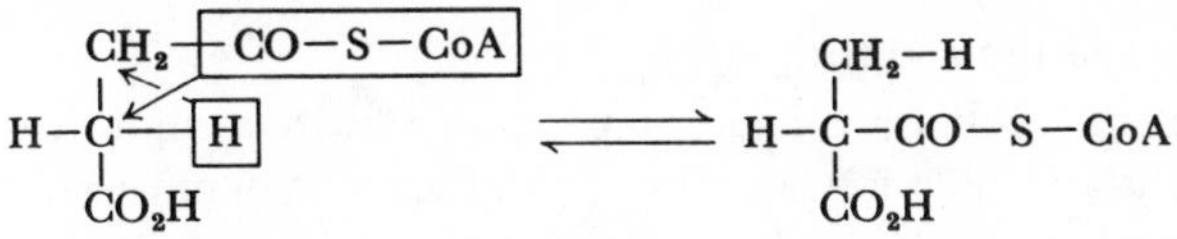

$$\underset{\mathbf{93}}{\text{H}-\overset{\text{CH}_2-\boxed{\text{CO}-\text{S}-\text{CoA}}}{\underset{\text{CO}_2\text{H}}{\text{C}}}-\boxed{\text{H}}} \rightleftharpoons \underset{\mathbf{94}}{\text{H}-\overset{\text{CH}_2-\text{H}}{\underset{\text{CO}_2\text{H}}{\text{C}}}-\text{CO}-\text{S}-\text{CoA}}$$

the hydrogen atom shown exchange positions[81] exactly, including stereochemistry. There is no exchange of the moving hydrogen atom with the solvent, and the exact mechanism of the reaction is not clear.

2. FURAN

A. Introduction

The earliest furan compound described is pyromucic acid (**1**), usually known as furoic or furan-2-carboxylic acid. It was obtained by Scheele in 1780 by the dry distillation of mucic acid. In 1832 furfural, or furan-2-carbaldehyde (**2**), was obtained by the action of sulphuric acid and manganese dioxide on sugar, but these furans were not related until 1860, when furfural was oxidized to furoic acid with silver oxide. Furan itself was obtained in 1870 by treating barium furoate with soda lime. In 1920 furfural became commercially available, cheaply and in large quantities, from the acid hydrolysis of oat hulls (husks). At this time the outlets for furfural were limited, but its cheapness and availability greatly stimulated the search for uses. Today furfural is a very valuable raw material and is the source of most furan derivatives.

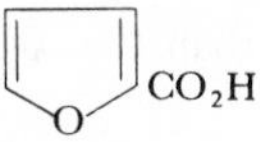

CO_2H CHO 4 3 5 2 1 β' β α' α

1 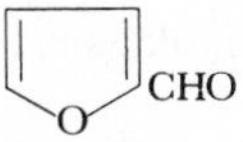 **2** 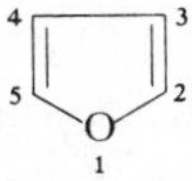 **3** 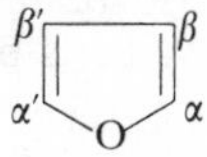 **4**

The positions in the furan ring are usually numbered as in **3**, but in older literature a less convenient lettering system (**4**) is sometimes used. The compound commonly known as furfural (**2**) could therefore be named furan-2-carbaldehyde, or furan-α-carbaldehyde. The trivial names of some groups derived from furan are shown.

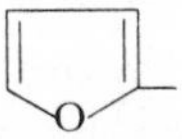

CH_2— CO—

2-Furyl 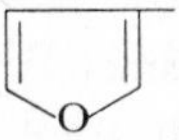 3-Furyl 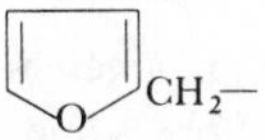 2-Furfuryl 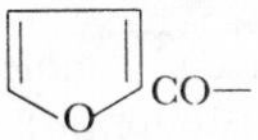 2-Furoyl

B. Physical properties and structure

Furan itself is a colourless liquid, b.p. 31·36°C, with a chloroform-like smell. It is only slightly soluble in water, but it is miscible with most organic solvents.

The arrangement of the atoms in the ring was proved by Baeyer in 1877 by the conversion of furfural both into furan and into pimelic acid (**5**), the structure of which had been established earlier.

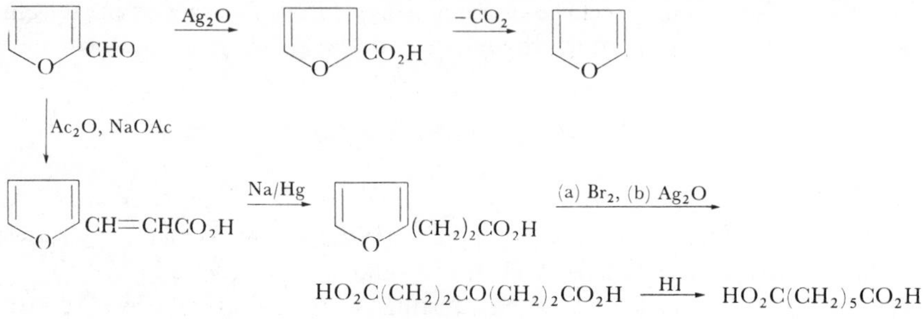

CHO $\xrightarrow{Ag_2O}$ CO_2H $\xrightarrow{-CO_2}$

Ac_2O, NaOAc

CH=CHCO$_2$H $\xrightarrow{Na/Hg}$ $(CH_2)_2CO_2H$ $\xrightarrow{(a)\ Br_2,\ (b)\ Ag_2O}$

$HO_2C(CH_2)_2CO(CH_2)_2CO_2H \xrightarrow{HI} HO_2C(CH_2)_5CO_2H$

5

Calculations using the heat of formation of liquid furan (62·4 kJ (14·9 kcal)/mole[82]) give values between 99·4 and 111·3 kJ (23·7 and 26·54 kcal)/mole for the resonance energy of the molecule. In good agreement, molecular-orbital calculations[83] based on the experimentally measured first electronic transition give a value of 109 kJ (26 kcal)/mole. However, calculations based on the heat of hydrogenation of furan give 71 kJ (17 kcal)/mole. The discrepancies arise in part from the different methods used in estimating the energy of the 'classical' non-resonating furan. It would appear that the correct value is probably 105 ± 4 kJ (25 ± 1 kcal)/mole, substantially less than that for benzene (150·5 kJ (36·0 kcal)/mole).

Microwave examinations have shown that the furan molecule is planar and

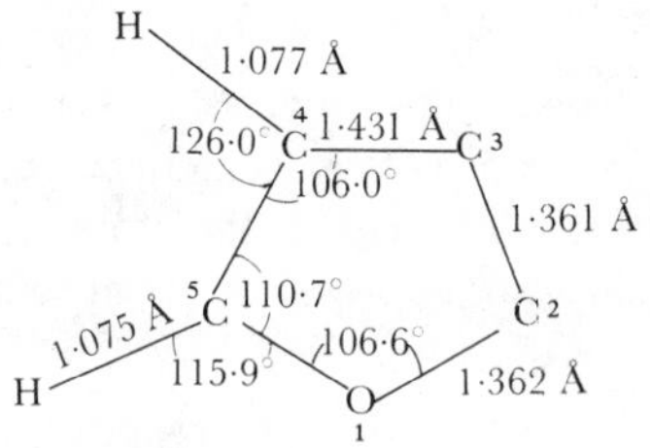

H 1·077 Å 126·0° C4 1·431 Å C3 106·0° 1·361 Å 1·075 Å C5 110·7° C2 106·6° 115·9° H O1 1·362 Å

Distances better than ±0·002 Å

have given[84] molecular parameters. The parameters[85] for tetrahydrofuran (**6**) and 2,5-dihydrofuran (**7**), some of which were determined by electron diffraction, are given for comparison. The results show that the carbon–oxygen bonds in furan are shorter than those of the reduced derivatives and also that the

1·54 Å (assumed) — (assumed) 1·54 Å — 111 ± 2° — 1·43 ± 0·03 Å

6

1·34 Å (assumed) — 110 ± 3° — 1·43 ± 0·03 Å

7

carbon–oxygen–carbon angle has contracted slightly in furan. The distance between carbon atoms 3 and 4 of furan is slightly longer than that of the aromatic carbon–carbon bond of benzene (1·40 Å) and that between carbon atoms 2 and 3, and 4 and 5 is very close to the ordinary ethylenic bond length, such as is assumed for 2,5-dihydrofuran (**7**). These results show that furan cannot be accurately represented by classical formula **8**, but that it is best considered as a resonance hybrid of which formulae **8a**–**8d** represent the more important hypothetical limiting structures. The furan molecule is far from a regular pentagon in shape, and the bond lengths suggest that structure **8** is the

8 **8a** **8b** **8c** **8d**

major contributor to the resonance hybrid. Of the other resonance structures shown, **8a** and **8b** are expected to be the most important for several reasons. The energy required to separate the charges is lower in these cases than in the other charged structures, as the distance involved in less. Structures **8a** and **8b** are completely conjugated and are therefore more stable than the 'cross-conjugated' **8c** and **8d**. However, calculation[86] of the π-electron densities for furan gives the result shown below, the reverse of what might have been expected from the above argument, but the localization energies[83] for electrophilic attack greatly favour substitution at positions 2 and 5.

π-Electron densities Localization energies (kJ(kcal)/mole)

In order for the furan ring to behave as an aromatic structure it must have six π-electrons in the ring. The oxygen atom must therefore provide two electrons for this purpose. The difference between the dipole moments of furan (0·71 D) and tetrahydrofuran (1·68 D) indicates a substantial drift of electrons away from the oxygen atom in furan, the mesomeric moment being the difference between the two. Contrary to many earlier statements, the oxygen atom is at the negative end of the dipole in both tetrahydrofuran and furan, and self-consistent field molecular orbital calculations of the dipole moment for furan give good agreement with that observed.[86] As oxygen is more electronegative than nitrogen, it provides the two electrons necessary for the aromatic sextet less easily, and, in consequence, furan is less aromatic than pyrrole.

The physical data therefore suggest that furan itself should behave chemically mainly as a diene ether which possesses unusually great resonance stabilization. This is in fact correct, but it should be specially noted that the introduction of substituents which can interact electronically with the ring system can alter the resonance energy of the final compound sufficiently to change the 'aromatic' and 'aliphatic' chemical characteristics of the ring.

C. Chemical properties

The ring of furan is opened moderately easily by acids, and thus an analogy to the very facile acid hydrolysis of enol ethers is apparent. 2,5-Addition to the double-bond system of the classical formula also occurs readily, leading to dihydrofuran derivatives. This type of addition is characteristic of 1,3-dienes.

Substitution of the hydrogen atoms of the furan ring can also take place. With very few exceptions, entering substituents always attack positions 2 and 5 of the ring preferentially. Substitution at positions 2 and 5 can take place either through addition, leading to a dihydro compound (**9**), followed by elimination,

RR′ —HR′ —H⁺ R⁺

9

or by direct displacement, as in the case of benzene. Where an addition compound has actually been isolated, as in the case of nitration (see p. 130), the course of the reaction is clear, but it may well be that the lability of the intermediate compounds has precluded their easy detection in a number of instances. This could result in a 2,5-addition–elimination sequence leading to a substitution being wrongly considered as a direct substitution. Substitution at positions 3 and 4 cannot involve the sole formation of 2,5-addition compounds as intermediaries. Substitution in these cases must either be similar to that of benzene or take a perhaps less expected course via the tetrahydrofuran (**10**). In some cases such intermediate tetrahydrofurans have been detected.

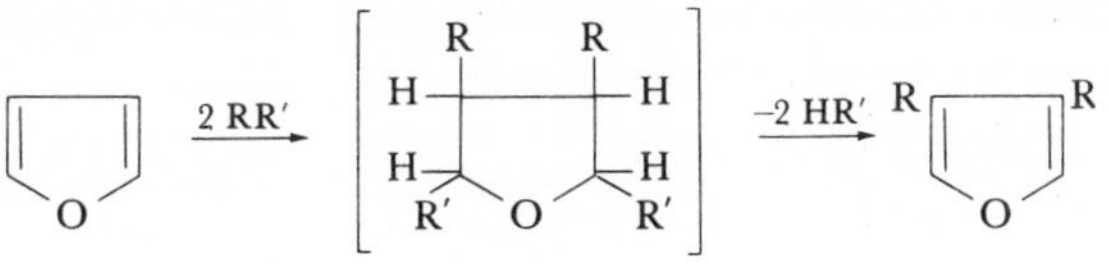

10

A number of furans have been converted into pyrroles (pp. 114, 115), pyridines (p. 273), and oxepines (p. 451).

The detailed chemical properties of furan are considered in the following three sections.

1. *Furan as an ether*

Enol ethers, in general, are very easily hydrolysed to the corresponding carbonyl compounds by dilute acids. Furan is not nearly so easily hydrolysed, because the initial attack of the proton involves the electrons of the oxygen atom, and these electrons contribute to the six π-electrons required for aromaticity. Their withdrawal reduces the resonance energy of the ring and converts the somewhat aromatic furan into the very much less stabilised cation. Deuterium exchange experiments[87] show that the cation **11** formed 10^3 times

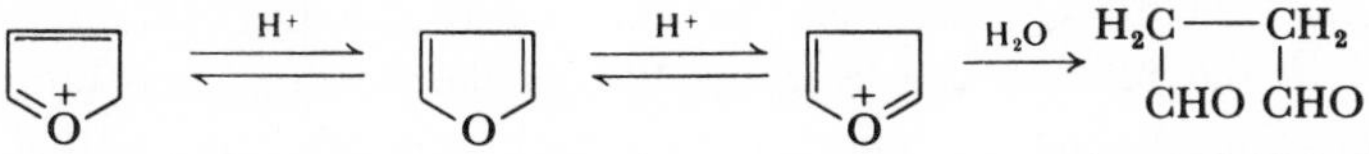

11 **12** Succindialdehyde

as rapidly as **12**, but hydrolysis to succindialdehyde presumably takes place through the latter. The mildest conditions necessary to effect this hydrolysis are such that polymerisation of the succindialdehyde occurs to a major extent.

2. *Addition reactions of furan*

Furan itself is not stable in the presence of air or oxygen and is usually stabilized by the addition of small quantities of hydroquinone. Aerial oxidation

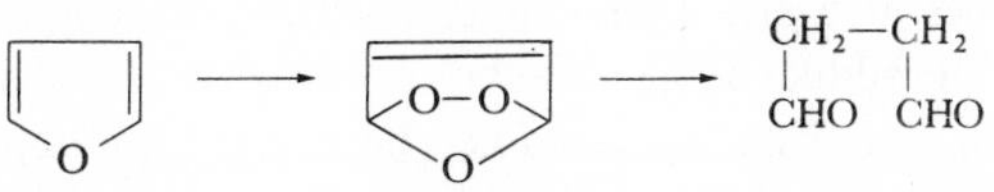

13

takes place by 2,5-addition, leading first to a peroxide (**13**), which has been isolated, and then through a free-radical polymerization to a resin. Hydrogenation of the peroxide gives succindialdehyde. Furan is completely oxidized by potassium permanganate.

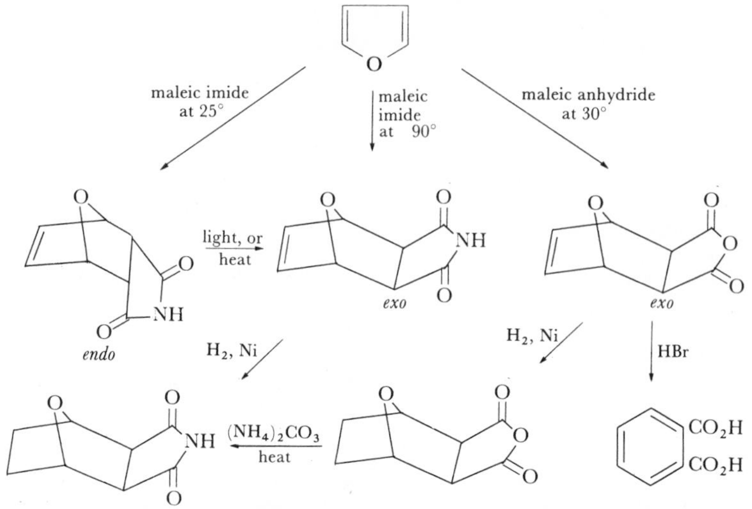

maleic imide at 25°; maleic imide at 90°; maleic anhydride at 30°; light, or heat; endo; exo; exo; H_2, Ni; H_2, Ni; HBr; $(NH_4)_2CO_3$ heat; NH; CO_2H; CO_2H

Furan behaves as a typical diene in the Diels–Alder reaction. With maleic anhydride it gives the *exo* adduct twice as fast as the less stable *endo* isomer, which quite quickly changes over via dissociation and recombination into the more stable *exo* form.[88] With maleic imide, and also the acid, the stereochemistry of the addition can be controlled by the reaction temperature. The maleic anhydride adduct, on boiling with hydrobromic acid, gives phthalic acid. This appears to be a general reaction for this type of compound. Dimethyl acetylenedicarboxylate also adds to furan, in a reaction catalysed by aluminium

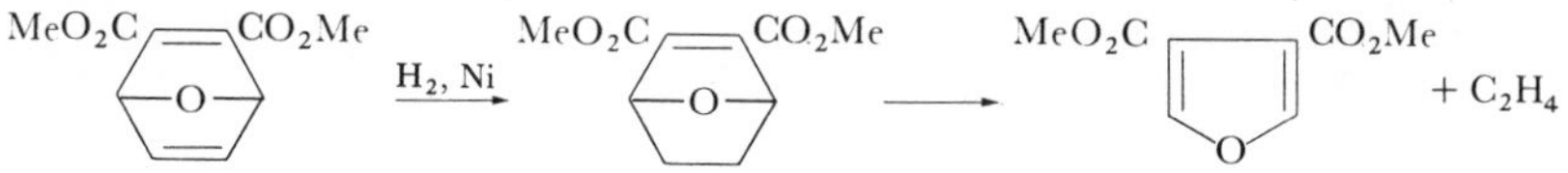

MeO_2C, CO_2Me (14) → H_2, Ni → MeO_2C, CO_2Me (15) → MeO_2C, CO_2Me (furan) + C_2H_4

chloride.[88a] The product (**14**) on catalytic reduction yields **15**, which on heating undergoes a reverse Diels–Alder reaction, forming dimethyl furan-3,4-dicarboxylate. The adduct **14** can also be converted to an oxepine (p. 451).

Under certain circumstances furan can behave as a nucleophilic reagent. With 2-acetylquinone[89] it yields **16**, and with acrolein or methyl vinyl ketone in the

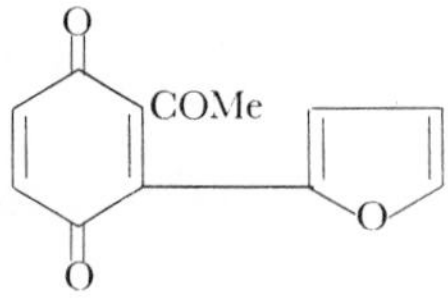

COMe (16); CH_2CH_2CHO (17); $OHCCH_2CH_2$, CH_2CH_2CHO (18)

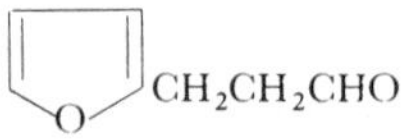

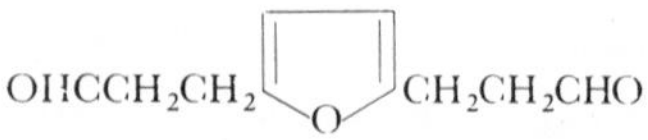

presence of sulphur dioxide reaction also occurs at position 2. The products from acrolein are **17** and **18** instead of normal Diels–Alder adducts. This type of behaviour also occurs with pyrroles (p. 101). Furan, however, behaves like 1-methylpyrrole with benzyne (p. 95), and with ethylene at high temperature and pressure it gives a Diels–Alder adduct which has been identified as shown below.

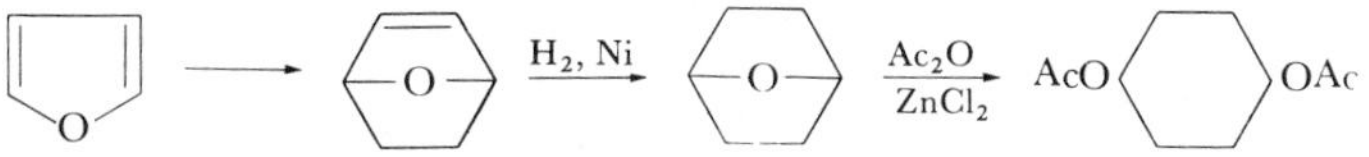

H_2, Ni

Ac_2O, $ZnCl_2$

AcO OAc

The photolysis of furan in benzene (contrast 1-methylpyrrole, p. 100) gives a 2,5-addition product (**19**) which undergoes a Cope rearrangement to an isomeric

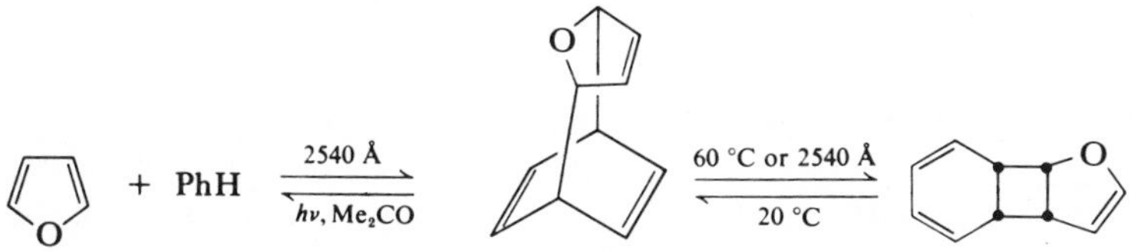

+ PhH 2540 Å / $h\nu$, Me_2CO

60 °C or 2540 Å / 20 °C

19

dihydrofuran.[90] Photolysis in the presence of benzophenone gives[91] a mixture of **20** and **21**. These, like the similar adduct[92] **22** from dimethylmaleic

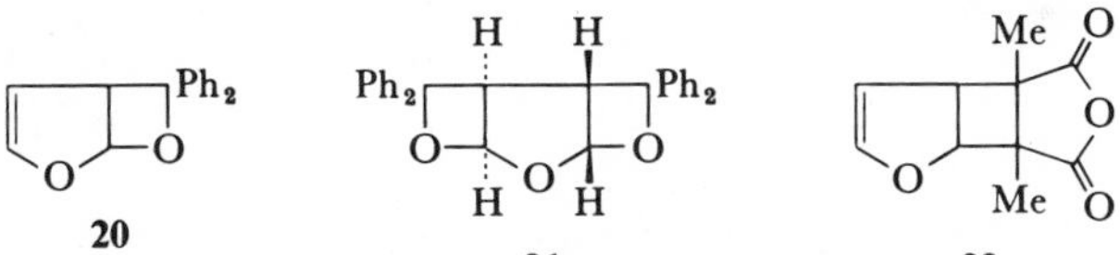

20 **21** **22**

anhydride (sensitised by benzophenone), are formed by $2\pi + 2\pi$ concerted cycloaddition processes (cf. oxetane formation, p. 71).

Furan can be oxidized[93] by bromine in methanol, or electrolytically in methanolic ammonium bromide, to a mixture of *cis*- and *trans*-2,5-dihydro-2,5-dimethoxyfuran (**24**), which has been separated by fractional distillation. Only the electrolytic oxidation is usually successful with furans possessing electron-attracting substituents. The exact mechanisms of these reactions have not been elucidated but clearly involve 2,5-addition. This dihydrodimethoxyfuran (**24**) is a cyclic acetal, and as such is hydrolysed by very dilute acid to maleicdialdehyde.

On hydrogenation the dihydro compound (**24**) yields the corresponding tetrahydrofuran (**25**), which on similar hydrolysis with very dilute acid gives succindialdehyde. This is the best available synthesis of this aldehyde. Excellent overall yields are obtained in contrast to those obtained by the direct acid hydrolysis of furan.

In a similar way[93] furan is attacked by lead tetraacetate in a mixture of acetic acid and anhydride, yielding **23**, which has been similarly hydrolysed to

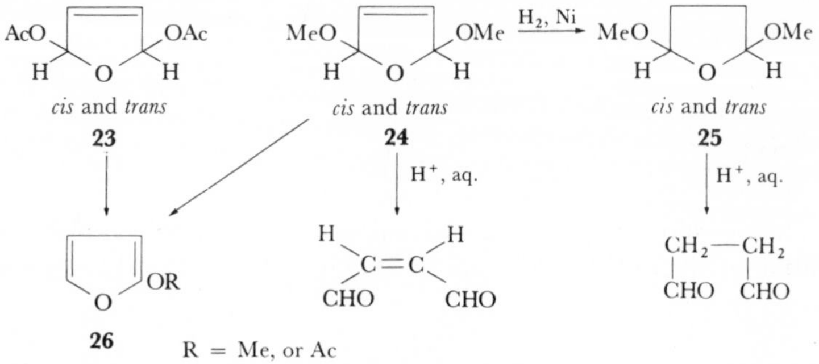

AcO, OAc, H, O, H — *cis* and *trans* **23**; MeO, OMe, H, O, H — *cis* and *trans* **24**; H_2, Ni → MeO, OMe, H, O, H — *cis* and *trans* **25**

24 → H^+, aq. → H, H, C=C, CHO, CHO

25 → H^+, aq. → CH_2—CH_2, CHO CHO

23, **24** → OR, O — **26** R = Me, or Ac

maleicdialdehyde and converted into succindialdehyde. An analogue of **23** is obtained from furan with dibenzoyl peroxide, presumably via a radical addition. Thermal decomposition (acid catalysed) of the diacetoxy (**23**) and dimethoxy (**24**) compounds yields derivatives (**26**) of the unknown 2-hydroxyfuran (p. 140).

Furan reacts with carbenes and nitrenes in alternative ways as shown below.

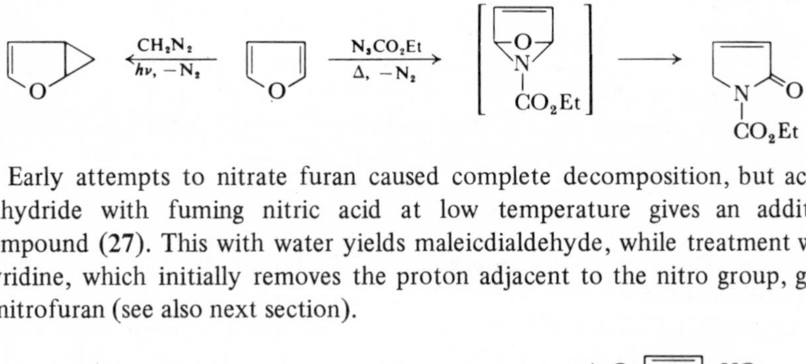

CH_2N_2, $h\nu$, $-N_2$; N_3CO_2Et, Δ, $-N_2$; O, N, CO_2Et; N, O, CO_2Et

Early attempts to nitrate furan caused complete decomposition, but acetic anhydride with fuming nitric acid at low temperature gives an addition compound (**27**). This with water yields maleicdialdehyde, while treatment with pyridine, which initially removes the proton adjacent to the nitro group, gives 2-nitrofuran (see also next section).

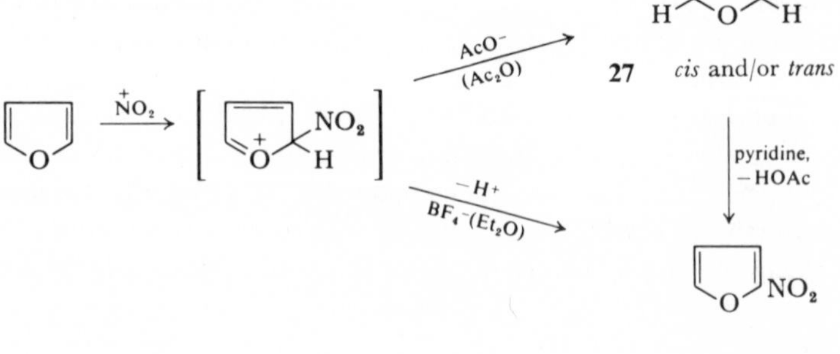

$\overset{+}{N}O_2$; O, +, NO_2, H; AcO^- (Ac_2O) → AcO, NO_2, H, O, H — **27** *cis* and/or *trans*; pyridine, −HOAc; $-H^+$ $BF_4^-(Et_2O)$ → O, NO_2

3. *Substitution reactions of furan*

The nitration of furan can occur through an addition–elimination sequence (above), but when nitronium tetrafluoroborate[94] is used, in which case the anion is non-nucleophilic, an intermediate compound is not observed.

Furan cannot be sulphonated under ordinary conditions owing to resinification, but with 1-proto-1-pyridium sulphonate (p. 234) in ethylene dichloride, a combination that does not attack toluene, a mixture of the 2-sulphonic (**28**) (41%) and 2,5-disulphonic acid (15%) is obtained. Oxidation to maleic acid with bromine water proves the position of the entering group, as a similar oxidation of furan-3-sulphonic acid gives sulphofumaric acid.

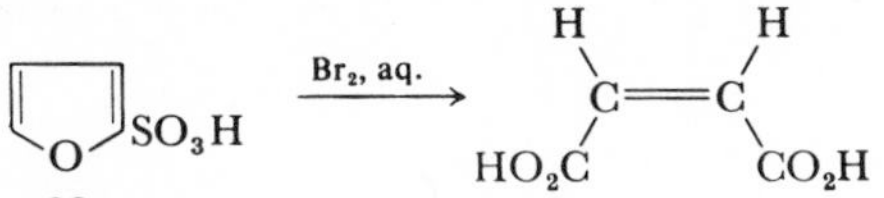

SO_3H **28** $\xrightarrow{Br_2,\ aq.}$ H(HO$_2$C)C=C(H)CO$_2$H Cl, Cl, Cl_2, Cl_2, O **29**

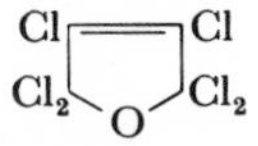

Furan reacts extremely rapidly with halogens, and the hydrogen halide liberated causes polymerization. Chlorination with 1·6 molecular proportions of chlorine at −40°C gives a mixture of the 2-chloro, 2,5-dichloro, and 2,3,5-trichloro derivatives, while increasing the proportion of chlorine yields tetrachlorofuran and its dichloride (**29**). 2-Bromofuran can be prepared directly in good yield from furan and dioxane dibromide at 0°C.[95]

Attempts to alkylate furan by the Friedel–Crafts reaction using alkyl halides with metallic halide catalysis cause extensive resinification, and no simple products have been isolated. This occurs presumably because of the strongly acidic nature of the catalysts and because alkylfurans are more easily polymerized than furan itself. However, boron tetrafluoride with propylene, or isobutylene with zinc chloride, gives 2-isopropyl and 2-*t*-butylfurans containing a small amount of the 3-isomers. It should be noted that no hydrogen halide is liberated in these alkylations.

Furan can be acylated very easily by both acid chlorides, or better acid

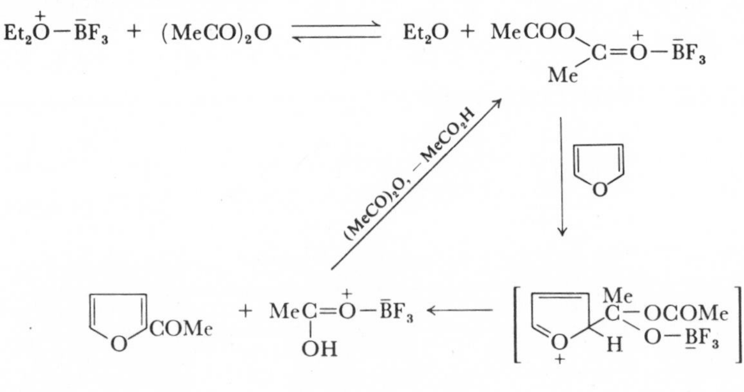

$$Et_2\overset{+}{O}-\bar{B}F_3 + (MeCO)_2O \rightleftharpoons Et_2O + MeCOO(Me)C=\overset{+}{O}-\bar{B}F_3$$

$(MeCO)_2O, -MeCO_2H$

COMe + $MeC(OH)=\overset{+}{O}-\bar{B}F_3$ ⟵ [Me, C−OCOMe, H, O−$\bar{B}F_3$, O, +]

anhydrides, in the presence of stannic chloride, but best results are obtained only when the reaction is carried out in such a manner that contact of the unreacted furan with the acidic catalyst is minimal. Traces of iodine and hydriodic acid can also act as catalysts in the anhydride condensations. Results are much better than in the alkylation of furan, as the acylfurans are relatively stable to acids. One of the best acetylation methods is to treat a mixture of furan and acetic anhydride at 0°C with boron trifluoride in ether.[96] The reaction proceeds by attack of the boron trifluoride–acetic anhydride complex on position 2 almost exclusively.

Furan, in contrast to thiophene, reacts with hydrogen cyanide and hydrogen chloride (Gattermann reaction) in the absence of metallic halides. Hydrolysis of the product (**30**) yields furfural in 35% overall yield. Furfural is better obtained

(a) HCN, HCl gives R = H, or
(b) Me_2NCHO, $POCl_3$, gives R = Me

$CH{=}\overset{+}{N}R_2$ Cl^-

30

H_2O

CHO

from furan by the Vilsmeier reaction, using phosphorus oxychloride and dimethylformamide,[97] which is so successful in the pyrrole series (p. 99).

Furan is inert to formaldehyde and dimethylamine under the usual Mannich conditions.

Furan reacts with phenyl radicals, obtained from benzenediazonium salts and alkali or by other methods, to give 2-phenylfuran (31–35% yield) containing no detectable (g.l.c.) amount of the 3-isomer; earlier claims that significant 3-arylation occurred have been disproved.[98]

D. Derivatives of furan

1. *Some reduced furans*

The furan ring is somewhat resistant to reduction except when carboxyl or phenyl groups are present as substituents. The hydrogenation of furan itself over Raney nickel at 80–140°C gives a high yield of tetrahydrofuran (**31**), and no intermediate dihydrofurans have been detected. Platinum catalysts behave similarly, but in the presence of acetic acid the ring is opened with the production of *n*-butyl alcohol.

CH_2—CH_2
CH_2OH CH_2OH

H_3PO_4, 270°
70 atm

31

$-CO, H_2$

CH_2OH

32

H_2, Ni

$HOCH_2C{\equiv}CCH_2OH \xleftarrow{2\ CH_2O} HC{\equiv}CH$

Tetrahydrofuran, which is also obtained industrially by the catalytic decomposition of tetrahydrofurfuryl alcohol (**32**) or by cyclization of butane-1,4-diol, is a valuable laboratory solvent, dissolving even polyvinyl chloride. It is a colourless liquid, b.p. 65°C, which easily forms an explosive peroxide with air. The intermediate radical (**33**) in the formation of the peroxide can be obtained thermally or by irradiation in the presence of a hydrogen acceptor and can be trapped as **34**.[99] At room temperature tetrahydrofuran combines with dicyanoacetylene to give **35**.[100]

$MeO_2CC{\equiv}CCO_2Me$

H; MeO_2C; CO_2Me; CN; CN

33 **34** **35**

Tetrahydrofuran is miscible with both water and most organic solvents. It is used in the laboratory as a solvent for lithium aluminium hydride reductions and for Grignard reactions. Grignard reaction type products can sometimes be obtained in better yield from the alkyl halide and sodium with aldehydes, ketones, or esters in tetrahydrofuran than by using a conventional Grignard reaction.[101] However, with butyllithium in hexane, tetrahydrofuran gives ethylene and the sodium salt of acetaldehyde[102] by an orbital-symmetry ($\pi 4s + \pi 2s$) permitted pathway. Tetrahydrofuran is also used commercially for

BuLi, $-$BuH; Li^+; $CH_2{=}CH_2$ + $CH_2{=}CH{-}O^-$ Li^+ → MeCHO

the production of hexamethylenediamine, which on condensation with adipic acid yields nylon. Another industrial application, which was used in Germany during the Second World War, is the catalytic vapour-phase dehydration of tetrahydrofuran to buta-1,3-diene. Tetrahydrofuran is also converted into adipic

$\xleftarrow{-H_2O}$ THF $\xrightarrow{HCl}$ $Cl(CH_2)_4Cl \xrightarrow{NaCN} NC(CH_2)_4CN \xrightarrow{H_2,\ Ni} H_2N(CH_2)_6NH_2$

acid, on an industrial scale, with carbon monoxide and nickel carbonyl. The catalytic oxidation of tetrahydrofuran yields γ-butyrolactone (**36**), while

1, 2, 3, 4, 5 → → $\xrightarrow{H_2O}$ $HO_2C(CH_2)_4CO_2H$

Adipic acid

chlorination at –30°C in the presence of light gives first 2-chloro- and then 2,5-dichlorotetrahydrofuran. If the chlorination temperature is near 0°C the 2-chlorotetrahydrofuran loses hydrogen chloride, forming **39**, which then adds more chlorine to give **37**. The chlorine atoms at position 2 in all these compounds are extremely reactive and are readily displaced by nucleophiles, such as ethanol, when **38** is obtained from **37**. This cyclic acetal (**38**) with sodium yields 2,3-dihydrofuran (**39**), which on strong heating undergoes an interesting rearrangement to cyclopropanecarbaldehyde. The double bond of

400° vapour → CHO

36 **37** **38** **39** **40**

2,3-dihydrofuran is very reactive and adds on water, yielding the semiacetal of 4-hydroxybutan-1-al (**40**). Alcohols, phenols, and acids add similarly. The resulting 'acetals' are easily hydrolysed by dilute acids to their hydroxylic

37 $\xrightarrow{RMgBr}$ $\xrightarrow{\text{powdered Na}}$ $RCH{=}CHCH_2CH_2OH$ (*cis and trans*) $\xrightarrow{H_2, Pt}$ $R(CH_2)_4OH$

precursors and **40**, and this procedure has been used for protecting hydroxyl groups. The chlorine atom at position 2 of compound **37** also reacts with Grignard reagents, and this is the basis of a very useful synthesis of olefines and a method of chain extension by four carbon atoms. 2,5-Dihydrofuran (**41**) is best obtained by the sequence indicated. The double bond is quite unlike that of 2,3-dihydrofuran and has typical olefinic properties.

$HC{\equiv}CH \xrightarrow{2\,CH_2O} HOCH_2C{\equiv}CCH_2OH \xrightarrow{H_2, Ni}$ $HOCH_2CH{=}CHCH_2OH$ (*cis*) $\xrightarrow[230\text{–}240°]{Al_2O_3}$ **41**; $HOCH_2CH{=}CHCH_2OH$ (*trans*) $\xrightarrow[230\text{–}240°]{Al_2O_3}$ $MeCH{=}CHCHO$

Tetrahydrofurfuryl alcohol (**42**), obtained from the catalytic hydrogenation of furfural, is another valuable industrial intermediate. The dihydropyran **43** possesses ^{14}C in approximately equal amounts at both the 2- and 6-positions

$Na^+ \bar{O}_2C(CH_2)_3CO_2^- Na^+$ ← molten NaOH — **42** (CH_2OH) — 270°, alumina gel → **43**

CH_2—CH_2, CH_2OH, C≡CH

44

when the side-chain carbon atom of **42** is labelled.[103] The corresponding chloride with sodamide in liquid ammonia yields the acetylene **44**. Tetrahydrofurans are difficult to dehydrogenate to furans, successful examples being the dehydrogenation of 2-methyltetrahydrofuran to 2-methylfuran by ruthenium on charcoal at 350–400°C, and of 3,4-diphenyltetrahydrofuran by sulphur at 210°C.[104]

2. *Alkyl- and arylfurans*

Many compounds of these types are known, but the majority have been obtained by indirect synthesis. 3-Alkylfurans unsubstituted at the 2- and 5-positions are often difficult to obtain because substitution of the furan ring always takes place preferentially at positions 2 and 5. 3-Methylfuran is now obtained[105] from the commercially available 2-methylallyl chloride (**45**). An

MeC($=CH_2$)CH_2Cl (**45**) —(a) Mg, Et_2O; (b) $CH(OEt)_3$→ MeC($=CH_2$)$CH_2CH(OEt)_2$ —perphthalic acid→ epoxide (Me, $CH_2CH(OEt)_2$) —0·1 M H_2SO_4, aq.→ 3-methylfuran

3-furoic acid (CO_2H) —(a) $SOCl_2$; (b) H_2, Pd/$BaSO_4$ (Rosenmund)→ 3-furaldehyde (CHO) —(a) NH_2NH_2; (b) NaOEt, EtOH, 200°→ 3-methylfuran (Me)

alternative synthesis involves the preparation and decarboxylation of 3-methylfuran-2-carboxylic acid.[106] 3-Methylfuran is acetylated by acetic anhydride and boron trifluoride etherate[96] and is nitrated in acetic anhydride solution, apparently only at the 2-position, but the Vilsmeier reaction (p. 99) gives a mixture of the 2- and 5-aldehydes in a 7 : 1 ratio.[107]

In general, the alkylfurans behave like furan itself to most reagents if allowance is made for their increased sensitivity to acids. The electron-donating character of the alkyl groups increases the basicity of the oxygen atom appreciably and the reactivity of the ring to electrophilic reagents. This results, for instance, in 2,5-di-*t*-butylfuran accepting a proton at position 2 in concentrated sulphuric acid to give a stable cation, and in sylvan (2-methylfuran) reacting more easily than furan with aqueous acids to give a useful synthesis of laevulinic aldehyde ($MeCOCH_2CH_2CHO$).

Alkylfurans usually react like the parent compound in the Diels–Alder reaction. A difference, however, occurs in the case of 2-vinylfuran (**46**), which gives a 1 : 1 adduct with maleic anhydride. This adduct is, however, 3*a*,4,5,6-tetrahydrobenzofuran-4,5-dicarboxylic anhydride (**47**), showing that of the two diene systems present in 2-vinylfuran the one involving the side chain is the more active.

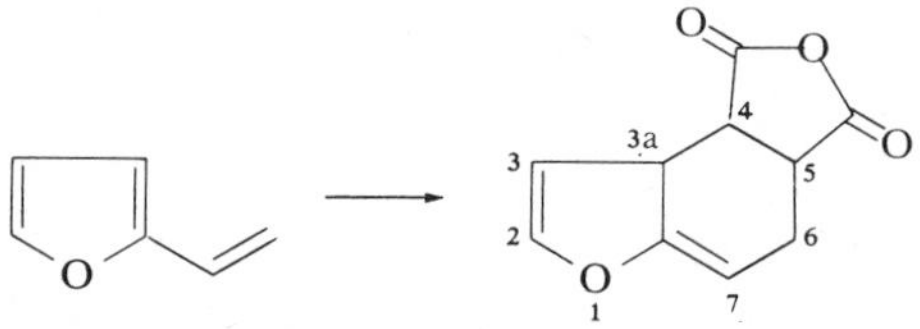

46 **47**

Substitution reactions of alkylfurans always involve free 2- or 5-positions, if present, and then the 3- or 4-positions. Oxidation of 2-methylfuran by potassium permanganate proceeds very easily with the production of acetic acid, and 2-methylfuran, unlike furan, undergoes the Mannich reaction at position 5.

2-Methylfuran, available directly from furfural (p. 145), can also be obtained from **48** with hot sodium ethoxide (Wolf–Kishner reaction). In this second synthesis 2-methylene-2,5-dihydrofuran (**49**) is also obtained. This compound (**49**) has a very different ultraviolet absorption spectrum from 2-methylfuran, into which it is converted by a trace of acid.

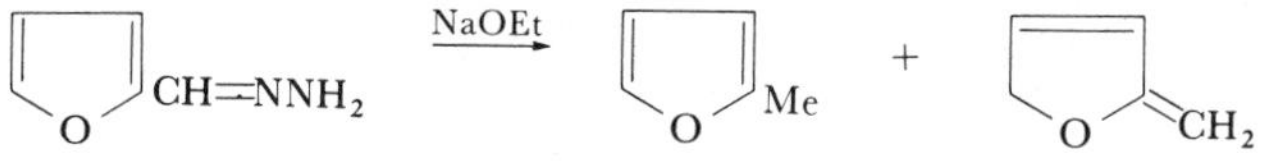

CH=NNH$_2$ —NaOEt→ Me + CH$_2$

48 **49**

The oxidation of 2,5-diphenylfuran with nitric and acetic acids opens the ring, yielding *cis*-1,4-diphenylbutene-1,4-dione (**50**), and ozonolysis gives the same ketone, along with some of the aldehyde **51**.[108] As the orientation in the ketone (**50**) is *cis*, it is virtually certain that the double bond moves to its final position before the ring opens. This implies that the oxidations proceed via 2,5-addition to the ring.

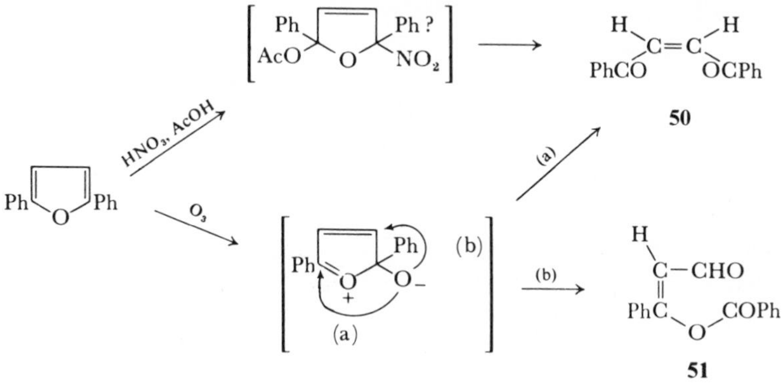

The photolysis of alkylfurans can cause rearrangements,[109] although the mechanism may not be clear.[38]

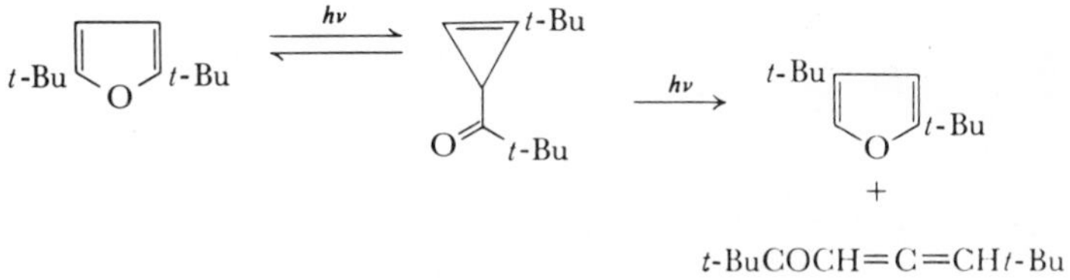

t-BuCOCH=C=CHt-Bu

3. *Halogen derivatives*

The monohalogen derivatives of furan are colourless liquids with a sweet odour, sparingly soluble in water but easily soluble in organic solvents. The 2- and 3-chlorofurans boil at 78°C and the 2- and 3-bromofurans boil at 102°C at atmospheric pressure. They discolour and resinify on long standing in the presence of air, and more rapidly with acids. They undergo the Diels–Alder reaction with maleic anhydride, and the products can be converted into phthalic acids.

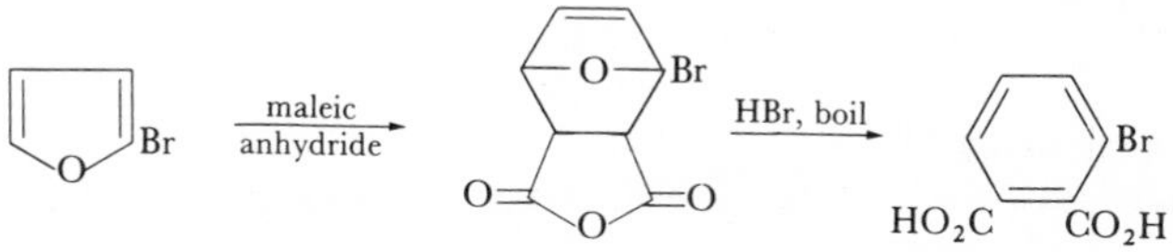

The halogen–carbon bond is quite stable in these compounds, which do not react with sodium hydroxide or cyanide, or with metallic sodium. 2-Bromofuran does, however, react with sodium methoxide or piperidine, and it gives a Grignard reagent with difficulty. Although 2-iodofuran behaves like iodobenzene with magnesium and gives a Grignard reagent with the usual properties, 3-iodofuran has not been induced to form a Grignard reagent. However, with butyllithium at −70°C it yields 3-furyllithium,[110] which is a useful intermediate for the preparation of 3-substituted furans. Copper is a valuable catalyst for the conversion of 3-iodofuran into 3-methoxy- and 3-cyanofuran with sodium methoxide and cuprous cyanide, respectively.[111]

Halogenated furans can sometimes be prepared by direct halogenation (p. 131), but more general synthetic methods are the treatment of furylmercuric salts with halogens and the decarboxylation of halogenated furancarboxylic

Cl_2 ; CO_2H ; Cl CO_2H ; heat in quinoline +Cu ; Cl

acids in quinoline. The carboxyl groups of furoic acids can also be replaced by bromine or iodine on treatment with the halogen in aqueous alkali.

HO_2C CO_2H ; Br_2, NaOH aq. ; Br Br

4. *Nitrofurans*

The 2- and 3-nitrofurans are solids, m.p. 29 and 27°C respectively, and the synthesis of 2-nitrofuran has been described (p. 130). 3-Nitrofuran has been obtained as shown. The nitration of furans which do not contain an electron-

Me CO_2H ; O_2N Me CO_2H poor yield + Me NO_2

NO_2 Me ; neutral $K_3Fe(CN)_6$; NO_2 CO_2H ; NO_2

attracting group is usually carried out under very mild conditions (cf. furan), otherwise resinification occurs. The entering group always prefers free 2- or 5-positions.

The presence of one nitro group, or of one other electron-attracting group, for example, carbonyl, usually permits further electrophilic substitution under conditions usual to benzene and it stabilizes the ring to acids. This is because the substituent withdraws electrons from the ring and the resulting deficiency in the aromatic sextet is made up by the oxygen atom. The aromaticity of ring is thus increased, as its electrons are less localized than before. At the same time the ring, denuded of electrons, is less easily attacked by a proton. 2-Nitrofurans are, however, easily attacked by strong alkalis. The furan ring is opened, presumably through semiacetal types of intermediates, with the formation of nitrites and a resin.

It is interesting that 3-methylfuran nitrates at position 2, while furan-3-

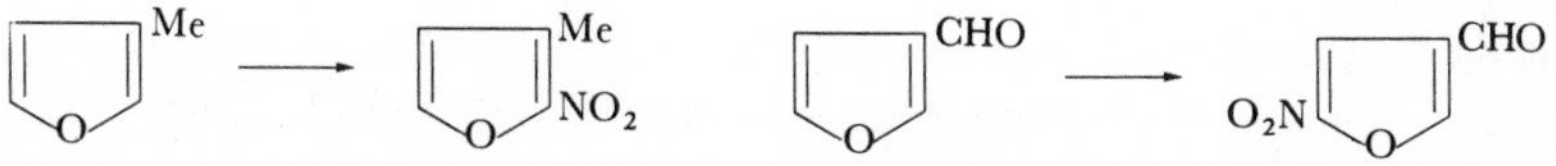

Me → Me NO$_2$ CHO → O_2N CHO

carbaldehyde yields the 5-nitro derivative. These results are in accord with the normal activating and deactivating effects of the respective groups on the *ortho* position in the benzene series. The effect of the methyl group is so marked that nitration of 3-methyl-2-furoic acid (**52**) also yields 3-methyl-2-nitrofuran (**54**),

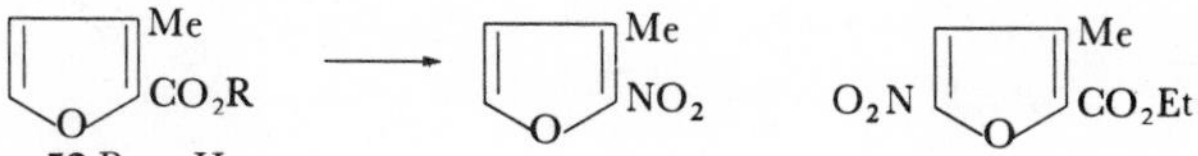

Me CO_2R → Me NO_2 O_2N Me CO_2Et

52,R = H
53,R = Et **54** **55**

the carboxyl group being eliminated. This type of elimination also occurs occasionally in the benzene series. The ester (**53**), however, gives **55** on nitration.

5-Bromo-2-nitrofuran, obtained by nitrating 5-bromo-2-furoic acid, has an activated bromine atom (cf. 4-bromonitrobenzene). The methyl group of

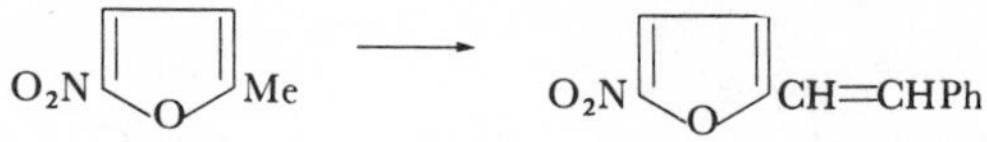

O_2N Me → O_2N CH=CHPh

56

2-methyl-5-nitrofuran is sufficiently active to give a low yield of **56** with benzaldehyde under conditions which do not affect 4-nitrotoluene.

The photolysis of 2-nitrofuran, and derivatives, proceeds by an interesting rearrangement.[112]

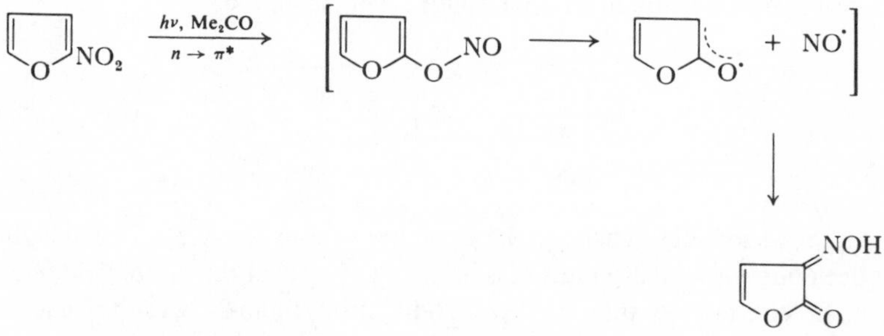

NO_2 $\xrightarrow[n \to \pi^*]{h\nu,\ Me_2CO}$ [O–NO → O· + NO·] → NOH O

5. *Hydroxyfurans*

In spite of a number of claims, it is certain that no simple free 2- or 3-hydroxyfuran has been prepared. Reactions which could lead to these compounds invariably give the tautomeric carbonyl derivatives or their decomposition products.

2-Acetoxy- and 2-methoxyfuran have been prepared by indirect means (p. 130), and with dilute acid and 2,4-dinitrophenylhydrazine the product isolated is the phenylhydrazone of 3-oxobutyric acid (cf. **57**). The initial protonation is the crucial step in the hydrolysis process, for 2-methoxyfuran in aqueous dimethylsulphoxide gives a mixture of products.[113]

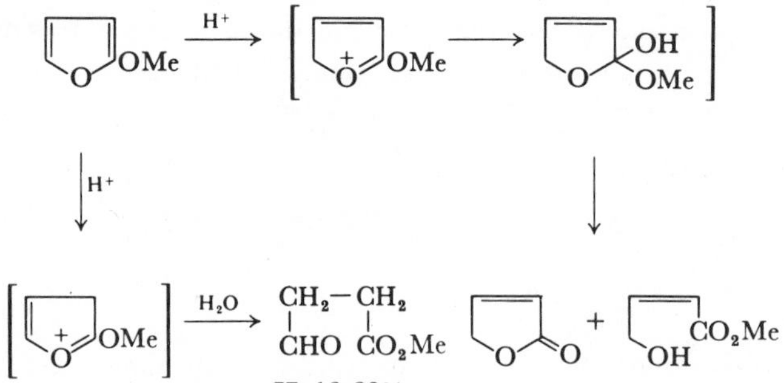

OMe H+ OMe OH OMe

H+

OMe H_2O CH_2-CH_2 CHO CO_2Me + CO_2Me OH

57, 16–23% 55–65% 16%

Attempted bromination or chlorination of 2-acetoxyfuran causes the elimination of acetyl halide, and the reaction with lead tetraacetate takes a similar course.

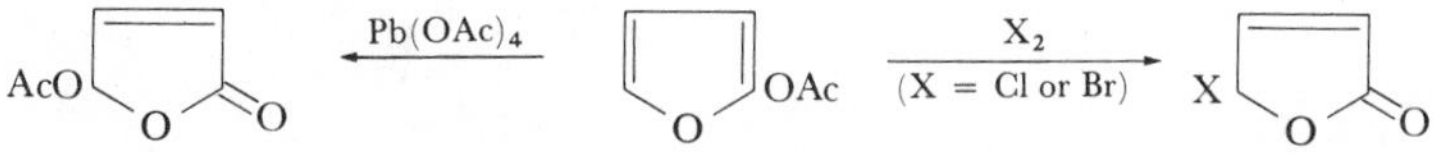

AcO O O $Pb(OAc)_4$ OAc X_2 (X = Cl or Br) X O O

The chemistry of α-angelica lactone (**58**) does not suggest that it exists in equilibrium with the tautomeric 2-hydroxy-5-methylfuran (**59**); only about $\frac{1}{3}$ mole of methane is liberated with methylmagnesium iodide. Succinic anhydride (**60**) likewise shows no properties expected of its enol formulation (**61**) and does not give 2,5-dichlorofuran with phosphorus pentachloride.[114]

Me O O **58** Me O OH **59** O O O **60** HO O OH **61**

Although some evidence for the existence of 3-hydroxy-2,4,5-triphenylfuran has been obtained, all attempts at isolation have yielded the tautomeric furanone (**62**). In contrast to this, diethyl 3,4-dihydroxyfuran-2,5-dicarboxylate (**63**), prepared from ethyl oxalate and ethyl diglycollate in the presence of sodium ethoxide, appears to exist only in the enol form. It gives a violet ferric chloride colour, and mono- and di-*O*-methyl ethers with methyl sulphate and alkali.

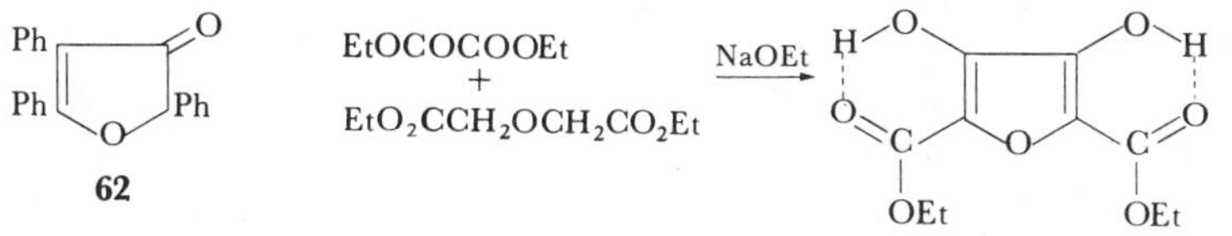
Ph
Ph
O
Ph
EtOCOCOOEt
+
EtO2CCH2OCH2CO2Et
NaOEt
H
O
H
C
OEt
OEt

62 **63**

Stabilization of the enol structure in this particular case is possible through hydrogen bonding. Compounds such as **62** are vinylogous esters and possess deactivated carbonyl groups which do not usually give reactions of ketones.

6. *Aminofurans*

A number of acyl and other derivatives of alkyl-2-aminofurans have been obtained, but attempted hydrolysis, and attempted reduction of 2-nitrofuran, disrupts the molecules with the liberation of ammonia. This suggests that tautomerization to the imine is followed by hydrolysis and polymerization under the reaction conditions.

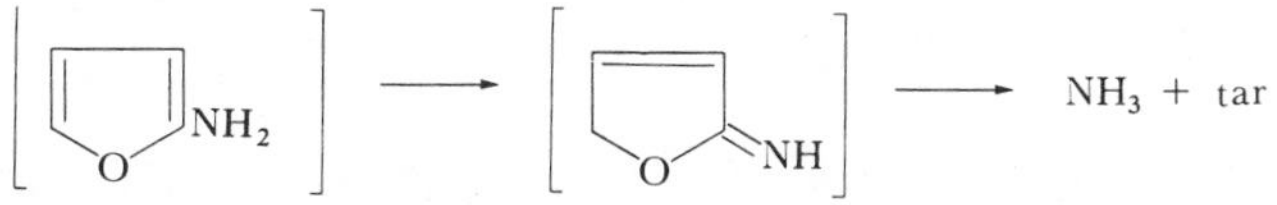
NH2
O
NH
NH3 + tar

3-Amino-2-methyl- and 3-amino-2,5-dimethylfuran, b.p. 51–52 and 55–56°C at 4 mm, respectively, have been obtained as indicated below. The hydrolysis of the formamido derivatives must be carried out under special conditions, and then excellent yields are obtainable. The amines are colourless liquids which smell like pyridine and discolour rapidly on standing in air. They benzoylate normally and give a positive carbylamine reaction. Both amines can be diazotized and coupled with alkaline 2-naphthol, although attempts to replace the diazo group by hydrogen and by the cyano group have failed. It is clear, therefore, that the amino group of 3-aminofurans (e.g., **64**) can react as such, although it is

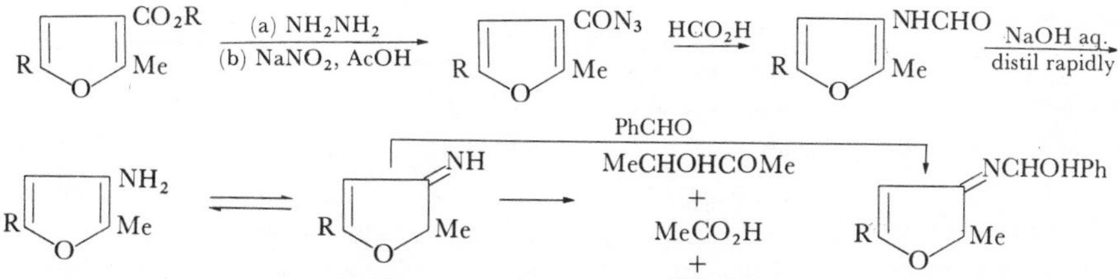
CO2R
R
Me
(a) NH2NH2
(b) NaNO2, AcOH
CON3
HCO2H
NHCHO
NaOH aq.
distil rapidly
NH2
NH
PhCHO
MeCHOHCOMe
+
MeCO2H
+
NH3
NCHOHPh

64 **65** **66** R = H or Me

probably in equilibrium with the ketimine form (**65**). This idea is supported by the fact that the Schiff's base (**66**) cannot be induced to dehydrate, and cleavage of the imine (**65**, R = Me) with aqueous alkali gives acetic acid, acetoin, and ammonia.

Ethyl 5-amino-2-furoate has been obtained, in contrast to 2-aminofuran, by catalytic or aluminium amalgam reduction of the corresponding nitro compound. Hydrolysis of the amino ester has not been carried out without complete decomposition, although prior acetylation permits a mild hydrolysis to 5-acetamido-2-furoic acid. This can be decarboxylated to 2-acetamidofuran.

An interesting synthesis of 2-dialkylaminofurans has been published.[115] A 3-keto acid is converted to the amide **67** by successive treatment with

CH_2—CH_2 R′CO CONR$_2$ **67** $\xrightarrow[HClO_4,\ <5\ °C]{Ac_2O}$ R′ AcO O $\overset{+}{N}R_2$ $\xrightarrow{AcOH}$ R′ O $\overset{+}{N}R_2$ **68** $\xrightarrow{Et_3N}$ R′ O NR$_2$ **69**

phosphorus trichloride and a secondary amine. Cyclisation below 5°C and the loss of acetic acid give the enamine cation (**68**), which is deprotonated to the furan **69**.

7. *Metallic compounds of furan*

Furan reacts more easily than thiophene with mercuric salts and initially forms 2-furylmercuric derivatives (e.g., **70**). Complete mercuration to 2,3,4,5-furan tetrakismercuriacetate is effected by excess mercuric acetate. Free

O HgCl **70**

i-Pr O CO_2H $\xrightarrow[HgCl_2]{NaOH,}$ i-Pr O HgCl

2- and 5-positions are more reactive than 3- and 4-positions, but both 3- and 4-positions can be successively substituted in 2,5-dimethylfuran. 2-Furoic acid reacts faster than furan with mercuric chloride, in the presence of sodium hydroxide, and gives furyl-2-mercuric chloride with the elimination of carbon dioxide. Under other conditions 2-furoic acid itself reacts with mercuric acetate, yielding a mixed salt (**71**), which on heating yields furan-3-mercuric acetate (**72**) through a cyclic intermediate.

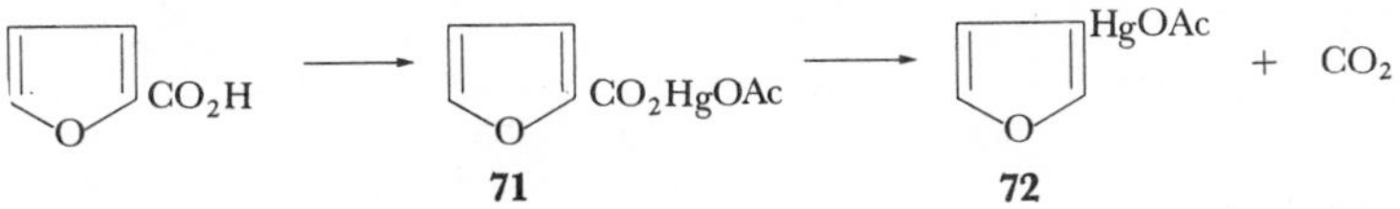

Although **72** with iodine yields 3-iodofuran, giving a valuable intermediate for the synthesis of 3-substituted furans, it does rearrange with some reagents.[111] Mercuration of furancarboxylic esters usually proceeds normally with the mercuration of the ring and retention of the ester group. The furan mercurials are useful synthetic compounds, as with acids the mercury is displaced by hydrogen, and with bromine and iodine halogenated furans are obtained.

Furan reacts with butyllithium, yielding 2-furyllithium, which has the usual properties of an aryllithium and yields 2-furoic acid with carbon dioxide. 2-Furyllithium (**73**) undergoes an interesting ring-opening sequence with trialkylboranes,[116] which gives *cis*-buten-1,4-diols (**74**). Some Grignard reagents, and 3-furyllithium, have been prepared from halogenated furans (pp. 137, 138).

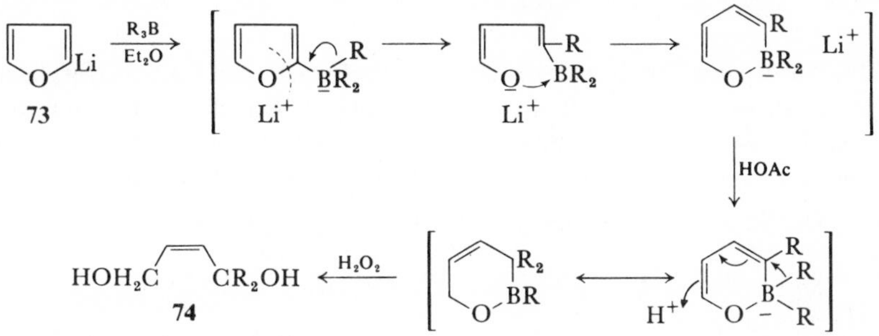

8. *Furfuryl alcohol*

This compound is a colourless liquid, b.p. 170°C, and is miscible with water and most organic solvents. It is obtained commercially by the catalytic hydrogenation of furfural and is stored in the absence of air and traces of acids to reduce resinification. It becomes immiscible with water on polymerization and has some uses in the plastics industry. Furfuryl alcohol behaves as a primary alcohol and a reactive furan. For example, it reduces aqueous permanganate in the cold, and it behaves like furan towards bromine and methanol. The ring is reduced on catalytic hydrogenation. Furfuryl alcohol can be used for synthesis in basic or neutral solution, but in acidic media it is very difficult to control the nature of the product.

Laevulinic acid (62–64%) can be obtained along with polymer by treating furfuryl alcohol with acids under controlled conditions so as to reduce resin formation. A likely mechanism for the internal oxidation–reduction is shown, although other possibilities are not excluded. This type of rearrangement and

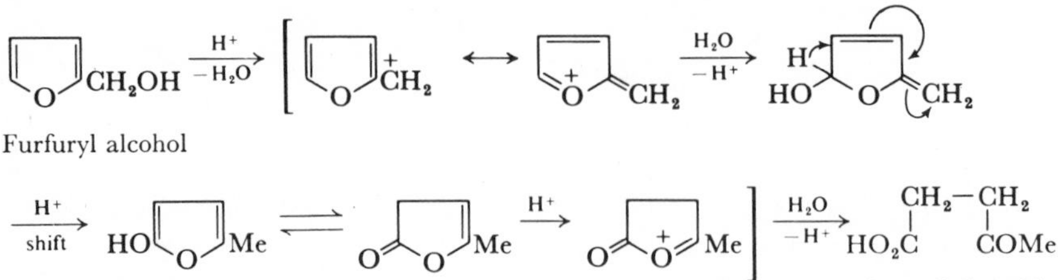

Furfuryl alcohol

Laevulinic acid

ring opening is quite general for furfuryl alcohols with a free 2- or 5-position. It also occurs with most furan derivatives with a vinyl side chain which can be hydrated to the furfuryl alcohol. Both **75** and **76**, which can be obtained from furfural and nitromethane under different conditions, yield the same product

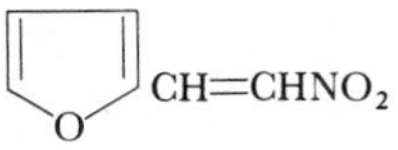

$CH{=}CHNO_2$ **75** $CHOHCH_2NO_2$ **76** $HO_2C(CH_2)_2CO(CH_2)_2NO_2$ **77**

(**77**) with aqueous hydrochloric acid. 2-Furylacrylic acid and furfurylidene-acetone behave similarly. 5-Hydroxymethylfurfural (**78**), obtained from hexoses with acids, yields laevulinic and formic acids on acid hydrolysis. The mechanism of this reaction is not certain, but it does not proceed through the initial formation of furfuryl alcohol and formic acid. The tracer study illustrated shows that the aldehyde group is the source of the formic acid.

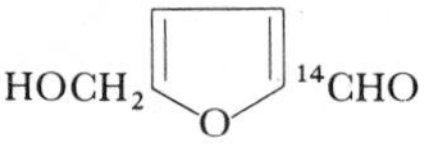

$$\text{HOCH}_2\text{-furan-}^{14}\text{CHO}\ (\mathbf{78}) \longrightarrow CH_3CO(CH_2)_2CO_2H + H^{14}CO_2H$$

Furfuryl alcohol is converted into the very reactive furfuryl chloride by thionyl chloride with ethereal pyridine at low temperatures. Furfuryl chloride is even more reactive than benzyl chloride, and attempts to form a Grignard reagent have yielded only *sym*-2-difurylethane. The chloride can be used as an alkyl halide in acetoacetic ester, and in many other syntheses when these are carried out in organic solvents; the expected products ensue. However, furfuryl chloride and aqueous potassium cyanide yield largely 5-methyl-2-furonitrile with some 2-furylacetonitrile. This may be due to an extended allylic rearrangement which can be compared with the acid ring opening of furfuryl alcohol; the intermediate **79** has been detected through its n.m.r. spectrum.[117] 2-Furyl-acetonitrile can, however, be made from the chloride and sodium cyanide in dimethylformamide, or from sodium cyanide and furfuryltrimethylammonium iodide.

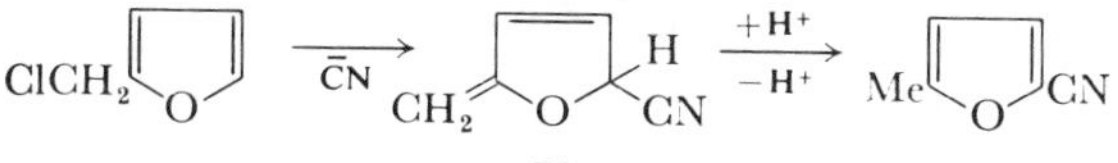

$ClCH_2$–furan $\xrightarrow{\bar{C}N}$ CH_2=(dihydrofuran)(H)(CN) **79** $\underset{-H^+}{\overset{+H^+}{\rightleftharpoons}}$ Me–furan–CN

An interesting reaction of furfuryl benzoate (**80**) is its thermolysis,[118] which yields 40% of the cyclobutenone **81**.

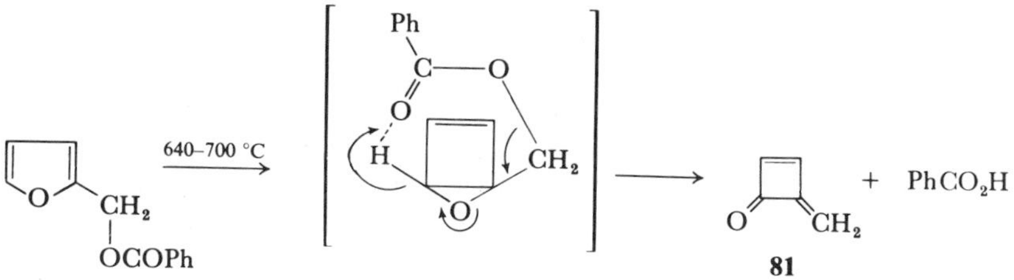

80 (furan–CH_2–OCOPh) $\xrightarrow{640–700\ °C}$ [intermediate] $\longrightarrow$ **81** + $PhCO_2H$

9. *Furan aldehydes and ketones*

Furfural (furan-2-carbaldehyde), b.p. 161·7°C, is a colourless liquid when freshly distilled and possesses an aromatic pungent odour. It is somewhat soluble in water (8% at 20°C) and is miscible with most common organic solvents, including aromatic hydrocarbons. It is only slightly soluble in saturated aliphatic hydrocarbons at ordinary temperatures and shows a remarkable power of extracting unsaturated hydrocarbons from mixtures with related saturated materials. This property enables otherwise very difficult separations of similar boiling materials to be effected, and about half the furfural produced is used for such separations in the vegetable oil and petroleum industries. Furfural reacts, like formaldehyde, with phenols to form resins, and these have industrial applications. The catalytic reduction of furfural on an industrial scale can lead to furfuryl alcohol or tetrahydrofurfuryl alcohol using nickel catalysts, and to 2-methylfuran using copper chromite.

Furfural has many reactions similar to those of benzaldehyde, but there are important differences which are noted below. It behaves like benzaldehyde towards the following reagents and gives the expected products: sodium bisulphite, hydrazine, and its derivatives, Grignard reagents, alkylzincs, chloro esters in the Reformatsky reaction, activated methylene groups in the Perkin (p. 124), Claisen, Knoevenagel, and aldol condensations, ammonia, orthoformic ester and ammonium chloride (giving the acetal), acetic anhydride and stannous chloride (giving the diacetate, p. 146), hydrogen over nickel (giving furfuryl alcohol), alkaline permanganate (giving 2-furoic acid), and so on. Like benzaldehyde, furfural undergoes the benzoin reaction with potassium cyanide and the Cannizzaro reaction with potassium hydroxide. The yields in both these reactions are, however, poor owing to the formation of unidentified coloured by-products. Alkaline oxidation converts furfural into 2-furoic acid, but aqueous acidic reagents usually cause ring rupture, with the formation of derivatives of maleic dialdehyde.

Furfural is much more stable to acidic reagents than furan because of the electron-attracting aldehyde group. Substitution in the ring, as in the case of furan, usually takes place at the unsubstituted 5-position. The ring is much more active than that of benzaldehyde and can be easily nitrated, acetylated, and

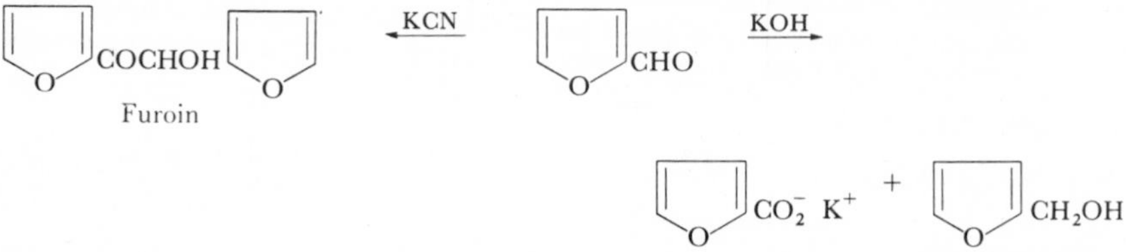

COCHOH ← KCN — CHO — KOH →

Furoin

CO_2^- K^+ + CH_2OH

alkylated (Friedel–Crafts). Bromination also causes 5-substitution,[119] but in the presence of aluminium bromide a mixture of mainly the 4- and 4,5-dibromo derivatives is formed in 20 and 40% yield, respectively.[120] There is no doubt that aluminium halides, particularly in large amount, can alter the preferred position for electrophilic attack from 5 to 4. The first example of attack at position 4 in furfural was the apparently exclusive formation of the 4-isopropyl derivative using isopropyl chloride and aluminium chloride.[121] The reaction does not proceed via substitution at the 5-position followed by rearrangement, as 5-isopropylfurfural is stable to aluminium chloride under the alkylation conditions.

CHO → i-Pr CHO

Furfural itself does not react with maleic anhydride, in common with furans possessing electron-attracting substituents. It behaves like furan, however, after conversion into the diacetate (**82**) or diethyl acetal[122] when the electron-attracting properties of the aldehyde group are neutralized. The diacetate also

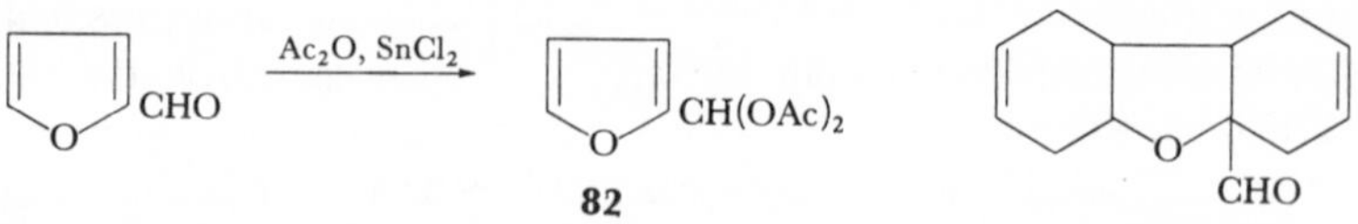

CHO —Ac_2O, $SnCl_2$→ $CH(OAc)_2$

82

CHO

83

reacts like furan towards bromine with methanol (cf. p. 129). Furfural itself can, however, react as a dienophile; with 2 moles of butadiene, **83** is formed.

Furfural reacts with many aliphatic and aromatic amines to form the expected anils, for example, **84** is formed from aniline and furfural alone. However, if the furfural is treated with aniline hydrochloride the ring is opened, yielding the purple glutaconic aldehyde derivative **85**, which dyes red on cotton, wool, and silk. This derivative can be converted into a pyridine (**86**).

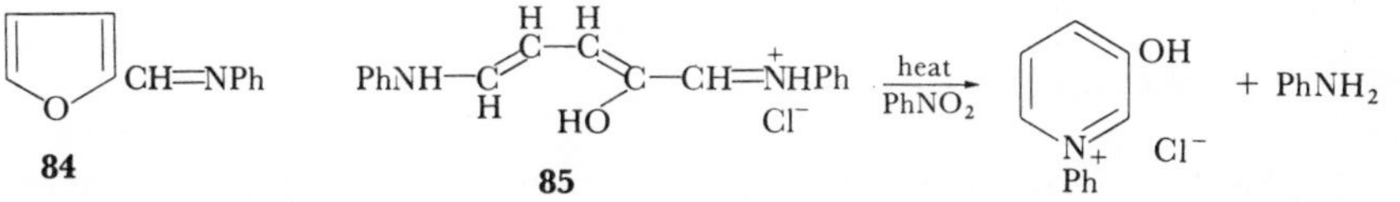

CH=NPh PhNH–C(H)=C(H)–C(H)=C(HO)–CH=$\overset{+}{N}$HPh Cl^- →(heat, $PhNO_2$) + $PhNH_2$

84 **85** **86**

Furan-3-carbaldehyde has been obtained from the catalytic hydrogenation of 3-furoyl chloride. It does not give a colour with aniline under acidic conditions.

2-Furyl methyl ketone is obtainable from furan by acetylation, and from 2-furoic acid derivatives by standard methods. 2-Furyl phenyl ketone can be prepared from benzene, 2-furoyl chloride, and aluminium chloride. Like furfural, both ketones react with aniline hydrochloride, and electrophilic substituents enter position 5. Many types of 2-furyl ketones yield pyridines (**87**) on heating with ammonia and ammonium chloride (p. 273). Replacing the ammonia by a secondary amine can give an enamine (**88**), which on distillation rearranges to the phenol (**89**).[123]

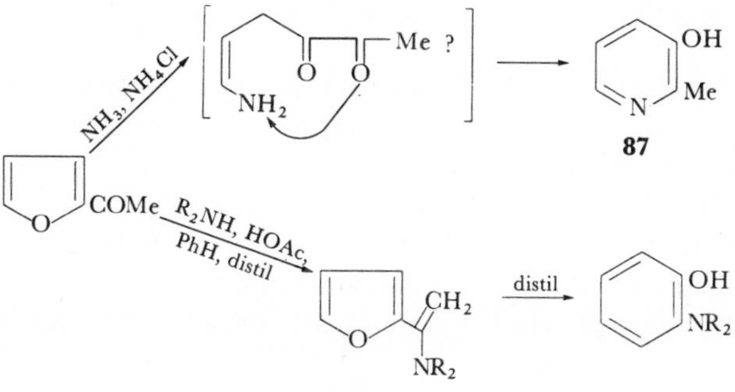

COMe; NH_3, NH_4Cl; NH_2; Me ?; OH; Me; N; R_2NH, HOAc, PhH, distil; CH_2; NR_2; distil; OH; NR_2

87 **88** **89**

10. *Furoic acids*

2-Furoic acid is a colourless crystalline solid of m.p. 133°C which sublimes easily and is sparingly soluble in water. It is very stable at room temperature, in contrast to many other furan derivatives, and is a stronger acid than benzoic acid. Furoic acid is obtainable in excellent yield by the oxidation of furfural by oxygen in the presence of alkali and a cuprous oxide–silver oxide catalyst.[124] 2-Furoyl chloride can be prepared with thionyl chloride and, like benzoyl chloride, is only slowly hydrolysed by water. It has the expected reactions with amines, alcohols, etc.

The electron-attracting carboxyl group of 2-furoic acid stabilizes the ring sufficiently to permit direct sulphonation, and both nitration with fuming nitric

R = H or CO_2H

acid or nitric and sulphuric acids, and chloromethylation take place at position 5. The nitration of 2-furoic acid, or of furan-2,5-dicarboxylic acid, yields first 5-nitro-2-furoic acid and then 2,5-dinitrofuran. This type of decarboxylation with the simultaneous introduction of a nitro group or halogen atom (p. 138) is common with furans, and a similar decarboxylation can occur with mercuric choride (see p. 142). 5-Bromo-2-furoic acid, along with some 2,3,4,5-tetrabromotetrahydrofuroic acid,[125] can be obtained from 2-furoic acid with bromine, but hot aqueous bromine causes decomposition to 'mucobromic acid' (**90**).

$$\longrightarrow OHCCBr{=}CBrCO_2H + {}^{14}CO_2$$

90

The carboxyl group of the furoic acid has been shown by tracer studies to give rise to the carbon dioxide.

Boiling 2-furoic acid in 40% sodium deuteroxide in deuterium oxide for 2·5 hours causes ca. 90% exchange of all the ring hydrogen atoms for deuterium;[126] this is a remarkably easy exchange.

Furoic esters do not decarboxylate on nitration, and in contrast to benzoic esters can be used in the Friedel–Crafts reaction when substitution can occur at both 4- and 5-positions. Methyl 2-furoate brominates at position 5, and if aluminium chloride is present the 4,5-dibromo derivative is formed.[120] It is not

91

affected by bromine and methanol, but electrolytic oxidation in methanolic ammonium bromide yields **91**. This ester may be converted into some comparatively inaccessible benzoic acids as shown.[127]

E. Synthetic methods

(1) Furan itself is best prepared in the laboratory by the decarboxylation of furoic acid in quinoline with a copper catalyst.[124] Industrially it is obtained by the catalytic decomposition of furfural in steam (90% yield).

$-CO_2$ $-CO$

CO_2H CHO

(2) Furfural is obtained commercially by the steam distillation of oat hulls (husks) and other naturally available material containing pentose residues in

CHOH—CHOH CHOH—CH ? HO ? CHO CHO

CH_2OH CHOHCHO CH_2OH CCHO HO

1–3M hydrochloric or sulphuric acid. Hydrolysis of the polysaccharides to the pentose is followed by the cyclization. The mechanism of the reaction is not clearly understood, although there is some evidence to support the sequence shown. A similar decomposition of hexoses yields 5-hydroxymethylfurfural.

(3) The cyclodehydration of 1,4-diketones, such as acetonylacetone (R = R‴ = Me, R′ = R″ = H), by acidic reagents, including sulphuric acid, zinc chloride, acetic anhydride, and phosphorus pentoxide, is a very general reaction for the

R′CH—CHR″ R′CH—CR″ ? R′ R″

RCO OCR‴ RCO CR‴ R R‴

HO

synthesis of furans and rarely fails. Some evidence is available for the mechanism shown, but it is not conclusive.

(4) A new synthesis,[58] which clearly has potentialities, starts from α-hydroxy ketones.

PhCO + CCO_2Me K_2CO_3 OH Ph CO_2Me MeOH, HCl Ph CO_2Me

PhCHOH CCO_2Me Ph CO_2Me Ph CO_2Me

(5) Feist's synthesis consists of treating an α-chloro ketone with an acetoacetic ester derivative in the presence of pyridine. Ammonia has been used instead of pyridine, but then some pyrrole is formed (p. 115) if the reactants are heated.

MeCO–CH_2Cl + CH_2CO_2Et–OCCH$_2$R →(pyridine)→ Me, CO_2Et, CH_2R (furan)

R = H or CO_2Et

(6) Furans can be obtained from certain 2-pyrones. Coumalic acid is obtained from malic acid with sulphuric acid. The malic acid, as a typical α-hydroxy acid, loses carbon monoxide and water and gives formylacetic acid, the enolic form of which is shown (**92**). Two moles condense to coumalic acid (**93**). This is then esterified, brominated with 1-bromopyridinium perbromide, and the product treated with base under carefully controlled conditions when furan-2,4-dicarboxylic acid is obtained in 90% yield.[128]

2 CO_2H–CH_2–CHOH–CO_2H → HO_2C…OH + OH…CO_2H (**92**) → HO_2C (2-pyrone) (**93**) →(2 stages)→

MeO_2C, Br (2-pyrone) → [MeO_2C, Br, CO_2^-, O^-] → HO_2C, CO_2H (furan)

(7) A new[129] synthesis starts with an acetylenic sulphonium salt. In ethanol this isomerises to an allene, which can add to the sodium salt of a 1,3-diketone or β-keto ester. Cyclisation and tautomerism now yield the furan.

HC≡C–$CH_2\overset{+}{S}Me_2$ →(EtOH)→ H_2C=C=CH$\overset{+}{S}Me_2$ + $^-$CHCOMe–OCMe →(H+)→ [H_2C=C(–C(COMe)H–C(=O)Me)–$CH_2\overset{+}{S}Me_2$] → Me, COMe, Me (furan)

F. Natural occurrence and compounds of special interest

Furan and sylvan (2-methylfuran) have been found in low-boiling wood oils. Furfuryl mercaptan (**94**) has a very disagreeable odour, but at very low concentrations its odour strongly resembles that of roasted coffee. It is a constituent of the oil from roasted coffee and is formed in the roasting process. A number of furans occur naturally, examples being **95**, which has been isolated

CH_2SH COi-Pr HO O Me Me

94 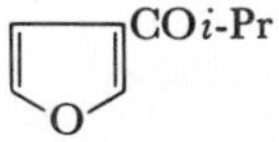 **95** 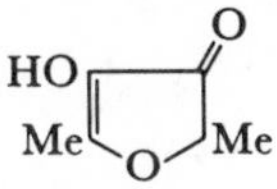 **96**

from *Perilla frutescens*, **96** which is a flavour component of pineapple,[130] and **97**, which has been obtained from sweet potatoes infected with *Ceratostomella fimbriata.*[131]

Furfural is the furan of greatest interest, and its chemistry has been discussed earlier. 'Furacin,' or 5-nitrofurfural semicarbazone (**98**), is a useful antibacterial.

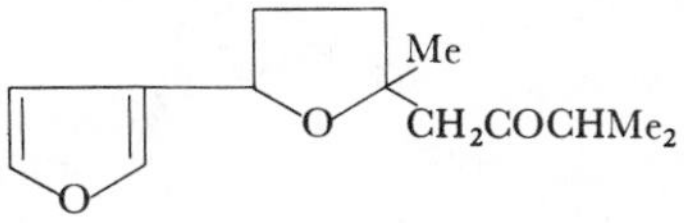

Me $CH_2COCHMe_2$ O_2N CH=NNHCONH$_2$

97 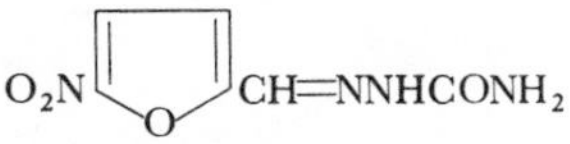 **98**

3. THIOPHENE

A. Introduction

The discovery of thiophene by Victor Meyer in 1882 was due to the substitution by an assistant of pure benzene, prepared from calcium benzoate, for the coal-tar benzene which was normally used in his lecture demonstration. On treating the pure benzene with isatin and concentrated sulphuric acid, the 'indophenine test,' which was supposed at that time to be specific for aromatic compounds, the expected blue colour which was normally formed with coal-tar benzene did not develop. Meyer soon came to the correct conclusion that coal-tar benzene contained an impurity responsible for the colour reaction. He found that the impurity was sulphonated more easily than benzene and that the product formed a crystalline lead salt. Purification of this, followed by desulphonation, yielded crude thiophene. Meyer recognized its aromatic character, as the dibromothiophene obtained on bromination, like bromobenzene, could neither be dehydrobrominated with alcoholic potash nor easily reduced like aliphatic compounds with the replacement of the bromine atoms by hydrogen. He then synthesized thiophene by heating sodium succinate with phosphorus trisulphide, a method which was widely used until the synthesis from butane and sulphur was put on a commercial basis.

Like pyrrole and furan, the atoms of the thiophene ring can be numbered (**1**) or lettered (**2**), but this second system is obsolete. The trivial names of

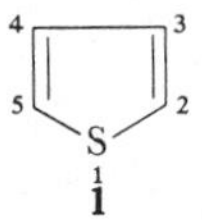

1 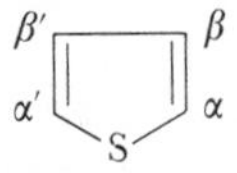 **2**  **3** 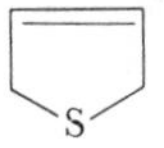 **4**  **5**

2-thiolene, 3-thiolene, and thiolane have been proposed for 2,3- (**3**) and 2,5-dihydrothiophene (**4**), and for tetrahydrothiophene (**5**) respectively, but as these names have not been generally accepted, the systematic names are used here. The common names for some groups derived from thiophene are shown.

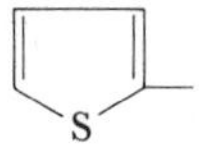

2-Thienyl 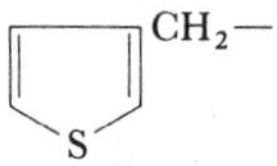 3-Thenyl 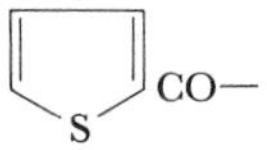 2-Thenoyl 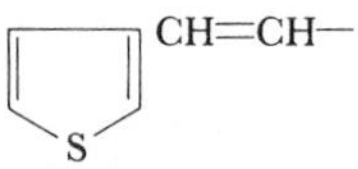 3-Thenal, or 3-Thenylidene

B. Physical properties and structure

Thiophene is a colourless liquid, b.p. 84·1°C and f.p. –38·3°C, which has an odour very similar to that of benzene. It is slightly denser (d_4^{25} 1·058) than water, with which it is immiscible, but it is miscible with most organic solvents.

Victor Meyer showed the arrangement of the atoms in the ring by two syntheses and proposed the same formula as is current today. The dimensions of

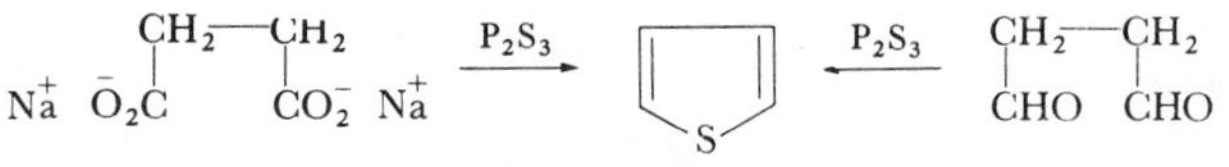

P_2S_3 P_2S_3

25–30% yield

the thiophene molecule have been determined with certainty by a study of the microwave spectrum;[132] no *a priori* assumptions had to be made in the calculation, and the results are in excellent agreement with those of earlier electron-diffraction studies. The carbon–sulphur–carbon angle is less than 105°, as might be expected by analogy with other compounds,[133] and is effectively the same as the bond angle in hydrogen sulphide (92·1 ± 0·2°). The carbon–sulphur distance is less than the normal single bond carbon–sulphur distance,

H 1·081 Å
4 1·423 Å
123·3° C C3
112·4°
1·370 Å
5
C 111·5° C2
1·078 Å
119·9° 92·2° 1·714 Å
H S
1

which is 1·82 Å. This information suggests that the molecule is not well represented by the classical structure (**1**), and that it is in a state of resonance to which other structures contribute. In agreement with this the resonance energy of thiophene, calculated from combustion data is 121–130 kJ (29–31 kcal)/mole. Although this is less than that of benzene (150·5 kJ (36·0 kcal)/mole), it is almost the same per ring atom, and it is higher than that of furan or pyrrole. This order of resonance energies is in agreement with the fact that sulphur is less electronegative than oxygen or nitrogen and is therefore more ready to release

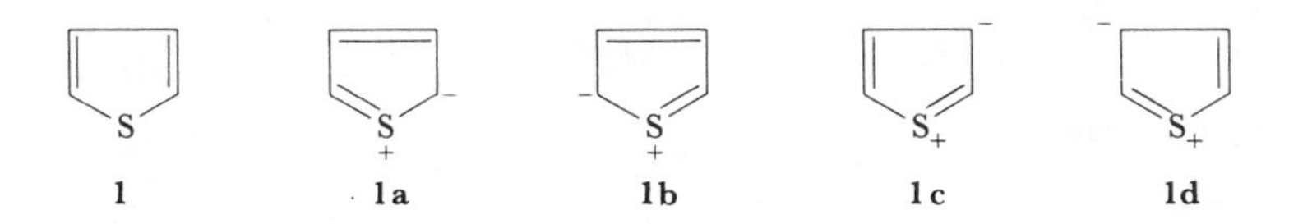

1 **1a** **1b** **1c** **1d**

electrons into the ring to form the sextet of π-electrons required for aromaticity. The major contributing resonance structures are **1** to **1d**.

Schomaker and Pauling suggested in 1939 that the sulphur–carbon bonds in thiophene did not consist of pure *p* orbitals, but that some of the 3*d* orbitals of the sulphur atom were used. This idea has been extended by Longuet-Higgins, who calculated the carbon–sulphur distance as 1·73 Å by the molecular-orbital method. More recently, Clark[134] has shown that if the contributions made by the 3*d*, 4*s*, and 4*p* orbitals of the sulphur are disregarded, then calculations by a self-consistent field molecular orbital method of the dipole moment, and hydrogen–hydrogen and ^{13}carbon–hydrogen coupling constants for thiophene, give quite a good fit with the observed data. However, he has also found[134] that if these three orbitals are included, the 4*s* and 4*p* orbitals having energies similar to that of the 3*d* orbital, a better fit is obtained. Including these orbitals makes only a small difference to the total energy calculated for the molecule, but it has a significant effect on the electron distribution in the ground state. It is therefore clear that these orbitals must be taken into consideration when calculations of reactivity, etc., for sulphur-containing molecules are made, and that structures 1*e*–1*i* must be given some weight when resonance formulae for thiophene are employed. Calculations[134] of the ground state

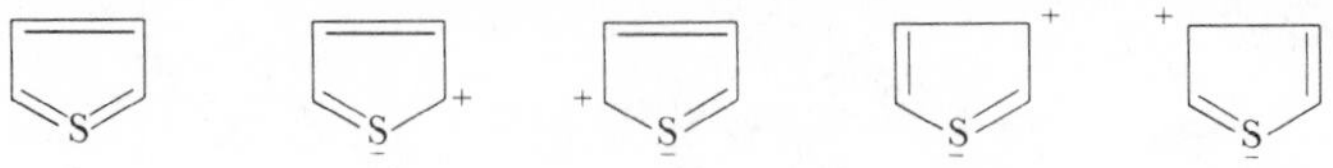

1e 1f 1g 1h 1i

π-electron distribution for thiophene show that positions 2 and 5 possess a small excess over positions 3 and 4. Localisation energy calculations[135] suggest that positions 2 and 5 will be much more easily attacked than positions 3 and 4, which are nevertheless more reactive than benzene to both electrophilic and radical reagents. This is exactly in accord with experiment, and the prediction that position 2 will be the more reactive to nucleophiles is still to be tested.

A most interesting observation (p. 155) is that thiophene can be alkylated on the sulphur atom;[136] the *S*-alkyl group is not in the ring plane.[137] This supports the idea that resonance structures with a negative charge on the sulphur, 1*f*–1*i*, have chemical significance. The difference between the dipole moments of thiophene (0·52 D) and tetrahydrothiophene (1·87 D), the sulphur being at the negative end of the dipole in both cases, shows that there is a substantial mesomeric moment, as in the furan series, and this is discussed later (p. 177).

C. Chemical properties

Thiophene resembles benzene, rather than furan or pyrrole, in many of its reactions, but it is much more reactive and less stable. It has very few properties which could be attributed to an enol ether or a 1,3-diene type of structure. Substitution at the reactive 2- and 5-positions of the molecule usually appears to be direct, as no intermediate 2,5-addition compounds have been isolated from nitration reactions, etc. (contrast furan). Thiophene is reported to form an ozonide. On autoxidation in the presence of light, thiophene is broken down to oxalic and sulphuric acids. The ring is also disrupted to hydrogen sulphide and other products in the presence of alumina at 200°C. Fluorination in the gas phase gives sulphur hexafluoride and fluorocarbons.

1. *Addition reactions*

Thiophene is stable to aqueous, but not to anhydrous, mineral acids. The ring is opened to some extent on boiling with 100% orthophosphoric acid, while 100% sulphuric acid, hydrogen fluoride, and stannic, ferric, and aluminium chlorides cause polymerisation to dark amorphous solids. Aluminium chloride is a very poor catalyst for Friedel–Crafts reactions involving thiophene and can be used for removing small quantities of thiophene from benzene.[138] Benzene, if a Friedel–Crafts reactant, should be thiophene-free for best results.

Under milder conditions orthophosphoric acid gives a trimer (**7**) which is not structurally analogous to pyrrole trimer (p. 94) but is formed by initial 2-protonation (**6**), as occurs with some pyrroles (p. 104), as outlined.

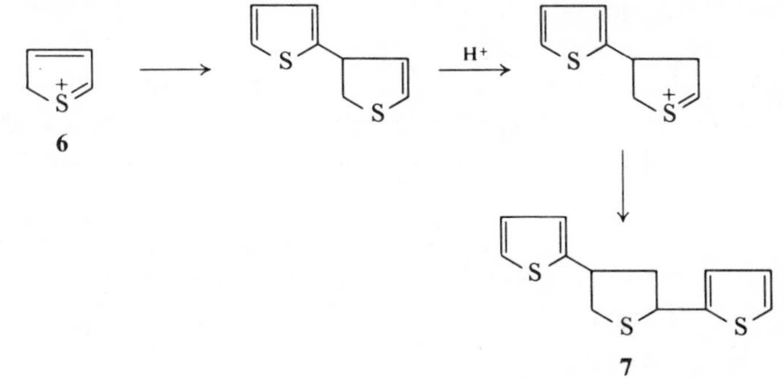

6

7

Thiophene, with either trimethyloxonium borofluoride or methyl iodide and silver perchlorate in inert solvents, gives the 1-methylthiophenium cation (**8**) which was isolated as the hexafluorophosphate. Its structure was proved as outlined. Both the ultraviolet absorption and nuclear magnetic resonance spectra of the ion, which is slowly solvolysed by water, suggest strongly that it should possess aromatic properties, but little data is available.[136] Corresponding benzo[*b*]thiophenium salts (**9**) have been much more fully investigated[137] and

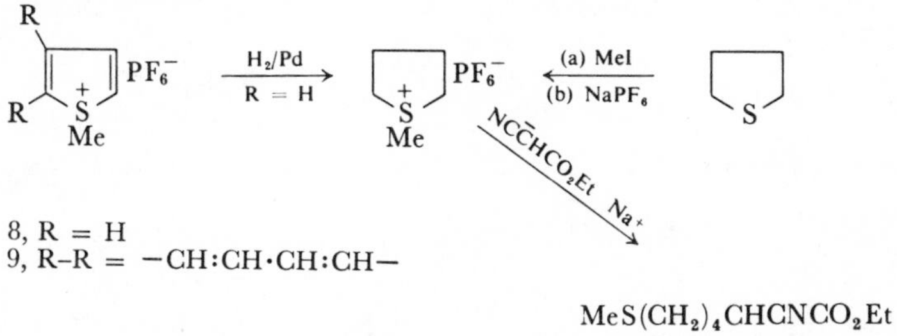

8, R = H
9, R–R = −CH:CH·CH:CH−

$MeS(CH_2)_4CHCNCO_2Et$

nuclear magnetic resonance studies have shown that the S–Me bond is not coplanar with the ring system. The compounds are powerful methylating agents, the original benzo[*b*]thiophene (p. 221) being reformed, and they are solvolysed in water or methanol. Nucleophiles open the 1-methyltetrahydrothiophenium ring as shown.[139]

The direct oxidation of thiophene probably yields first the 1-oxide and then the 1,1-dioxide, which polymerize *in situ.* The 1-oxide (**11**) has been obtained[140] in solution by treating the *trans*-tetrahydrothiophene (**10**) with

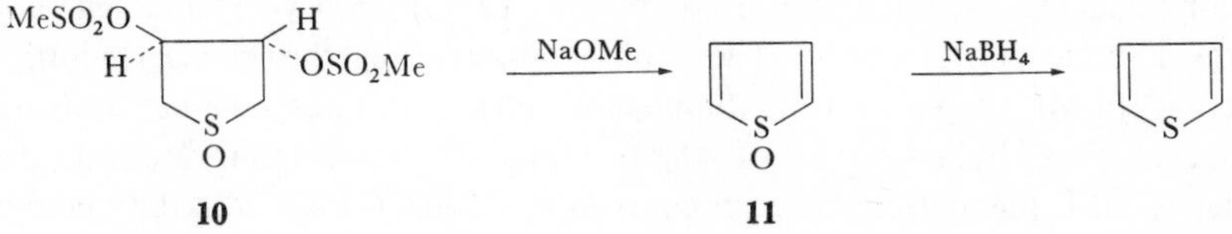

10

11

cold sodium methoxide or *t*-butoxide. All attempts to isolated it gave a dimer, but reduction to thiophene confirms its structure.

Thiophene 1,1-dioxide has also been synthesised by the indirect route[141] outlined below. The first stage is now known to be a concerted reversible process.[142] The 1,1-dioxide (**14**) has not been isolated in a pure condition, as it

+ SO_2 → **12** (a) Br_2, (b) Me_2NH → Me_2N ... NMe_2 (a) MeI, (b) Ag_2O → NMe_2 ... (a) MeI, (b)Ag_2O → **14** → [SO_2 adduct] $-SO_2$ → **13**; **13** $3H_2$ → ; **14** Me_2NH → NMe_2; light → **15**; **16**

reacts rapidly with itself, with the loss of sulphur dioxide yielding **13**, probably through the intermediary of an adduct of the Diels–Alder type, as shown.

The dioxide (**14**) can be extracted from water by chloroform, and its ultraviolet absorption spectrum closely resembles those of the 1-oxide **11** and cyclopentadiene but not those of compounds **12** and **15**. Cyclopentadiene also forms a dimer (**16**) readily. These observations, and the Michael addition reaction shown, suggest that the thiophene oxides do not possess much aromatic character. The hybridisation of the orbitals of the sulphur atom is not understood. Diphenylthiophene, in contrast to thiophene, is oxidized by peroxy acids to an isolable stable 1,1-dioxide, which on heating gives a dimer of analogous structure to **13**.

Although thiophene is inert to maleic anhydride, it does combine[143] with the extremely reactive dienophile dicyanoacetylene and with tetrafluorobenzyne[144] to give **18** and **19**, respectively, with the extrusion of sulphur; presumably initial additions across the 2,5-positions occur (e.g., **17**). Thermal decomposition of ethyl azidoformate (N_3CO_2Et) in thiophene gives 21% of the

17  18 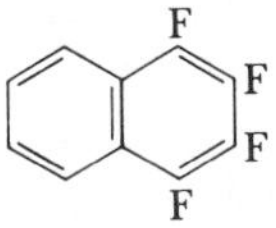 19  20

pyrrole (**20**) and other products. This suggests that 2,5-addition of the nitrene $:NCO_2Et$ is followed by loss of the sulphur atom.[145] In the presence of light, thiophene adds to dimethylmaleic anhydride in exactly the same way as furan (p. 129), and 2,5-dimethylthiophene forms **21** with benzophenone.[146] Photolysis of ethyl diazoacetate in thiophene initially gives a carbene ($:CHCO_2Et$), which yields **22**, converted by acid to **23**.

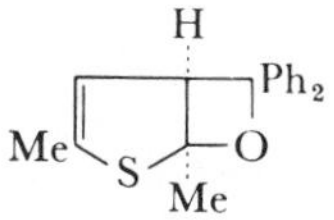

21 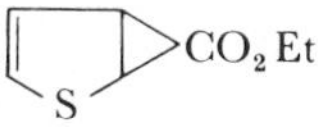 22 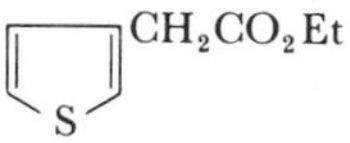 23 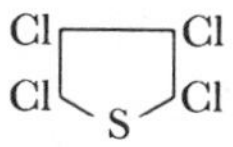 24

The reduction of thiophene gives a number of products according to the conditions (see pp. 160–161), and at 40°C with chlorine a maximum of 13% of 2,3,4,5-tetrachlorotetrahydrothiophene (**24**), along with 2-chloro- and 2,5-dichlorothiophene, is obtained. Two of the six possible geometric isomers of **24** have been characterized. Further substitution of **24** is possible, and octachlorotetrahydrothiophene can be obtained from thiophene and chlorine in the presence of iodine.

2. *Substitution reactions*

Thiophene, while much more reactive than benzene, is much more stable than pyrrole or furan to acids. This enables it to undergo many substitution reactions under comparatively mild conditions. There is a great tendency for the attacking group to enter 2- or 5-positions in preference to 3- or 4-positions, although there are some notable exceptions in substituted derivatives. There is a much larger tendency for existing substituents to be displaced by incoming groups than in the benzene series.

The sulphonation of thiophene, a reaction used widely for its separation from benzene, occurs readily with 95% sulphuric acid at 30°C. The product is largely the 2-sulphonic acid (69–76%), and better yields (86%) are claimed if 1-proto-1-pyridinium sulphonate (pp. 98 and 234) is used. Some thiophene-3-sulphonic acid is also formed, and further reaction leads to disulphonic acids. Desulphonation of the acids with superheated steam gives back thiophene.

Thiophene reacts with isatin in sulphuric acid to give a deep blue colour, which is due to the formation of indophenine. This reaction is used for the

Isatin

Indophenine

detection of thiophene in coal-tar distillates and is sensitive to 0·025%. This colour reaction is positive for some substituted thiophenes but has not been systematically investigated.

TABLE 1. Chlorothiophenes from the chlorination of thiophene (2100 g)

Derivative	Amount (g)	% yield based on chlorine
2-Chloro-	1055	36·8
3-Chloro-	3·3	0·1
2,3-Dichloro-	63	3·3
2,4-Dichloro-	64	3·3
2,5-Dichloro-	519	27·1
3,4-Dichloro-	113	5·9
2,3,4-Trichloro-	120	7·3
2,3,5-Trichloro-	2·5	0·2

The chlorination of thiophene proceeds very rapidly even in the dark at –30°C and gives substitution as well as addition products. In acetic acid at 25°C attack only appears to occur at position 2, and $1{\cdot}25 \times 10^7$ as fast as a similar attack on benzene. The substitution products obtained from 25 moles each of thiophene and chlorine after dehydrochlorination with mixed sodium–

potassium hydroxide are given in Table 1. The alkali dehydrochlorinates tetrachlorotetrahydrothiophene to a mixture of the 2,4- (44%), 2,5- (2%), and 3,4-dichlorothiophenes (56%). The chlorination of thiophene with sulphuryl chloride also gives 2-chlorothiophene (43%).

Bromine reacts rapidly with thiophene, but addition products have not been characterized. Treatment of the product with alkali gives largely 2,5-dibromothiophene. Refluxing thiophene with *N*-bromosuccinimide is the best preparative method for 2-bromothiophene (77% yield). The direct iodination of thiophene in the presence of mercuric oxide or nitric acid[147] gives an excellent yield of 2-iodothiophene.

Attempts to nitrate thiophene under conditions normal for the benzene series invariably result in the destruction of the ring, but nitration at 10°C in acetic acid–acetic anhydride gives over 80% of the 2-nitro compound; a little of the 3-nitro isomer is also formed. Only a weak secondary isotope effect was observed[148] in the nitration of 2-^3H-thiophene with benzoyl nitrate in methyl cyanide solution at –2°C, so that the loss of the proton (or $^3H^+$) is not rate determining, as is also the case for benzene.

The direct alkylation of thiophene with olefines in the presence of catalysts, such as phosphoric acid, boron trifluoride, and 80% sulphuric acid, has been investigated only comparatively recently. Ethylene fails to react under the conditions used, propylene reacts slowly, and isobutylene reacts rapidly. It is interesting that both the 2- and 3-positions are attacked to about the same extent by the very reactive, and therefore relatively unselective, cations derived from the olefines. In contrast to this the acylation of thiophene by acyl halides, or anhydrides in the presence of stannic chloride or phosphoric acid, which cause less polymerization than aluminium chloride, gives exclusively 2- or 5-acylthiophenes when these positions are free. Acylation may also be effected by merely boiling thiophene, a carboxylic acid, and phosphorus pentoxide in benzene, and by the Vilsmeier reaction (p. 99).

The chloromethylation of thiophene proceeds very easily, yielding 2-chloromethylthiophene at 10°C and the 2,5-bis derivative at 50°C. Thiophene is

$$\text{(thienyl)}CH_2N{=}CH_2\cdot HCl \ (\mathbf{25}) \xrightarrow{MeOH} \text{(thienyl)}CH_2\overset{+}{N}H_3\ Cl^- \ (\mathbf{26}) + CH_2(OMe)_2$$

reactive enough to combine with formaldehyde and ammonium chloride, in a type of Mannich reaction, to give **25**. With methanol, **25** yields thenylamine hydrochloride (**26**). Thiophene is very easily mercurated (see p. 166).

Phenyl radicals, from aniline and pentyl nitrite or *N*-nitrosoacetanilide, with thiophene give a mixture of the 2- (93%) and 3-phenyl (7%) derivatives,[149] and other radicals also show a very strong preference for the 2-position.[150]

D. Derivatives of thiophene

1. *Some reduced thiophenes*

Thiophene is reduced by sodium in a mixture of methanol and liquid ammonia at –40°C to a mixture of compounds.[151] In contrast to thiophene, but like tetrahydrothiophene, both dihydro compounds form methiodides.

Na, MeOH, NH_3 → + + butanethiols (mixture)

2,3-Dihydrothiophene, b.p. 112°   2,5-Dihydrothiophene, b.p. 122°

2,5-Dihydrothiophene is quite stable, and with hydrogen peroxide gives the 1,1-dioxide, which has also been obtained from butadiene and sulphur dioxide (p. 156, **12**). 2,3-Dihydrothiophene is very reactive and polymerizes on standing. On oxidation with hydrogen peroxide it also gives a 1,1-dioxide which has been obtained from 2,5-dihydrothiophene 1,1-dioxide by irradiation with ultraviolet light (p. 156).

Tetrahydrothiophene (**27**), sometimes known as thiolane, is a colourless liquid, b.p. 121·1°C, with a slightly puckered ring and can be obtained as indicated. In contrast to thiirane, thietane, and tetrahydrofuran, it is not affected by butyllithium, and it has the typical properties of an aliphatic sulphide. On oxidation with hydrogen peroxide it gives the 1,1-dioxide, and it reacts with methyl iodide, forming the sulphonium iodide. Tetrahydrothiophene can be dehydrogenated to thiophene (32%) with platinized charcoal.

CH_2—CH_2 / CH_2Cl CH_2Cl —Na_2S→ **27** ←H_2S, Al_2O_3, 400°— CH_2—CH_2 / CH_2OH CH_2OH or tetrahydrofuran

Catalytic hydrogenation of thiophene under most conditions usually removes the sulphur atom, and butane is the major product. Nickel catalysts are rapidly

—Raney Ni→ Me—CH_2—CH_2—Me + NiS

poisoned by thiophene, although Raney nickel is very useful for desulphurisations (see index).[152] The desulphurisation of many thiophene derivatives, often readily synthesised, to the corresponding butanes by boiling with Raney nickel has been widely used as a synthetic procedure.[153] Hydrogenation in the presence of molybdenum disulphide, or over a very large excess of palladium on charcoal, does, however, reduce thiophene to tetrahydrothiophene.

2. *Alkyl- and arylthiophenes*

Many alkylthiophenes have been obtained by cyclization of the appropriate succinic or laevulinic acids, and the Wolf–Kishner reduction (cf. p. 103) of acylthiophenes has been widely applied. The direct alkylation of thiophene by olefines is mentioned on p. 159, and thiophene in the presence of zinc chloride combines with benzyl alcohol, giving **28** and **29**, and with benzaldehyde giving

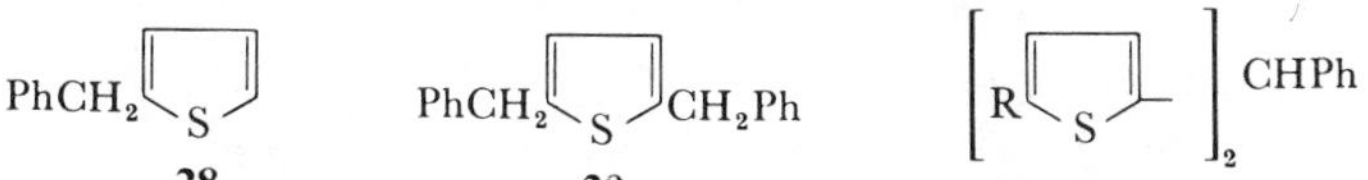

$PhCH_2$ S

28

$PhCH_2$ S CH_2Ph

29

[R S]$_2$ CHPh

30, R = H
31, R = Me

30. 2-Vinylthiophene has been obtained by two methods, and behaves like 2-vinylfuran towards maleic anhydride and quinone.

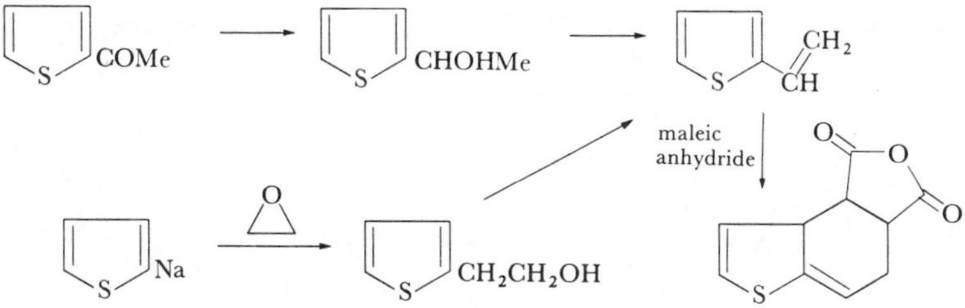

S COMe → S CHOHMe → S $CH=CH_2$

S Na —(ethylene oxide)→ S CH_2CH_2OH

maleic anhydride

Alkylthiophenes are very easily substituted, and substituents enter positions 2 or 5 first when possible. In the case of 3-methylthiophene substitution can occur at either the 2- or 5-position. Except in the case of metallation, most substituents enter the 2-position predominantly, showing that the major effect of the methyl group is at this position (Table 2).

Elemental bromine usually substitutes all the available ring positions before attacking the side chain, whatever the conditions, but side-chain bromination does occur first if *N*-bromosuccinimide is used in the presence of peroxides. The direct oxidation of methylthiophenes gives very poor yields of thiophenecarboxylic acids, methyl groups at position 2 being attacked preferentially. 2-Methylthiophene combines with benzaldehyde in the presence of zinc chloride to give **31**, showing that the ring is more reactive than the methyl group, but 2,5-dimethylthiophene couples with 2,4-dinitrophenylbenzenediazonium salts to give[155] a mixture of the expected **32** and of **33**. The formation of the latter

TABLE 2. Isomer formation in the monosubstitution of 3-methylthiophene

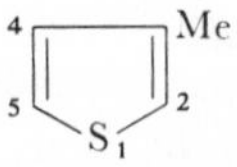

Reaction	Isomer ratio	
	2-	5-
Formylation[154] (Vilsmeier)	86·5%	13·5%
Acylation	80%	20%
Aminomethylation	100%	0%
Halogenation	100%	0%
Nitration	Major	Minor
Metallation	0%	100%

product, and of similar compounds from 2,3,5-tri- and 2,3,4,5-tetramethylthiophenes, from the supposedly electron-rich 5-methyl groups is noteworthy.

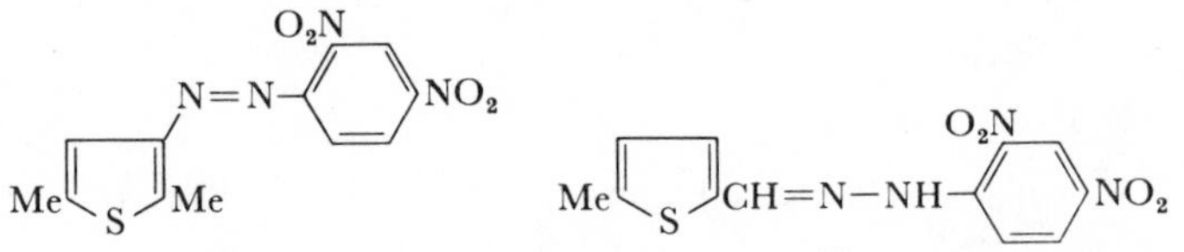

32

33

2- and 3-Phenylthiophenes have been prepared by cyclization of the appropriate keto acids with phosphorus pentasulphide, and the 2-isomer has also been obtained in 20% yield from benzenediazonium chloride, sodium hydroxide, and thiophene. 2-Phenylthiophene is attacked by electrophilic reagents at position 5, and in benzene solution on exposure to ultraviolet light it is isomerized in good yield to 3-phenylthiophene, the labelled (*) carbon atom remaining with the phenyl group.[156] The mechanism of the transformation (cf. p. 137) has not been established.[38]

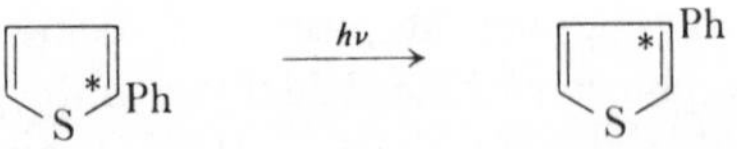

Tetraphenylthiophene (thionessal, **34**) was the first thiophene obtained by synthesis and is easily prepared from stilbene and sulphur. It boils without decomposition at 460°C. Vigorous oxidation breaks the ring, while hydrogen peroxide yields a stable 1,1-dioxide (**35**), in contrast to thiophene itself.

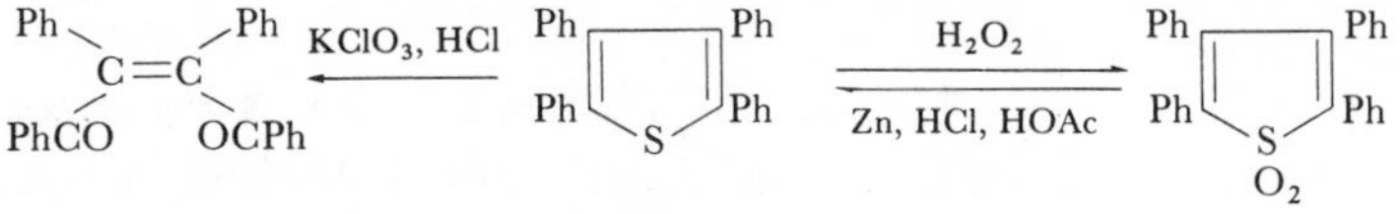

3. *Halogenated thiophenes*

2-Chlorothiophene is a colourless liquid, b.p. 128°C. It is reported not to form a Grignard reagent, but with sodium in benzene 2-thienyl sodium (**36**) is formed. It is remarkable that if the reaction is carried out in ether, 2-chloro-, 2-bromo-, and 2-iodothiophene yield the 5-(2-halothienyl)sodium (e.g., **37**). 2-Chlorothiophene undergoes the usual substitution reactions of thiophene, but more slowly.

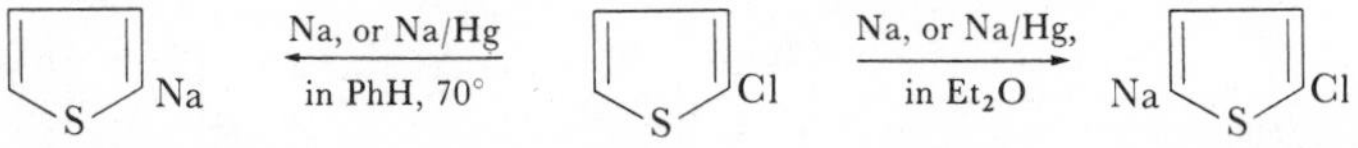

3-Bromothiophene can be obtained by brominating thiophene in chloroform to the 2,3,5-tribromo stage and reducing out the more reactive bromine atoms, or by treating 2-bromothiophene with sodium in liquid ammonia, in which case an interesting rearrangement[157] to the 3-isomer occurs; a benzyne intermediate is probably involved.

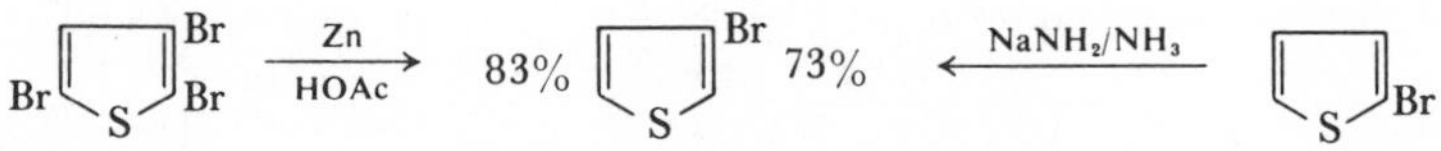

Although 3-bromo- and 3-iodothiophene do not react with magnesium under ether, they do so in the presence of reacting methyl iodide, and the 2-isomers form Grignard reagents under the usual conditions. 2-Bromothiophene is converted into thiophene-2-carboxylic acid by aqueous–alcoholic potassium cuprocyanide at 200°C and is reduced by sodium amalgam to thiophene. Bromine atoms are also easily displaced by chlorine atoms, or by nitro groups, for example, **38** gives **39**.

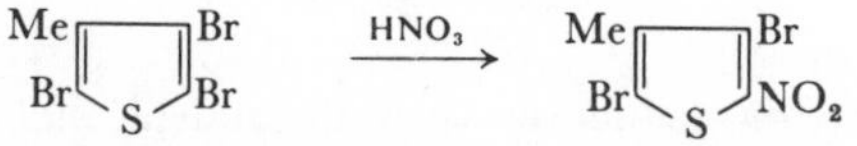

4. *Nitrothiophenes*

2- and 3-Nitrothiophene are pale yellow solids of m.p. 46·5°C and 78–79°C, respectively. 2-Nitrothiophene is obtained by the nitration of thiophene in acetic anhydride, and 3-nitrothiophene is best prepared indirectly through nitration of thiophene-2-sulphonic acid, or better the sulphonyl chloride. These last two

HNO_3 → [NO_2 … SO_2Cl, 10%] + O_2N … SO_2Cl, 90% → (a) H_2SO_4 (b) steam distil → NO_2

nitrations are specially noteworthy, as in very few other instances in the thiophene series does a substituent enter position 4 predominantly when position 5 is unsubstituted. Nitration of 2-nitrothiophene gives 56% of 2,4- and 44% of 2,5-dinitrothiophene, while the nitration of 3-nitrothiophene is reported to give 2,4-dinitrothiophene.

Cl → HNO_3, Ac_2O → O_2N … Cl → HNO_3, H_2SO_4 → O_2N … NO_2, Cl

The nitration of 2-chlorothiophene eventually gives 2-chloro-3,5-dinitrothiophene. The chlorine atom of this compound is much more easily displaced by anionic reagents (e.g., ^-OH) than that of chloro-2,4-dinitrobenzene and has been displaced by hydroxyl, alkoxyl, and amino groups.

2-Nitrothiophene undergoes an interesting ring opening to give **42** with dimethylamine in the presence of air.[158] The intermediates **40** and **41** are thought to be formed en route.

$Me_2\overset{+}{N}H$ … S … NO_2^- (**40**) ⇌ Me_2N … $\overset{+}{S}H$ … NO_2^- → Me_2N … SH … NO_2 (**41**) → $[O_2]$ → [Me_2N … NO_2, S–]$_2$ (**42**)

5. *Aminothiophenes*

2-Nitrothiophene on reduction with tin and alcoholic hydrogen chloride yields 2-aminothiophene stannichloride. An alternative synthesis[159] is from the *cis*-thioether **43**, which yields the hydrochloride **44**; attempted Hofmann degradation of thiophene-2-carboxyamide gives only thiophene-2-carboxylic acid. The free base is best obtained from the salts by treatment with strong bases under nitrogen; it has m.p. 12–13°C, and b.p. 77–79°C at 11 mm. It very rapidly decomposes in air but can be diazotised and coupled with β-naphthol.

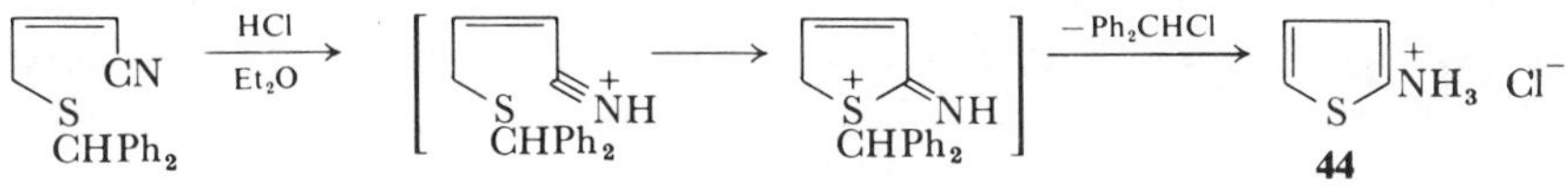

HCl, Et_2O; $-Ph_2CHCl$

43 CN, S, $CHPh_2$ → [S, NH, $CHPh_2$ → S, NH, $CHPh_2$] → **44** S, $\overset{+}{N}H_3$ Cl^-

The *N*-acetyl derivative can be obtained by acylation but is most easily prepared by the Beckmann rearrangement of methyl 2-thienyl ketone oxime.

3-Aminothiophene has been obtained both by the reduction of 3-nitrothiophene and by the Hofmann degradation of thiophene-3-carboxyamide. It decomposes more readily than 2-aminothiophene, and its salts also polymerize. However, with acetic anhydride and benzoyl chloride it gives stable acyl derivatives. The 'instability' of the aminothiophenes is probably due both to their tending to tautomerize to reactive products and to the very great ring activation produced by the electron-donating characteristics of the amino group. The presence of electron-attracting substituents greatly stabilises aminothiophenes. The 3-amino-2-carboxylic acid, for example, is a stable substance and with amyl nitrite and hydrogen chloride[160] gives a diazonium salt, which, on heating decomposes probably via a benzyne-type of intermediate in the same way as diazotised anthranilic acid.

3-Dialkylaminothiophenes react as enamines with dimethyl acetylenedicarboxylate to give thiepines (p. 454).

6. *Hydroxythiophenes*

Both 2- and 3-hydroxythiophenes have been obtained from the thienylmagnesium bromides, either by treatment with oxygen or by successive reaction with butyl borate and hydrogen peroxide.[161] Both are soluble in alkali, give deep red ferric chloride colours, couple with benzenediazonium chloride, and decompose on standing at room temperature. 3-Hydroxythiophene[162] is a pale yellow liquid, b.p. 39–40°C at 0·01 mm. Its infrared absorption spectrum in liquid film suggests that approximately equal amounts of bonded hydroxyl and unsaturated carbonyl groups are present. The compound is therefore best considered as a tautomeric mixture of **45** and **46**. Nuclear magnetic resonance

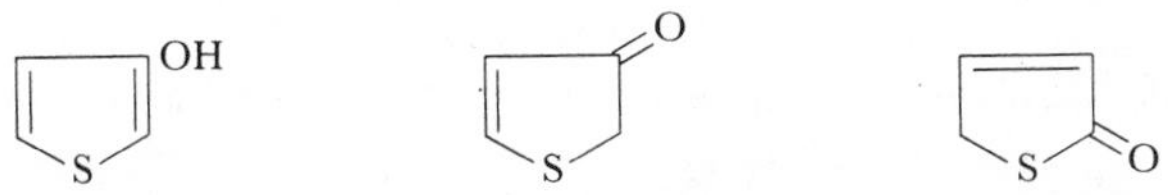

OH, S — **45**; O, S — **46**; S, O — **47**

studies[161] show that 2-hydroxythiophene exists almost exclusively as **47**. 2-Methoxythiophene is a colourless liquid, b.p. 153–155°C and has been obtained both from 2-iodothiophene, sodium methoxide, and cupric oxide and from 2-hydroxythiophene with methyl sulphate and alkali.

7. *Metallic derivatives of thiophene*

The mercuration of thiophenes under various conditions has been widely studied. Thiophene reacts with mercuric chloride in the presence of some sodium acetate to give mono (**48**) and bis (**49**) derivatives. This reaction can be used for removing thiophene from benzene, and reaction proceeds, very slowly if at all, in the presence of carbonyl, nitro, or other electron-attracting groups. Steam distillation in the presence of hydrochloric acid decomposes the mercurichlorides to the parent thiophenes. Thiophene combines with mercuric acetate, giving **50** or **51** according to the conditions, and if the 2- and 5-positions are blocked the 3,4-bismercuriacetate is formed; carboxyl groups are displaced in this reaction with the elimination of carbon dioxide. Unlike the chloride, mercuric acetate attacks 2- and 3-nitrothiophenes, forming **52** and **53**, respectively. Treatment of **51** with iodine in potassium iodide gives **54**. This type of replacement of mercury by halogen is typical. Most of the mercury derivatives of thiophenes have melting points suitable for characterization purposes.

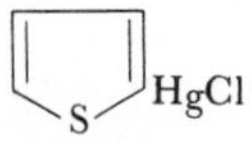

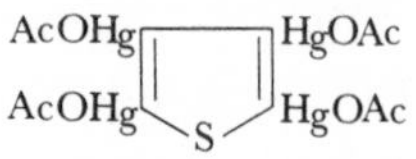

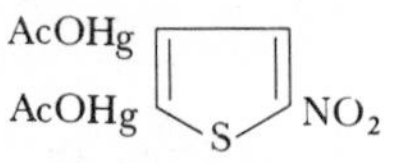

The Grignard reagents derived from thiophene have the usual reactions and are mentioned on pp. 163–165. However, lithium and sodium derivatives of thiophene are often more convenient than the Grignard reagents, as they can be prepared without using a halogenated thiophene. Butyllithium and thiophene yield 2-thienyllithium, while, surprisingly, sodium amalgam and 2-halothiophenes in ether yield 2-halo-5-thienyl sodium (see p. 163).

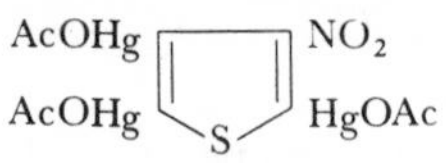

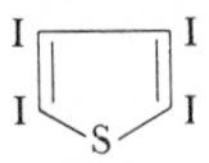

3-Thienyllithium can be obtained from 3-bromothiophene and butyllithium at –70°C, but at higher temperatures rearrangements occur. These metal derivatives undergo typical Grignard-type reactions with carbon dioxide, carbon disulphide, ethylene oxide, aldehydes, ketones, etc. and are particularly useful for preparing 3-substituted thiophenes.

8. *Thiophene alcohols, aldehydes, ketones, and acids*

2-Thenyl alcohol (**55**) is a colourless liquid, b.p. 207°C. In the presence of acids it polymerizes, like furfuryl alcohol, forming resins. It can be obtained by several general methods, including the treatment of 2-thienylmagnesium iodide with formaldehyde. 2-Thenyl chloride is obtained from the alcohol with thionyl chloride in the usual way, and in contrast to furfuryl chloride, gives a Grignard

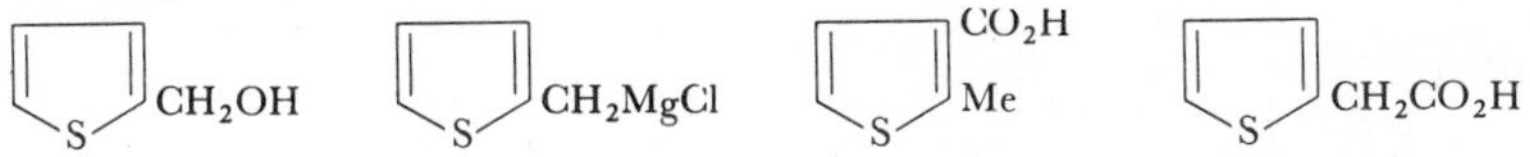

CH_2OH CH_2MgCl CO_2H Me CH_2CO_2H

55 **56** **57** **58**

derivative. Treatment of this (**56**) with water gives the expected 2-methylthiophene, but carbonation yields a mixture of **57** and **58** in a 1 : 2 ratio;[163] 3-thenylmagnesium bromide on carbonation similarly yields a 2 : 3 mixture of 3-thienylacetic acid and 3-methylthiophene-2-carboxylic acid. Benzylmagnesium chloride also behaves anomalously in some circumstances, for instance with formaldehyde, 2-methylbenzyl alcohol is formed.

Thiophene-2-carbaldehyde, b.p. 198°C, resembles benzaldehyde greatly in smell and chemical behaviour. It also undergoes typical Cannizzaro and benzoin reactions. 2,2′-Thenoin (**59**) can be oxidized to 2,2′-thenil (**60**), which with

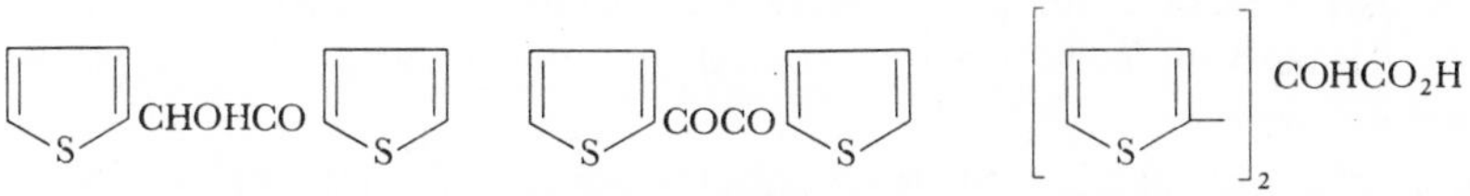

CHOHCO COCO $COHCO_2H$

59 **60** **61**

alkali yields 2,2′-thenilic acid (**61**). Thiophene-2-carbaldehyde is best obtained from thiophene with dimethylformamide (cf. pp. 99 and 132) and phosphorus oxychloride.[164]

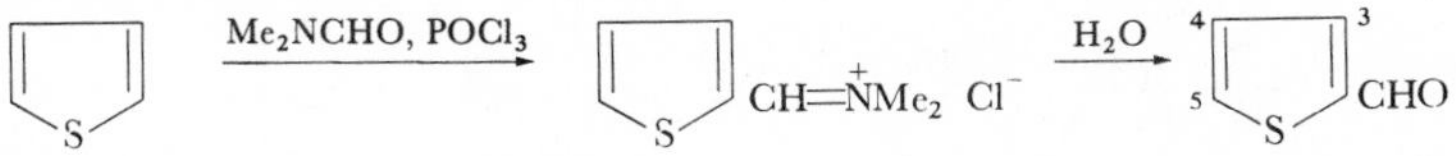

$Me_2NCHO, POCl_3$ $CH{=}\overset{+}{N}Me_2\ Cl^-$ H_2O CHO

Nitration of the aldehyde diacetate in acetic anhydride, and bromination of the aldehyde in chloroform, gave 5-substituted derivatives, while bromination of the aldehyde in the presence of aluminium chloride,[165] and nitration in sulphuric acid, caused 4-substitution. Complex formation or protonation must be the cause of electrophilic attack occurring at the 4-position.

2-Acetylthiophene, or 2-thienyl methyl ketone, b.p. 214°C, is very similar to acetophenone in colour, odour, and chemical properties. On nitration it yields a mixture of the 5- and 4-nitro derivatives, the former being present in greater proportion. 2-Acetylthiophene is best prepared in the laboratory from thiophene, acetic anhydride, and 85% orthophosphoric acid.

Thiophene-2-carboxylic acid, m.p. 129–130°C, greatly resembles benzoic acid in most of its properties, although it is a stronger acid (pK_a 3·53). It is, however, nitrated much more easily, and the product consists mostly of the 5-nitro derivative. This shows that the directing effect of the carboxyl group

CO_2H → O_2N CO_2H + O_2N CO_2H + NO_2

minor products

does not overcome that of the ring. Thiophenecarboxylic acids can be decarboxylated by boiling with copper in quinoline, by copper bronze,[165] or by displacement of the carboxyl by a mercuriacetate group and subsequent removal of the mercury.

Cl Cl CO_2H —$Hg(OAc)_2$ in 50% HOAc aq.→ Cl Cl HgOAc —NaCl, HCl steam distil→ Cl Cl

80% overall yield

A useful method for the conversion of an aromatic carboxyl to a methyl group has been successfully employed in the thiophene series.[166]

MeO_2C CO_2H —$SiHCl_3$→ The acid anhydride? —$SiHCl_3$, R_3N→ MeO_2C CH_2SiCl_3 —base→ MeO_2C Me

E. Synthetic methods[167]

(1) Thiophene can be prepared in the laboratory by distilling sodium succinate with phosphorus trisulphide. This method was first used by Victor

$Na^+ \ ^-O_2C-CH_2-CH_2-CO_2^- \ Na^+$ —P_2S_3→ thiophene ←P_2S_3— $CHO-CH_2-CH_2-CHO$

Meyer. By replacing the sodium succinate by salts of keto acids, such as laevulinic acid, alkyl- and arylthiophenes may be obtained. A variation of this

$$\underset{MeCO}{CH_2}-\underset{CO_2H}{CH_2} \xrightarrow{P_2S_3} \text{2-methylthiophene}$$

synthesis is to heat a 1,4-dialdehyde or diketone with phosphorus trisulphide. This synthesis is similar to those used in the furan and pyrrole series but is limited by the availability of the intermediates.

(2) Thiophene is available commercially from the continuous cyclization of butane, butene, or butadiene with sulphur. The contact time is about 2 seconds at 566°C, and the redistilled product has a purity of 99%. It contains traces of carbon disulphide and benzene. This method can be applied to other

$$MeCH_2-CH_2Me \xrightarrow{S} \text{thiophene} + \text{etc.}$$

hydrocarbons, for instance to isoprene (2-methylbutadiene), which yields 3-methylthiophene.

(3) A very general synthesis is that of Hinsberg. X and Y can be hydrogen, alkyl, aryl, hydroxyl, methoxyl, or carboxyl. It is notable that oxalic acid undergoes the reaction. A mechanism for the synthesis, which is consistent both with ^{18}O tracer studies and the formation of a half-ester as the final product, has been put forward.[168]

X Y, O O + $EtO_2C\bar{C}H$—SCH_2CO_2Et (from EtO_2CCH_2—SCH_2CO_2Et, −H⁺) ⟶ intermediate (H, X, Y, C, O, O, EtO_2C, $SCH_2C(=O)OEt$) $\xrightarrow{-EtO^-}$ (H, X, Y, C, O, O, EtO_2C, $SCH_2C=O$) $\xrightarrow{-H^+}$ X, Y, EtO_2C, S, O, $CH_2CO_2^-$ $\xrightarrow[H^+]{-H_2O}$ X, Y, EtO_2C, S, CO_2H

(4) A synthesis, of general application, starts with vinyl ketones or the related β-dialkylaminoketones.[169] A variation using ethoxymethylene ketones (EtOCH:CHCOR) gives an intermediate (**62**, R = EtO) at a higher stage of oxidation which only needs acid treatment to give the thiophene.

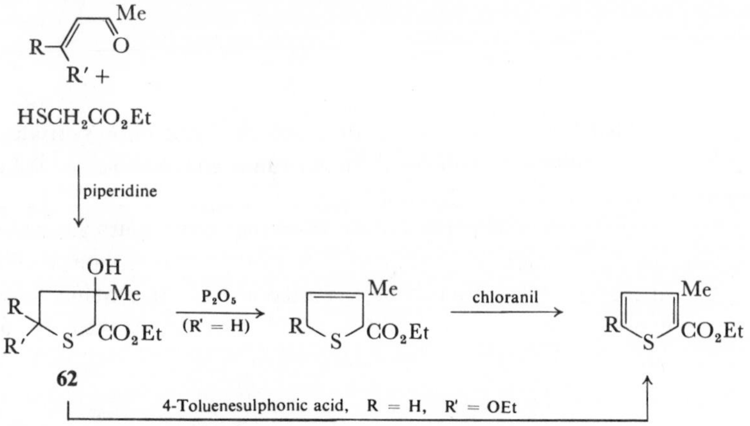

R′ + $HSCH_2CO_2Et$ —piperidine→ **62** —P_2O_5 (R′ = H)→ chloranil→

4-Toluenesulphonic acid, R = H, R′ = OEt

(5) Diacetylenes combine with hydrogen sulphide under weakly basic conditions to give thiophenes. The substituents R and R′ can be hydrogen, alkyl aryl, carboxyl, or substituted ethinyl. This synthesis seems to be limited only by the availability of the diacetylene and may be the route by which thiophenes are formed in plants.[170]

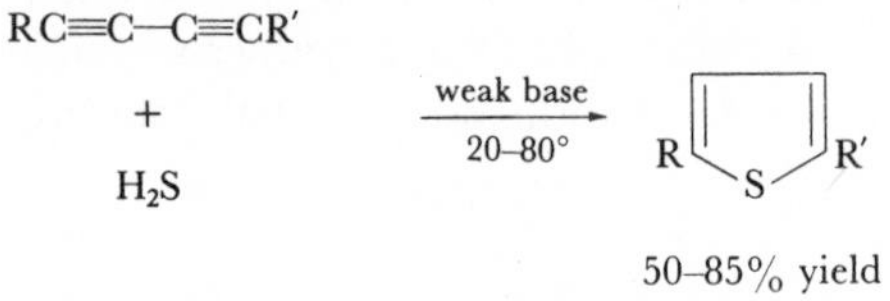

RC≡C—C≡CR′ + H_2S —weak base, 20–80°→

50–85% yield

(6) Acetylene combines with sulphur, yielding thiophene, but better results are obtained with dimethyl acetylenedicarboxylate.

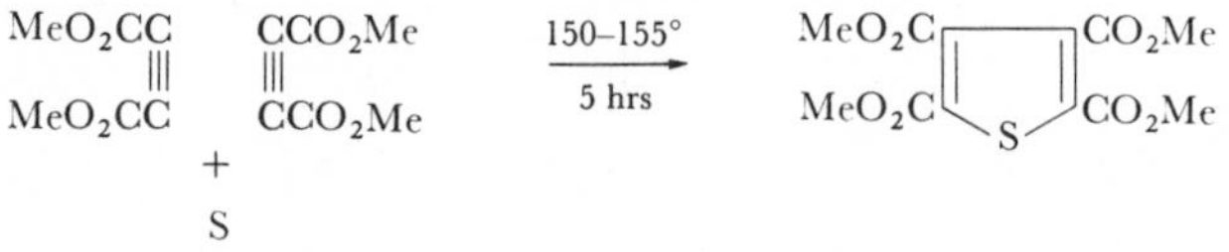

$MeO_2CC{\equiv}CCO_2Me$ + S —150–155°, 5 hrs→

(7) A rather unusual synthesis of 2,4-diphenylthiophene has been reported.

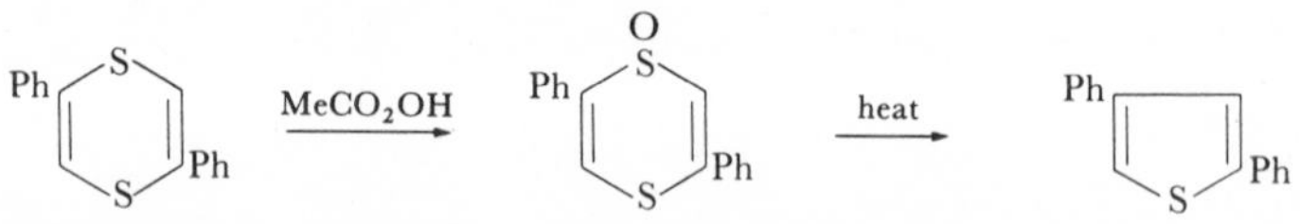

—$MeCO_2OH$→ —heat→

F. Natural occurrence and compounds of special interest

There is no evidence that thiophene or tetrahydrothiophene occurs as such in nature, although they are constituents of coal tar and shale oil. All the possible methylthiophenes have also been isolated from the same sources. α,α-Terthienyl

(63) has been isolated from the flowers of the Indian marigold. It is not impossible that it could have been formed from a polyacetylene. Another thiophene **(64)**,

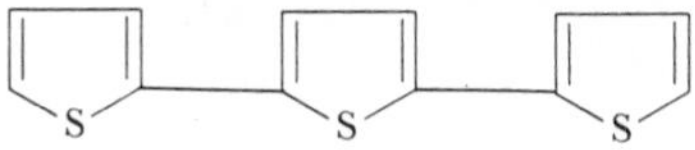

63

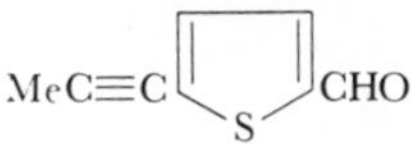

MeC≡C ... CHO

64

isolated from *Daedalea juniperina* cultures, is interesting in that it contains an acetylene link.

Biotin **(67)**, a tetrahydrothiophene, is the most important naturally occuring thiophene derivative so far recognized. It is a growth factor for yeast, and by using this property as an assay method Kögl isolated a very small quantity of the compound. In 1936 Kögl reported finding a richer source of the material in egg yolk, and from 250 kg of dried duck egg yolks he isolated 1·1 mg of biotin methyl ester. Coenzyme R, a well recognized growth factor for several species of

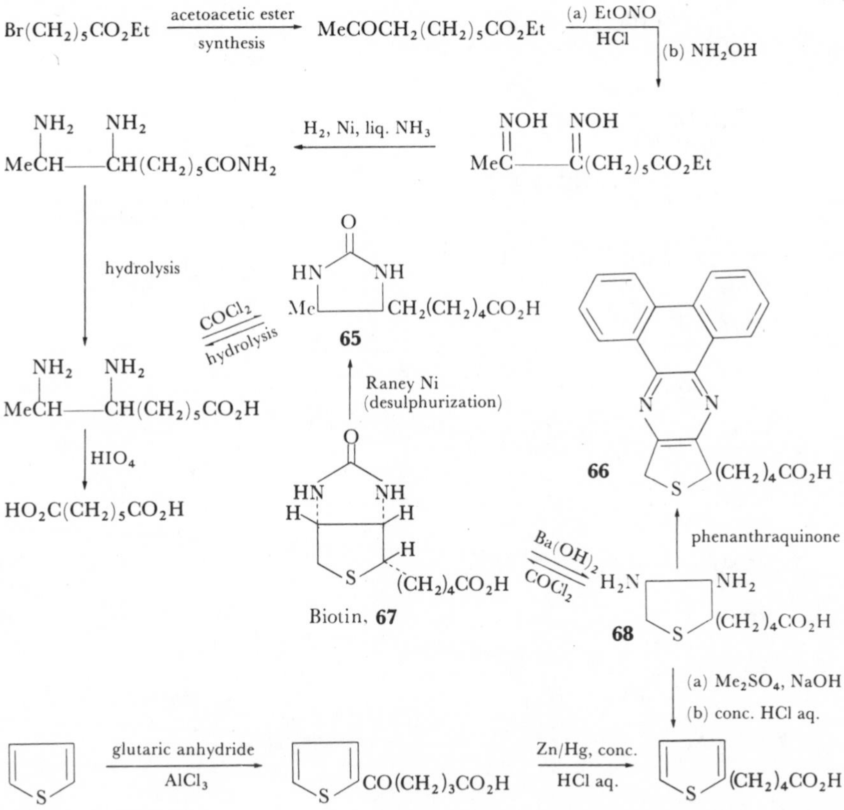

$Br(CH_2)_5CO_2Et$ —acetoacetic ester synthesis→ $MeCOCH_2(CH_2)_5CO_2Et$ —(a) EtONO, HCl; (b) NH_2OH→ $MeC(=NOH)C(=NOH)(CH_2)_5CO_2Et$

—H_2, Ni, liq. NH_3→ $MeCH(NH_2)CH(NH_2)(CH_2)_5CONH_2$ —hydrolysis→ $MeCH(NH_2)CH(NH_2)(CH_2)_5CO_2H$ —HIO_4→ $HO_2C(CH_2)_5CO_2H$

$COCl_2$ / hydrolysis

HN, NH, O, Me, $CH_2(CH_2)_4CO_2H$

65

Raney Ni (desulphurization)

HN, NH, H, H, H, S, $(CH_2)_4CO_2H$

Biotin, **67**

$Ba(OH)_2$ / $COCl_2$

H_2N, NH_2, S, $(CH_2)_4CO_2H$

68

phenanthraquinone

N, N, S, $(CH_2)_4CO_2H$

66

(a) Me_2SO_4, NaOH
(b) conc. HCl aq.

Thiophene —glutaric anhydride, $AlCl_3$→ S, $CO(CH_2)_3CO_2H$ —Zn/Hg, conc. HCl aq.→ S, $(CH_2)_4CO_2H$

69

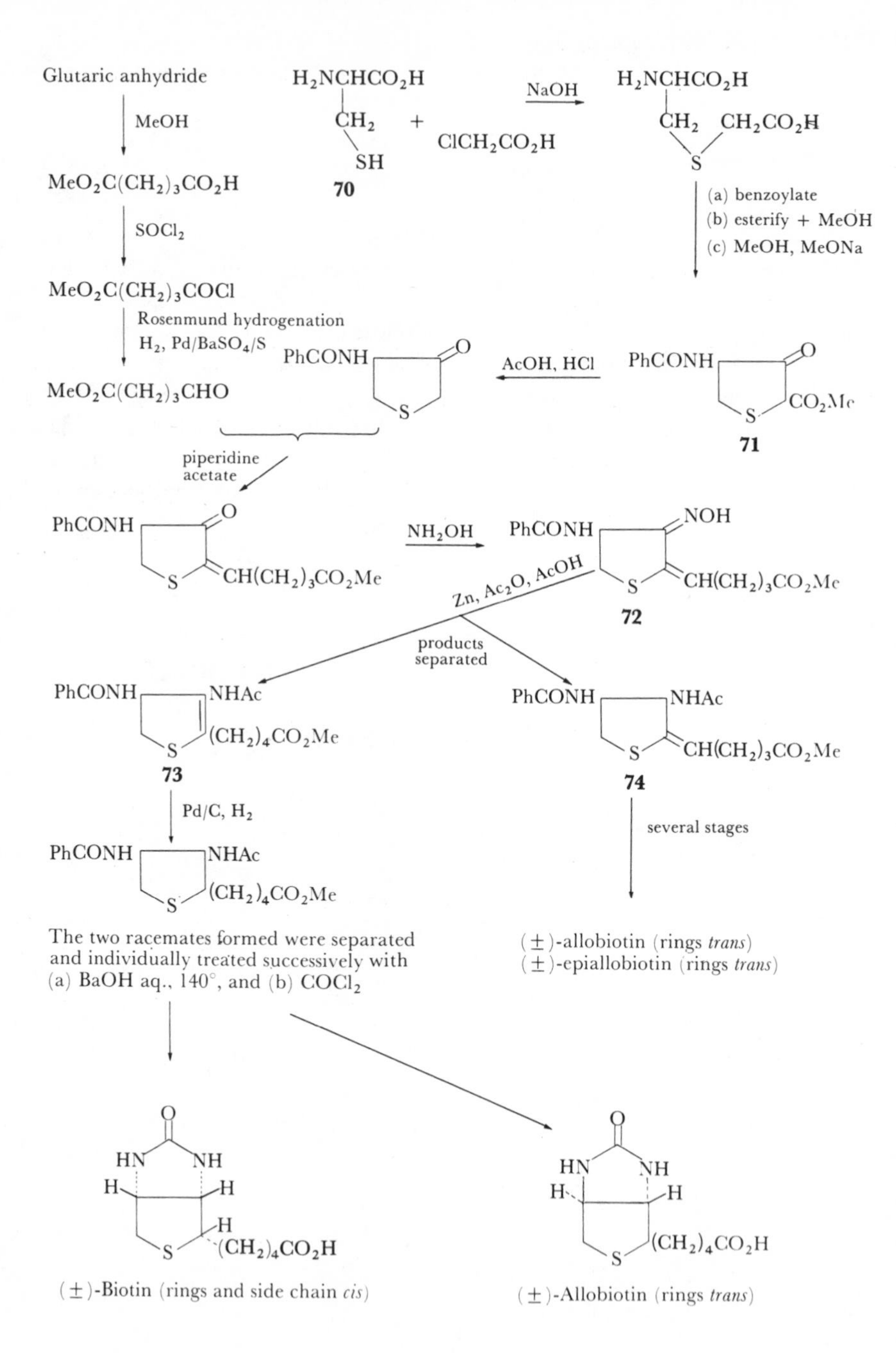

Glutaric anhydride
MeOH
$MeO_2C(CH_2)_3CO_2H$
$SOCl_2$
$MeO_2C(CH_2)_3COCl$
Rosenmund hydrogenation
H_2, Pd/$BaSO_4$/S
$MeO_2C(CH_2)_3CHO$
H_2NCHCO_2H
CH_2
SH
70
$ClCH_2CO_2H$
NaOH
H_2NCHCO_2H
CH_2 CH_2CO_2H
S
(a) benzoylate
(b) esterify + MeOH
(c) MeOH, MeONa
PhCONH
O
S
AcOH, HCl
PhCONH
O
CO_2Me
S
71
piperidine acetate
PhCONH
O
S
$CH(CH_2)_3CO_2Me$
NH_2OH
PhCONH
NOH
S
$CH(CH_2)_3CO_2Me$
72
Zn, Ac_2O, AcOH
products separated
PhCONH
NHAc
S
$(CH_2)_4CO_2Me$
73
Pd/C, H_2
PhCONH
NHAc
S
$(CH_2)_4CO_2Me$
PhCONH
NHAc
S
$CH(CH_2)_3CO_2Me$
74
several stages
The two racemates formed were separated and individually treated successively with (a) BaOH aq., 140°, and (b) $COCl_2$
(±)-allobiotin (rings *trans*)
(±)-epiallobiotin (rings *trans*)
O
HN
NH
H
H
H
S
$(CH_2)_4CO_2H$
(±)-Biotin (rings and side chain *cis*)
O
HN
NH
H
H
S
$(CH_2)_4CO_2H$
(±)-Allobiotin (rings *trans*)

legume nodule bacteria (*Rhizobium* sp.), was shown to be identical with biotin in 1940, and a little later that year biotin was also shown to be identical with a growth factor required by man and present in liver. From this source du Vigneaud isolated crystalline biotin.

The degradations which proved the structure of biotin are shown on the chart (p. 171). It is of interest that a quinoxaline (**66**) is obtained from the diamine (**68**) and phenanthraquinone instead of the expected dihydroquinoxaline. The Raney nickel desulphurization of biotin (**67**) to desthiobiotin (**65**) is a characteristic reaction of thiophenes (p. 160), and the modified Hofmann exhaustive methylation of **68**, which gives **69**, are worthy of note.

Biotin has three asymmetric carbon atoms, and so eight optically active forms (four racemates) of the molecule can exist. All these have been obtained by synthesis. Only one of the optically active forms, that which is identical to natural (+)-biotin is biologically active. In confirmation of deductions from chemical studies, X-ray examinations[171] have shown that the ring junction is *cis*, that the orientation of the side chain in respect to it is also *cis*, and that the absolute configuration is as shown.

Three quite similar syntheses of biotin have been effected, but only that of the Merck group is outlined here. L-Cysteine (**70**) was the starting material, but racemization took place at the cyclization stage, which gave **71**. The next point of interest is the reduction of the oxime (**72**), as two isomeric products were formed, separated, and their structures elucidated. Both of these compounds were treated similarly, although the sequence is only shown on the chart for the one which gave biotin. On hydrogenation both gave two racemates, all of which were subjected to the reactions indicated. (±)-Biotin (rings *cis*) and (±)-allobiotin (rings *trans*) were obtained from **73**, while **74** gave (±)-allobiotin and (±)-epiallobiotin (rings *trans*). The fourth racemate (±)-epibiotin, which is isomeric with (±)-biotin at position 2, has been obtained by another synthesis. Biotin acts as a coenzyme in carboxylation and decarboxylation *in vivo*, the 1′-carboxylic acid (**75**) being the intermediate compound. Biotin is also involved in the synthesis of long-chain fatty acids.

Biocytin, ε-*N*-biotyllysine (**76**), has been obtained from yeast and by synthesis. It is as active as biotin in promoting the growth of some organisms.

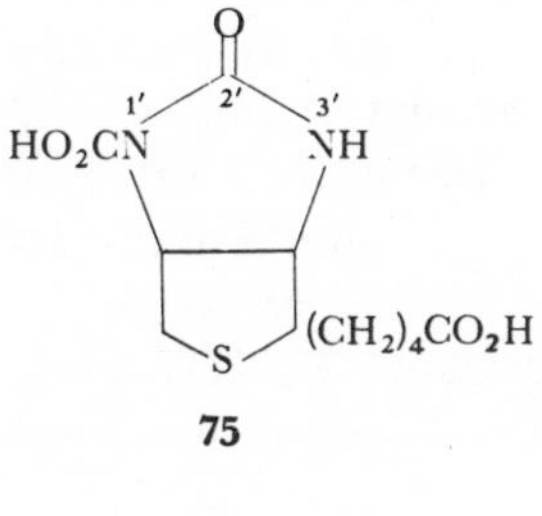

O; 1′, 2′, 3′; HO_2CN; NH; S; $(CH_2)_4CO_2H$

75

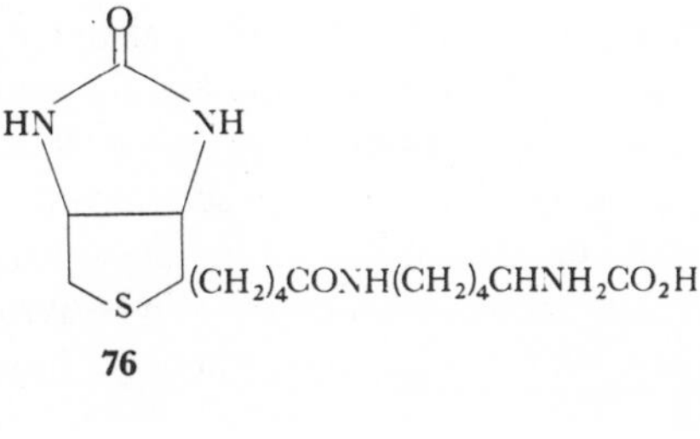

O; HN; NH; S; $(CH_2)_4CONH(CH_2)_4CHNH_2CO_2H$

76

4. A SUMMARY COMPARISON OF THE AROMATIC PROPERTIES OF PYRROLE, FURAN, AND THIOPHENE

The meaning of the term 'aromatic' is in practice determined by common usage and is extremely difficult to define precisely in a generally acceptable way. Sir Robert Robinson's statement, that aromatic compounds show reduced unsaturation and a tendency to retain the type, gives an indication as to how aromatic compounds may be expected to *behave* in chemical reactions, and it is from this standpoint the present discussion starts.

In the Diels–Alder reaction furan generally behaves as a normal diene, pyrrole does not behave as a normal diene but undergoes addition reactions of the Michael type, and thiophene does not react. Exceptions are of course known, with the substituted heterocycles and with extremely reactive dieneophiles. With aqueous mineral acid the same order of reactivity is shown, as furan readily polymerizes, pyrrole forms recognizable salts which then polymerize, while thiophene is inert.

Furan forms 2,5-addition compounds readily, and compounds of this sort are often isolable intermediates in reactions leading, eventually, to substituted furans. It is even likely that these addition compounds are formed in most reactions which introduce substituents at position 2 or 5 of the furan ring. Intermediate 2,5-addition compounds have so far not been detected in the electrophilic substitution of pyrrole, and it is known that the nitration of thiophene is essentially similar to that of benzene.

The reactivities of furan, pyrrole, and thiophene, and a number of derivatives, to several electrophilic reagents have been measured recently.[172] Direct comparisons of reactivity towards trifluoroacetic anhydride in dichloromethane, which is reactive enough to introduce the trifluoroacetyl group without the necessity of a Lewis acid catalyst, and summary results in a series of bromination experiments which give approximate values for comparative brominations, are given in Table 3. It is also known that *N*-methylation approximately doubles the reactivity of the pyrrole ring to trifluoroacetylation. All the molecules are very much more easily attacked by electrophiles than benzene. It is also clear that the effect of existing substituents on electrophilic attack is similar to that in the benzene series.[172,173] Comparisons of reactivity to radical attack have not been made, but relative rates of nucleophilic displacement of bromine by piperidine from 2-bromo-5-nitrofuran and 2-bromo-5-nitrothiophene, and 1-bromo-4-nitrobenzene are in the ratio $8{\cdot}9 \times 10^4 : 4{\cdot}7 \times 10^2 : 1$, respectively. 2,5-Dinitro-1-methylpyrrole does react readily with piperidine to give 5-nitro-2-(1-pyrrolidinyl) pyrrole, but no rate data are available.[174]

The photolysis[175] of ethyl diazoacetate in furan and thiophene gives, presumably via the formation of ethoxycarbonylcarbene (**1**), the adducts **2** and

TABLE 3. Comparison of reactivity to electrophilic attack at position 2 for pyrrole, furan and thiophene

Heterocycle	Relative rates for Trifluoroacetylation	Relative rates for Bromination	Partial rate factors for position 2 (benzene 1)
Pyrrole	$5{\cdot}3 \times 10^7$	$5{\cdot}9 \times 10^8$	3×10^{18}
Furan	$1{\cdot}4 \times 10^2$	$1{\cdot}2 \times 10^2$	6×10^{11}
Thiophene	1	1	5×10^9

3. Treatment with methanolic hydrogen chloride, in the case of **2**, opens the rings to give the aldehyde **5**, while the thiophene derivative **3** is aromatised to the 3-substituted thiophene **6**. The product from the copper-catalysed ther-

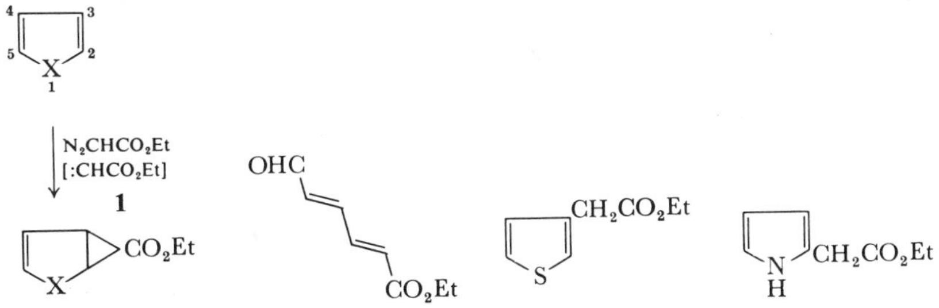

N_2CHCO_2Et [$:CHCO_2Et$]

1

CO_2Et

2, X = O
3, X = S
4, X = NCO_2Me

OHC ... CO_2Et **5**

CH_2CO_2Et (S) **6**

CH_2CO_2Et (NH) **7**

molysis of ethyl diazoacetate in 2-methylfuran is the methyl ketone corresponding to the aldehyde **5**,[176] while in pyrrole it is the 2-substituted pyrrole **7**.[177] A bridged intermediate is likely also in these cases, for the adduct **4** can be obtained from methyl 1-pyrrolecarboxylate.[178] Pyrrole with sodium hydrogen sulphite is stated[179] to give the pyrrolidine **8**; comparable reactions are not known with furan and thiophene. Presumably protonation at position 3 (p. 94) is followed by addition of the counterion.

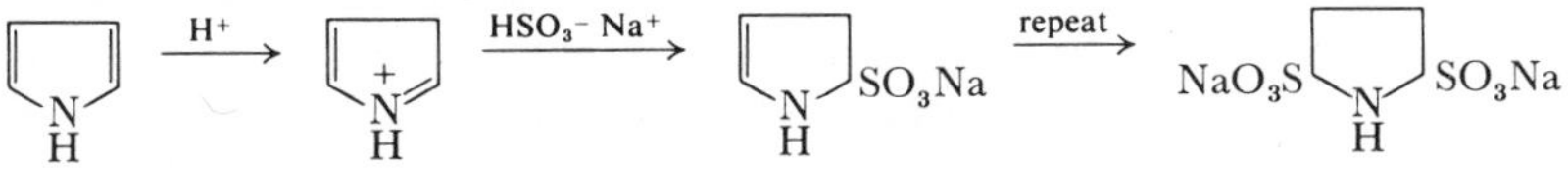

H^+ → HSO_3^- Na^+ → SO_3Na → repeat → NaO_3S ... SO_3Na

8

Only thiophene can accept an additional substituent on the heteroatom, leading to isolable products bearing a formal positive charge on the sulphur (see p. 155). They are very reactive compounds, there is little evidence to suggest that they are 'aromatic,' and the hybridization of the sulphur orbitals has not been ascertained. From the point of view of *behaviour* in chemical reactions, as defined by *retention of type*, one can conclude that aromaticity increases in the order furan, pyrrole, and thiophene. Although not long ago this gradation was related more or less satisfactorily to the bond lengths and resonance energies of these heterocycles, the new and much more precise data discussed below make these older comparisons of trends unsatisfactory.

In the last few years the shapes of furan, pyrrole, and thiophene have been determined with great accuracy, with the surprising result that these compounds are very much more similar than might have been imagined from a consideration of earlier structural information. A noticeable gradation in bond lengths is not shown. The percentage contraction in length of the bond between the heteroatom and the adjacent carbon atom, as compared with the corresponding length in saturated aliphatic analogues, is remarkably constant (Table 4).

TABLE 4. Comparison of bond lengths in furan, pyrrole, and thiophene

c
b
X a

	Bond length (Å)				
	in ring				
Heteroatom	*c*	*b*	*a*	X–C saturated	% contraction of X–C in heterocycle
N	1·429	1·371	1·383	1·47	5·9
O	1·431	1·361	1·362	1·43	4·8
S	1·423	1·370	1·714	1·82	5·8

Although the resonance energy for pyrrole is somewhat uncertain it must be of the same order of those of furan and thiophene, which numerically approach that of benzene (150·5 kJ (36·0 kcal)/mole). If the resonance energies of these compounds are divided by the number of atoms constituting the rings, and contributing to the π-electron sextet, one finds the surprising result that, *per atom*, thiophene is at least as stabilized as benzene and the other compounds only somewhat less (Table 5); again there is no clear-cut gradation as was once thought to be the case.

Two of the six π-electrons required for the aromaticity[180] of these heterocycles must come from the heteroatom. According to resonance theory this means that the charged structures **9a**–**9d**, as well as the uncharged structure **9**, are contributors to the resonance hybrid.

TABLE 5. Resonance energies and dipole moments of furan, pyrrole, and thiophene

	Resonance energy in kJ/mole (kcal/mole)						
	A	B	λ_{max} (nm)	ϵ_{max} (nm)	Dipole moment (D)	Dipole moment of tetrahydro derivative (D)	Mesomeric moment (D)
Furan	99–107 (23·7–25·6)	108·5 (26)	200	10000	0·71	1·68	0·97
Pyrrole	88–100 (21–24)	103 (24·7)	210	5100	1·80	1·57	2·73
Thiophene	121–129·5 (29–31)	142 (34)	235	4300	0·52	1·87	1·35

A, from heats of combustion; B, from molecular orbital calculations.

9 **9a** **9b** **9c** **9d**

X = O, NH, or S

Attempts have been made to calculate and to estimate, from observed data, the magnitude of the ring current induced in these heterocycles by an applied magnetic field. The only certain conclusion, as yet, is that the magnitude of the induced current for all these heterocycles is less than that for benzene.[181]

The (vector) difference between the dipole moment of a heterocycle and its tetrahydro derivative, the mesomeric moment (Table 5), gives an indication of the extent of electron release from the heteroatom to the ring. For the furan and thiophene, and their tetrahydro derivatives, the heteroatoms are at the *negative* end of the dipoles,[182,183] in contrast to what was often supposed earlier. In the case of pyrrole the nitrogen atom is at the positive end, while for tetrahydropyrrole it is at the negative end, and assuming its direction is along the axis of the lone pair and that the bonds round the nitrogen are tetrahedrally disposed, an estimate of the mesomeric moment can be made. The nitrogen atom is far more ready to give up its electrons than either oxygen or sulphur, and this is reflected in the relatively very high reactivity of the pyrrole ring to electrophilic attack. Charged structures **9a–9d** contribute much more greatly, therefore, to the resonance picture for pyrrole than for furan or thiophene. The localisation energies for electrophilic substitution are lowest at position 2 for all three heterocycles.

A comparison of the acid dissociation constants (Table 6) for the carboxylic acids from these three heterocycles is interesting. The 3-carboxylic acids are all stronger than the corresponding 2-carboxylic acids, and the order of decreasing acidities, for structurally comparable acids, is furan > thiophene > pyrrole. This

TABLE 6. Acid dissociation constants.[182]

Acid	pK_a
Pyrrole-2-carboxylic	4·45
Pyrrole-3-carboxylic	5·07
Furan-2-carboxylic	3·16
Furan-3-carboxylic	3·95
Thiophene-2-carboxylic	3·53
Thiophene-3-carboxylic	4·10
Benzoic	4·21

fits exactly the situation anticipated from the mesomeric moments of the heterocycles and also agrees with the supposition that the charged structures **9a** and **9b** are more important than **9c** and **9d** as there is less energy expended in charge separation.

The molar refractivities (M_D) for furan, pyrrole, and thiophene increase in this order, and this gives an order of increasing polarisability for the compounds (Table 7). It is reflected by the steady decrease in the ionisation potentials of

TABLE 7. Ionisation potentials and electronegativities

Compound	M_D[184]	Ionisation potentials[185] (eV)	Compound	Ionisation potentials (eV)	Atom	Electro-negativities[186]
Furan	18·43	8·99	H_2O	12·56	O	3·50
Pyrrole	21·7	8·40	NH_3	11·2	N	3·07
Thiophene	24·36	9·12	H_2S	10·42	S	2·44
					C	2·50

water, ammonia, and hydrogen sulphide, which is in line with the decreasing electronegativities of the hetero atoms of these compounds. However, the ionisation potentials of the heterocycles themselves, measured by electron impact, follows the order pyrrole < furan < thiophene. This confirms what is indicated by dipole moment and theoretical studies – that the contribution by the heteroatom to the aromatic six π-electron system is not a simple function of the polarisability and electronegativity of the heteroatom. This last order also reflects decreasing reactivity of the heterocycles to positively charged reagents, as might be anticipated.

Although general agreement on the precise meaning of the term 'aromatic character' and the best means of assessing it differ, it can be concluded without doubt that the 'aromatic character' of the series furan,[187] pyrrole, and thiophene[187] increases in this order. Furan has many properties of a rather stable enol ether or butadiene, pyrrole behaves as an enamine with a great tendency to retain its type, and thiophene possesses properties more akin to those of benzene.

5. HETEROCYCLIC ANALOGUES WITH OTHER HETEROATOMS

In the last few years a great deal of interest has arisen in novel heterocyclic systems with heteroatoms other than nitrogen, oxygen, or sulphur, and the whole area is being developed rapidly. Only a few examples are described here.

A. Selenium and Tellurium

Selenophene (**1**),[188] b.p. 110°C at 758 torr and m.p. –38°C, can be synthesised by many of the methods applicable to thiophene, but employing intermediates containing selenium instead of sulphur. Microwave spectra of the compound and deuteriated derivatives have shown that the molecule is planar and have given molecular dimensions. Selenophene resembles thiophene in many respects, undergoing electrophilic substitution at position 2 and giving the 2-lithium derivative with butyllithium. In acidic media the protons at position 2 are exchanged 5 x 10^4 faster than those at position 3, and about 10 times faster than those at position 2 of thiophene.[188]

MeCH=CHMe + Se —(580 °C)→ **1** [1·433 Å; 1·370 Å; 1·855 Å; 114·6°; 111·6°; 87·8°; Se] —((a) BuLi, (b) CO_2)→ selenophene-2-carboxylic acid (CO_2H at 2; ring numbered 1 (Se)–5)

Tellurophene (**2**) has been synthesised[189] from sodium telluride and butadiyne in methanol, oxygen and water being excluded. It has b.p. 90°C at 100 torr, m.p. ca. –36°C, and is easily decomposed by acids. Halogens add at

butadiyne + Na_2Te → **2** (Te, ring numbered 1–5) —(Ac_2O, $SnCl_4$)→ 2-acetyltellurophene ($COCH_3$)

position 1, electrophiles attack at position 2, and a 2-lithium derivative is obtained with butyllithium.

The 'aromaticities' of thiophene, selenophene, tellurophene, and furan decrease in this order, according to a number of criteria.[187] At the same time their reactivity to phosgene and dimethylformamide, in a modified Vilsmeier reaction (p. 99), which leads to the 2-aldehydes, goes up in the ratio 1 : 3·6 : 36·8 : 107, respectively.[190]

B. Phosphorus[191]

1-Methylphosphole (4) has been synthesised[192] as outlined. It was obtained 87% pure as a colourless liquid, b.p. 82–83°C at 317 torr (slight decomposition), containing residual pentane. Its basicity is very low and protonation causes decomposition, but nevertheless reaction with methyl iodide occurs. The product is a dimer,[193] but if ring substituents are present dimerisation does not occur and compounds such as **5** can be obtained (cf. 1,1-dimethylindolium and 1-methylthiophenium salts, pp. 195 and 155).

+ $MePCl_2$ ⟶ **3** (P⁺ Me Cl, Cl^-) $\xrightarrow{H_2O}$ (P⁺ Me O⁻) $\xrightarrow{Br_2}$ Br, Br (P⁺ Me O⁻) $\xrightarrow{SiHCl_3}$

Br, Br (P Me) $\xrightarrow[\text{pentane}]{\text{KO}t\text{-Bu}}$ **4** (P Me; positions 1–5) **5** (Me, Me, P⁺ Me_2, I^-)

The u.v. spectrum is similar to that of 1-methylpyrrole and quite unlike that of divinylphosphines. The proton and phosphorus resonances in the n.m.r. spectra have been examined, and show without doubt that there is considerable electron delocalisation in the ring.[193]

1,3,5-Triphenylphosphole (7) is formed[194] in 50–65% yield from 1,4-diphenylbutadiene and phenylphosphonous dichloride, presumably via an intermediate similar to **3**, which loses hydrogen chloride. It is bright yellow, has m.p. 187–189°C, and absorbs at much longer wavelengths in the ultraviolet than 1,2,5-triphenylpyrrole. It does not combine with maleic anhydride, under conditions which give a Diels–Alder adduct with the oxide (**6**), but with dimethyl acetylenedicarboxylate, addition and loss of phosphorus must occur, for the product is **8**. X-Ray studies[195] show that the bond lengths for **7** are close to those expected of a butadiene, but it cannot be concluded from this that there is little electron delocalisation in the phosphole ring, because the steric and

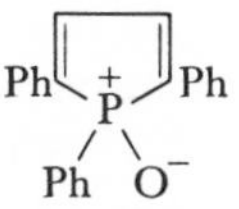

6

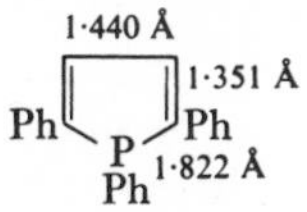

7

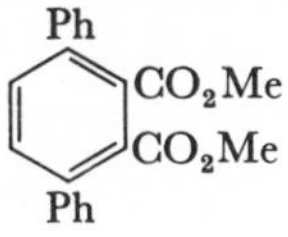

8

electronic properties of the three vicinal phenyl groups will have a large effect on the heterocyclic ring.

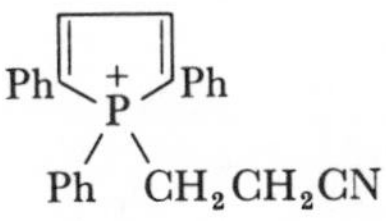

9

$Ph_3\overset{+}{P}CH_2CH_2CN$

10

There is an abnormally low inversion barrier for the phosphorus atom in some phospholes, suggesting significant delocalisation of the electrons of this atom in spite of indications to the contrary mentioned above. The positive charge of the cation **9** destabilises the molecule compared with the cation **10**, for reaction of the corresponding hexafluoroantimonates with sodium methoxide splits off the aliphatic side chain from only the former, giving **7**.[196]

Triphenylphosphine combines easily with both dimethyl acetylenedicarboxylate[197] and dicyanoacetylene[198] to give the 1,1,1-triphenylphospholes (e.g., **11**). This ester (**11**), the structure of which has been established from its nuclear magnetic resonance spectrum, isomerizes on standing to the diene **12**. The rearrangement is assisted by methanol.

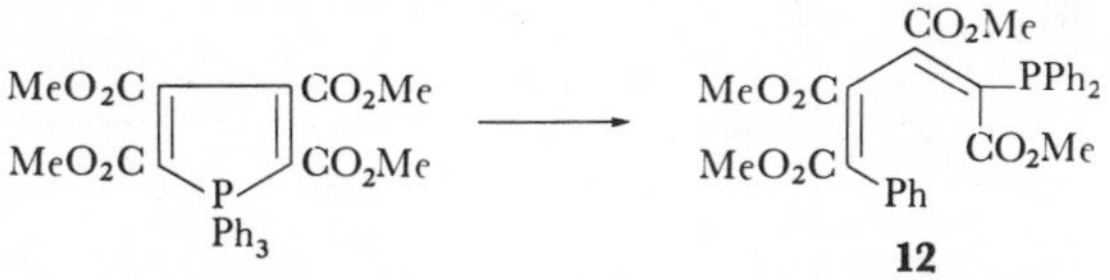

11

12

C. Silicon

[31Si]-1-Silacyclopenta-2,4-diene (**13**) is obtained from silicon atoms recoiling from the $^{31}P(n, p)^{31}Si$ nuclear transformation in gaseous phosphine and butadiene.[199] A few more complex derivatives have been prepared.[200]

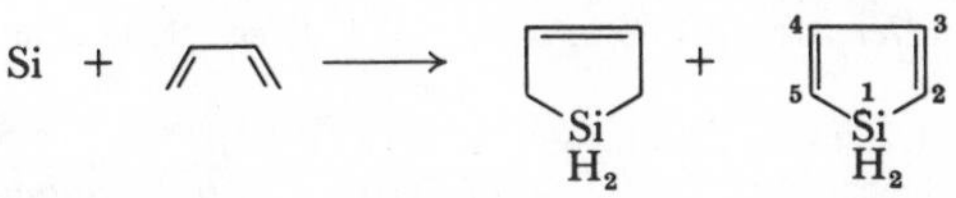

13

D. Boron

Simple boroles have not yet been obtained, but pentaphenylborole (**14**) has been synthesised[201] in two ways. In spite of the great steric shielding by the phenyl groups, which must be roughly perpendicular to the plane of the

Ph Ph Ph Ph $SnMe_2$ —$PhBCl_2$→ Ph Ph Ph Ph B Ph ←$PhBBr_2$— Ph Ph Ph Ph Li Li

14

H_2O_2, ^-OH → PhCOCHPhCHPhCOPh

O_2 → Ph Ph Ph Ph O

$PhC\equiv CPh$ → Ph Ph Ph B Ph Ph Ph Ph

heterocyclic ring, the compound is very reactive. The borole ring could be considered 'antiaromatic', as it possesses only four π-electrons.

E. Aluminium

The first derivative of aluminole to be synthesised, 1,2,3-triphenylbenzo-[*b*]aluminole (**15**), readily loses aluminium.[202]

PhC≡CPh —$AlPh_3$→ Ph Ph Al Ph **15**

H_2O → $PhCH{=}CPh_2$ + PhH

I_2 → I CPh ClPh + PhI

GENERAL BIBLIOGRAPHY

PYRROLE

A. Gossauer, *Die Chemie der Pyrrole, Organische Chemie in Einzeldarstellungen*, Vol. 15, Springer, Berlin, 1974.

H. D. Springall and I. T. Millar, *Sidgwick's Organic Chemistry of Nitrogen*, Oxford University Press, London, 1966.

A. H. Corwin, in *Heterocyclic Compounds*, Vol. I, ed. R. C. Elderfield, Wiley, New York, 1950.

E. Baltazzi, and L. I. Krimen, *Chem. Rev.*, **63**, 511 (1963).

R. A. Jones, 'Physicochemical Properties of Pyrroles,' *Adv. Heterocycl. Chem.*, **11**, 383 (1970).

G. P. Gardini, 'The Oxidation of Monocyclic Pyrroles,' *Adv. Heterocycl. Chem.*, **15**, 67 (1973).

FURAN

A. P. Dunlop and F. N. Peters, *The Furans*, Am. Chem. Soc. Monograph No. 119, Reinhold, New York, 1953.

F. M. Dean, *Naturally Occurring Oxygen Ring Compounds*, Butterworths, London, 1963.

R. C. Elderfield and T. N. Dodd in *Heterocyclic Compounds,* Vol. I, ed. R. C. Elderfield, Wiley, New York, 1950.

D. G. Jones and A. W. C. Taylor, *Q. Rev.*, **4**, 195 (1950).

A. P. Dunlop, *Furfural from Agricultural Sources*, Royal Institute of Chemistry Lectures, No. 4, 1956.

R. D. Topsom, *Rev. Pure Appl. Chem.*, **14**, 127 (1964).

P. Bosshard and C. H. Engster, 'Development of the Chemistry of Furans, 1952–1963,' *Adv. Heterocycl. Chem.*, **7**, 377 (1966).

THIOPHENE

H. D. Hartough, *Thiophene and its Derivatives*, Interscience, New York, 1952.

H. D. Hartough and S. L. Meisel, *Compounds with Condensed Thiophene Rings*, Interscience, New York, 1954.

F. F. Blicke in *Heterocyclic Compounds*, Vol. I, ed. R. C. Elderfield, Wiley, New York, 1950.

S. Gronowitz, *Adv. Heterocycl. Chem.*, **1**, 1 (1963).

REFERENCES

1. B. Bak, D. Christensen, L. Hansen and J. Rastrup-Andersen, *J. Chem. Phys.*, **24**, 720 (1956); cf. C. W. N. Cumper, *Trans. Farad. Soc.*, **54**, 1266 (1958).
2. T. J. Thomas, R. W. Roth, and J. G. Verkade, *J. Amer. Chem. Soc.*, **94**, 8854 (1972); G. Marino, *J. Heterocycl. Chem.*, **9**, 817 (1972), and papers cited.
3. D. T. Clark, *Tetrahedron,* **24**, 4689 (1968), and papers therein quoted.
4. G. da Re and R. Scarpati, *Rend. Accad. Sci. Fis. Mat. Soc. Naz. Sci. Napoli,* **31**, 88 (1964).
5. R. C. Lord and F. A. Miller, *J. Chem. Phys.*, **10**, 328 (1942).
6. P. Hikasai and D. McLeod, *J. Amer. Chem. Soc.*, **95**, 27 (1973).
7. D. L. Currell and L. Zechmeister, *J. Amer. Chem. Soc.*, **80**, 205 (1958).
8. E. B. Whipple, Y. Chiang, and R. L. Hinman, *J. Amer. Chem. Soc.*, **85**, 26 (1963).
9. R. A. Jones, *Adv. Heterocycl. Chem.*, **11**, 838 (1970).
10. G. P. Bean, *Chem. Comm.*, 421 (1971).
11. S. F. Dyke, *Enamines*, Cambridge Chemistry Texts, Cambridge, 1973.
12. G. F. Smith, *Adv. Heterocycl. Chem.*, **2**, 287 (1963).
13. P. de Mayo and S. T. Reid, *Chem. Ind.* (*Lond.*), 1576 (1962); see also D. L. Lightner and L. K. Low, *Chem. Comm.*, 625 (1972).

13a. M. G. Barlow, R. N. Hazeldine, and R. Hubbard, *Chem. Comm.*, 301 (1969).

14. R. M. Acheson, *Adv. Heterocycl. Chem.*, **1**, 125 (1963),
15. G. Wittig and B. Reichel, *Chem. Ber.*, **96**, 2851 (1963), and earlier papers; E. Wolthuis, D. V. Jagt, S. Mels, and A. De Boer, *J. Org. Chem.*, **30**, 190 (1965).
16. R. C. Bansal, A. W. McCullock, and A. G. McInnes, *Can. J. Chem.*, **48**, 1472 (1970).

16a. C. Rivas et al., *Acta Cient. Venez.*, **22**, 145 (1971).

17. E. P. Papadopoulos and N. F. Haidar, *Tetrahedron Lett.*, 1723 (1968).

18. C. F. Hobbs, C. K. McMillin, E. P. Papadopoulos, and C. A. Vander Werf, *J. Amer. Chem. Soc.*, **84**, 43 (1962); H. Heaney and S. U. Ley, *J.C.S. Perkin I*, 499 (1973).
19. G. P. Bean, *J. Org. Chem.*, **32**, 228 (1967).
20. H. J. Anderson and L. C. Hopkins, *Can. J. Chem.*, **42**, 1279 (1964).
21. P. S. Skell and G. P. Bean, *J. Amer. Chem. Soc.*, **84**, 4660 (1962); see also C. E. Loader and H. J. Anderson, *Can. J. Chem.*, **49**, 1064 (1971).
22. G. S. Reddy, *Chem. Ind. (Lond.)*, 1426 (1965).
23. A. P. Terent'ev and A. U. Dombrowskii, *J. Gen. Chem. (U.S.S.R.)*, **21**, 281 (1951).
24. A. R. Cooksey, K. J. Morgan, and D. P. Morrey, *Tetrahedron*, **26**, 5101 (1970) and earlier papers.
25. R. M. Silverstein, E. E. Ryskiewicz, C. Willard, and R. C. Koehler, *J. Org. Chem.*, **20**, 668 (1955); see also G. F. Smith, *J. Chem. Soc.*, 3842 (1954).
26. R. A. Jones and R. L. Laslett, *Australian J. Chem.*, **17**, 1056 (1964).
27. E. P. Papadopoulos, *J. Org. Chem.*, **38**, 667 (1973); **37**, 351 (1972).
28. J. W. Harbuck and H. Rapoport, *J. Org. Chem.*, **37**, 3618 (1972).
29. V. Bocchi, L. Chierichi, and G. P. Gardini, *Tetrahedron*, **26**, 4073 (1970) and earlier papers.
30. M. Bellas, D. Bryce-Smith, and A. Gilbert, *Chem. Comm.*, 263 (1967).
31. R. C. Jones and C. W. Rees, *J. Chem. Soc. (C)*, 2249 (1969).
32. G. L. Closs and G. M. Schwartz, *J. Org. Chem.*, **26**, 2609 (1961).
33. C. B. Hudson and A. V. Robertson, *Tetrahedron Lett.*, 4015 (1967).
34. A. J. Birch, R. W. Richards, and K. J. S. Stafford, *Australian J. Chem.*, **22**, 1321 (1969).
35. M. J. T. Robinson, private communication.
36. J. M. Patterson, L. T. Burka and M. R. Boyd, *J. Org. Chem.*, **33**, 4033 (1968).
37. J. M. Patterson and L. T. Burka, *Tetrahedron Lett.*, 2215 (1969).
38. A. C. Day and J. A. Barltrop, *Chem. Comm.*, 177 (1975).
39. D. J. Chadwick, *Chem. Comm.*, 790 (1974).
40. C. O. Bender and R. Bonnett, *J. Chem. Soc. (C)*, 2526 (1968).
41. R. Bonnett, I. A. D. Gale, and G. F. Stevenson, *J. Chem. Soc.*, 1518 (1965).
42. N. L. Weinberg and E. A. Brown, *J. Org. Chem.*, **31**, 4054 (1966).
43. C. A. Grob and H. Utzinger, *Helv. Chim. Acta*, **37**, 1256 (1954).
44. J. M. Tedder and B. Webster, *J. Chem. Soc.*, 3270 (1960).
45. J. M. Tedder and B. Webster, *J. Chem. Soc.*, 1638 (1962).
46. C. A. Grob and P. Ankli, *Helv. Chim. Acta*, **32**, 2010, 2023 (1949).
47. R. Kuhn and G. Osswald, *Chem. Ber.*, **89**, 1423 (1956).
48. R. S. Atkinson and E. Bullock, *Can. J. Chem.*, **41**, 625 (1963).
49. R. J. Motekactis, D. H. Heindert, and A. E. Martell, *J. Org. Chem.*, **35**, 2504 (1970).
50. R. L. Hinman and S. Theodoropulos, *J. Org. Chem.*, **28**, 3052 (1963).
51. H. J. Anderson and H. Nagy, *Can. J. Chem.*, **50**, 1961 (1972), and earlier papers.
52. J. K. Groves, H. J. Anderson, and H. Nagy, *Can. J. Chem.*, **49**, 2427 (1971).
53. Y. Badar, W. J. S. Lockley, T. P. Toube, B. C. L. Weedon, and L. R. G. Valadon, *J.C.S. Perkin I*, 1416 (1973).
54. H. J. Anderson and S. H. Lee, *Can. J. Chem.*, **43**, 409 (1965).
55. P. Hodge and R. W. Rickards, *J. Chem. Soc.*, 2543 (1963), 459 (1965).
56. M. K. A. Khan, K. J. Morgan, and D. P. Morrey, *Tetrahedron*, **22**, 2095 (1966).
57. M. K. A. Khan and J. K. Morgan, *Tetrahedron*, **21**, 2197 (1965).
58. J. B. Hendrickson, R. Rees, and J. F. Templeton, *J. Amer. Chem. Soc.*, **86**, 107 (1964).

59. H. S. Broadbent, W. S. Burnham, and R. M. Sheeley, *J. Heterocycl. Chem.*, **5**, 757 (1968).
60. P. Rosenmund and K. Grübel, *Angew. Chem.*, **80**, 702 (1968).
61. W. G. Terry, A. H. Jackson, G. W. Kenner, and G. Kornis, *J. Chem. Soc.*, 4389 (1965).
62. S. Hauptmann, M. Weissenfels, M. Scholz, E. M. Werner, H. J. Köhler, and J. Weisflog, *Tetrahedron Lett.*, 1317 (1968).
63. K. E. Schulte, J. Reisch, and H. Walker, *Chem. Ber.*, **98**, 98 (1965).
64. B. Bak, D. Christensen, W. B. Dixon, L. Hansen-Nygaard, J. R. Andersen, and M. Schottländer, *J. Mol. Spectrosc.*, **9**, 124 (1962).
65. S. Murahashi, T. Shimamura, and I. Moritani, *Chem. Comm.*, 931 (1974).
66. M. Gorman, R. L. Hamill, R. P. Elander, and J. Mabe, *Biochem. Biophys. Res. Comm.*, **31**, 294 (1968).
67. K. Tanaka, K. Kariyone, and S. Umio, *Chem. Pharm. Bull. (Tokyo)*, **17**, 622 (1969).
68. Y. Morinoto, M. Hashimoto, and K. Hattori, *Tetrahedron Lett.*, 209 (1968).
69. E. E. Garcia, *Chem. Comm.*, 78 (1973).
70. P. E. Sorvet, *J. Med. Chem.*, **15**, 97 (1971).
71. B. Witkop, *Experientia*, **27**, 1121 (1971).
71a. A. W. Johnson, *Chem. Soc. Reviews*, **4**, 1 (1975).
72. L. E. Webb and E. B. Fleischer, *J. Amer. Chem. Soc.*, **87**, 667 (1965).
73. J. H. Mathewson and A. H. Corwin, *J. Amer. Chem. Soc.*, **83**, 135 (1961).
74. L. P. Vernon and G. R. Seely (eds.), *The Chlorophylls*, Academic Press, New York, 1960.
75. H. Brockmann, *Angew. Chem. Int. Ed.*, 7, 221 (1968).
76. R. B. Woodward, W. A. Ayer, J. M. Beaton, F. Bickelhaupt, R. Bonnett, P. Buchschacher, G. L. Closs, H. Dutler, J. Hannah, F. P. Hanck, S. Itô, A. Langemann, E. Le Goff, W. Leimgruber, W. Lwowski, J. Sauer, Z. Valentia, and H. Volz, *J. Amer. Chem. Soc.*, **82**, 3800 (1960).
77. M. T. Cox, T. T. Howarth, A. H. Jackson, and G. W. Kerner, *J.C.S. Perkin I*, 2478 (1973).
78. L. Smith, *Vitamin B_{12}*, 3rd ed., Methuen, London 1965; J. M. Wood and D. G. Brown, *Structure and Bonding*, **11**, 47 (1972); H. A. O. Hill, in *Inorganic Biochemistry*, Vol. 2, ed. G. L. Eichhorn, Elsvier, Amsterdam 1973, p. 1067.
79. D. Crowfoot Hodgkin, J. Kamper, J. Lindsey, M. MacKay, J. Pickworth, J. H. Robertson, C. B. Shoemaker, J. G. White, R. J. Prosen and K. N. Trueblood, *Proc. R. Soc. (Lond.) Ser. A*, **242**, 228 (1957).
80. A. Eschenmoser, *Q. Rev.*, **24**, 366 (1970); R. B. Woodward, *J. Pure Appl. Chem.*, **17**, 579 (1968); **25**, 283 (1971); **33**, 145 (1973).
81. R. M. Acheson, *Acc. Chem. Res.*, **4**, 177 (1971).
82. G. B. Guthrie, D. W. Scott, W. N. Hubbard, C. Katz, J. P. McCullough, M. E. Cross, K. D. Williamson, and G. Waddington, *J. Amer. Chem. Soc.*, **74**, 4662 (1952).
83. D. S. Sappenfeld and M. Kreevoy, *Tetrahedron*, **19**, Suppl. 2, 1957 (1963).
84. B. Bak, D. Christensen, W. B. Dixon, L. Hansen-Nygaard, J. R. Andersen, and M. Schottländer, *J. Mol. Spectrosc.*, **9**, 124 (1962).
85. J. Y. Beach, *J. Chem. Phys.*, **9**, 54 (1941).
86. D. T. Clark, *Tetrahedron*, **24**, 3285 (1968), and references cited.
87. J. L. Goldfarb, J. B. Volkenstein, and L. I. Belankij, *Angew. Chem. Int. Ed.*, 7, 519 (1968).
88. F. A. C. Anet, *Tetrahedron Lett.*, 1219 (1962).

88a. A. W. McCullough, D. G. Smith, and A. G. MacInnes, *Can. J. Chem.,* **51**, 4125 (1973).
89. C. H. Eugster and P. Bosshard, *Helv. Chim. Acta,* **46**, 815 (1963).
90. J. Berridge, D. Bryce-Smith, and A. Gilbert, *Chem. Comm.*, 964 (1974).
91. J. Leitich, *Tetrahedron Lett.*, 1937 (1967); R. Evanega and E. B. Whipple, *ibid.*, 2163 (1967).
92. G. O. Schenck, W. Hartmann, S. Mannsfeld, W. Metzner, and C. H. Krauch, *Chem. Ber.,* **95**, 1642 (1962).
93. N. Elming, *Adv. Org. Chem.,* **2**, 67 (1960); J. Froborg, G. Magnusson, and S. Thóren, *J. Org. Chem.,* **40**, 122 (1975).
94. G. Oláh, S. Kuhnand, and A. Mlinkó, *J. Chem. Soc.,* 4257 (1956).
95. A. P. Terent'ev, L. I. Belen'kii, and L. A. Yanovskaya, through *Chem. Abstr.,* **49**, 12327d (1955).
96. P. A. Finan and G. A. Fothergill, *J. Chem. Soc.*, 2723 (1963).
97. V. J. Traynelis, J. J. Miskel, and J. R. Sowa, *J. Org. Chem.,* **22**, 1269 (1957).
98. L. Benati, N. La Barba, M. Tiecco, and A. Tundo, *J. Chem. Soc. (B)*, 1253 (1969).
99. G. Ahlgren, *J. Org. Chem.,* **38**, 1369 (1973).
100. R. J. Bushby, *J.C.S. Perkin I*, 271 (1974).
101. P. J. Pearce, D. H. Richards, and N. F. Scilly, *J.C.S. Perkin I*, 1655 (1972).
102. R. B. Bates, L. M. Kroposki, and D. E. Potler, *J. Org. Chem.,* **37**, 560 (1972).
103. W. J. Gensler, J. E. Stouffer, and R. G. McInnes, *J. Org. Chem.,* **32**, 200 (1967).
104. P. Bosshard and C. H. Eugster, *Adv. Heterocycl. Chem.,* 7, 377 (1966).
105. J. W. Cornforth, *J. Chem. Soc.*, 1310 (1958).
106. D. M. Burness, *Org. Synth.,* **39**, 46 (1959).
107. D. J. Chadwick, J. Chambers, H. E. Hargraves, G. D. Meakins, and R. L. Snowden, *J.C.S. Perkin I*, 2327 (1973).
108. H. M. White, H. O. Colomb, Jr., and P. S. Bailey, *J. Org. Chem.,* **30**, 481 (1965).
109. E. E. van Tamelen and T. H. Whitesides, *J. Amer. Chem. Soc.,* **90**, 3894 (1968).
110. S. Gronowitz and G. Sörlin, *Acta Chem. Scand.,* **15**, 1419 (1961).
111. S. Gronowitz and G. Sörlin, *Ark. Kemi,* **19**, 515 (1963).
112. R. Hunt and S. T. Reid, *J.C.S. Perkin I*, 2527 (1972).
113. J. E. Garst and G. L. Schmir, *J. Org. Chem.,* **39**, 2920 (1974).
114. G. W. Wheland, *Advanced Organic Chemistry*, Wiley, New York, 1949, p. 646.
115. G. U. Boyd and K. Heatherington, *J.C.S. Perkin I*, 2523 (1973).
116. A. Suzuki, N. Miyaura, and M. Itoh, *Tetrahedron,* **27**, 2775 (1971).
117. S. Divall, M. C. Chun, and M. M. Joullié, *Tetrahedron Lett.,* 777 (1970).
118. W. S. Trahanovsky and M. Park, *J. Amer. Chem. Soc.,* **95**, 5412 (1973).
119. Z. N. Nazarova, through *Chem. Abstr.,* **49**, 6214f (1953).
120. D. J. Chadwick, J. Chambers, G. D. Meakins, and R. L. Snowden, *J.C.S. Perkin I*, 1766 (1973).
121. H. Gilman, N. O. Calloway, and R. R. Burtner, *J. Amer. Chem. Soc.,* **57**, 906 (1935).
122. L. Mavoungou-Gomès, *Bull. Soc. Chim. Fr.,* 1753 (1967).
123. L. Birkoffer and G. Daum, *Chem. Ber.,* **95**, 183 (1962).
124. J. B. Cloke and F. J. Pilgrim, *J. Amer. Chem. Soc.,* **61**, 2667 (1939).
125. T. J. Mabry, *J. Org. Chem.,* **28**, 1699 (1963).
126. D. J. Chadwick, J. Chambers, G. D. Meakins, and R. L. Snowden, *J.C.S. Perkin I*, 201 (1973).
127. N. Clauson-Kaas and P. Nedenskov, *Acta Chem. Scand.,* 9, 27 (1953).
128. D. L. Dave, I. D. Entwistle, R. A. W. Johnstone, *J.C.S. Perkin I*, 1130 (1973).
129. J. W. Betty, P. D. Howes, and C. J. M. Stirling, *J.C.S. Perkin I*, 65 (1973).

130. D. W. Henry and R. M. Silverstein, *J. Org. Chem.*, **31**, 2391 (1966).
131. P. K. Bose, *J. Indian Chem. Soc.*, **37**, 653 (1960).
132. B. Bak, D. Christensen, L. Hansen-Nygaard, and J. Rastrup-Andersen, *J. Mol. Spectrosc.*, **7**, 58 (1961).
133. S. C. Abrahams, *Q. Rev.* (London), **10**, 407 (1956).
134. D. T. Clark, *Tetrahedron*, **24**, 2663 (1968); see also M. H. Palmer, R. H. Findlay and A. J. Gaskell, *J.C.S. Perkin II*, 420 (1974).
135. D. T. Clark, *Tetrahedron*, **24**, 2567 (1968).
136. G. C. Brumlik, A. I. Kosak, and R. Pitcher, *J. Amer. Chem. Soc.*, **86**, 5360 (1964).
137. R. M. Acheson and D. R. Harrison, *J. Chem. Soc.* (*C*), 1764 (1970).
138. D. H. Johnson, *J. Chem. Soc.* (*C*), 2275 (1967).
139. A. Winkler and I. J. Gosselck, *Tetrahedron Lett.*, 2433 (1970).
140. M. Procházka, *Collect. Czech. Chem. Comm.*, **30**, 1158 (1965).
141. W. J. Bailey and E. W. Cummins, *J. Amer. Chem. Soc.*, **76**, 1932, 1940 (1954).
142. W. L. Mock, *J. Amer. Chem. Soc.*, **88**, 2857 (1966); also S. D. McGregor and D. M. Lemal, *ibid.*, **88**, 2858 (1966).
143. R. Helder and H. Wynberg, *Tetrahedron Lett.*, 605 (1972).
144. D. D. Callander, P. L. Coe, and J. C. Tatlow, *Chem. Comm.*, 143 (1966).
145. K. Hafner and W. Kaiser, *Tetrahedron Lett.*, 2185 (1964).
146. T. Nakano, C. Rivas, C. Perez, and K. Tori, *J.C.S. Perkin I*, 2322 (1973).
147. A. R. Butler and A. P. Sanderson, *J.C.S. Perkin II*, 1214 (1974).
148. B. Östman, *Ark. Kemi*, **19**, 499 (1963).
149. C. M. Camaggi, R. Leardini, M. Tiecco, and A. Tundo, *J. Chem. Soc.* (*B*), 1683 (1970).
150. C. M. Camaggi, R. Leardini, and A. Tundo, *J.C.S. Perkin I*, 271 (1974); A. P. Manzara and P. Kovacie, *J. Org. Chem.*, **39**, 504 (1974).
151. S. F. Birch and D. T. McAllan, *J. Chem. Soc.*, 2556 (1951).
152. H. Hauptmann and W. F. Walter, *Chem. Rev.*, **62**, 347 (1962); G. R. Pettit and E. E. van Tamelen, *Org. React.*, **12**, 356 (1962).
153. A. I. Meyers, *Heterocycles in Organic Synthesis*, Wiley-Interscience, New York, 1974.
154. P. Linda, A. Lucarelli, G. Marino, and G. Savelli, *J.C.S. Perkin II*, 1610 (1974).
155. S. T. Gore, R. K. Mackie, and J. M. Tedder, *Chem. Comm.*, 272 (1974).
156. H. Wynberg and H. van Driel, *Chem. Comm.*, 203 (1966).
157. M. G. Reinecke, H. W. Adickes, and C. Pyun, *J. Org. Chem.*, **36**, 2690 (1971).
158. G. Guanti, C. Dell'Erba, G. Leandri, and S. Thea, *J.C.S. Perkin I*, 2357 (1974).
159. D. L. Eck and G. W. Stacy, *J. Heterocycl. Chem.*, **6**, 147 (1969).
160. M. G. Reineke and R. H. Walter, *Chem. Comm.*, 1044 (1974).
161. A. Hörnfeldt and G. Gronowitz, *Ark. Kemi*, **21**, 239 (1963).
162. M. C. Ford and D. Mackay, *J. Chem. Soc.*, 4985 (1956).
163. E. Campaigne and C. J. Collins, *J. Heterocycl. Chem.*, **2**, 136 (1968).
164. E. Campaigne and W. L. Archer, *J. Amer. Chem. Soc.*, **75**, 989 (1953).
165. D. J. Chadwick, J. Chambers, G. D. Meakins, and R. C. Snowden, *J.C.S. Perkin I*, 1766, 2079 (1972).
166. R. A. Benkesler and D. F. Ehler, *J. Org. Chem.*, **38**, 3660 (1973).
167. D. E. Wolf and K. Folkers, *Org. React.*, **6**, 410 (1951).
168. H. Wynberg and H. J. Kooreman, *J. Amer. Chem. Soc.*, **87**, 1739 (1965).
169. B. D. Tilak, H. S. Desai, and S. S. Gupte, *Tetrahedron Lett.*, 1609 (1964); private communication from Professor B. D. Tilak.
170. F. Bohlmann, M. Wotschokowsky, U. Hinz, and W. Lucas, *Chem. Ber.*, **99**, 984 (1966).

171. W. Traub, *Nature,* **178**, 649 (1956); J. Trotter and J. A. Hamilton, *Biochemistry J.*, 713 (1966).
172. G. Marino, *Adv. Heterocycl. Chem.,* **13**, 235 (1971).
173. R. N. McDonald and J. M. Richmond, *Chem. Comm.*, 333 (1974).
174. G. Doddi, P. Mencarelli, and F. Stegel, *Chem. Comm.*, 273 (1975).
175. G. O. Schenck and R. Steinmetz, *Annalen,* **668**, 19 (1963).
176. C. W. Rees and C. E. Smithen, *Adv. Heterocycl. Chem.,* **3**, 63 (1964).
177. P. S. Skell and G. P. Bean, *J. Amer. Chem. Soc.,* **84**, 4655 (1962).
178. F. W. Fowler, *Chem. Comm.*, 1359 (1969).
179. A. Treibs and R. Zimmer-Galler, *Annalen,* **664**, 140 (1963).
180. J. W. Armit and R. Robinson, *J. Chem. Soc.,* **127**, 1604 (1925).
181. D. W. Davies, *Chem. Comm.*, 258 (1965).
182. G. Marino, *J. Heterocycl. Chem.,* **9**, 817 (1972).
183. T. J. Barton, R. W. Roth, and J. G. Verkade, *J. Amer. Chem. Soc.,* **94**, 8854 (1972).
184. C. P. Smyth and W. S. Walls, *J. Amer. Chem. Soc.,* **54**, 3230 (1932); H. Kofod, L. E. Sutton, and J. Jackson, *J. Chem. Soc.,* 1467 (1952); R. Keswani and H. Freiser, *J. Amer. Chem. Soc.,* **71**, 218 (1949).
185. P. Linda, G. Marino, and S. Pignataro, *Ris. Sci.,* **39**, 666 (1969).
186. A. L. Allred and E. G. Rochow, *J. Inorg. Nucl. Chem.,* **5**, 264, 269 (1957).
187. F. Fringuelli, G. Marino, A. Taticchi, and G. Grandolini, *J.C.S. Perkin II,* 332 (1974).
188. N. N. Magdesieva, *Adv. Heterocycl. Chem.,* **12**, 1 (1970).
189. F. Fringuelli and A. Taticchi, *J.C.S. Perkin I*, 199 (1972).
190. G. Marino, S. Clementi, F. Fringuelli, P. Linda, G. Savelli, and A. Taticchi, *J.C.S. Perkin II*, 1097 (1973).
191. F. G. Mann, *Heterocyclic Derivatives of Phosphorus, Arsenic, Antimony and Bismuth*, 2nd ed., Wiley-Interscience, New York, 1970.
192. L. D. Quin, J. G. Bryson, and C. G. Moreland, *J. Amer. Chem. Soc.,* **91**, 3308 (1969).
193. L. D. Quin, S. G. Borleske, and J. F. Engel, *J. Org. Chem.,* **38**, 1858, 1954 (1973).
194. I. G. M. Campbell, R. C. Cookson, M. B. Hocking, and A. N. Hughes, *J. Chem. Soc.,* 2184 (1965).
195. W. P. Ozbirn, R. A. Jacobson, and J. C. Clardy, *Chem. Comm.*, 1062 (1971).
196. W. B. Farnham and K. Mislow, *Chem. Comm.*, 469 (1972).
197. J. B. Hendrickson, R. E. Spenger, and J. J. Sims, *Tetrahedron,* **19**, 707 (1963).
198. G. S. Reddy and C. D. Weiss, *J. Org. Chem.,* **28**, 1822 (1963).
199. P. P. Gaspar, R-J. Hwang, and W. C. Eckelman, *Chem. Comm.*, 242 (1974).
200. T. J. Barton, A. J. Nelson, and J. Clardy, *J. Org. Chem.,* **37**, 895 (1972), and references cited.
201. J. J. Fisch, N. K. Hoto, and S. Kojima, *J. Amer. Chem. Soc.,* **91**, 4575 (1962).
202. J. Eisch and W. C. Kaska, *J. Amer. Chem. Soc.,* **88**, 2976 (1966) and earlier papers.

CHAPTER IV

FUSED RING SYSTEMS INVOLVING PYRROLE, FURAN, AND THIOPHENE RINGS

A benzene ring may be fused onto the carbon atoms of each of these three heterocyclic compounds in two different positions. Six possible structures

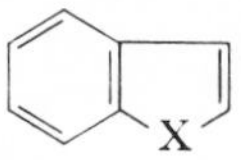

Indole X = NH
Benzofuran X = O
Benzo[*b*]thiophene X = S

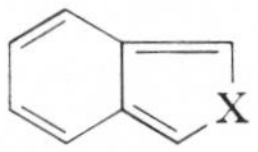

Isoindole
Isobenzofuran
Benzo[*c*]thiophene

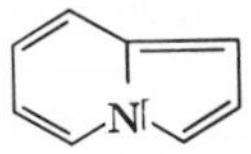

Indolizine

emerge, and derivatives of all are known. In the case of pyrrole, where the heteroatom is trivalent, yet another structure can be written, that of indolizine. Of these seven ring systems that of indole requires most attention.

1. INDOLE

A. Introduction

Indole was first prepared in 1866 by heating oxindole with zinc dust, and in recent years it has become commercially available. The indole ring system is found in many naturally occurring compounds of great chemical and biochemical

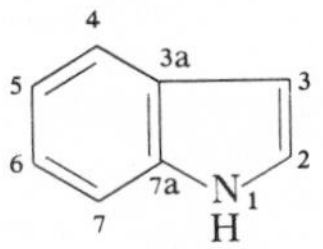

4 3a 3 5 6 7a N 1 2 7 H

1

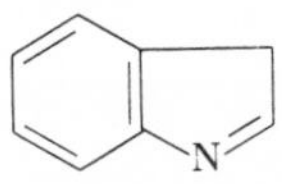

3*H*-Indolenine

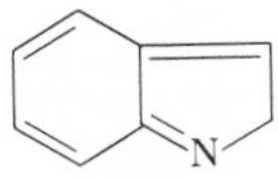

2*H*-Indolenine

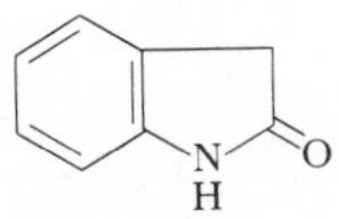

Oxindole

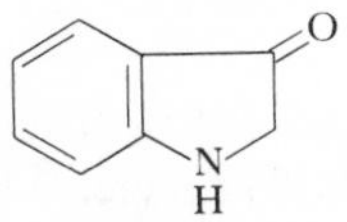

Indoxyl

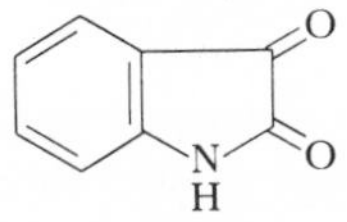

Isatin

interest. Three specific simple examples are the essential amino acid tryptophan, the dyestuff indigo, and the plant growth hormone indole-3-acetic acid.

The positions of the indole ring are usually numbered as shown (**1**), although occasionally α- and β- are used to indicate the 2- and 3-positions, respectively, as in the case of pyrrole. Derivatives of both indolenines have been prepared, and the common names for some important indoles are shown.

B. Physical properties and structure

Indole is a colourless crystalline solid, m.p. 52°C, which boils with some decomposition at 254°C. It is remarkably volatile and is easily soluble in most organic solvents. It may be crystallized from water and is sparingly soluble in the cold. When pure it has a pleasant, but very persistent, smell, and it is used as a perfume base.

The formula of indole, as usually written (**1**), was proposed by Baeyer in 1869 and was based on the syntheses outlined.

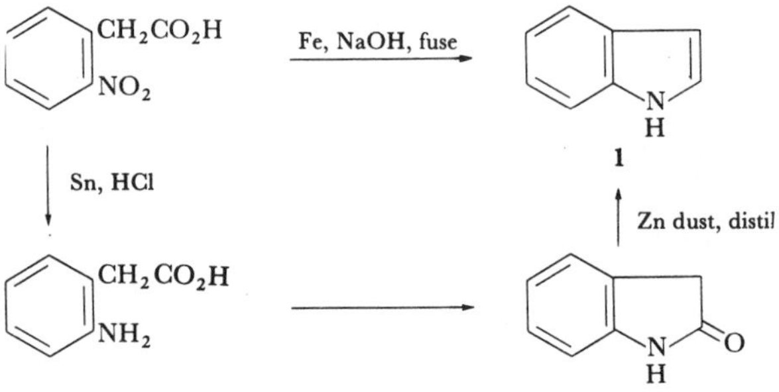

CH_2CO_2H, NO_2 —(Fe, NaOH, fuse)→ N H **1**

↓ Sn, HCl ↑ Zn dust, distil

CH_2CO_2H, NH_2 ——→ N H, O

The molecular parameters for indole-3-acetic acid have been determined[1] by X-ray methods and may be compared with those of pyrrole (p. 91). The π-electron densities suggest that, in conformity with experiment, the 3-position will be the most susceptible to electrophilic attack.

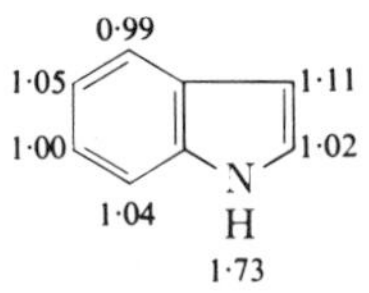

0·99, 1·05, 1·00, 1·04, 1·11, 1·02, N H, 1·73

π-Electron densities

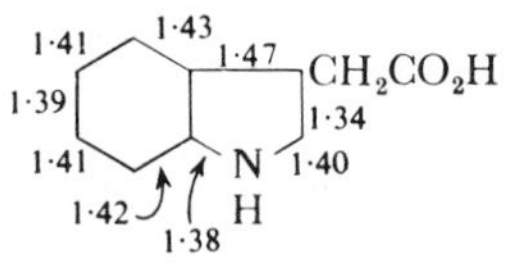

1·41, 1·43, 1·47, CH_2CO_2H, 1·39, 1·34, 1·41, N, 1·40, 1·42, H, 1·38

Indole-3-acetic acid, bond lengths in Å

The resonance energy of indole, calculated from its heat of combustion, is 190–204 kJ (47–49 kcal)/mole. The increase in resonance energy (ca. 96 kJ (ca. 24 kcal)/mole) over that of pyrrole (96–113 kJ (23–27 kcal)/mole) is almost identical to the difference between the resonance energies of naphthalene

(254·5 kJ (61 kcal)/mole) and benzene (150·5 kJ (36 kcal)/mole) and shows the similar effect of adding the additional ring. Of the resonance structures which can be written for indole (compare pyrrole), the uncharged structure **1** is probably the most important, as it involves full conjugation in both rings; the benzene ring can, of course, be written in its other Kekulé form **(1a)**. Of the ionic structures possible only those with a negative charge in the nitrogen-containing ring will be

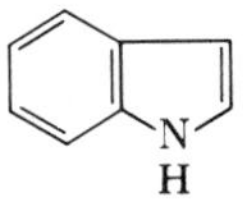

1 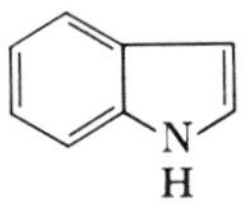 **1a** 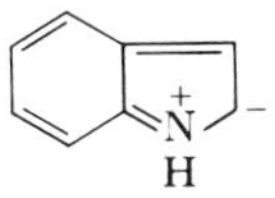 **1b**

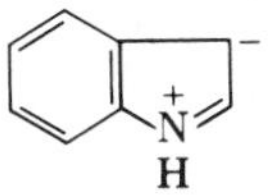

1c 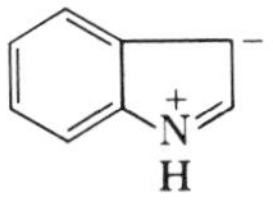 **1d** 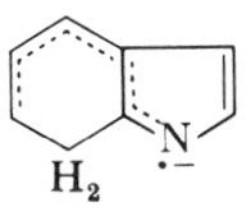 **2**

considered. Only one structure **(1b)**, involving an *ortho*-quinonoid benzene ring, can provide a negative charge at position 2. In the case where the negative charge is at position 3 the benzene ring can have all the uncharged resonance forms possible for benzene itself; two are shown (**1c** and **1d**). It therefore appears, in agreement with the molecular-orbital calculation, that one effect of fusing the benzene and pyrrole rings is to alter the position of greatest reactivity to electrophilic reagents of the pyrrole ring from position 2 to position 3. Various assumptions are of course involved in this, but the chemical reactions of indole agree with this conclusion. The non-benzenoid part of the indole system, taken in isolation, can be regarded as an enamine, and on this basis the high reactivity to electrophiles at position 3 is readily understandable. The status of the indolenine tautomer of indole corresponds exactly to the pyrrole analogy. Although both tautomers of 2-ethoxyindole are known (p. 201) there is no evidence suggesting that indole itself tautomerizes to a detectable extent, and the formation of indolenines (p. 195; cf. pyrrolenines, p. 90) requires only the addition of the electrophilic reagent at the very reactive 2- or 3-position of the indole ring, followed in some cases by the subsequent expulsion of a proton from position 1, and not the prior tautomerization of the original indole to an indolenine. The e.s.r. spectrum of the radical anion of indole, formed[2] by sodium in an argon matrix at 4°K, shows that it has the surprising structure **2**, where resonance in the carbocyclic ring has been interrupted.

C. Chemical properties

The chemical properties of indole are, in general, quite similar to those of pyrrole if allowance is made for the marked stabilising influence of the benzene ring and its effect in directing electrophilic reagents to position 3. Indole, like pyrrole, gives a red colour with a pine splint dipped in hydrochloric acid and a positive Ehrlich reaction (see p. 194).

1. *The direct oxidation of indole*

The nitrogen-containing ring of many substituted indoles can be opened by the action of peroxy acids, which give mixtures of products,[3] ozone, or sodium periodate in methanol[4] and is also opened enzymically *in vivo*. In the case of indole itself (**1**), some oxidation to 2-formamidobenzaldehyde (**3**) takes place, but the major product is indigo (p. 210).

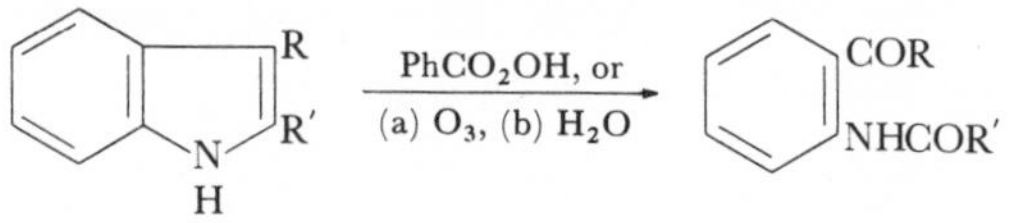

$PhCO_2OH$, or (a) O_3, (b) H_2O

1, R=R′=H **3**, R=R′=H

2. *Addition reactions*

Indole, like pyrrole, is a very weak base but accepts a proton in strongly acid solution as an enamine to give the cation **4**. Deuterium exchange under these conditions takes place most rapidly at position 1, then at position 3, and occasionally at position 2.[5] The cation attacks another molecule of indole to yield the crystalline dimer (**5**),[6] and a trimer is also formed. Electron-donating groups in the five-membered ring assist protonation, while it is inhibited or prevented by electronegative substituents.

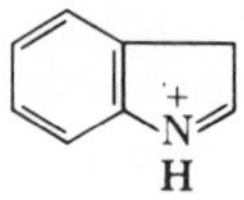

4 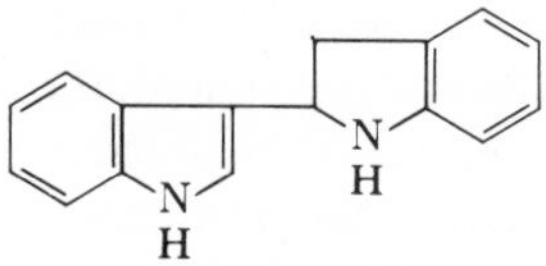 **5**, Indole dimer

Indole itself in the presence of acid is easily reduced by zinc[7] or hydrogen[8] over palladium or platinum to 2,3-dihydroindole, or indoline (**6**), and it is likely that the cation **4** is the species actually reduced. The reduction of derivatives can be difficult. More powerful hydrogenation yields octahydroindole (**7**) and, finally, 2-ethylcyclohexylamine (**8**). 2,3-Dihydroindole is a colourless liquid, b.p. 230°C, and on catalytic dehydrogenation (Pd/C) or oxidation with manganese dioxide gives indole; 2,3-dichloro-4,5-dicyanobenzoquinone is valuable for

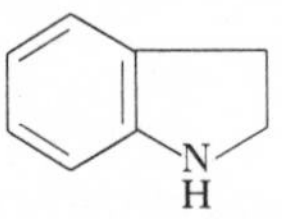

6 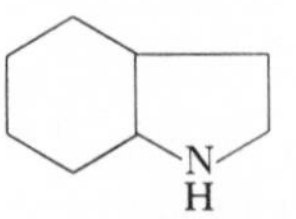 **7**  **8** 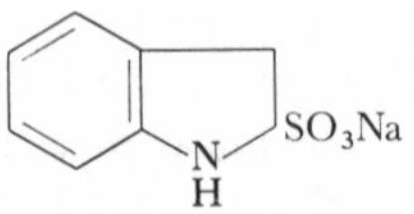 **9**

aromatizing complex 2,3-dihydroindoles.[9] 1,2-Dihydroindole has the typical properties of an aromatic–aliphatic secondary amine.

Indole also adds sodium bisulphite, presumably via the ion **4**, forming **9**, which acetylates on the nitrogen atom.[10] Electrophilic substitution, such as nitration, now takes place successively at positions 5 and 7. Subsequent treatment with sodium hydroxide splits out the substituents in the five-membered ring and gives a useful synthesis of substituted indoles. Photochemical

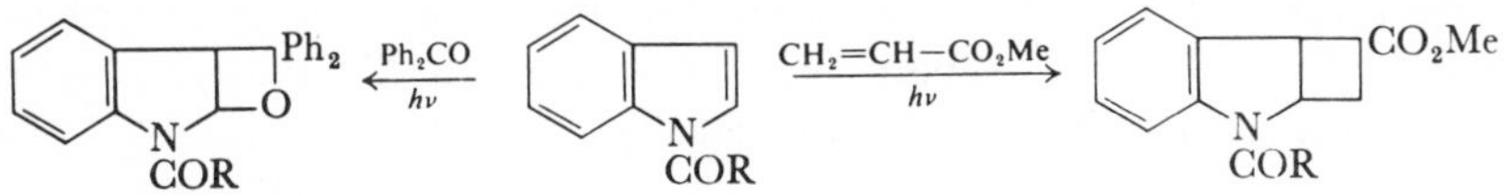

$$\xleftarrow[h\nu]{Ph_2CO} \qquad \xrightarrow[h\nu]{CH_2{=}CH{-}CO_2Me}$$

additions across the 2,3-positions of 1-acylindoles have recently been carried out,[11] and similar thermal reactions with aceylenes lead to azepines (p. 448).

3. *Substitution reactions*

The hydrogen at position 1 of indole is appreciably acidic (cf. pyrrole, p. 96) and can be displaced by metallic sodium, potassium hydroxide at 125–130°C, or Grignard reagents. The sodium salt, often prepared with sodamide in liquid ammonia, with methyl iodide gives mainly 1-methylindole with small amounts of 2- and 3-methylindole, while the potassium derivative, obtained from potassium hydroxide in dimethyl sulphoxide, gives high yields of 1-alkyl derivatives when treated with the alkyl bromide or iodide at room temperature.[12] Another very convenient alkylation or acetylation method is to form the sodium salt with sodium hydride in dimethyl sulphoxide and add the appropriate halide.[13] Indole Grignard reagents,[14] which are largely ionic, give mixtures of 1- and 3-substituted indoles depending on the conditions, and this can be compared with similar reactions, but at position 2, in the pyrrole series.

In general the electrophilic substitution of indoles yields 3-derivatives. If this position is blocked the substituent enters position 2, otherwise it enters the benzene ring. However, chlorination of indole by sodium hypochlorite gives the

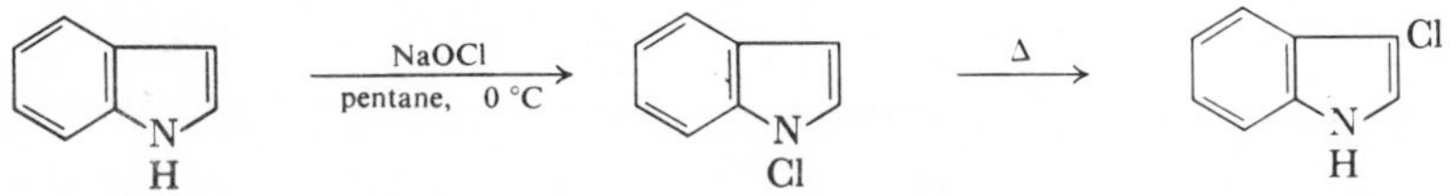

$$\xrightarrow[\text{pentane, } 0\,^\circ C]{NaOCl} \qquad \xrightarrow{\Delta}$$

1-chloro derivative, which has been detected in solution from its spectra, and which easily isomerises to 3-chloroindole.[15] 3-Halogenated indoles are obtained using 2,4,4,6-tetrabromocyclohexa-2,5-dienone,[16] iodine, or sulphuryl chloride, which also gives 2,3-dichloroindoles.[17] Nitrosation of substituted indoles, such as tryptophan (p. 209) can occur[18] at position 1 under the correct conditions, but indole itself is attacked[19] at position 3.

Ehrlich discovered that indoles on treatment with hydrochloric acid and 4-dimethylaminobenzaldehyde give red to violet colours, a test which still bears his name. The colour does not develop if the indole has substituents at both 2- and 3-positions. The product from the reaction of 2-methylindole and the aldehyde is a 3-indolenine (**10**) which has been isolated as the perchlorate; a similar 2-indolenine is obtainable from 3-methylindole. The colour is associated with the fact that, as in the aminotriphenylmethane dyes, the positive charge is not restricted to the nitrogen atom of the indole ring, as in **10**, but can also be written on the other nitrogen atom (**10a**). The compound is best considered as a resonance hybrid. Similar red-purple compounds are formed in the pyrrole series where attack occurs at position 2 of the ring.

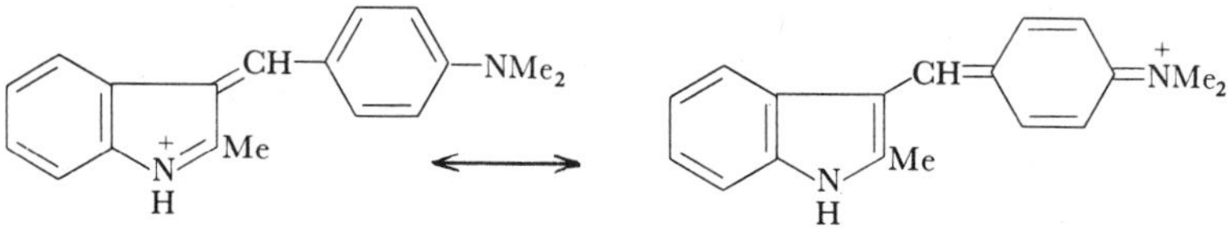

CH—⟨⟩—NMe_2 ... Me ... $\overset{+}{N}$ H ⟷ CH=⟨⟩=$\overset{+}{N}Me_2$ ... Me ... N H

10 **10a**

The attempted nitration of indole under ordinary acidic conditions leads to decomposition of the ring system. 3-Nitroindole can, however, be obtained from indole, sodium ethoxide, and ethyl nitrate (cf. 2-nitropyrrole, p. 98). The sulphonation of indole with 1-proto-1-pyridinium sulphonate (p. 234) in refluxing pyridine[20] gives the 3-sulphonic acid. Indole can be mercurated by mercuric acetate to a bismercuriacetate, which is probably the 1,3-derivative.

Indole is attacked at position 3 by oxalyl chloride, ethyl diazoacetate, formaldehyde and dimethylamine (Mannich reaction), and propiolactone with the respective formation of **11**, **12**, **13**, and **14**. α-Acetamidoacrylic acid (**15**)

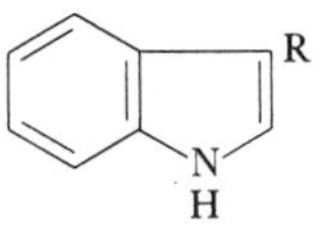

R
11, COCOCl
12, CH_2CO_2Et
13, CH_2NMe_2
14, $CH_2CH_2CO_2H$

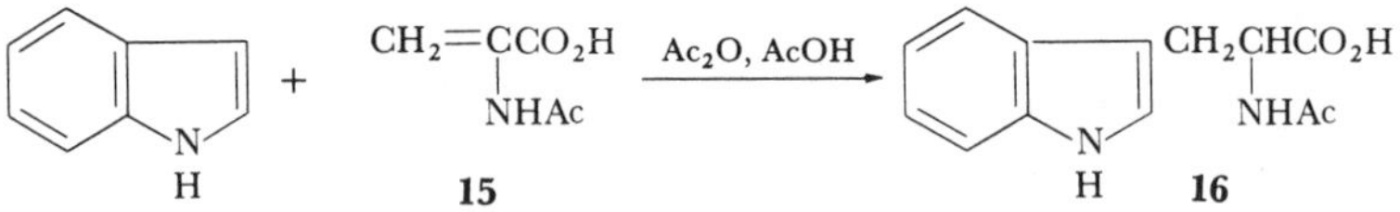

Indole (N H) + $CH_2{=}C(NHAc)CO_2H$ (**15**) $\xrightarrow{Ac_2O,\ AcOH}$ 3-($CH_2CH(NHAc)CO_2H$)indole (**16**)

also reacts at position 3, through a Michael type of addition yielding acetyltryptophan (**16**). It is therefore particularly interesting that acrylonitrile, which reacts with indole (anion) in the presence of base at position 1 and under neutral conditions at position 3 in the normal way to give **17** and **18**, in methanol in the presence of light gives **19**. An exiplex is involved.[21] Indole is

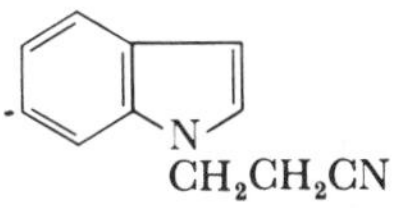
N
CH_2CH_2CN

17

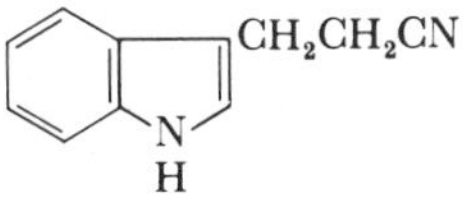
CH_2CH_2CN
N
H

18

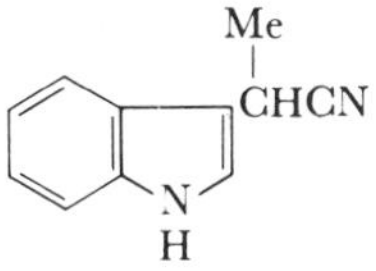
Me
CHCN
N
H

19

attacked by acetic anhydride (cf. pyrrole, p. 98) at 180–200°C, giving 1-acetylindole with some 1,3-diacetylindole. At higher temperatures the second compound is the main product, and it is easily hydrolysed by dilute alkali to 3-acetylindole.

A kinetic study of the reaction of 4-nitrobenzenediazonium chloride with indole and its 3-deuterium derivative has shown that there is a slow attack of the diazonium cation on the neutral indole molecule at position 3, followed by a rapid expulsion of a proton or deuteron. This substitution therefore proceeds by the same mechanism invoked in the nitration of benzene.

Treatment of indole with methyl iodide converts it successively into the 3-methyl and 2,3-dimethyl derivatives followed by **20** and **21**. It is not possible to stop the methylation efficiently until ionic products are formed. It is particularly interesting that compounds **20** and **21** nitrate at position 5 in 88 and 77% yield, respectively.[22] This is the first example of a positively charged group directly attached to a benzene ring directing an attacking nitronium cation to the *para* position.

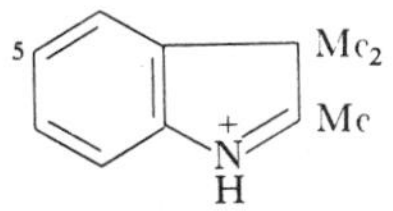
5
Me$_2$
Me
+
N
H

20

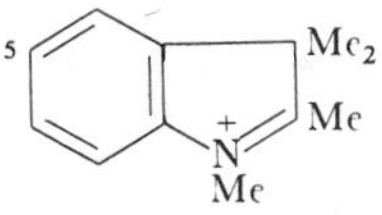
5
Me$_2$
Me
+
N
Me

21

1,1-Dimethylindolium hydroxide (**23**), convertible into the perchlorate, has been obtained indirectly from isatin.[23] Reduction with lithium aluminium hydride gave the alcohol (**22**), which was treated as indicated. The hydroxide

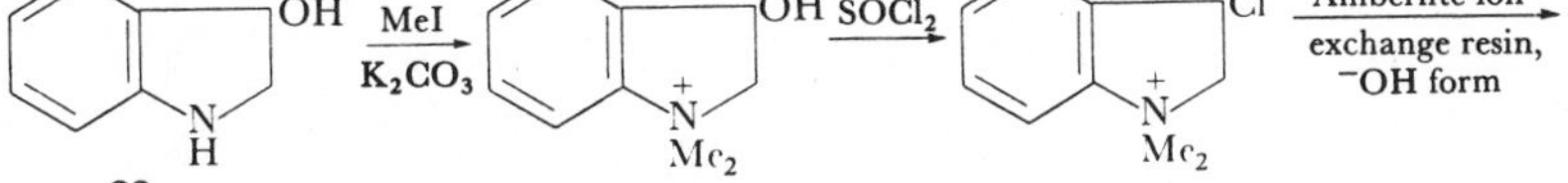
OH
MeI
K_2CO_3
N
H
OH
$SOCl_2$
+
N
Me$_2$
Cl
Amberlite ion
exchange resin,
$^-$OH form
+
N
Me$_2$

22

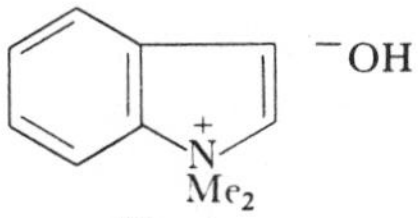
$^-$OH
+
N
Me$_2$

23

(**23**) possessed an ultraviolet absorption spectrum very similar to those of styrene and indene. It was remarkably easily demethylated to 1-methylindole, standing for 1 hour with a solution of sodium hydroxide at room temperature being sufficient.

The alkylation of indoles with sodium alkoxides at high temperatures proceeds only at position 3. The mechanism of this substitution is probably the same as that of a similar substitution in the pyrrole series (p. 103).

Indole with benzyl radicals gives[24] small amounts of 1- (1·67%) and 3-benzylindole (2·15%), and 1-methylindole with benzoyl peroxide gives mainly the 3-benzoyloxy derivative.[24a]

D. Derivatives of indole

In general, substituents attached to the carbocyclic and heterocyclic rings of indole have properties normal to the corresponding benzene and pyrrole derivatives, respectively.

1. *Alkyl- and arylindoles*

Alkylindoles can be prepared by direct synthesis or by the direct or indirect alkylation of indole. Rearrangements often occur in the further alkylation of 3-substituted indoles; for instance, cyclisation of the tritiated alcohol (**24**) to the tetrahydrocarbazole (**26**) gave a product in which the tritium was almost equally distributed between the 1 and 4 positions.[25] This shows that the spiro compound **25** must be an intermediate and that rearrangement of this leads to **26**.

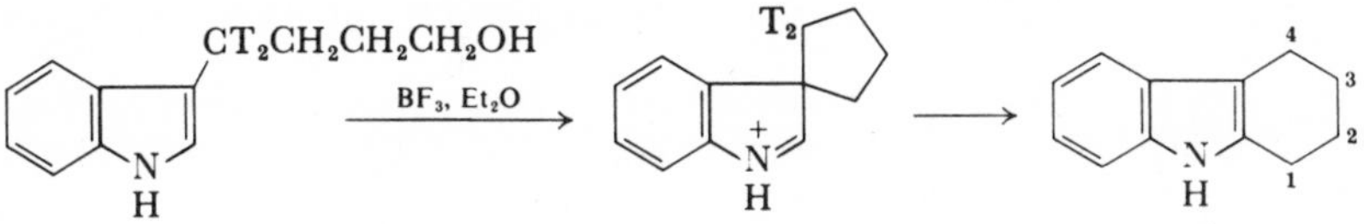

$CT_2CH_2CH_2CH_2OH$ BF_3, Et_2O T_2 N H   N H   N H   4 3 2 1

24   **25**   **26**

In a similar rearrangement, heating the 3*H*-indoline **27** or **28** with polyphosphoric acid, which presumably causes initial *N*-protonation, gives[26] an equilibrium mixture of these indolenines in 1 : 3·5 ratio, respectively. The nitration of 2-methyl- and 2,3-dimethylindole in sulphuric acid solution gives the 5-nitro derivatives. Presumably the 3-protonated indoles are actually nitrated, for 1-acetyl-2,3-dimethylindole, which is much less ready to accept a proton, yields the 6-nitro derivative.

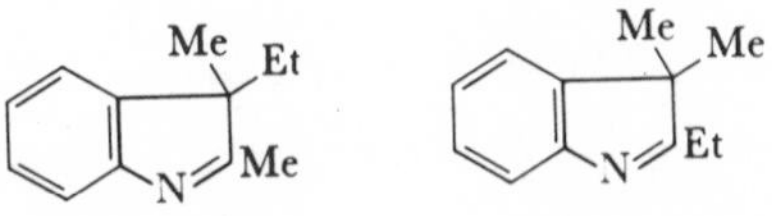

Me Et N Me   Me Me N Et

27   **28**

3-Methylindole (skatole) is the major odoriferous constituent of human faeces. It is also a valuable perfume base and can be obtained by Fischer's synthesis (p. 205). 2-Methylindole can also be obtained by Fischer's synthesis and is converted into a carbazole (**29**) on heating with methyl vinyl ketone,

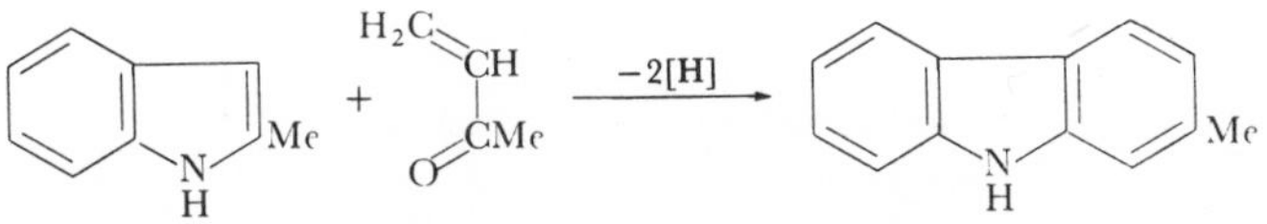

29

indicating that the 2-methyl group must be somewhat activated. Both 2- and 3-methylindole are easily oxidised by ferric chloride in ether to red dyestuffs.

2,3-Dimethylindole on bromination in aqueous media yields much 2-hydroxymethyl-3-methylindole (**30**). The reaction scheme[27] below accounts for this, and other similar transformations are known in the indole series. The same indole emits a green light when oxidised by oxygen or hydroperoxide[28] in

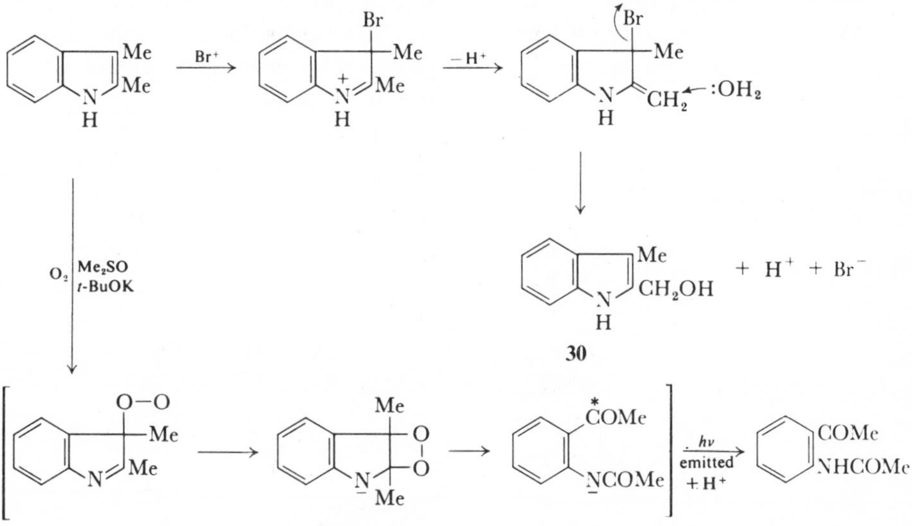

30

31

strongly basic media. Opening a dioxetane (**31**) ring is thought to be the crucial stage in light production (see also p. 385).

3-Methyl-2-phenylindole (**32**) is of interest because, like many other 2-arylindoles, it gives a stable ozonide (**33**), and on autoxidation in boiling light petroleum it yields a hydroperoxide (**34**).

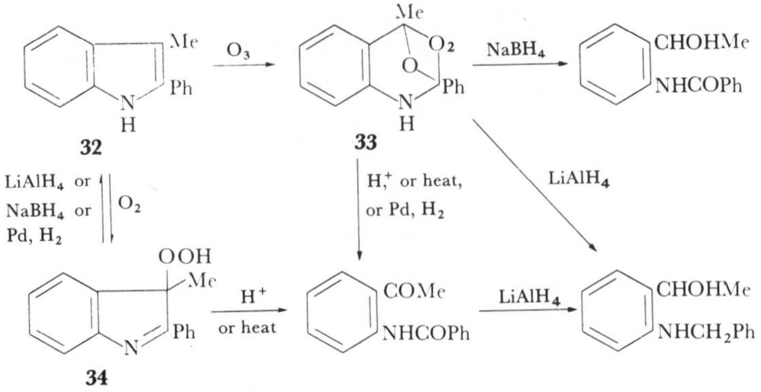

Gramine (**35**), obtained from indole with formaldehyde and dimethylamine by the Mannich reaction, forms a crystalline methiodide (**37**). The aliphatic amino groups of **35** and **37** can be readily displaced by suitable nucleophiles, and kinetic studies have shown that **36** is an intermediate.[29] 1-Methylgramine methiodide (**38**) does undergo similar reactions, but much more slowly, by a direct S_N2 displacement of trimethylamine. Syntheses of 3-indoleacetonitrile (**39**) and of tryptophan (**40**) are outlined. The nitrile is found in cabbages and other plants and on hydrolysis yields indole-3-acetic acid. This last is a plant growth hormone known as auxin, and is also found in human urine.

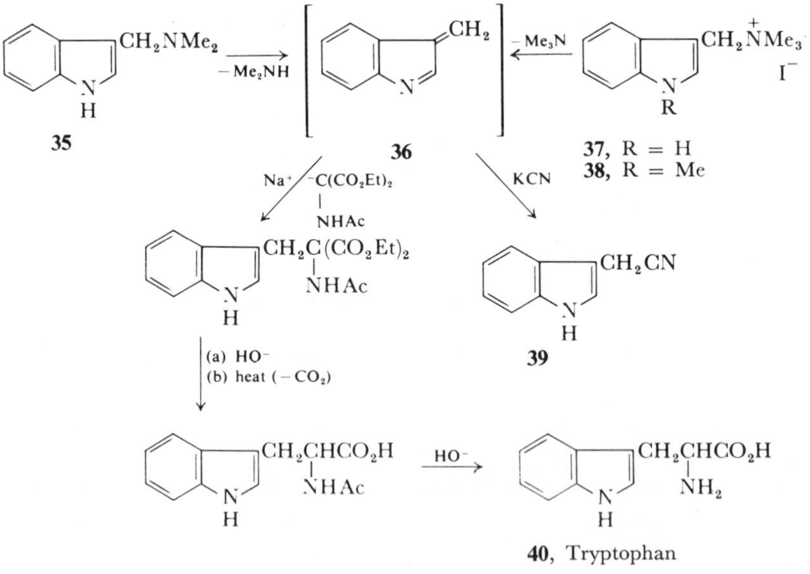

37, R = H
38, R = Me

40, Tryptophan

1-Methylindole with butyllithium gives the 2-lithium derivative, which has normal reactions (cf. 1-methylpyrrole, p. 104).

2. *Indole-3-carboxyl derivatives*

Indole-3-carbaldehyde (**42**) is best obtained from indole by the Vilsmeier–Haack reaction using *N,N*-dimethylformamide–phosphorus oxychloride, which is also very successful with pyrrole (p. 99). The substitution even works when electron-withdrawing substituents are present at the 2-position. Indole-3-carbaldehyde can also be obtained, along with 3-chloroquinoline, from indole with chloroform and alkali. This synthesis proceeds through the conversion of chloroform into dichlorocarbene, which may combine with indole or the anion (**41**), but precisely how this occurs has not been determined experimentally (cf. p. 100). A cyclopropane intermediate (e.g., **43**), however, must be on the pathway to 3-chloroquinoline.

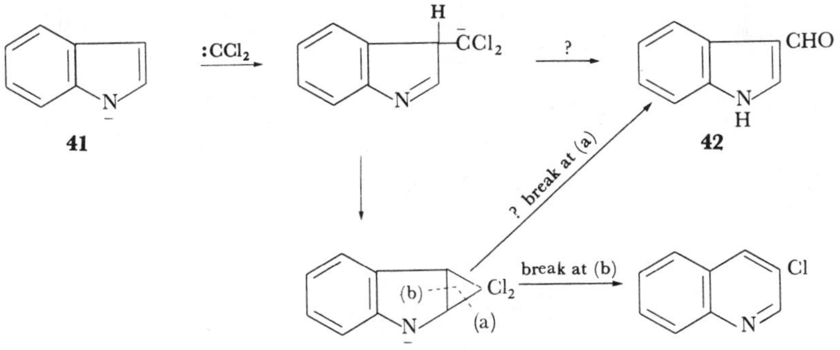

:CCl$_2$ H $\bar{C}Cl_2$? CHO N H
41 **42**
? break at (a)
(b) Cl_2 (a) break at (b) Cl
43 3-Chloroquinoline

Indole-3-carbaldehyde (**42**) has only some of the properties of an aromatic aldehyde. Its low solubility in ether and its ultraviolet absorption spectrum show

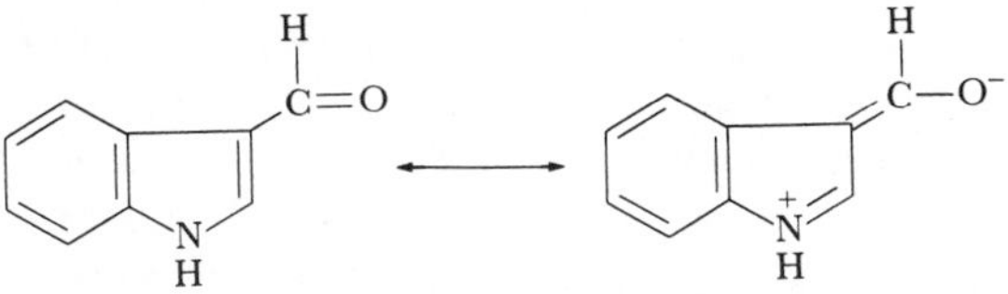

H C=O N H H C—O$^-$ N$^+$ H
42 **42 a**

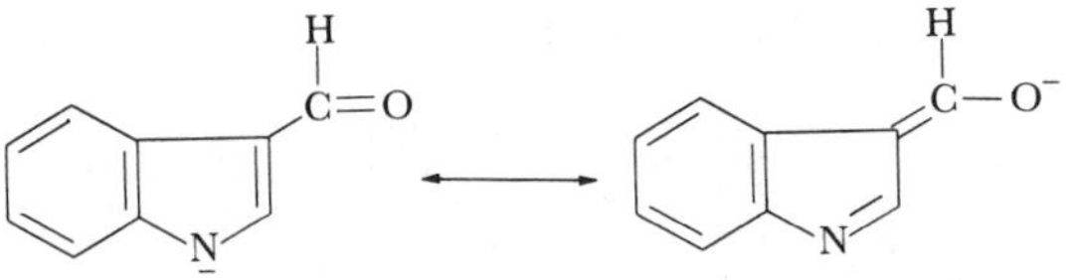

H C=O N H C—O$^-$ N
44 **44 a**

that the molecule is very zwitterionic in character, and it can be considered as a 'vinylogous amide.' The conclusion that structure **42a** is an important contributor to the resonance hybrid which represents the aldehyde is confirmed by infrared absorption spectrum studies It also agrees with the fact that a cyanohydrin is not formed from the aldehyde and hydrogen cyanide. Indole-3-carbaldehyde does not react under the ordinary conditions of the Cannizzaro reaction, although 1-methylindole-3-carbaldehyde undergoes the reaction normally. It therefore appears that the hydrogen atom at position 1 is connected with the failure, and its removal under the strongly alkaline conditions would leave the resonant anion (**44**–**44a**), which could conceivably be too stable to undergo further transformation. Indole-3-carbaldehydes, perhaps for similar reasons, do not undergo the benzoin reaction.

Many indolecarboxylic acids have been prepared by Fischer's synthesis (p. 205). The 3-carboxylic acid, along with a small proportion of the 2-isomer, can also be obtained by heating the Grignard or potassium derivative of indole with carbon dioxide. Decarboxylation of these acids also occurs readily, the 3-carboxyl group being the more labile.

2. *Oxindole, or 2-indolinone*

Oxindole (**45**) is a colourless solid, m.p. 126–127°C, and has been obtained in a variety of ways, including the reduction of isatin and of 2-nitrophenylacetic acid. The most generally useful synthesis is outlined.

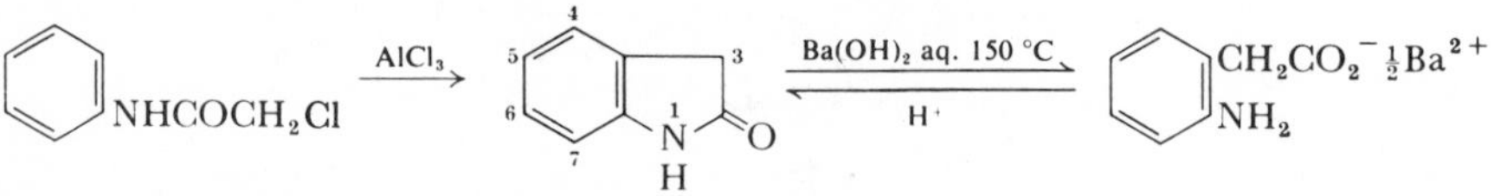

$C_6H_5NHCOCH_2Cl$ $\xrightarrow{AlCl_3}$ **45** $\underset{H^+}{\overset{Ba(OH)_2\ aq.\ 150\ °C}{\rightleftharpoons}}$ $C_6H_4(NH_2)CH_2CO_2^-\ \tfrac{1}{2}Ba^{2+}$

4, 5, 6, 7, 1, 3, N, H, O

45

Oxindole under normal conditions is probably best represented as **45**, owing to the resemblance of its ultraviolet absorption spectrum to those of its 1-methyl and 1,3,3-trimethyl derivatives. This conclusion is supported by infrared absorption spectrum studies. Nevertheless, 2 moles of methane are evolved with methylmagnesium iodide. This is almost certainly due to a direct attack of the Grignard reagent on **45** giving the resonant dianion **46**. It is not necessary to postulate tautomerization of **45** to **47** before reaction with the Grignard reagent. This is because conversion of **45** into **47** would involve the ionization of a

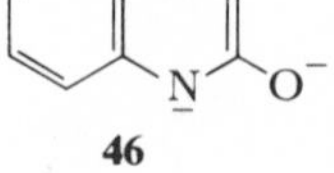

N, O^-

46

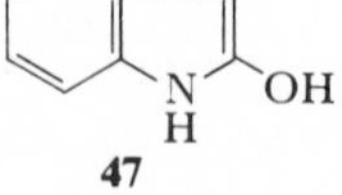

N, H, OH

47

proton from position 3, and Grignard reagents react with protons irrespective of their sources. The sodium and silver salts of oxindole with methyl iodide usually give *N*- and *O*-alkyl derivatives, respectively. Lithium aluminium hydride does not reduce oxindole, probably because of proton removal to give **46**, but it does reduce 1-substituted oxindoles to the corresponding 2,3-dihydroindoles.[30] Borane in tetrahydrofuran does, however, reduce oxindole to 2,3-dihydroindole.[31] Oxindole has a reactive methylene group at position 3 which reacts with aldehydes ketones and so forth in the usual way.

2-Ethoxyindole (**48**), obtained from oxindole with triethyloxonium fluorborate, possesses an N–H absorption band in the infrared. It is particularly interesting,[31a] as on sublimation it gives the isomeric indolenine (**49**), which shows no N–H absorption. Both compounds give the same nuclear magnetic resonance spectrum in carbon tetrachloride where equilibrium is established. The ratio of **48**/**49** is about 1 : 1.5 under these conditions.

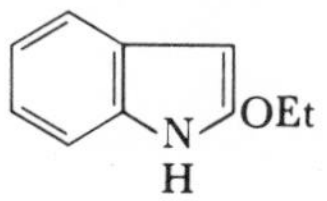

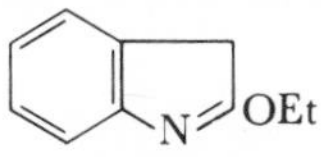

Oxindole is weakly basic, forms a water-soluble hydrochloride, and is nitrated at position 5. With phosphorus oxychloride it yields 2-chloroindole,[17] which is rather unstable. Surprisingly, 3-chloroindole yields oxindole on acid hydrolysis, and this may take place through initial protonation as shown.[17]

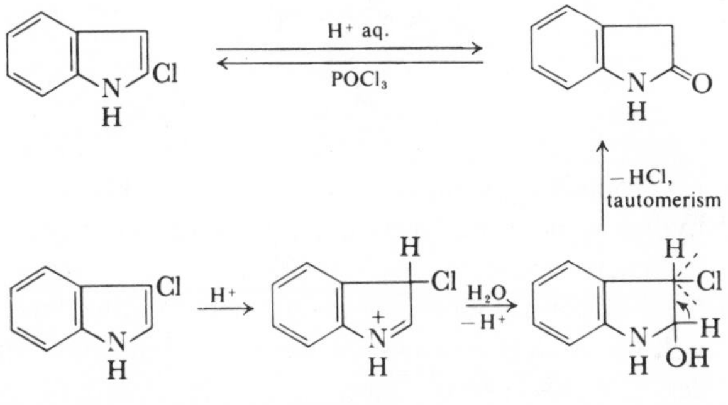

The oxidation of indoles by *N*-bromosuccinimide, or bromine, can lead to oxindoles, and occasionally rearrangements occur as in the example given below.[32] An ester shift is not common.[33]

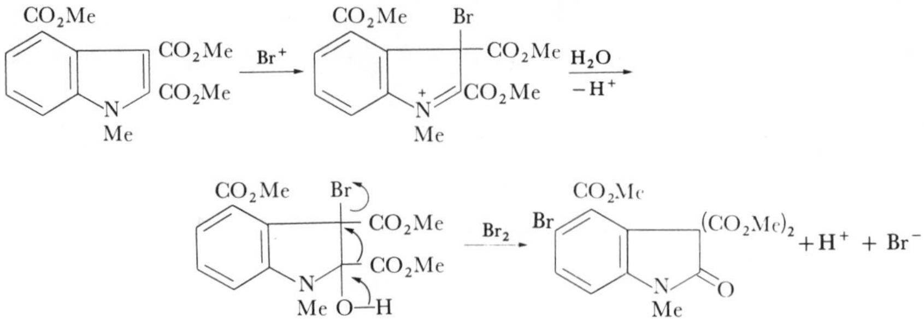

4. *Indoxyl, or 3-indolinone*

3-Indolinone, commonly called indoxyl (**53**), is an unstable yellow compound which is very easily oxidised to indigo, (p. 210) and can be obtained as shown.[34] The 1-acetyl derivative (**50**) exists entirely as the keto tautomer, but the possibility that 3-hydroxyindole is in tautomeric equilibrium with 3-indolinone itself has not been established. The *O*-sulphate (**51**), 'indoxyl sulphate,' occurs as salts in human urine.

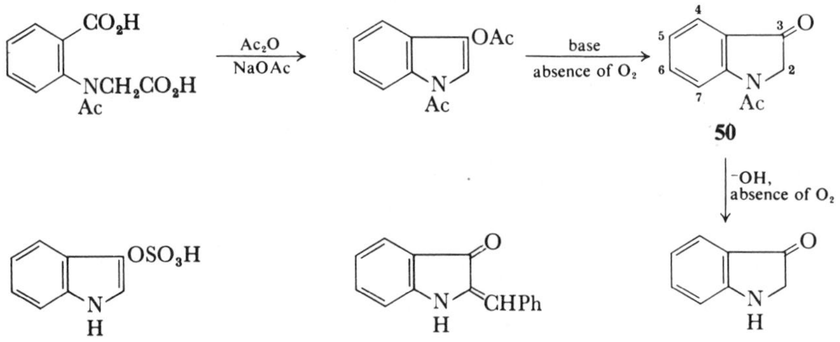

51 **52** **53**

3-Indolinone is soluble in alkali, gives a red ferric chloride, and is resinified by concentrated hydrochloric acid. Treatment with dimethyl sulphate and sodium carbonate gives 3-methoxyindole, but if methyl iodide is used 2,2-dimethyl-3-indolinone is the product. 3-Indolinone has a highly reactive 2-methylene group which reacts with benzenediazonium salts, nitrous acid, nitroso compounds, and many aldehydes and ketones in acetic acid. Benzaldehyde gives **52**, and compounds of this type are called indogenides.

5. *Isatin*

Isatin, systematically named 2,3-indolindione (**54**), crystallizes in red needles, m.p. 200–201°C. It was first obtained by the oxidation of indigo and has been synthesized by many methods. The most widely used synthesis starts from aniline, and a useful synthesis of 1-substituted isatins involves oxalyl chloride.

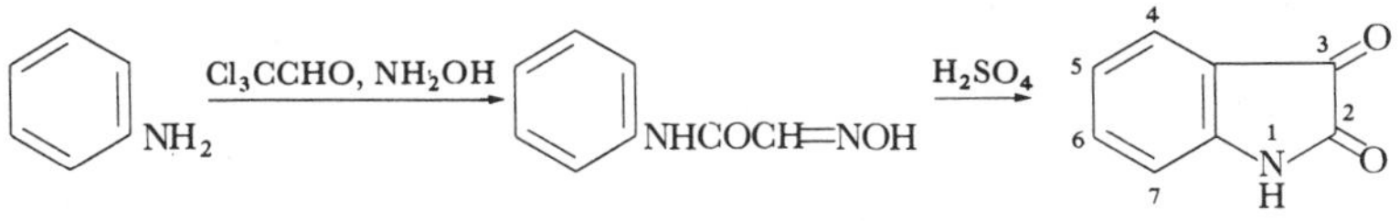

Cl_3CCHO, NH_2OH → NHCOCH=NOH → H_2SO_4 → **54**

ClCOCOCl → NRCOCOCl → $AlCl_3$ →

The structure of crystalline isatin has been determined from an X-ray examination which shows that it exists in the lactam form. In solution, however, several ionic and molecular species can be present, depending on the hydrogen ion concentration, but the infrared absorption spectrum of a chloroform solution shows that under these conditions only the lactam form is present.

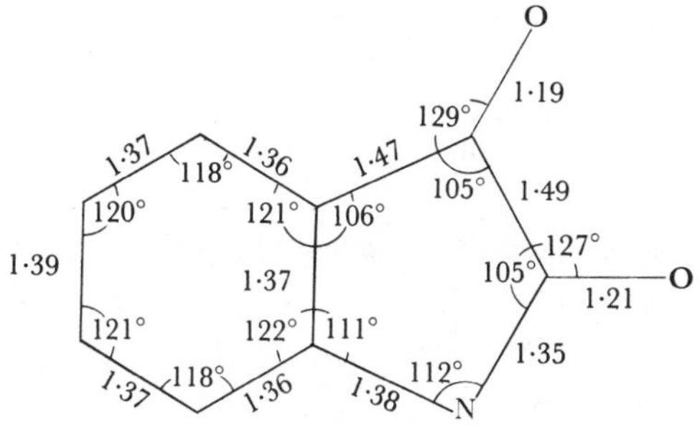

Isatin, distances in Å

Isatin is basic enough to dissolve in concentrated hydrochloric acid, and it forms a stable perchlorate. It also has an acidic hydrogen and forms sodium and silver salts. The alkylation of these gives 1-alkyl and 2-alkoxy derivatives, respectively. The anion derived from isatin is therefore best represented as a resonance hybrid of **55** and **55a** with a very small contribution from the *ortho*-quinonoid **55b**.

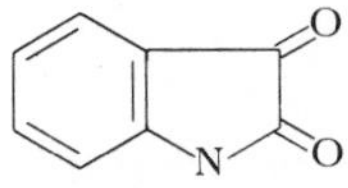

55 **55 a** **55 b**

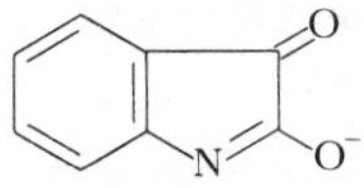

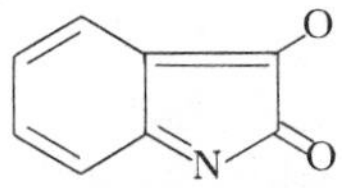

The two carbonyl groups of isatin have widely differing properties. The one at position 3 is ketonic and undergoes typical reactions of ketones. For instance, with hydroxylamine it gives **56**, with acetone it gives **57** or **58** according to the conditions, and it undergoes the Reformatsky reaction normally. The indophenine reaction, which also involves condensation at position 3, is discussed on

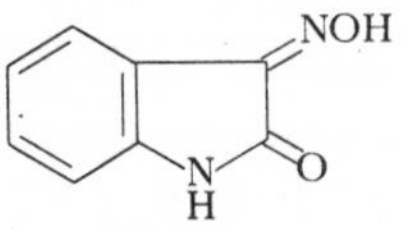

NOH
N H O
56

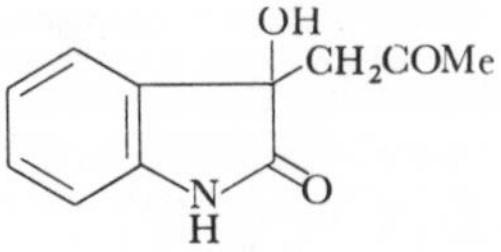

OH
CH_2COMe
N H O
57

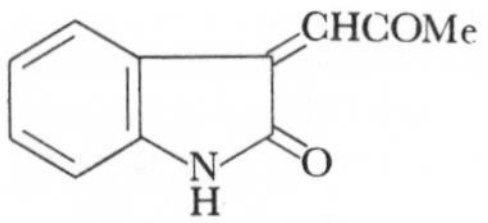

CHCOMe
N H O
58

p. 158. Reduction of isatins with lithium aluminium hydride usually gives the corresponding indole as the major product. The carbonyl group at position 2 is amidic in character and is hydrolysed by alkali to give sodium isatinate (**59**).

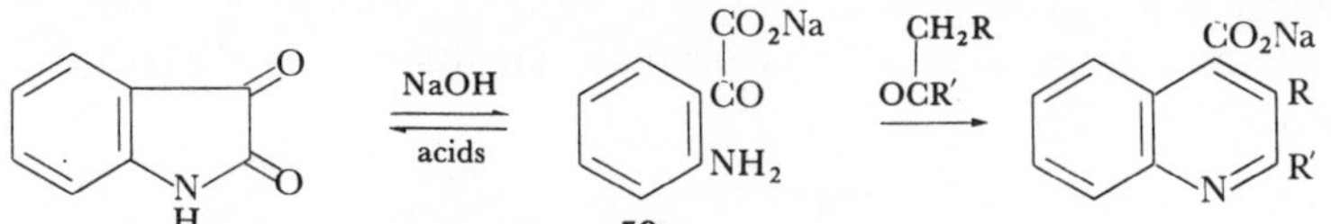

O
N H O
NaOH
acids
CO_2Na
CO
NH_2
59
CH_2R
OCR′
CO_2Na
R
N R′

Acidification immediately reprecipitates isatin. Sodium isatinate condenses with suitable ketones to give quinoline-4-carboxylic acids, a quinoline synthesis (Pfitzinger) of very general application (p. 314).

The oxidation of isatin with chromic acid gives isatoic anhydride (**60**), which is hydrolysed to anthranilic acid by dilute acid. Anthranilic acid can also be obtained by direct oxidation.

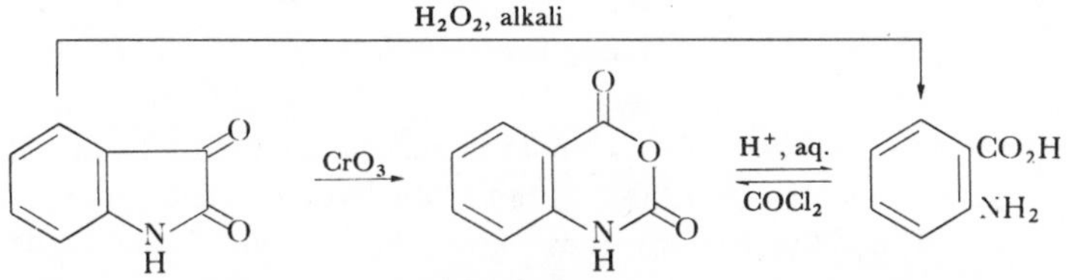

H_2O_2, alkali
O
N H O
CrO_3
O
O
N H O
60
H^+, aq.
$COCl_2$
CO_2H
NH_2

1-Methylisatin 3-thiosemicarbazone appears to be an extremely valuable prophylactic agent against smallpox, and this is the first example of the prophylaxis of a virus disease by a synthetic chemical.[35]

6. *Carbazole*

Carbazole (**62**), or dibenzopyrrole, can be isolated from the anthracene fraction of coal tar. When pure it exhibits no fluorescence and has m.p.

245–246°C. It can be synthesized by boiling 2-nitrodiphenyl with triethyl phosphite, and by the dehydrogenation of 1,2,3,4-tetrahydrocarbazole obtained via the Fischer indole synthesis from cyclohexanone phenylhydrazone. Alternative syntheses, which probably proceed through nitrene or carbene intermediates, are the photolysis of 2-azidobiphenyls (**61**) and the thermolysis of 1-arylbenzotriazoles (**63**) which are obtained by diazotising 2-aminobiphenylamines.

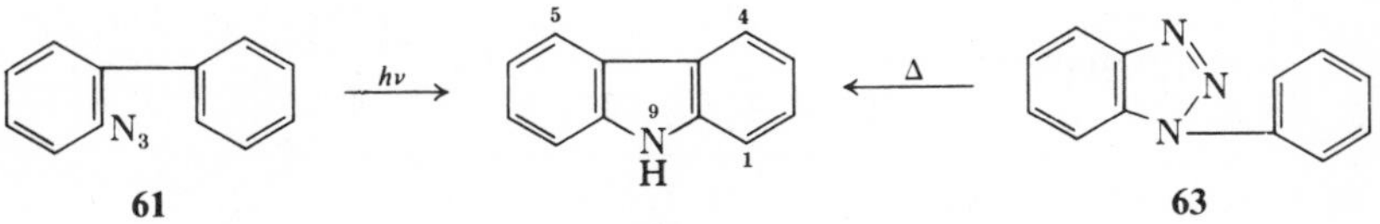

Like pyrrole, the nitrogen atom of carbazole can be alkylated by alkyl halides in the presence of sodamide, and vinylated by acetylene in the presence of sodium hydroxide. *N*-Vinylcarbazole can be polymerized alone or as a copolymer to give commercially useful products. The benzene rings protect the 'pyrrole' ring of carbazole, so much that it gives a mixture of 3,6-dinitrocarbazole and higher nitration products with nitric–concentrated sulphuric acid. This is in agreement with molecular-orbital calculations, which show that the highest π-electron density is at positions 3 and 6. The resonance energy of carbazole [309 kJ (74 kcal)/mole] is only slightly greater than that of biphenyl [297 kJ (71 kcal)/mole] or of two benzene molecules (2 x 150·5 kJ (36 kcal)/mole).

E. Synthetic methods

A large number of methods of indole synthesis are available (see both books listed in the General Bibliography p. 225), but only the more important are mentioned here.

(1) The most widely used synthesis of indole derivatives is due to Fischer,[36] and consists of heating ketone or aldehyde phenylhydrazones (e.g., **64**), with anhydrous zinc chloride, boron trifluoride, polyphosphoric acid, or other acid catalysts. It should be particularly noted that indole itself cannot be obtained from acetaldehyde phenylhydrazone by this method, although the synthesis is otherwise very general and can also be applied to the preparation of 1-alkyl- or 1-arylindoles. Indole has, however, been obtained by the decarboxylation of indole-2-carboxylic acid, prepared by cyclizing pyruvic acid phenylhydrazone.

The mechanism of Fischer's synthesis has been the subject of much discussion and experiment, but that originally proposed by Robinson and Robinson, with minor additions, is compatible with all the available data and is generally accepted. The basicities of the various compounds in the scheme are expected to increase steadily, with the exception of the final indole (**68**). The position of the nitrogen atom which is eliminated as ammonia has been established by ^{15}N tracer studies in which the fate of both the nitrogen atoms of the original phenylhydrazone have been followed, but the way in which the hydrogen atoms move is not certain. It may be that the addition of a proton to the more basic nitrogen of **65** to give **66** is the significant step. All the intermediates shown, except **66**, for particular indoles have either been trapped as derivatives or

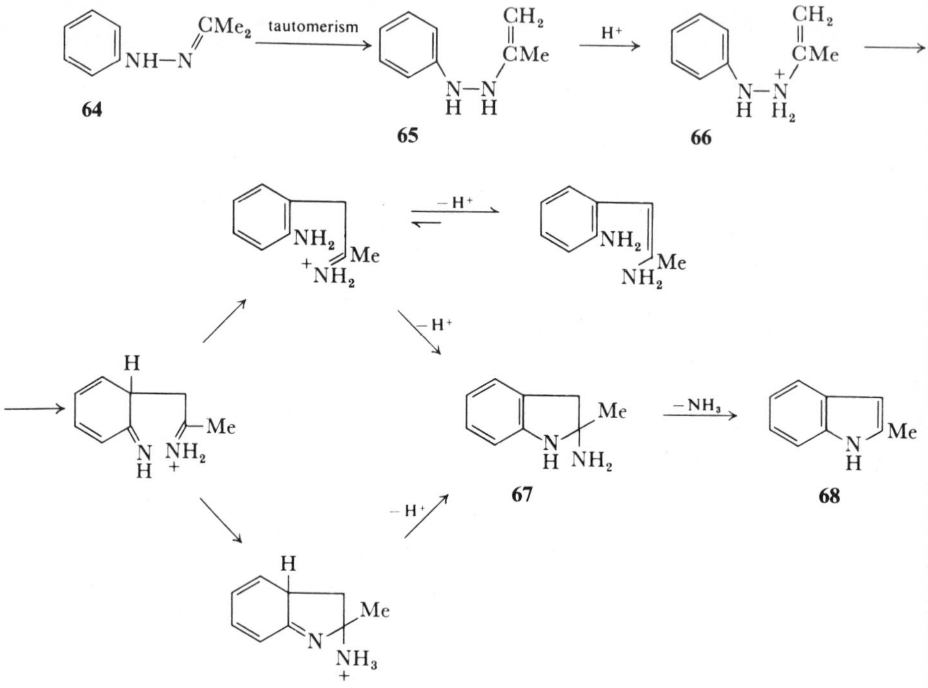

actually have been isolated and characterised, a recent example[37] being a derivative of **67**.

Another possibility is the addition of protons to both nitrogen atoms of **65** followed by a 'benzidine' type of rearrangement to the indole. One difficulty, however, is that indoles can be obtained in good yield by heating suitable phenylhydrazones in an inert solvent, such as tetralin in the absence of acid, and even in the presence of a trace of alkali.[38] It is not impossible, however, that indole formation occurs by a different mechanism under these different conditions. It has been clearly established that for the synthesis under neutral conditions the reaction is intramolecular, as crossed products are not formed.

The Japp–Klingemann reaction provides a useful route to some of the phenylhydrazones required for the Fischer synthesis. This involves the attack of a diazonium salt on a β-keto acid or ester, under basic conditions.

$$MeCOCH(CH_2Ph)CO_2^- \, Na^+ \xrightarrow{PhN_2^+ \, Cl^-} MeCOC(CH_2Ph){=}NNHPh + CO_2$$

$$MeCOCH(CH_2Ph)CO_2Et \xrightarrow{PhN_2^+ \, Cl^-} PhNHN{=}C(CH_2Ph)CO_2Et + MeCO_2H$$

The carboxyl or the acyl group of the β-keto compound is respectively eliminated with the formation of the hydrazone.

(2) The best modification of the Madelung synthesis involves the cyclization of a 2-acylaminotoluene with potassium *t*-butoxide or other very strong bases at 300°C, and both indole and 2- and 3-substituted indoles have been made.[39]

Me, NHCOR ⇌ CH_2^-, CR, $\bar{N}$, O → R, $\bar{N}$, O^- → ($2H^+$, $-H_2O$) → N, H, R

(3) Indoles may be obtained (Bischler's synthesis) by heating α-halo or α-hydroxy ketones with arylamines. The reaction is acid catalysed, and many different indoles have been obtained from aniline and *N*-methylaniline.

XCHR′, O=CR $\xrightarrow{PhNH_2}$ PhNHCHR′, O=CR $\xrightarrow{PhNH_2}$ HNPh, CHR′, N=CR $\xrightarrow{-PhNH_2}$ R′, R, N, H

(4) The Reissert synthesis gives good results, but has been used less.

Me, NO_2 $\xrightarrow[(CO_2Et)_2]{NaOEt}$ CH_2, $COCO_2Et$, NO_2 $\xrightarrow{Zn, \, HOAc}$ N, H, CO_2Et

(5) A new synthesis[40] from anilines gives yields of 41–85%.

PhNHCl —MeSCHR′COR″, −70 °C→ PhNH–$\overset{+}{S}$(Me)–CHR′COR″ —Et_3N, warm to 20 °C→ 3-SMe-3-R′-2-R″-3H-indole —$LiAlH_4$→ 3-R′-2-R″-indole

$PhNH_2$ —t-BuOCl, −70 °C→ PhNHCl

(6) Heating 2-aminoarylketones with α-bromoketones appears to be a useful general synthesis.[41]

2-(R′NH)C$_6$H$_4$COR″ + $BrCH_2COR'''$ —Me_2NCHO, Δ→ 1-R′-3-R″-2-COR‴-indole

(7) Photolysis of azides leads to indoles via an intermediate nitrene.[42]

2-N_3C$_6$H$_4$CHO —MeCOR→ 2-N_3C$_6$H$_4$CH=CHCOR —hν, −N_2→ 2-COR-indole (NH)

(8) The Nenitzsecu synthesis of 5-hydroxyindole-3-carboxylic esters, or 3-acyl-5-hydroxyindoles, has received much attention recently,[43] as 5-hydroxyindoles have powerful biological effects.

p-benzoquinone + R′C(NHR″)=CHCOR‴ → HO–C$_6$H$_3$(OH)–C(COR‴)=C(R′)NHR″ —(a) −2[H], (b) cyclisation→ 5-oxo-indoleninium (N$^+$–R′, 2-R″, 3-COR‴) —+2[H]→ 5-HO-1-R′-2-R″-3-COR‴-indole

F. Natural occurrence and compounds of special interest

Indole itself occurs free in jasmines, orange blossoms, citrus fruit, etc. It is formed by the decomposition of tryptophan residues in proteins. Many derivatives of tryptophan are of great biological interest.

1. *Tryptophan and derivatives*

L-Tryptophan (**69**), a very widely distributed α-amino acid essential to man and many living organisms, is a constituent of many proteins. It is decomposed when proteins are hydrolysed by acids, but not when alkalis are employed. Tryptophan has been synthesised from indole-3-carbaldehyde by the oxazolone method using alkaline hydrolysis (p. 374), from gramine (p. 198), and from indole by a very convenient route (p. 194). Resolution can be achieved by separating the brucine salts of acetyltryptophan.

$OP_2O_6H_3$ + $CH_2CH(NH_2)CO_2H$ → $CH_2CH(NH_2)CO_2H$ + $H_4P_2O_7$

N H   4 5 7 N H

69

The biosynthesis of tryptophan by the mould *Neurospera crassa* occurs by condensation of serine pyrophosphate with indole, and it has also been shown that anthranilic acid can be converted into indole microbiologically. A possible chemical sequence is outlined and has been followed using a labelled carboxyl group.

$^{14}CO_2H$ NH_2 + H O H, H H, $H_3O_6P_2O$ $CH_2OPO_3H_2$, OH OH → $^{14}CO_2H$ N H, H O H, H H, $CH_2OPO_3H_2$, OH OH →

5-Phosphoribosyl-1-pyrophosphate

$^{14}CO_2H$, $HOC{-}(CHOH)_2OPO_3H_2$, CH, N H → N H + $^{14}CO_2$ + $OHCCH(OH)CH_2OPO_3H_2$

Tryptophan undergoes many transformations in animals and is decarboxylated to tryptamine (**70**), hydroxylated at position 5, and decarboxylated to 5-hydroxytryptamine (**71**). This compound is known as serotonin and is of immense physiological and biochemical interest.[44] The *N,N*-dimethyl derivative of serotonin is bufotenine (**72**), a hallucinogen obtainable from *Amanita spp.*

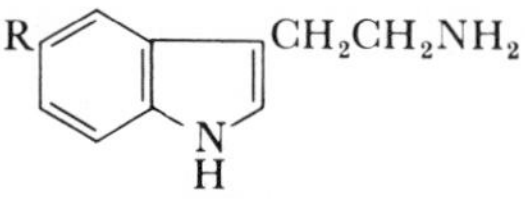

$CH_2CH_2NH_2$ HO $CH_2CH_2NMe_2$ R $CH_2CH_2NMe_2$

70, R = H
71, R = OH

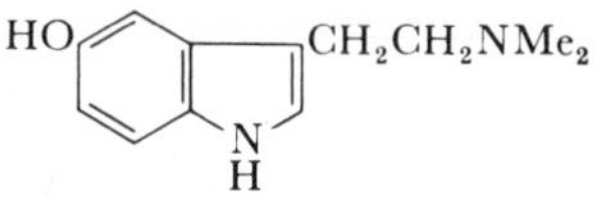

72

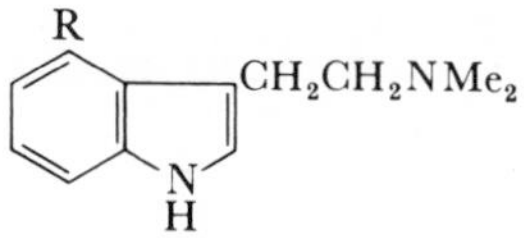

73, R = OH
74, R = OPO_3H_2

4-Hydroxy-*N,N*-dimethyltryptamine, psilocin (**73**), and its phosphate psilocybin (**74**) are even more powerful hallucinogens,[45] and are synthesised in the plant from tryptophan via *N,N*-dimethyltryptamine.[45a]

2. *Indigo*

Indigo (**75**) does not occur as such in plant sources but rather, as *O*-derivatives of indoxyl (p. 202), such as the β-glucoside in *Indigofera tinctoria* and the 5-ketogluconate in the woad plant of the ancient Britons.[46] Hydrolysis to indoxyl and aerial oxidation, which proceeds through the indoxyl radical, gives indigo. Indigo has been used as a dye since antiquity, as has its 6,6′-dibromo derivative, better known as Tyrian Purple, which can be obtained from Mediterranean and other molluscs.[47]

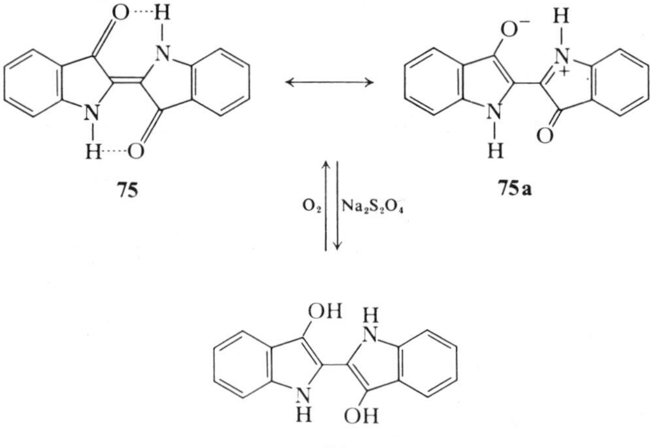

75 ⟷ **75a**

O_2 ⇅ $Na_2S_2O_4$

76

Robinson pointed out that the rings in indigo were very likely to be *trans*; this has been confirmed by X-ray measurements, and the compound is best considered as a resonance hybrid (**75** and **75a**). For dyeing purposes indigo is reduced to the yellow leucoindigo (**76**) by sodium hydrosulphite. The fabric is treated with a solution of this in aqueous sodium carbonate, and the indigo is regenerated by aerial oxidation.

There are many syntheses of indigo, the first of which dates from 1870. A very efficient commercial synthesis giving a 90% overall yield is outlined below.

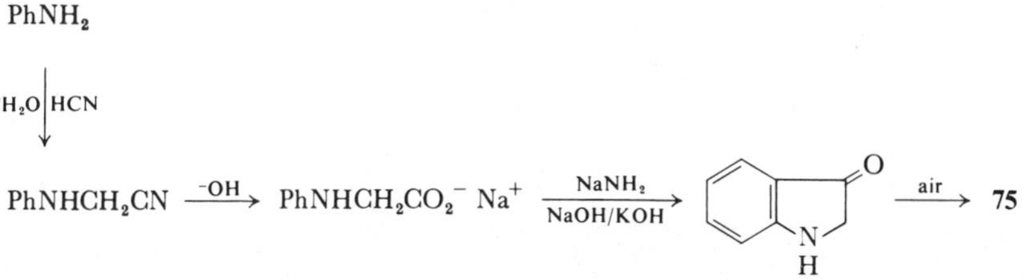

$PhNH_2 \xrightarrow{CH_2O \mid HCN} PhNHCH_2CN \xrightarrow{^-OH} PhNHCH_2CO_2^- Na^+ \xrightarrow[NaOH/KOH]{NaNH_2}$ (indoxyl) $\xrightarrow{air}$ **75**

Isoindigo (**78**) and indirubin (**77**) are the deep red isomers of indigo obtained from isatin with oxindole and indoxyl, respectively, but they are of no value as dyestuffs. Thioindigo is derived from thionaphthene and is considered on p. 224.

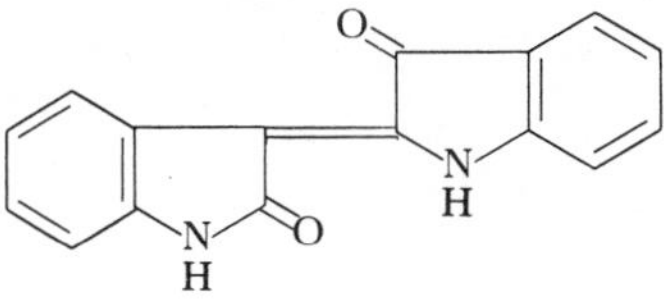

77 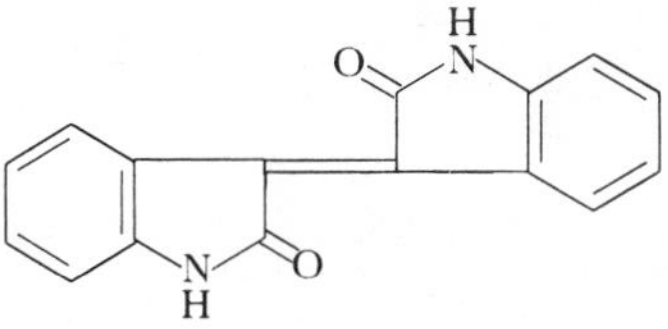 **78**

3. *Betanidin*

The structure (**79**) of this deep red beetroot pigment, which is synthesised in the plant from 3,4-dihydroxyphenylalanine,[48] defied investigators for many

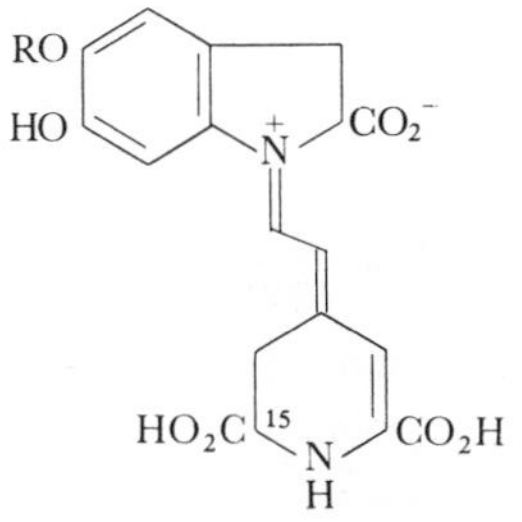

79, R = 1-β-glucopyranoside
80, R = H

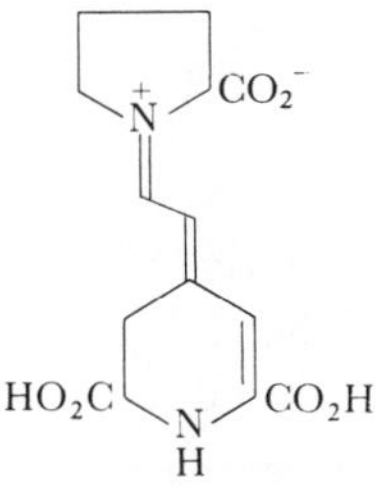

81

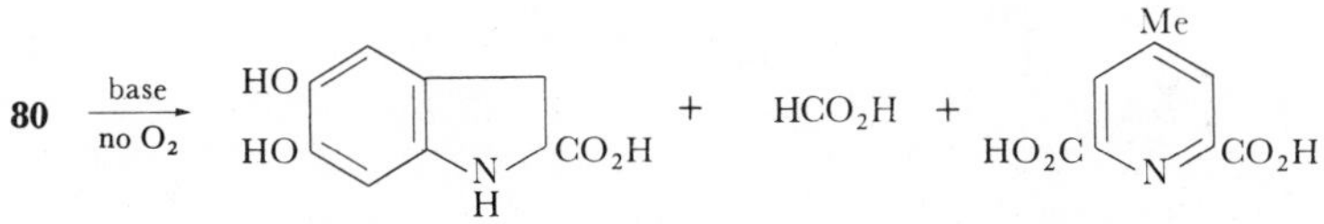

80 $\xrightarrow[\text{no } O_2]{\text{base}}$ (5,6-dihydroxyindoline-2-carboxylic acid) + HCO_2H + (4-methylpyridine-2,6-dicarboxylic acid)

years and was elucidated[49] through the use of modern chromatographic and spectral methods. Scission of the glucose residue yields betanidin (**80**), which is obtained along with its epimer at C_{15}, isobetanidin. Treating betanidin with *S*-proline and base gives indicaxanthin (**81**), a yellow pigment present in certain cacti.[50]

2. ISOINDOLES

Although isoindoles, or benzo[*c*]indoles, have attracted much attention recently,[51] isoindole itself (**4**) was first isolated in 1972. It was obtained[52] by the thermolysis of the carbonate ester **1**, or **2**,[53] and collected on a cold finger at −196°C. It decomposes rapidly in air at room temperature, in contrast to the

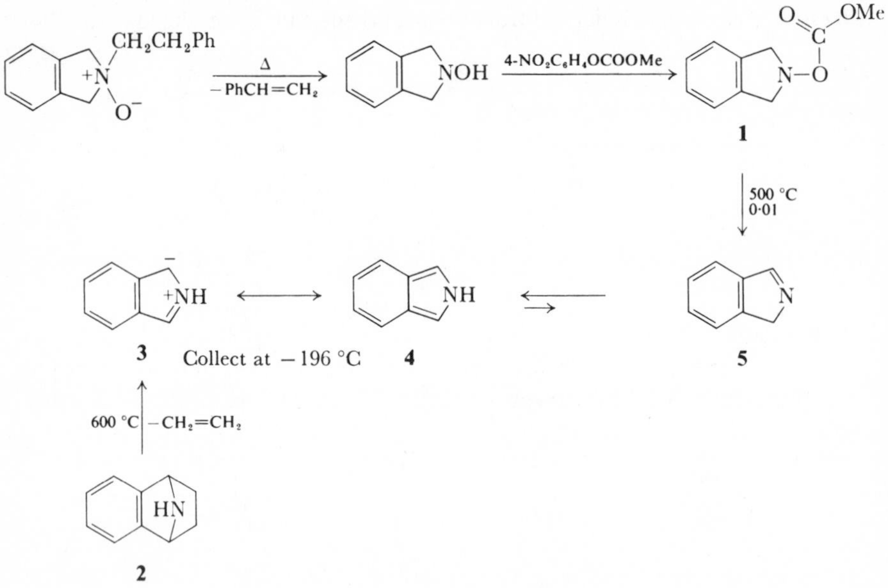

CH_2CH_2Ph — Δ, $-PhCH{=}CH_2$ → NOH — $4\text{-}NO_2C_6H_4OCOOMe$ → **1**

500 °C, 0·01 → **5**

3 Collect at −196 °C **4**

600 °C, $-CH_2{=}CH_2$

2

much more stable 2-alkylisoindoles (e.g., **7**), substituted isoindoles, and benzoisoindoles. The i.r. and n.m.r. spectra show the N–H group, and the u.v. spectrum is very similar to that of 2-alkylisoindoles. Little of the 1*H*-tautomer (**5**) can be present, as is also the case with 1-phenylisoindole (**6**).[54] Isoindoles usually give

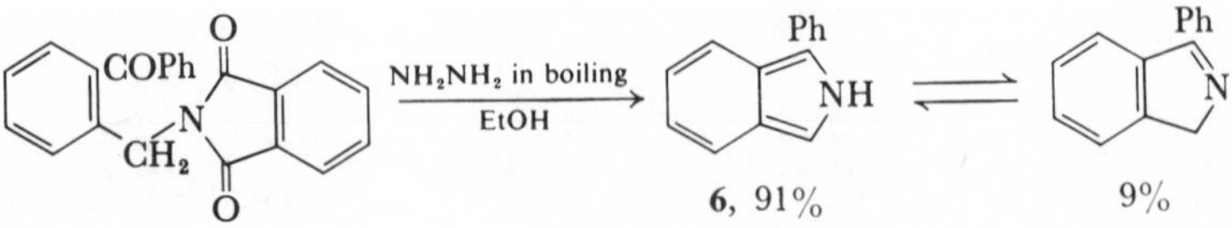

NH_2NH_2 in boiling EtOH →

6, 91% 9%

blue colours with Ehrlich's reagent, resinify with acid, and undergo Diels–Alder reactions with maleic anhydride across the 1,3-positions.

The most important charged resonance structure is **3**, if aromaticity in the carbocyclic ring is to be preserved, in conformity with electron-density calculations. The hydrogen atoms at positions 1 and 3 of 2-methylisoindole (**7**) are rapidly exchanged for deuterium in deuterium oxide tetrahydrofuran solution, presumably by a protonation–deprotonation sequence.[55]

CH_2Cl, Cl $\xrightarrow{MeNHCH_2CN}$ NMe, CH_2CN, Cl $\xrightarrow{KNH_2, NH_3}$ NMe

7

2-Methylisoindole (**7**) is a colourless crystalline solid, m.p. 90–91°C, which does not react with methyl iodide. It has been synthesised in several ways, one of which is outlined.[56]

Dihydroisoindole (**8**) is easily prepared and has the properties of a typical secondary amine.

CH_2Br, CH_2Br + $H_2NSO_2C_6H_4Me$ $\xrightarrow{NaOEt}$ $NSO_2C_6H_4Me$ $\xrightarrow{HBr}$ NH

8

The phthalocyanines are by far the most valuable derivatives of isoindole and are extremely stable blue dyestuffs. Phthalocyanin itself (**12**) can be obtained by heating phthalonitrile (**10**) or 2-cyanobenzamide (**9**) with magnesium oxide. The bright blue magnesium derivative (**11**) is formed and is converted into free phthalocyanin (**12**) by concentrated sulphuric acid at −3°C. The phthalocyanin ring system is very similar to that of porphyrin and is remarkably stable. It is, however, broken down by hot concentrated sulphuric acid to phthalimide and phthalic acid. A number of metal derivatives of phthalocyanin are known, and, of these, copper phthalocyanin is particularly stable and can be sublimed unchanged at ca. 580°C *in vacuo*. It is unaffected by boiling hydrochloric acid or molten potassium hydroxide and is precipitated unchanged from its sulphuric acid solution by water.

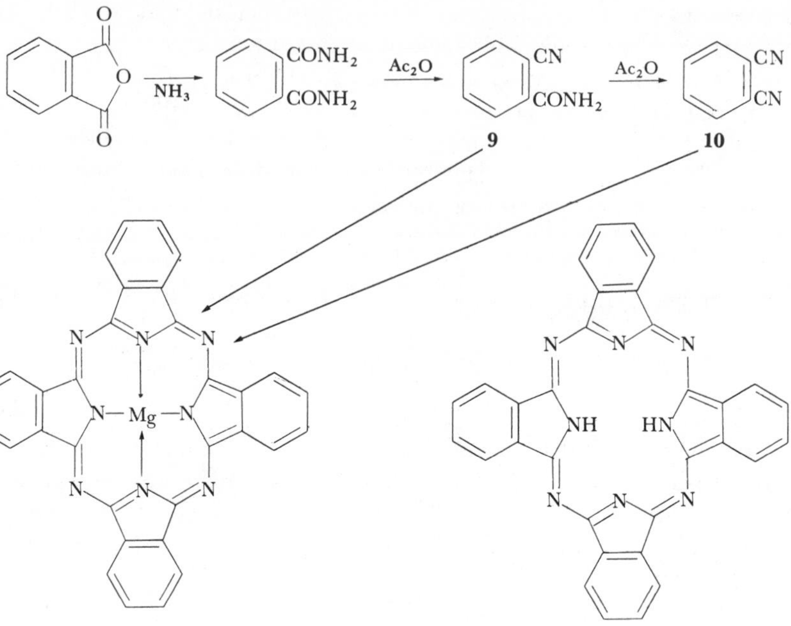

$\xrightarrow{NH_3}$ CONNH$_2$ / CONNH$_2$ $\xrightarrow{Ac_2O}$ CN / CONH$_2$ **9** $\xrightarrow{Ac_2O}$ CN / CN **10**

11 **12**

3. INDOLIZINE AND CYCLAZINE

Indolizine (**1**), previously called pyrrocoline, is isoelectronic with indole. A number of charged resonance structures can be written, but only the two involving a fully aromatic pyridine ring are shown. Of these **1a** is the more important, as the charge separation is less than in **1b**. This qualitative conclusion is in agreement with the results of π-electron density molecular-orbital calculations.[57]

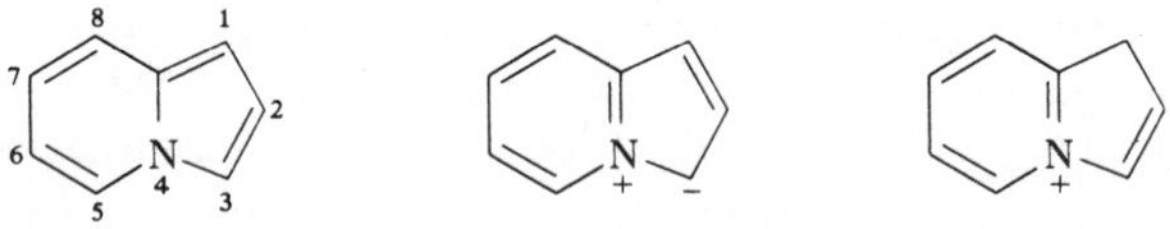

1 **1a** **1b**

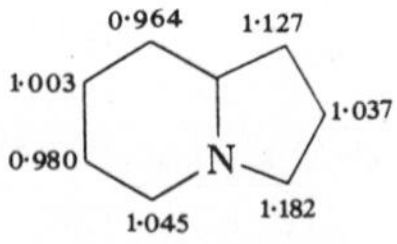

0·964 1·127 1·003 1·037 0·980 1·045 1·182

π-Electron densities

Indolizines, although very weak bases, usually[58] protonate at position 3. Deuterium oxide at 50°C causes proton exchange at this position, and also at position 1, but more slowly.[55] Indolizines are substituted by electrophilic reagents preferentially at position 3 with great ease, examples being the 4-nitrobenzenediazonium cation and even hot ethyl acetoacetate, which gives 3-acetoacetylindolizine.[59] This confirms the results of localization energy calculations,[60] which indicate that electrophilic reagents should attack position 3 first, then position 1.

Indolizine is a colourless solid, m.p. 74°C, with a blue fluorescence and an odour resembling naphthalene. It can be synthesized[61] from **2** in 35% overall yield. Another synthesis starts from 2-methylpyridine, and this type of cyclization has been used widely for the preparation of indolizines.[62] Indolizine-1- and indolizine-3-carboxylic acids decarboxylate just above their melting points.

$CH_2(CH_2)_2OH$ Ac_2O $CH(OAc)(CH_2)_2OAc$ pyrolyse
+N O⁻
2
N
(a) NaOH (b) distil + CaO
Me $BrCH_2COCO_2Et$ Me NaHCO$_3$ CO_2Et
N +NCH_2COCO_2Et N
Br⁻

Cycl[3,2,2]azine (**3**), a derivative of indolizine, has a particularly interesting structure which possesses 10 π-electrons and conforms to Hückel's rule. It was first prepared[63] as outlined, and is a stable yellow aromatic compound of m.p. 64°C. It protonates,[64,65] nitrates, and acetylates under Friedel–Crafts conditions[66] at position 1 in accordance with the results of localization-energy calculations.[63] Nuclear magnetic resonance studies show that the molecule can sustain a large ring current.[65,67]

N Me (a) n-BuLi (b) Me_2NCHO N CH_2CHO AcOH, hot
6 5 7 4a 7a N 4 1 3 2a 2
23% yield
3

4. BENZOFURAN

A. General properties

Benzofuran (**1**), sometimes called coumarone, is a colourless liquid, b.p. 173°C, which may be isolated from coal tar. Its structure has not been investigated in detail, but it has aromatic properties and is best treated as a resonance hybrid to which **1** is the major contributor; minor contributions are made by charged structures such as **1a** and **1b**. Benzofuran is much more stable to chemical attack than furan.

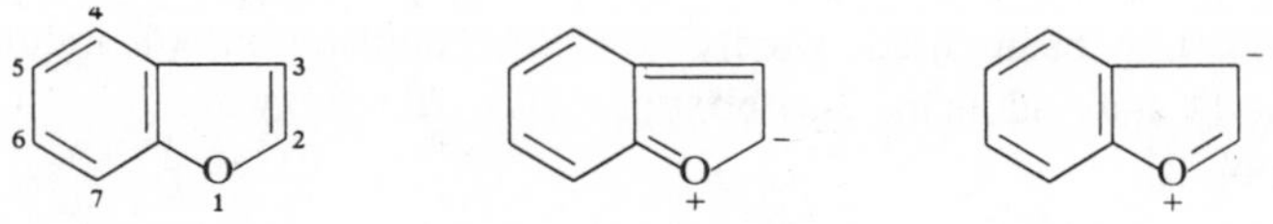

1 **1a** **1b**

The heterocyclic ring of benzofuran is much less easily opened than that of furan, but it can be split by the two routes outlined. Although much more stable

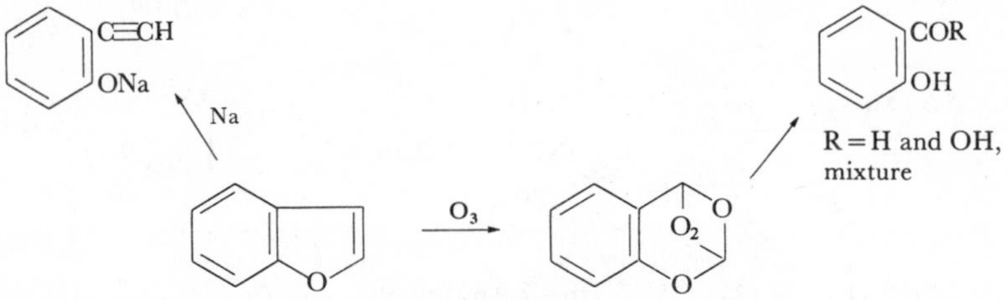

C≡CH ONa Na O_3 O_2 O O COR OH R = H and OH, mixture

than furan to acids, benzofuran is resinified by concentrated sulphuric acid; the 2-sulphonic acid can be obtained with 1-protopyridium sulphonate. The direct nitration of benzofuran with nitric and acetic acids gives the 2-nitro derivative, and other electrophilic reagents, such as acetic anhydride with boron trifluoride, also attack position 2 virtually exclusively. This is in very marked contrast to indole and thianaphthene, where attack takes place predominantly at position 3, and has been attributed to the much greater electronegativity of oxygen than that of nitrogen or sulphur. Proceeding further, and rather naively, if the electrons of the oxygen atom did not take part in the reaction with the electrophilic reagent, then benzofuran could be treated as styrene. The formation of addition products, as occurs readily with bromine and chlorine, is then expected and the addition of the electrophilic nitronium cation would occur at the position, in this case position 2, which would give the more stabilized cation (**2**). Why the electronegativity of the oxygen should be the controlling factor, if in fact it is

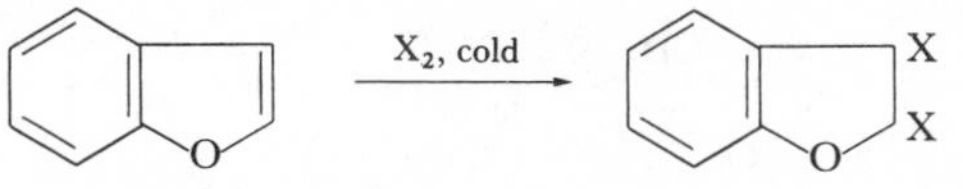

X_2, cold

X=Cl or Br

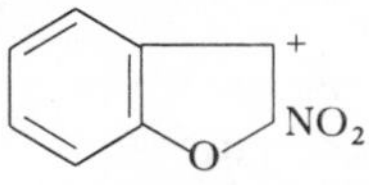

2

so, in the case of benzofuran is not clear, but it may be that the bonding of the oxygen in the heterocyclic ring is not comparable to the bonding of the oxygen atom of anisole. Phenyl radicals also attack[68] position 2 mainly (75·9%), and also positions 4 (17·5%) and 7 (6·6%). Some benzofurans form stable 2,3-epoxides with perbenzoic acid.

Benzofuran is easily reduced by hydrogen over palladium, or sodium and alcohol, to the 2,3-dihydro derivative, which is also known as coumaran. Coumaran behaves as a typical aryl alkyl ether and is dehydrogenated by sulphur to benzofuran.

Coumaran-2-one (**3**) is the lactone of 2-hydroxyphenylacetic acid. Its chemical reactions do not indicate appreciable tautomerism to **4**, and are those expected of a lactone.

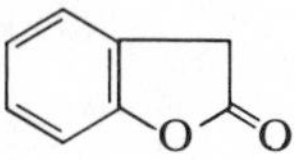

3 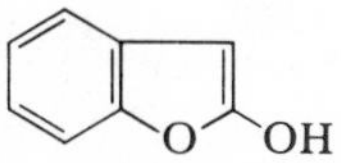 **4** 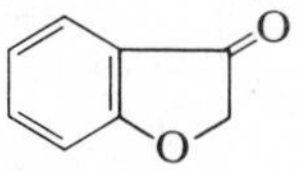 **5**

Coumaran-3-one is a stable colourless crystalline solid, m.p. 97°C. It appears to exist entirely in the keto form (**5**) according to bromine titration, and it is a strong reducing agent. The carbonyl group behaves normally towards hydroxylamine, and the methylene group is sufficiently reactive to condense with benzaldehyde in the presence of hydrochloric acid. Coumaran-3-one with methyl sulphate and acetic anhydride gives 3-methoxy- and 3-acetoxybenzofuran, respectively. This does not provide evidence for tautomerization of **5** to the hydroxy tautomer (c.f. **4**), as the conjugate anion (**6**) and cation (**7**) are the same for both of these structures and are probably the reacting species.

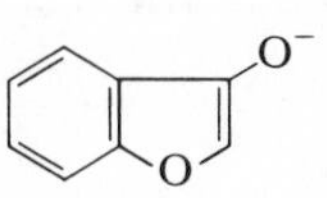

6 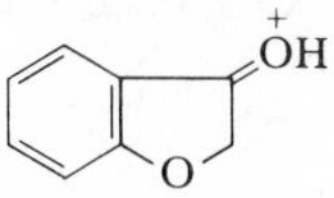 **7**

B. Synthetic methods

(1) The vapor phase synthesis outlined gives high conversions.

OH; $CH_2{=}CH_2$, 170–180 °C; CH_3, CH_2, O; Al_2O_3/Fe_2O_3, CoS, 650 °C; O; + H_2

(2) A useful laboratory synthesis starts with coumarin and is generally applicable.

O, O; Br_2; Br, Br, O, O; KOH; CH, CCO_2^- 2K^+, Br, O^-

O; CO_2H; Distil, CaO, $-CO_2$; O

(3) An internal Claisen reaction may be used to bring about ring closure. When the sodium derivative of ethyl salicylate is used the product is ethyl

COR; ONa; $BrCH_2CO_2Et$; COR; CH_2CO_2Et; O; NaOEt; R; CO_2Et; O

coumaran-3-one-2-carboxylate (**8**), which can be decarboxylated giving the best synthesis[69] of coumaran-3-one (**9**).

O; CO_2Et; O; O; O

8 **9**

(4) An interesting new route[70] involves a Claisen rearrangement.

OH; $ClCH_2CCl{=}CH_2$, K_2CO_3/Me_2CO; Cl; O; Δ, $PhNEt_2$; Cl; OH; conc. HCl; O; Me

5. DIBENZOFURAN

Dibenzofuran, or diphenylene oxide (**1**), is made in 20% yield by the pyrolysis of phenol over lead oxide. It is very much more stable to acids than benzofuran, it is very resistant to ether-cleaving agents, and with chlorosulphonic acid it yields the 2-sulphonic acid followed by the 2,8-disulphonic acid. Halogenation also takes place at these positions. Nitration in acetic anhydride[71] gives a

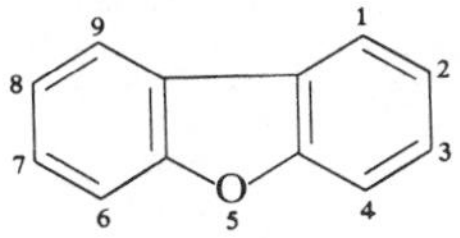

1

mixture of the 1-, 2-, and 3-nitro derivatives in 1 : 2 : 2 ratio. The free radical $\cdot CH_2CO_2H$ attacks the 1-, 2-, 3-, and 4-positions in 53, 0, 15, and 32% ratio respectively,[72] in agreement with localization-energy calculations.[72] Reaction with butyllithium occurs at position 4, giving the 4-lithium derivative.

5-Methyldibenzofuran tetrafluoroborate (**2**) is an even more powerful methylating agent than trimethyloxonium tetrafluoroborate. It methylates dibenzothiophene on the suphur atom.[73]

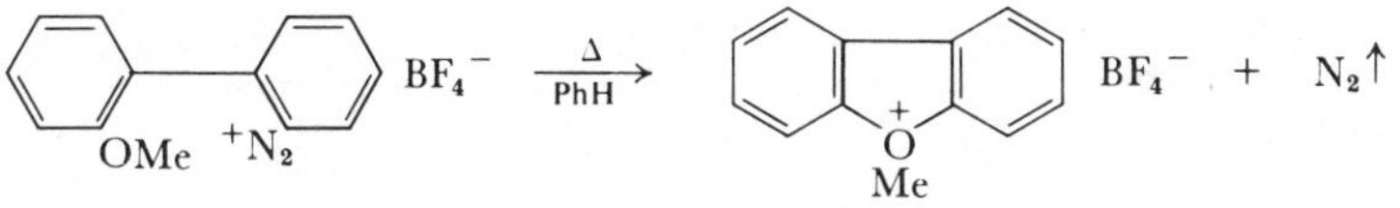

2

6. ISOBENZOFURANS

Isobenzofuran (**3**), or benzo[*c*]furan, has been obtained[74] as a highly reactive compound, m.p. 20°C, by pyrolysis of the adduct (**1**) *in vacuo*, and it is collected on a cold finger. It polymerises rapidly at room temperature, and undergoes Diels–Alder additions with *N*-methylmaleic imide across the 1,3-positions to give both *exo* and *endo* adducts. The reactivity is related to the fact that resonance in the carbocyclic ring is restricted unless the heterocyclic ring develops charges, as in the resonance form **2**. Few alkyl- and arylisobenzofurans are known, although derivatives of 1,3-dihydroisobenzofuran, for example, phthalic anhydride, are very numerous.

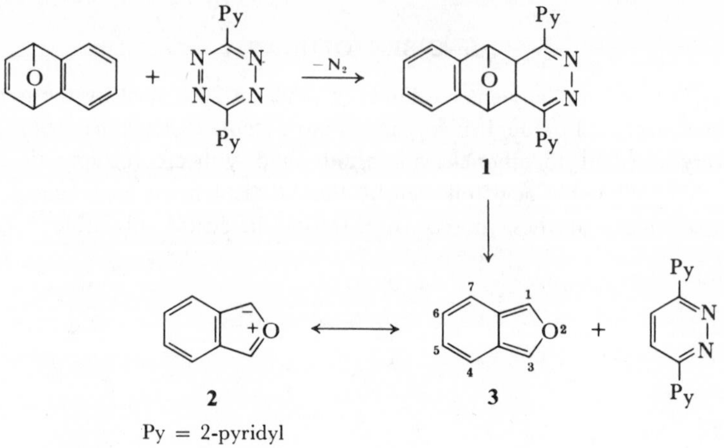

Py = 2-pyridyl

7. BENZO[*b*]THIOPHENE AND RELATED COMPOUNDS

A. Physical properties and structure

Benzo[*b*]thiophene (**1**), also known as thianaphthene, was first obtained by a difficult synthesis in 1893, and in 1902 it was isolated from coal tar. It is the major sulphur-containing impurity in technical naphthalene. Benzothiophene itself has little commercial value, but derivatives in the form of thioindigo dyes have great value and have been much investigated.

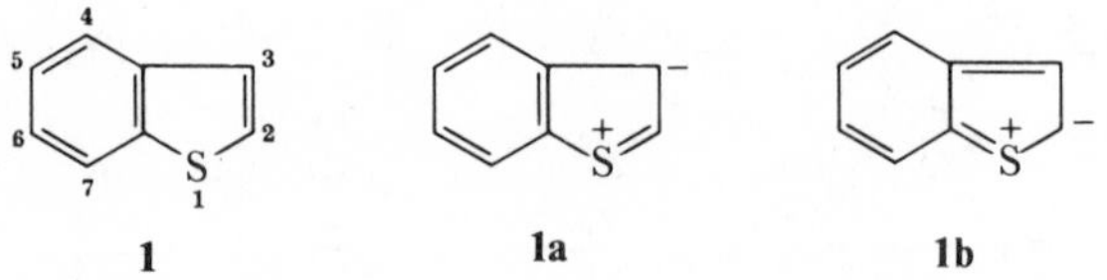

Benzothiophene is a colourless solid, m.p. 31·4°C, b.p. 220°C at 760 torr, and has a dipole moment of 0·62 D. Its structure has not been investigated by physical means, but chemical properties indicate that it is best represented as a resonance hybrid of **1** and the charged structures **1a** and **1b** in order of decreasing importance.

B. Chemical properties and derivatives

Benzo[*b*]thiophene is more stable and less reactive than thiophene. The sulphur-containing ring can be opened in several ways, the most generally useful being a Raney nickel desulphurisation (cf. p. 160). It does not react with methyl iodide alone, but in the presence of silver tetrafluoroborate or on methylation with trimethyloxonium tetrafluoroborate *S*-alkyl derivatives such as **3** can be obtained.[75] The stereochemistry and bonding of the sulphur atom are particularly interesting, as the n.m.r. spectrum of the *S*-ethyl compound (**4**) shows that the methylene hydrogen atoms are nonequivalent. The *S*-alkyl bond is therefore not coplanar with the ring system. It is difficult to assess the

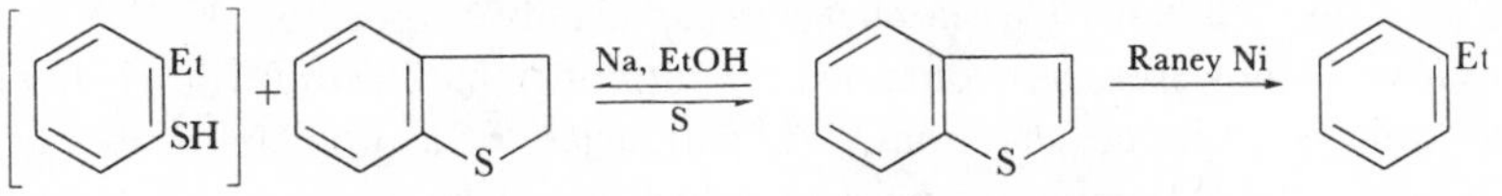

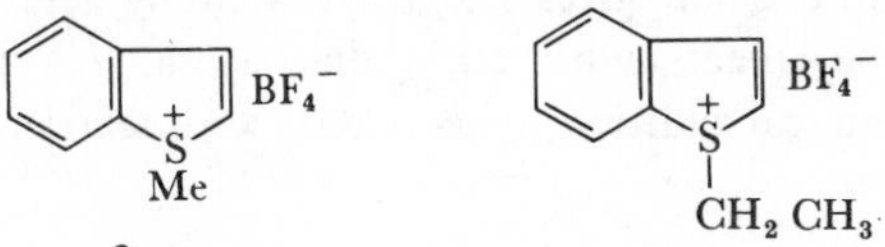

aromaticity of the heterocyclic ring in these compounds, but they are powerful alkylating agents, for example, they slowly alkylate methanol giving back the parent benzo[*b*]thiophene.

With hydrogen peroxide (30%) in acetic acid benzo[*b*]thiophene gives the 1,1-dioxide (**5**), or sulphone, which is much more stable than the corresponding compound from thiophene (p. 155) and only undergoes a similar Diels–Alder reaction with itself at 220°C. The 2,3-bond of the 1,1-dioxide (**5**) has the

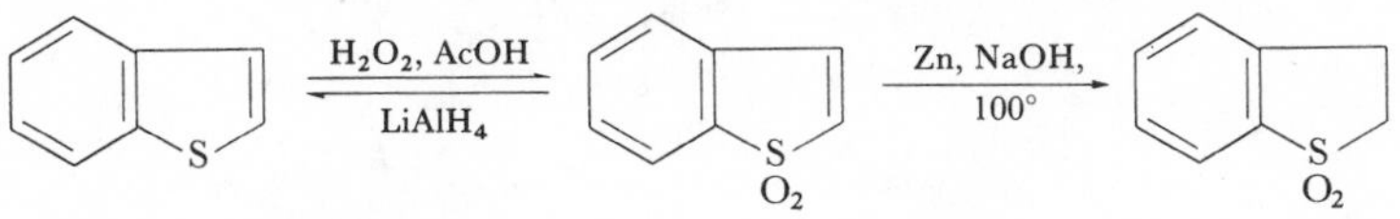

interesting property of adding thiophenol in different directions according to whether ionic or radical catalysis is employed, and nitration occurs at position 6.

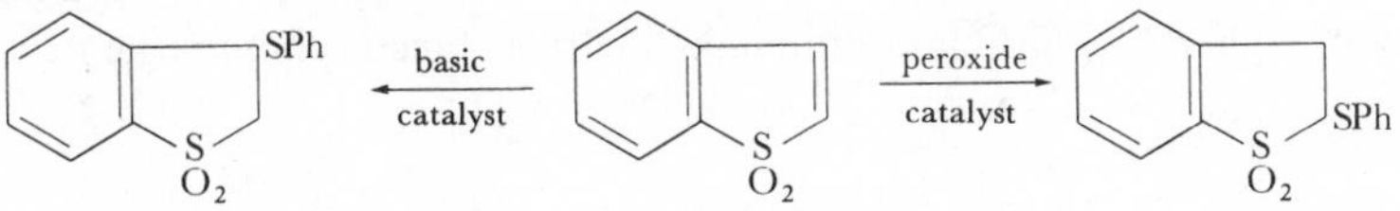

The sulphonation, bromination (*N*-bromosuccinimide), chloromethylation, acylation (Friedel–Crafts), alkylation with olefines and polyphosphoric acid,[76]

and nitration of benzo[*b*]thiophene occur mainly at position 3. The 3-nitro derivative is, however, best obtained by nitrating the 3-carboxylic acid, when this group is displaced and isomer formation less of a problem.[77] Reduction gives the unstable 3-amino compound, which is usually isolated as a derivative.

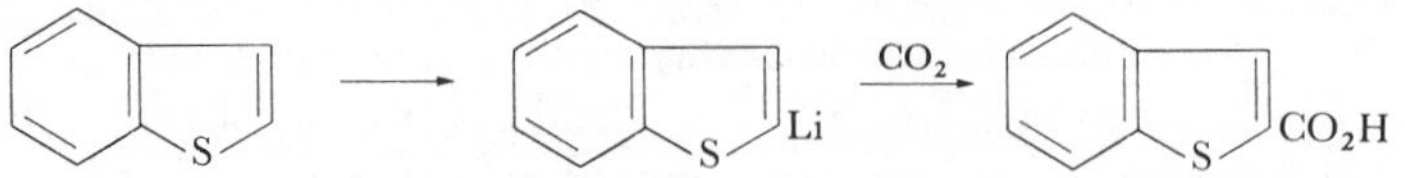

The metallation of benzo[*b*]thiophene with sodium or butyllithium occurs at position 2, and carbonation gives the corresponding acid (**6**). Benzo[*b*]thiophene, like benzene, reacts with ethyl diazoacetate to give **7**, and it also undergoes photochemical addition to acetylenes with rearrangement.[78,79]

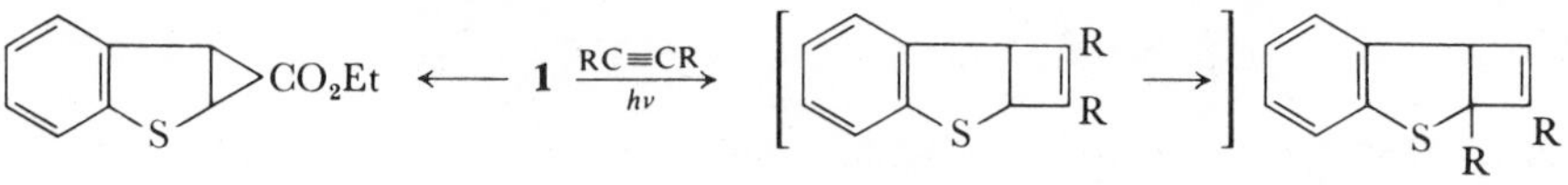

R = Ph or CO_2Me

5-Hydroxybenzothiophene is of special interest, as on bromination it yields **8**. By analogy with 2-naphthol, bromination only at position 4 would be expected, and the result shows the lesser effect of fusing a thiophene ring, in comparison with a benzene ring, onto the phenol system.

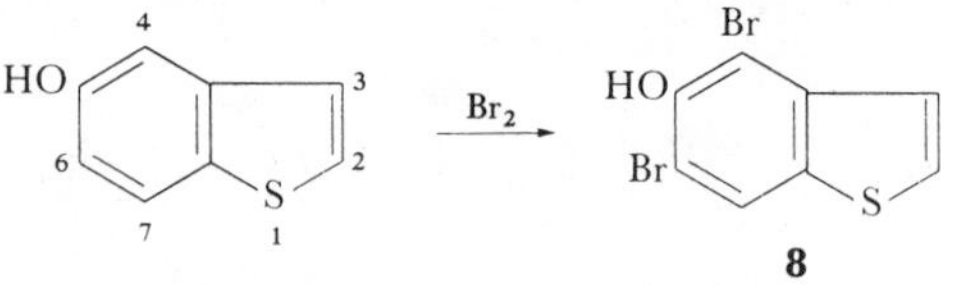

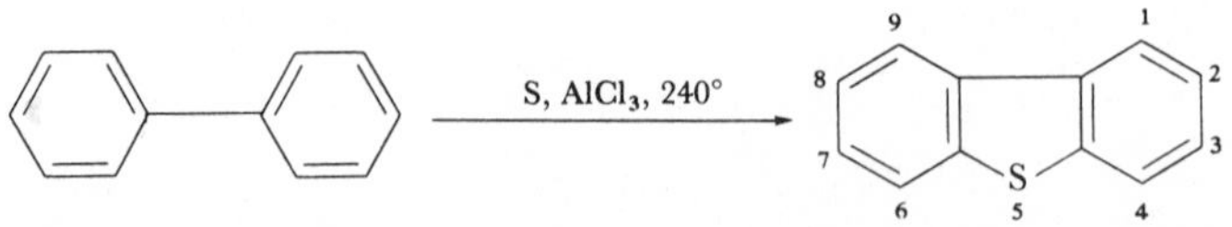

Dibenzothiophene (**9**) can be obtained[80] from diphenyl and has m.p. 99–100°C. Oxidation with hydrogen peroxide in acetic acid first gives the 5-oxide, followed by the 5,5-dioxide. Both these compounds are reduced to dibenzothiophene by lithium aluminium hydride. Electrophilic reagents attack positions 2 and 8 of dibenzothiophene as might be expected, and metallation usually takes place at position 4. Phenyl radicals, from benzoyl peroxide, attack positions 1, 2, 3, and 4 in approximately 3 : 1 : 2 : 3 ratio.

C. Synthetic methods[80]

(1) Benzothiophene is available commercially and is possibly prepared from ethylbenzene in the vapour phase. It can also be obtained in 71% yield from thiophenol and acetylene at 600–650°C.

Et; H_2S; Cr_2O_3/Al_2O_3; S; Cr_2O_3/Al_2O_3; 475°; Et; SH

(2) The best method[80] of synthesising benzothiophenes is shown; other routes similar to those for benzofurans [p. 218 (3)] have been used.

PhLi —$[SCH_2CH(OMe)_2]_2$ / $-LiSCH_2CH(OMe)_2$→ $PhSCH_2CH(OMe)_2$ —polyphosphoric acid→ S

PhSH —$BrCH_2CH(OMe)_2$→

(3) Another synthesis involves the Diels–Alder reaction.

O; O; O; + H_2C; HC; S; O; O; O; S; decarboxylation; dehydrogenation; S

(4) The displacement of nitro groups *ortho* to cyano (leading to 3-amino compounds) or aldehyde groups in a benzene ring by a suitable sulphur nucleophile, followed by cyclisation gives a versatile synthesis.[81]

CHO; NO_2; $HSCH_2CO_2Me$; base, 0 °C DMF; $-HNO_2$; [CHO; SCH_2CO_2Me]; S; CO_2Me

D. Benzo[*c*]thiophene

Benzo[*c*]thiophene (**10**), or isothianaphthene, has m.p. 50–51°C, an odour like that of naphthalene and is stable at –30°C under nitrogen for only one day.

With maleic anhydride it gives the expected adduct (**11**), which is desulphurized to a naphthalene (**12**) by sodium hydroxide.[82]

vapour phase
Pd/C, 330°
heat
NaOH

10 **11** **12**

E. Compounds of special interest

Thioindigo (**14**), which has been shown by X-ray measurements to have the *trans* structure, is of special interest, as many of its derivatives are valuable fast dyestuffs varying in colour from yellow-orange to violet-black. Two and a half million pounds of thioindigo dyes were sold in the United States in 1958. For a full discussion of the immense amount of work on this subject the reader is referred to Hartough and Meisel's monograph on condensed thiophenes (see General Bibliography).

Thioindigo itself is obtainable by the mild oxidation of the sulphur analogue of indoxyl (**13**), sometimes called 3-hydroxybenzothiophene.

$K_3Fe(CN)_6$

13 **14**

Asymmetrical thioindigos can also be made (X = O or NH, or S when the two benzene rings have different substituents), as can analogues of isoindigo and

indirubin by methods similar to those used in the indigo series. The reduction of thioindigos to weakly acidic, alkali-soluble, colourless dihydro compounds by sodium hydrosulphite in the dyeing vat, followed by aerial oxidation on the fibre, constitutes the dyeing process, as in the case of indigo dyes.

GENERAL BIBLIOGRAPHY

INDOLE

W. J. Houlihan (ed.), *Indoles*, Part I and Part II, Wiley-Interscience, New York, 1972.

R. J. Sundberg, *The Chemistry of Indoles*, Academic Press, New York, 1970.

S. P. Hisemath and R. S. Hosmane, 'Nuclear Magnetic Resonance Spectra of Indoles,' *Adv. Heterocycl. Chem.*, **15**, 278 (1973).

ISOINDOLE

J. D. White and M. E. Mann, *Adv. Heterocycl. Chem.*, **10**, 113 (1969).

F. H. Moser and A. L. Thomas, *Phthalocyanine Compounds*, Am. Chem. Soc. Monograph No. 157, Reinhold, New York, 1963.

INDOLIZINE

W. L. Mosby, *Heterocyclic Compounds with Bridgehead Nitrogen Atoms*, Part I (1961) and Part II, 1962, Interscience, New York.

BENZOFURANS

A. Mustapha, *Benzofurans*, Wiley-Interscience, New York, 1974.

BENZO[*b*] THIOPHENES

H. D. Hartough and S. L. Meisel, *Compounds with Condensed Thiophene Rings*, Interscience, New York, 1954.

B. Iddon and R. M. Scrowston, *Adv. Heterocyclic Chem.*, **11**, 178 (1970).

BENZO[*c*] THIOPHENES

B. Iddon, *Adv. Heterocycl. Chem.*, **14**, 331 (1972).

DIBENZO[*bd*] THIOPHENE

J. Ashby and C. C. Cook, *Adv. Heterocycl. Chem.*, **16**, 181 (1974).

REFERENCES

1. I. L. Karle, K. Britts, and P. Gum, *Acta Crystallogr.*, **17**, 496 (1964).
2. P. Hikasai and D. McLeod, *J. Amer. Chem. Soc.*, **95**, 27 (1973).
3. E. Brandeau, S. David, and J. C. Fischer, *Tetrahedron*, **30**, 1445 (1974).
4. L. J. Dolby and D. L. Booth, *J. Amer. Chem. Soc.*, **88**, 1049 (1966).
5. R. L. Hinman and J. Lang, *J. Amer. Chem. Soc.*, **86**, 3796 (1964); B. C. Challis and F. A. Long, *ibid.*, **85**, 2524 (1963).
6. G. F. Smith, *Adv. Heterocycl. Chem.*, **2**, 287 (1963).
7. L. J. Dolby and G. W. Gribble, *J. Heterocycl. Chem.*, **3**, 124 (1966).
8. A. Smith and J. H. P. Utley, *Chem. Comm.*, **427** (1965); R. Kuhn and I. Butula, *Angew. Chem.*, **80**, 189 (1968).
9. E. Walton, F. W. Holly, and S. R. Jenkins, *J. Org. Chem.*, **33**, 192 (1968).
10. J. Thesing, G. Semler, and G. Mohr, *Chem. Ber.*, **95**, 2205 (1962).
11. D. R. Julian and G. D. Tringham, *Chem. Comm.*, 13 (1973); D. R. Julian and R. Forto, *Chem. Comm.*, 311 (1973).
12. H. Heaney and S. V. Ley, *J.C.S. Perkin I*, 499 (1973).
13. R. J. Sundberg and H. F. Russell, *J. Org. Chem.*, **38**, 3324 (1973).
14. R. H. Heacock and S. Kaspárek, *Adv. Heterocyclic Chem.*, **10**, 43 (1969).
15. M. De Rosa, *Chem. Comm.*, 482 (1975).
16. V. Caló, F. Ciminale, L. Lopez, F. Naso, and P. E. Todesco, *J.C.S. Perkin I*, 2567 (1972).
17. J. C. Powers, *J. Org. Chem.*, **31**, 2627 (1966).

18. R. Bonnett and R. Holleyhead, *J.C.S. Perkin I*, 962 (1974); H. F. Hodson and G. F. Smith, *J. Chem. Soc.*, 3546 (1957).
19. B. C. Challis and A. J. Lawson, *J.C.S. Perkin II*, 918 (1973).
20. G. F. Smith and D. A. Taylor, *Tetrahedron*, 669 (1973).
21. K. Yamasaki, T. Matsuura, and I. Sacto, *Chem. Comm.*, 944 (1974).
22. K. Brown and A. R. Katritzky, *Tetrahedron Lett.* 803 (1964).
23. R. L. Hinman and J. Lang, *J. Org. Chem.*, **29**, 1449 (1964).
24. J. Hutton and W. A. Waters, *J. Chem. Soc.*, 4253 (1965).
24a. T. Kanaoka, M. Aiwa, and S. Hariya, *J. Org. Chem.*, **36**, 458 (1971).
25. A. H. Jackson, B. Naidoo, and P. Smith, *Tetrahedron,* **24**, 6119 (1968); A. H. Jackson and P. Smith, *Chem. Comm.*, 264 (1967).
26. Y. Kanaoka, K. Miyashita, and O. Yonemitsu, *Chem. Comm.*, 1365 (1969).
27. W. I. Taylor, *Proc. Chem. Soc.*, 247 (1962).
28. N. Sugiyama, M. Akutagawa, and H. Yamamoto, *Bull. Chem. Soc. Jap.,* **41**, 936 (1968).
29. E. Baciocchi and A. Schiroli, *J. Chem. Soc.* (*B*), 401 (1968).
30. C. B. Hudson and A. V. Robertson, *Australian J. Chem.,* **20**, 1699 (1967).
31. F. J. McEvoy and G. R. Allen, *J. Org. Chem.,* **38**, 3350 (1973).
31a. J. Harley-Mason and T. J. Leeney, *Proc. Chem. Soc.*, 368 (1964).
32. R. M. Acheson and R. W. Snaith, *J. Chem. Soc.*, 3229 (1964).
33. R. M. Acheson, *Accounts Chem. Res.,* **4**, 177 (1971).
34. K. N. Kilminster and M. Sainsbury, *J.C.S. Perkin I*, 2264 (1972).
35. D. J. Bauer, L. St. Vincent, C. H. Kempe, and A. W. Downie, *Lancet,* **ii**, 494 (1963).
36. B. Robinson, *Chem. Rev.,* **69**, 227 (1969); **63**, 373 (1963).
37. T. P. Forrest and F. M. F. Chan, *Chem. Comm.*, 1067 (1972).
38. J. T. Fitzpatrick and R. D. Hiser, *J. Org. Chem.,* **22**, 1703 (1957).
39. R. L. Augustine, A. J. Gustavsen, S. F. Wanat, I. C. Pattison, K. S. Houghton, and G. Koletar, *J. Org. Chem.,* **38**, 3004 (1973).
40. P. G. Gossman, D. P. Gilbert, and T. J. van Bergen, *Chem. Comm.*, 201 (1974).
41. C. D. Jones and T. Suarez, *J. Org. Chem.,* **37**, 3623 (1972); C. D. Jones, *J. Org. Chem.*, **37**, 3624 (1972).
42. R. J. Sundberg, H. F. Russell, W. V. Ligon, and L-S. Lin, *J. Org. Chem.,* **37**, 719 (1972); P. Gemerand and H. W. Moore, *Chem. Comm.*, 358 (1973).
43. G. R. Allen, *Org. React.*, Vol. 20, Wiley, New York, 1973.
44. J. D. Barchas and E. Usdin (eds.), *Serotonin and Behavior*, New York, 1973.
45. H. P. Weber and T. J. Petcher, *J.C.S. Perkin II*, 942, 946 (1974) and papers therein cited.
45a. S. Agurell and G. Nilson, *Acta Chem. Scand.,* **22**, 1210 (1968).
46. M. C. Elliott and B. B. Stowe, *Plant Physiol.,* **48**, 498 (1971) and earlier papers.
47. J. T. Baker, *Endeavour,* **33**, 11 (1974).
48. H. E. Miller, H. Rösler, A. Wohlpart, H. Wyler, M. E. Wilcox, H. Frohofer, T. J. Mabry, and A. S. Dreiding, *Helv. Chim. Acta,* **51**, 1470 (1968).
49. M. E. Wilcox, H. Wyler, T. J. Mabry, and A. S. Dreiding, *Helv. Chim. Acta,* 48, 252 (1965).
50. H. Wyler, M. E. Wilcox, and A. S. Dreiding, *Helv. Chim. Acta,* **48**, 361 (1965).
51. J. D. White and M. E. Mann, *Adv. Heterocycl. Chem.,* **10**, 113 (1969).
52. R. Bonnett, R. F. C. Brown, and R. G. Smith, *J.C.S. Perkin I*, 1432 (1973); R. Bonnett and R. F. C. Brown, *Chem. Comm.*, 393 (1972).
53. J. Bornstein, D. E. Remy, and J. E. Shields, *Chem. Comm.*, 1149 (1972); G. M. Priestley and R. N. Warrener, *Tetrahedron Lett.*, 4295 (1972).

54. D. F. Veber and W. Lwowski, *J. Amer. Chem. Soc.,* **86**, 4152 (1964).
55. M. Engewald, M. Mühlstädt, and C. Weiss, *Tetrahedron,* **27**, 4171 (1971).
56. B. Jaques and R. G. Wallace, *Chem. Comm.*, 397 (1972).
57. A. Galbraith, T. Small, R. A. Barnes, and V. Boekelheide, *J. Amer. Chem. Soc.,* **83**, 453 (1961).
58. W. L. F. Armarego, *J. Chem. Soc.*, 4226 (1964); M. Fraser and D. H. Reid, *ibid.*, 1421 (1963).
59. F. N. Stepanov and N. I. Grineva, *Zh. Obshch. Khim.,* **32**, 1529 (1962); *J. Gen. Chem. U.S.S.R.,* **32**, 1515 (1962).
60. V. Boekelheide and T. Small, *J. Amer. Chem. Soc.,* **83**, 462 (1961).
61. V. Boekelheide and W. Feely, *J. Org. Chem.,* **22**, 589 (1957).
62. D. R. Bragg and D. G. Wibberley, *J. Chem. Soc.*, 3277 (1963).
63. R. J. Windgassen, W. H. Saunders, and V. Boekelheide, *J. Amer. Chem. Soc.,* **81**, 1459 (1959).
64. F. Gerson, E. Heilbronner, and H. Zimmermann, *Helv. Chim. Acta,* **46**, 1940 (1963).
65. V. Boekelheide, F. Gerson, E. Heilbronner, and D. Meuche, *Helv. Chim. Acta,* **46**, 1951 (1963).
66. V. Boekelheide and T. Small, *J. Amer. Chem. Soc.,* **83**, 462 (1961).
67. L. M. Jackman, Q. N. Porter, and G. R. Underwood, *Australian J. Chem.,* **18**, 1221 (1965).
68. P. Spagnolo, M. Tiecco, A. Tundo, and G. Martelli, *J.C.S. Perkin I*, 556 (1972).
69. D. C. Schroeder, P. O. Corcoran, C. A. Holden, and M. C. Mulligan, *J. Org. Chem.,* **27**, 586 (1962).
70. W. K. Anderson and E. J. La Voie, *Chem. Comm.*, 174 (1974).
71. M. J. S. Dewar and D. S. Urch, *J. Chem. Soc.*, 345 (1957).
72. P. L. Southwick, M. W. Munsell, and E. A. Barktus, *J. Amer. Chem. Soc.,* **83**, 1358 (1961).
73. A. J. Copson, H. Heaney, A. A. Logun, and R. P. Sharma, *Chem. Comm.*, 315 (1972).
74. R. N. Warrener, *J. Amer. Chem. Soc.,* **93**, 2347 (1971).
75. R. M. Acheson and D. R. Harrison, *J. Chem. Soc. (C)*, 1764 (1970).
76. J. Cooper and R. M. Scrawston, *J.C.S. Perkin I*, 414 (1972).
77. K. J. Armstrong, M. Martin-Smith, N. M. D. Brown, G. C. Brophy, and S. Sternhell, *J. Chem. Soc. (C)*, 1766 (1969).
78. W. H. F. Sasse, P. J. Collins, and D. B. Roberts, *Tetrahedron Lett.,* **54**, 4791 (1969).
79. D. C. Neckers, J. H. Dopper, and A. Wynberg, *Tetrahedron Lett.,* **34**, 2913 (1969).
80. B. D. Tilak, *Tetrahedron,* **9**, 76 (1960).
81. J. R. Beck, *J. Org. Chem.,* **37**, 3224 (1972).
82. R. Mayer, H. Kleinert, S. Richter, and K. Gewald, *Angew. Chem.,* **74**, 118 (1962); B. D. Tilak, H. S. Desai, and S. S. Gupta, *Tetrahedron Lett.,* 1956 (1966).

CHAPTER V

HETEROCYCLIC ANALOGUES OF BENZENE WITH ONE HETEROATOM

The most important ring system of this class, that of pyridine, can be derived theoretically from benzene by the replacement of a carbon and a hydrogen atom by one of nitrogen. This change leaves the aromatic structure of the ring essentially unaltered. A similar replacement of carbon and hydrogen atoms by

Pyridine   Pyrylium cation   1,2-Pyran   1,4-Pyran   1-R-Thia(IV)benzene

one of oxygen cannot be envisaged because of valency considerations unless the oxygen atom subsequently bears a positive charge. This is the case with the pyrylium cation, which has aromatic properties reminiscent of those of certain benzenes. Reduction of the pyrylium cation could lead to the nonaromatic 1,2- and 1,4-pyrans, and derivatives of both are known. Sulphur analogues of the pyrylium salts and pyrans, which are called thiapyrylium salts and thiapyrans, respectively, and some unstable tetravalent sulphur-containing compounds, the thia(IV)benzenes, have been prepared recently.

1. PYRIDINE

A. Introduction

Pyridine was discovered by Anderson in 1849, and he isolated this base in a pure state from bone oil. Anderson also obtained pure picoline (methylpyridine) and lutidine (dimethylpyridine) from the same source. These pyridines do not occur free in bones but are formed thermally during the distillation.

Compounds containing the pyridine ring are widely distributed in nature. Certain of them, examples being vitamin B_6 and the nicotinamide adenine dinucleotide phosphates, are of the greatest biochemical importance. The pyridine ring also features in the structures of many drugs, dyes, and alkaloids.

Pyridine is an aromatic compound that is very similar to benzene in general structure. The asymmetry of the ring, however, greatly increases the number of structural isomers possible in comparison with the benzene series; there is only one monomethylbenzene, but there are three monomethylpyridines. Pyridine has 19 possible methyl substitution products, whereas benzene gives rise to only 12. Pyridine is also a tertiary base and has many properties characteristic of such compounds.

Because of the difficulty in introducing substituents into the pyridine ring by electrophilic reagents, although pyridine itself is available in quantity, many syntheses involving building the ring have been developed. This situation is in direct contrast to that of the benzene series, where interest in syntheses of the ring has long been lacking owing to the ease with which substitutions and transformations of existing substituents may be carried out. In very recent years the position has altered noticeably with the development of the chemistry of the pyridine 1-oxides and of radical and nucleophilic reagents which can give high yield reactions with the pyridine nucleus.

The positions of substituents round the pyridine ring are numbered (**1**) or occasionally lettered (**2**). The cation derived from pyridine is called pyridinium, and hexahydropyridine is usually known as piperidine. A comparatively large number of pyridines possess trivial names.

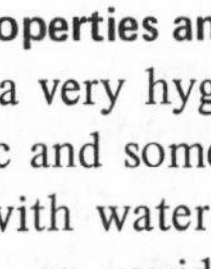

4 5 3 6 2 N 1 γ β' β α' α N $\overset{+}{N}$ H N H

1 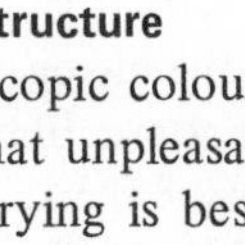 **2** 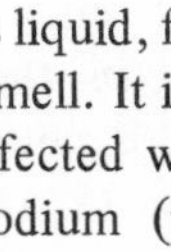 Pyridinium 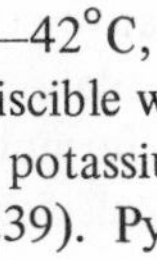 Piperidine

B. Physical properties and structure

Pyridine is a very hygroscopic colourless liquid, f.p. –42°C, b.p. 115°C, with a characteristic and somewhat unpleasant smell. It is miscible with most organic solvents and with water. Drying is best effected with potassium hydroxide or barium oxide, as pyridine reacts with sodium (p. 239). Pyridine is a very powerful solvent for many types of organic compounds, and it often dissolves salts to give conducting solutions.

The cyclic structure usually written for pyridine (**1**) was suggested in 1869 by Körner, and the arrangement of the atoms was confirmed by a synthesis from pentamethylenediamine.

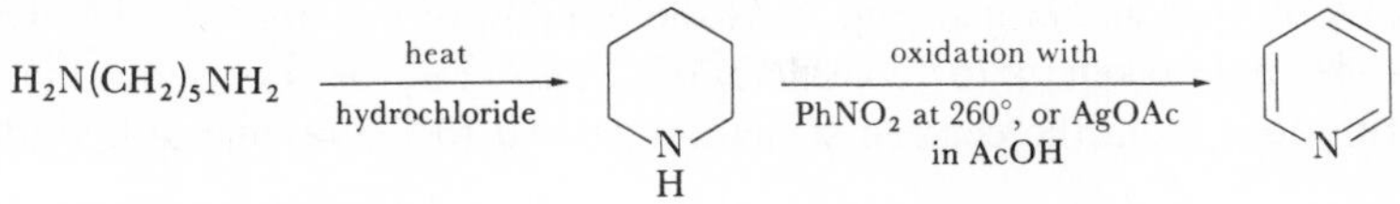

$$H_2N(CH_2)_5NH_2 \xrightarrow[\text{hydrochloride}]{\text{heat}} \text{piperidine (N–H)} \xrightarrow[\text{PhNO}_2\text{ at }260°\text{, or AgOAc in AcOH}]{\text{oxidation with}} \text{pyridine (N)}$$

The literature contains a number of values for the resonance energy of pyridine, but the most reliable result, based on 1963 combustion data,[1] is 124 ± 8 kJ (31·9 ± 2·0 kcal)/mole. This is slightly less than the accepted value for benzene [150·5 kJ (36 kcal)/mole].

The dimensions of the pyridine molecule have been calculated from microwave measurements.[2] The pure sp^2 hybridization of the electrons of the carbon atoms which is present in benzene is slightly disturbed, as the bond angles

Dimensions of pyridine

are not exactly 120°. A comparison of the carbon–carbon bond lengths of benzene, pyridine, and pyrrole shows that all the carbon–carbon bonds in the first two compounds are in the 1·39–1·40 Å region. In pyrrole the carbon–carbon bonds show a much larger alternation in length (p. 91), which indicates less resonance than in pyridine or benzene. As the pyridine ring is not quite symmetrical, although planar, it is reasonable that the resonance energy of pyridine is of the same order but a little less than that of benzene.

π-Electron density calculations for pyridine show a drift of electrons from the 2- and 4-positions, and to a lesser extent from the 3-position, to the nitrogen atom, although the magnitude varies with the method employed and the assumptions made. This is in accord with the dipole moment (2·20 D) of pyridine, the negative pole of which is at the nitrogen atom.

From all these data it is clear that an alternation of double and single bonds, as suggested by formula **1**, is not actually the case in pyridine. A resonance hybrid of the Kekulé formulae (**1**, **1a**), which suffices to describe the structure in

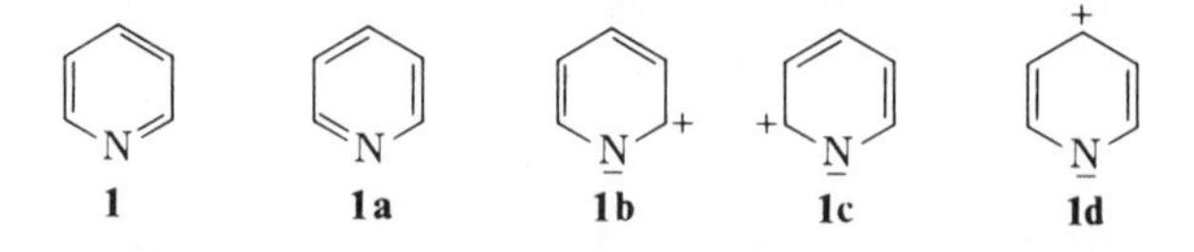

the case of benzene, in inadequate, as the buildup of electrons on the nitrogen atom is not taken into consideration. This can be done by including contributions from the charged structures **1b**–**1d** to the resonance hybrid, and then the chemical properties of pyridine can be adequately accounted for. Of the charged structures, those involving the least charge separation should be favoured. This is in agreement with the preferential attack of nucleophilic reagents at position 2 of the ring. Localisation energy calculations[3] give the same result and also suggest correctly that radicals will attack mainly position 2 and electrophiles position 3 in both pyridines and the pyridinium cation. In spite of the importance of charged and other structures to the resonance picture of pyridine, it is normally convenient to represent pyridine, like benzene, by the Kekulé structure (**1**).

C. Chemical properties

Pyridine is a very stable compound with a great deal of aromatic character. In many respects it greatly resembles benzene. At the same time it contains a tertiary nitrogen atom which possesses many properties of a typical tertiary amine. Pyridine is a weak monoacid base and forms stable salts with strong acids. It can even be used as a solvent for some chromic acid oxidations although it is oxidised to carbon dioxide and ammonia by sodium hypochlorite at pH 12–13.[4] It should be particularly noted that reactions taking place under strongly acidic conditions may involve the pyridinium cation, which is isoelectronic with benzene, while comparable reactions in the benzene series involve the benzene molecule. It is also very important to note that, in contrast to the case of pyrrole (p. 93), the lone pair of electrons on the nitrogen atom employed in salt formation is not needed to make up the sextet of electrons essential for aromaticity, and, in consequence, in the pyridine series salt formation does not destroy the aromatic properties of the ring.

1. *Pyridinium salts*

Pyridine (pK_a 5·23) is a much weaker base than ammonia (pK_a 9·23) or trimethylamine (pK_a 9·90) but forms stable, water-soluble salts with many acids; the mercurichloride ($PyHHgCl_3$), chloroaurate ($PyHAuCl_4$), and chloroplatinate ($Py_2H_2PtCl_6$), in contrast, are very sparingly soluble.

Pyridine is a very useful solvent-base in a number of reactions. It is used with piperidine in the Knoevenagel condensation of aromatic aldehydes with malonic acid, the products being cinnamic acids. It can be employed for removing hydrogen halides from compounds under anhydrous conditions. It is also valuable in the acylation of phenols, alcohols, and amines by acid chlorides, acid anhydrides, and mixed anhydrides. The acid derivative combines with the

PhCOCl → $\overset{+}{N}$–COPh Cl^- (**2**) —ROH→ PhCOOR + $\overset{+}{N}$–H Cl^- (**3**)

pyridine to give a quaternary salt which can be isolated in some cases (e.g., **2**), or detected kinetically[5] as an intermediate. This salt then reacts with the hydroxy or amino group yielding the acyl derivative; the liberated acid is taken up as the pyridine salt (**3**). Quaternary salts such as **2** are immediately decomposed by water to pyridine hydrochloride and the organic acid but can be attacked by nucleophiles with ring opening (pp. 233–234).

Pyridine is easily converted into quaternary salts by reagents such as dimethyl sulphate and methyl iodide. 1-Methylpyridinium iodide (**5**) is a colourless hygroscopic solid easily soluble in water, and it behaves as a strong electrolyte in aqueous solution. In conformity with its ionic properties it is virtually insoluble in ether or hydrocarbon solvents. It can be demethylated in high yield on long boiling in dimethylformamide,[6] and on heating in a sealed tube the N–Me bond is split homolytically to give mainly the hydriodides of 2- and 4-methylpyridine.[7] In 95% acetone solution the iodide **5** is present in two distinct types of ion pair, one concentration dependent and the other (ions separated) not.[8] Although aqueous solutions do not accurately obey Beer's law in the ultraviolet, electrical conductivity measurements show that the compound is, apparently, completely ionised. Consequently, it seems likely that a small amount of the pseudoiodide (**7**)[7] is present in equilibrium. As both the conductivity, allowing for the additional ions present, and the ultraviolet absorption spectrum of the iodide (**5**) are unchanged by the addition of sodium hydroxide, the 1-methylpyridinium hydroxide (**6**) must also be completely ionised within the limitations of the experiment. Oxidation of this hydroxide (**6**) by alkaline potassium

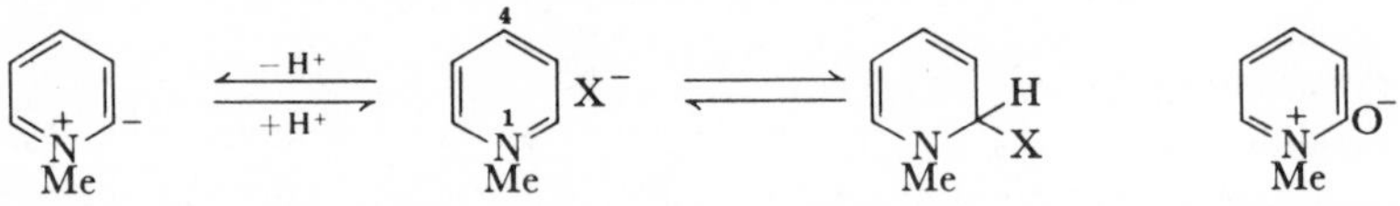

4 ⇌ ($-H^+$ / $+H^+$) **5**, X = I; **6**, X = OH ⇌ **7**, X = I; **8**, X = OH — **9**

ferricyanide does, however, give 1-methyl-2-pyridone (**9**) in good yield. As oxidations with ferricyanide ions usually proceed by the abstraction of hydrogen and not by the addition of oxygen, a case can be made for the supposition that a small proportion of the pseudohydroxide (**8**) is in equilibrium with the ionic hydroxide (**6**) in aqueous solution. Boiling **5** with aqueous sodium hydroxide gives only methylamine as identifiable product.

The salt **5** in deuterium oxide undergoes base-catalysed exchange of only the 2 and 6 hydrogen atoms for deuterium. An intermediate of type **4** must be involved, and if a 1-substituent is absent, the presence of acid is necessary and a deuteron is added to position 1 before exchange can occur.[9] The exchange is greatly facilitated by electron-attracting substituents at position 3.

Nucleophiles can attack pyridinium salts to give stable 1,2- and 1,4-dihydropyridines, which are considered on p. 239.

The infrared absorption spectra of solutions of iodine in pyridine suggest[10] that *N*-iodopyridinium iodide (cf. **5**) is obtained and similar compounds are formed with bromine and chlorine. The catalytic effect of pyridine on halogenations, such as the bromination of benzene is understandable, as the pyridinium cation could be a more ready source of positive halogen than the halogen molecule itself.

Pyridine is oxidised easily to the *N*-, or 1-oxide, a type of internal pyridinium salt, which is considered later (p. 249).

2. *The ring opening of pyridinium salts*

When the group attached to the nitrogen atom is strongly electron attracting, the ring is comparatively easily opened by nucleophilic reagents. 2,4-Dinitrophenylpyridinium chloride (**10**) is a colourless crystalline solid formed from pyridine and 2,4-dinitrochlorobenzene at 100°C; the reaction is reversed at 200°C. With water at 150°C the chloride yields pyridine hydrochloride and 2,4-dinitrophenol, but with cold aqueous alkali the deep red compound **12** is formed. It is probable that the reaction occurs through addition of hydroxyl anion to give **11** and subsequent electrocyclic ring opening. Treatment of **12**

Cl^- N⁺ Ar

10

Ar = 2,4-$(NO_2)_2C_6H_3$—

HO^-

OH N Ar

11

N Ar OH

12

$\xrightarrow[HCl]{PhNH_2}$ $ArNH_2$ + NHPh Cl^- $\overset{+}{N}H$ Ph

13

with dilute acid and aniline yields 2,4-dinitroaniline and the purple glutaconic aldehyde derivative hydrochloride (**13**), which is highly resonance stabilised. Pyridine with cyanogen bromide yields 1-cyanopyridinium bromide (**16**), which with aniline also yields **13** under suitable conditions, and this reaction can be used to detect nucleophilic pyridines in small quantities on chromatograms. If **13** is treated with triethylamine, the corresponding base (**13a**) is formed, and this cyclises by a first order process, the characteristics of which are consistent with an electrocyclic reaction,[11] giving **14**, followed by a rapid acceptance of a proton to form 1-phenylpyridinium chloride and aniline.

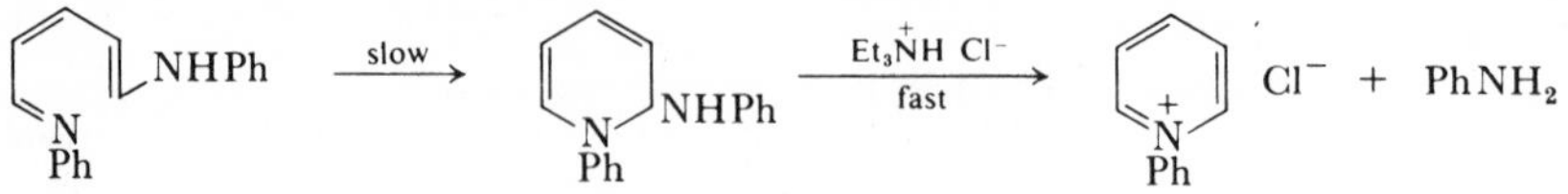

13a **14** **15**

The formation[12] of the open-chain compound **17** from the pyridinium salt **2** and the sodium salt of ethyl cyanoacetate can be explained in the same way, and electrocyclic openings of suitable 1,2-dihydropyridine rings can be postulated in other cases.

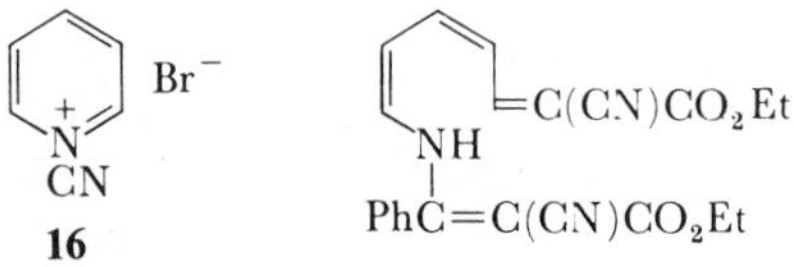

16 **17**

In methylene chloride, pyridine and sulphur trioxide give the addition compound 1-pyridiniumsulphonate (**18**), which is a useful mild sulphonating agent yielding the pyridinium salt of the sulphonic acid produced, very strongly acid conditions being avoided. In a similar way the adduct from 2-methyl-pyridine and nitronium tetrafluoroborate is a valuable nonacidic nitrating agent. The sulphonate **18** is hydrolysed by sodium hydroxide to the sodium salts of sulphamic acid and of glutaconic aldehyde (**19**, enol form); the product (**20**) from pyridine and thiophosgene likewise gives[13] the isothiocyanate **21**.

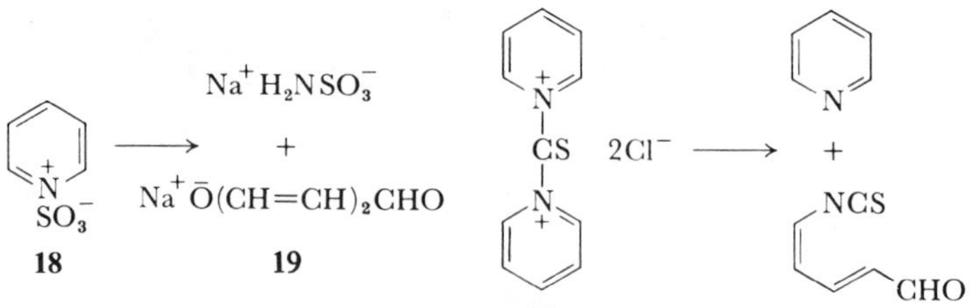

18 **19** **20** **21**

Azulene (**22**) can be obtained in 60% overall yield from pyridine, via the salt **10**, as indicated.

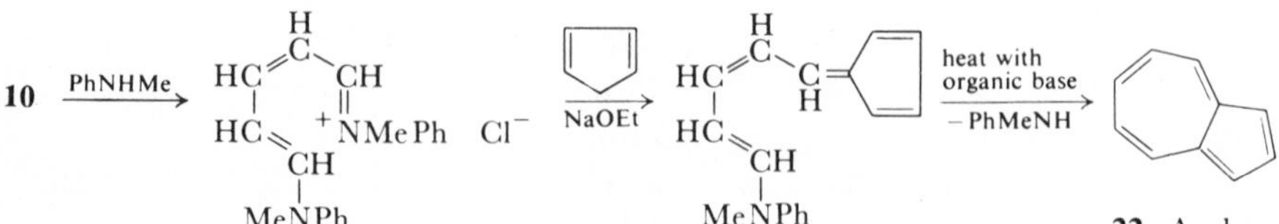

22, Azulene

2. *The photolysis of pyridine*

Pyridine, and simple derivatives, slowly turn yellow in light and this has long been attributed to ring opening to glutaconic derivatives. The photolysis of pyridine has recently[14] been shown to give a 'Dewar pyridine,' 2-azabicyclo-[2.2.0]hexa-2,5-diene (**23**), which reverts completely to pyridine at room temperature. The activation energy for the reversion is high [62 kJ (14·8 kcal)/mole], enabling the n.m.r. spectrum of **23** to be obtained. This pyridine valence tautomer (**23**) can be trapped by carrying out the photolysis in aqueous sodium borohydride, when **24** is formed and can be isolated. Irradiation of aqueous pyridine gives the aldehyde **25**, of unknown stereochemistry, by the route shown. It reverts to pyridine in the dark. Irradiation of pyridine in an argon matrix at 8°K gives[15] cyclobutadiene (**26**) and hydrogen cyanide,

N 2537 Å $t_{1/2}$ 2·5 min/15 °C N NaBH$_4$ H$_2$O NH H$_2$, Ni NH$_2$ CH$_2$

23 **24** +

dark $h\nu$ H$_2$O NH

NH$_2$ CHO NH OH + HCN

25 **26**

(a) NaBH$_4$ (b) NaBH$_4$/nickel boride

$H_2N(CH_2)_5OH$

presumably via the intermediary of **23**. Irradiation[15a] of the perfluoroalkyl-pyridine **26a** gives a mixture of a different type of Dewar pyridine (**26b**) and the prismanes **26c** and **26d**, all of which have been isolated. The formation of **26b**, which can cyclise to **26d**, is easy to follow. The prismane **26c**, however, is probably formed via **26e** (cf. 23), isomerising to **26f** and finally giving **26c**. Thermolysis of **26d**, which can only take place in one way, gives **26i**, while **26c** gives a mixture of **26g** and **26h**. Identification of the products was possible from their ^{19}F n.m.r. spectra. The photolysis of pyridines can therefore give a very complicated situation.

Ultrasonic waves[16] cleave pyridine to acetylene and hydrogen cyanide, but the mechanism of this transformation has not been established.

$X = CF_2CF_3$, $Y = CF_3$, $Z = CF(CF_3)_2$

3. *Substitution reactions*

Molecular-orbital calculations (p. 230) show that the π-electron density at position 3 of the pyridine ring is higher than that at positions 2 or 4. This is a consequence of nitrogen being more electronegative than carbon, and the nitrogen atom therefore causes a drift of electrons in its direction. This drift denudes the 2- and 4-positions of electrons by a resonance interaction (**27**), and the effect is very strong. It can be compared with the similar drift of electrons from the 2- and 4-positions of the ring in nitrobenzene (**28**). There is also the inductive effect of the nitrogen atom in pyridine which is expected to reduce the electron density at position 3 somewhat (however, see p. 230), although very little in comparison with the reduction at positions 2 and 4. This has been detected chemically in certain reactions of 3-methylpyridine (p. 248) and of 3-bromopyridine (p. 256).

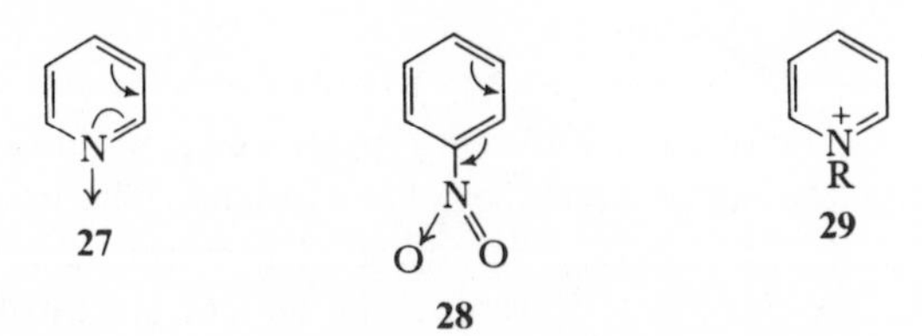

27 **28** **29**

A general comparison of the reactions of pyridine and nitrobenzene is often made. The analogy is helpful if not pressed too hard and is satisfactory for nucleophilic substitutions. Under acid conditions or attack by electrophilic reagents the two compounds are not strictly comparable. Pyridine can be attacked at the most electronegative position, the nitrogen atom, to give a pyridinium cation (**29**) where the positive charge is actually on the ring. This positive charge greatly accentuates the electronegative properties of the nitrogen atom in withdrawing electrons from the ring so that attack on the cation, if it takes place appreciably, does so very slowly and at position 3.

Nitrobenzene, in contrast to pyridine, is such an extremely weak base that it can be considered 'non-basic' under most conditions. Even if the nitro group does accept a cation to some extent under the forcing conditions of nitration, there will still be a substantial equilibrium concentration of uncharged nitrobenzene present, and this will be the species nitrated.

Experiments have established that for some pyridines with electron-donating substituents (2,6-dimethoxy) the cation (cf. **29**) is the species actually nitrated, and for some with electron-attracting substituents (e.g., 2,6-dichloro) which reduce the electron availability to an attacking electrophile, the free base is actually attacked. It can be concluded that in either case electrophilic substitution is highly discouraged.

An assessment of the electrophilic reactivity of the pyridine ring, as opposed to the situation where the nitrogen carries a positively charged atom or group, has been made[17] from a consideration of the rates of pyrolysis of a series of acetates (**30**). The transition state (**31**) is stabilized, or destabilized, by the nature of the aromatic ring (R).

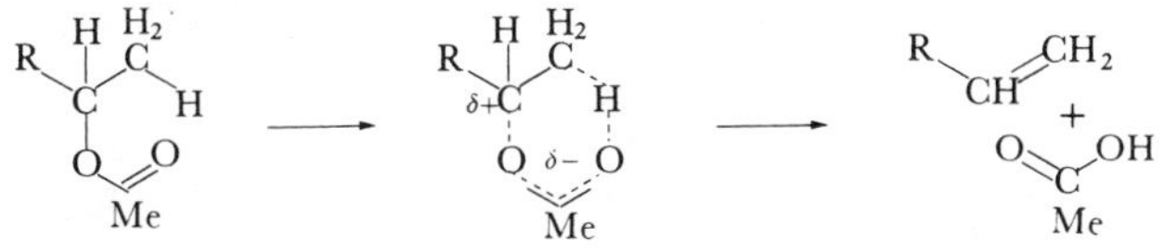

30 → **31** → $RCH{=}CH_2$ + $MeCO_2H$

The observed rates for phenyl, and the three pyridyl groups were:

phenyl > 2-pyridyl > 3-pyridyl > 4-pyridyl

and this gives the order of relative abilities of these aromatic groups to stabilize an adjacent carbonium ion.

The Friedel–Crafts reaction fails with pyridine. 3-Nitropyridine (4·5%) and some 2-nitropyridine (0·5%) are obtained under optimum conditions from pyridine, 100% sulphuric acid, and sodium and potassium nitrates at 300°C. At 450°C only 2-nitropyridine (2%) is obtained (cf. bromination at 500°C). Pyridine with nitronium borofluoride yields 1-nitropyridinium borofluoride. The sulphonation of pyridine does not proceed significantly unless a trace of mercuric sulphate is present, in which case a 70% yield of the 3-sulphonic acid is obtained at 230°C. An intermediate, similar to that (**32**) formed in the relatively facile electrophilic mercuration shown, is probably involved.

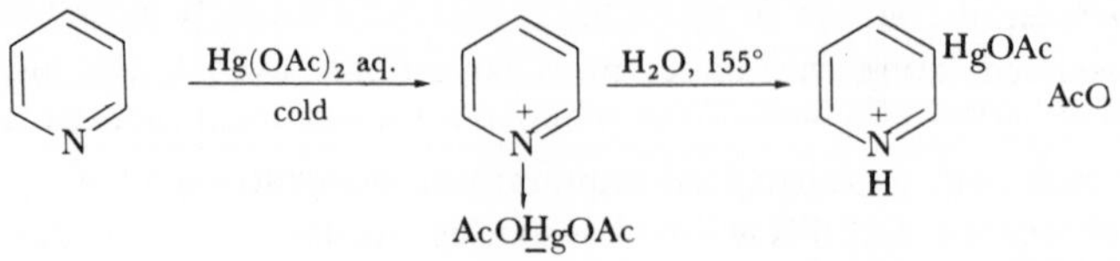

Hg(OAc)$_2$ aq. / cold ; H_2O, 155° ; HgOAc ; AcO ; H ; AcOHgOAc

32

Halogenation of pyridine with excess aluminium,[18] sulphur, or thionyl chlorides[19] gives the 3- and 3,5-derivatives in fair yield. Bromination in concentrated sulphuric acid[20] or in the vapour phase at about 300°C gives 3-bromo- and 3,5-dibromopyridine. In the vapour at about 500°C, substitution appears to proceed through a radical mechanism, and the products are the 2-bromo- (48%) and 2,6-dibromopyridines. Similar results are obtained on chlorination. Pyridine reacts with thionyl chloride to form 1-(4-pyridyl)pyridinium chloride (**33**) in good yield. This product is of special interest, as on hydrolysis with water or ammonia it gives 4-pyridone or 4-aminopyridine, respectively.

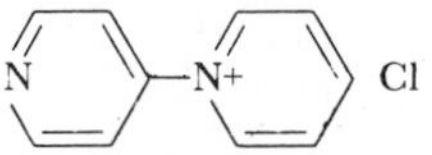

N ; N+ ; Cl

33

The reaction of pyridine at 105°C with benzoyl peroxide, lead tetrabenzoate, phenyliodosobenzoate, or *N*-nitrosoacetanilide, all of which act as a source of phenyl radicals gives in every case a mixture of 2-phenyl- (55%), 3-phenyl- (31%), and 4-phenylpyridine (14%), and methyl radicals similarly give a mixture of 2- (62·7%), 3- (20·3%) and 4- (17·0%) methylpyridines.[21]

In acetic acid solution, where the pyridinium cation is present, phenyl radicals show an even greater preference[22] for position 2. Localisation energy calculations account for these results if the parameters are well chosen.[3] Phenyl radicals attack nitrobenzene mainly at positions 2 (60–70%) and 4 (40–30%), with up to 10% at position 3, so the situation is rather different from that of pyridine.

Nucleophilic attack takes place easily at the 2- and 4-positions, the former usually being favoured, to give an anion (e.g., **34**). This can either lose a hydride ion to reform the aromatic system and lithium hydride or accept a proton to give a dihydropyridine, which can be trapped or aromatised. An example is the reaction of pyridine with phenyllithium, which usually yields 2-phenylpyridine (**36**), but the intermediate can be trapped[23] as the acetyl derivative **35**. Grignard reagents can behave similarly, but lithium alkyls can be employed to synthesise 2-, 2,6-di-, and 2,4,6-trialkylpyridines by using progressively more vigorous conditions.[24]

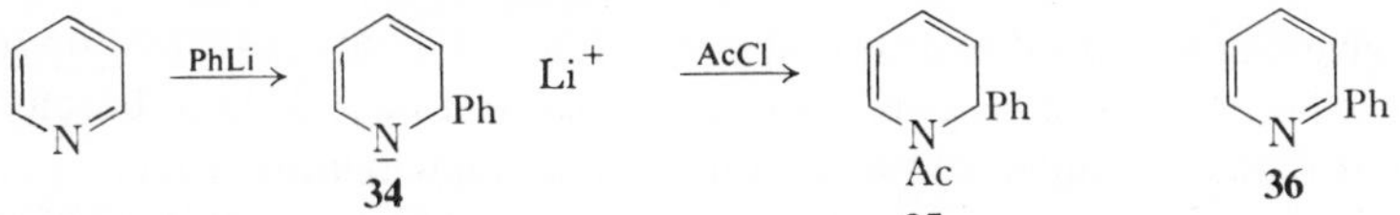

Sodamide in dimethylaniline converts pyridine to 2-aminopyridine (p. 257), further reaction giving the 2,6-diamino derivative, while reaction with sodium hydroxide yields 2-pyridone (p. 261). Pyridine reacts with sodium to give a number of products, depending on the conditions. A radical (**37**) is probably formed first, and if tetrahydrofuran or hexamethylphosphoramide is employed as the solvent, dimerisation to **39** and subsequent loss of sodium hydride occur as 4,4′-bipyridine (**41**) is formed. The quaternary salt of the last compound, 4,4′-bis-dimethylpyridinium chloride, is the highly toxic weed killer ‘paraquat.’ Pyridine also reacts with zinc and acetic anhydride to give **40**, which dissociates reversibly on heating to give the yellow radical **38**.[25] 2,2′-Dipyridyl along with some 2-(2-pyridyl)pyrrole is formed on boiling pyridine with degassed Raney nickel.[26]

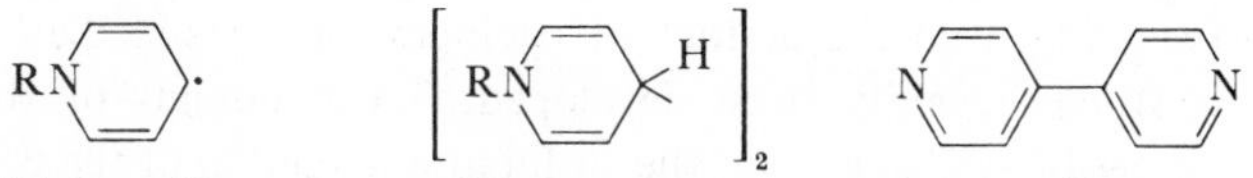

37, R = Na
38, R = Ac

39, R = Na
40, R = Ac

41

D. Derivatives of pyridine

1. *Reduction of pyridines and reduced pyridines*

As compounds such as 2-pyridone, which can be regarded as a dihydropyridine (1,2-dihydro-2-oxopyridine) or, less correctly, as a pyridine (2-hydroxypyridine), are dealt with elsewhere, comparatively few simple dihydropyridines[27] remain for consideration. The reduction of pyridine itself theoretically

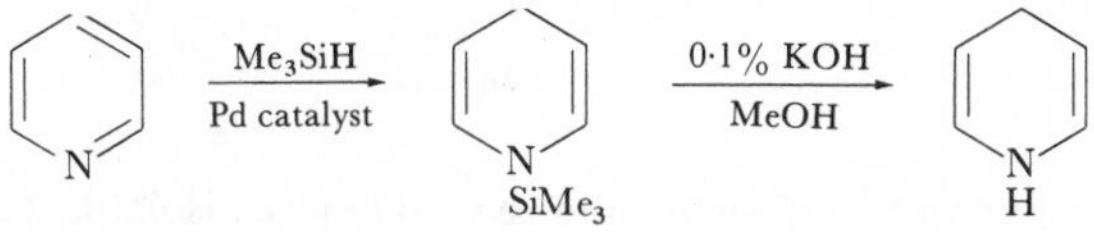

Me_3SiH, Pd catalyst → $SiMe_3$; 0·1% KOH, MeOH → **42**

could give five possible dihydro compounds, but so far only 1,4-dihydropyridine (**42**) has been obtained.[28] It is extremely sensitive to air. Sodium borohydride does not reduce pyridine, and with lithium aluminium hydride reduction cannot be controlled unless forcing conditions are used and then piperidine is formed. The lithium–ethanol–liquid ammonia reduction of 4-methylpyridine followed by quenching with methyl iodide gives[28a] 1,4-dihydro-1,4-dimethylpyridine. Using sodium–ethanol–liquid ammonia 2-methylpyridine appears to give dihydropyridines as the first stage, for treatment of the product with acids gives cyclohexenone. A few highly substituted dihydropyridines unsubstituted on the nitrogen atom have been prepared, by reductive procedures, but most have been

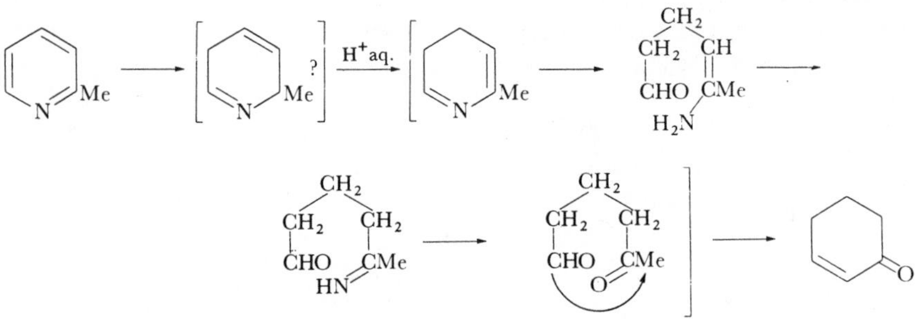

obtained as intermediates in the Hantzsch pyridine synthesis and oxidised *in situ* to the corresponding pyridines (p. 271). A new type of synthesis of 1,4-dihydropyridines (e.g., **44**) from the adduct of cyclopentadiene and diethyl azodicarboxylate (**43**) has recently been developed.[29] Compounds of this type are usually most easily available from the dithionite reduction of the corresponding pyridinium salts (e.g., **46**) which proceeds via intermediates[30] of type **45**.

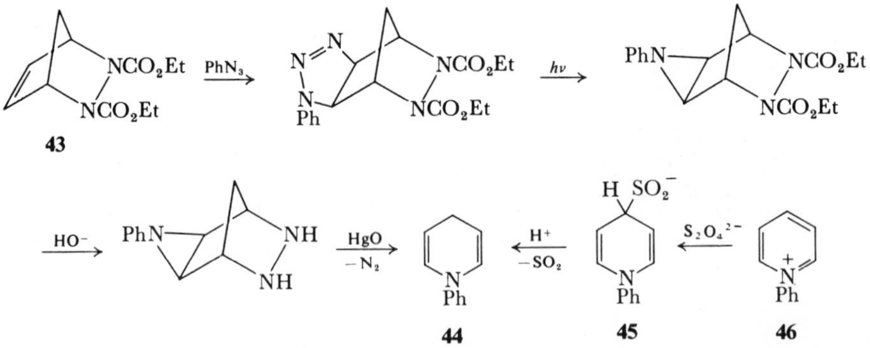

Borohydride reduction[31] of such salts (e.g., **47**) gives isolable 1,2-dihydropyridines (**48**), providing that the *N*-substituent is not very large when hydride addition at position 4 also occurs[32]; however, if a proton source is present, further reduction to a 1,2,3,6-tetrahydropyridine can occur. The pyridine nucleotide enzymes (p. 276) are pyridines that are involved in oxidation–

Me Ph N Me 47 —$\bar{B}H_4$→ 48 —H^+→ 49 —$\bar{B}H_4$→

reduction processes *in vivo* through reversible conversion to their 1,4-dihydro derivatives.

Ac N CH_2Ph 50 ⇌ (H^+) ⇌ (H_2O, $-H^+$) HO 51

It should be noted that both 1,2- and 1,4-dihydropyridines can behave as enamines,[33] examples being the protonation of **48** to give **49**, the addition of water to the dihydropyridine **50**, yielding **51**,[34] and the surprising reaction of dimethyl acetylenedicarboxylate with **52** to give **53**, the amide group also being eliminated via a six-membered transition state.[35] 1,2-Dihydropyridines can undergo normal Diels–Alder reactions.

$CONH_2$ CH_2Ph **52** → → —NHCO → **53**

$E = CO_2Me$

Many anions, including carbanions, can add like the hydride ion at the 2- or 4-positions of 1-substituted pyridinium salts to give dihydropyridines. An example is the cyanide anion, which adds to 1-methylpyridinium salts at position 2 (cf. **7**). However, rearrangement to the thermodynamically more stable 4-cyano-1,4-dihydropyridines (cf. **45**) subsequently takes place.[36]

The electrolytic reduction of the pyridinium cation **54** gives a remarkably stable radical (**55**), which distils *in vacuo* as an emerald green oil. It is very sensitive to oxygen, and magnetic measurements suggest that it contains 60–100% of the radical.[37] Anodic oxidation of 2,6-dimethoxypyridine in

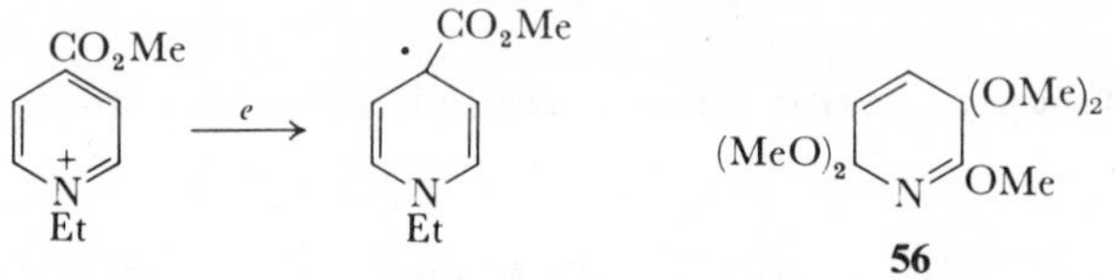

CO_2Me CO_2Me $(OMe)_2$ $(MeO)_2$ OMe N Et e Et

54 **55** **56**

methanol gives **56**, one of the very few examples of 2,5-dihydropyridines.[38] Pyridine does not undergo the Diels–Alder reaction, but combines with dimethyl acetylenedicarboxylate (p. 330).

1,2,3,6-Tetrahydropyridine (**58**)[38a] has properties typical of an olefine and an aliphatic secondary amine. It can be obtained by the reduction of pyridine with aluminium hydride, and from the 4-pyridone derivative (**57**). A derivative (**59**) is present in betel nut. In aqueous solution 2,3,4,5-tetrahydropyridines

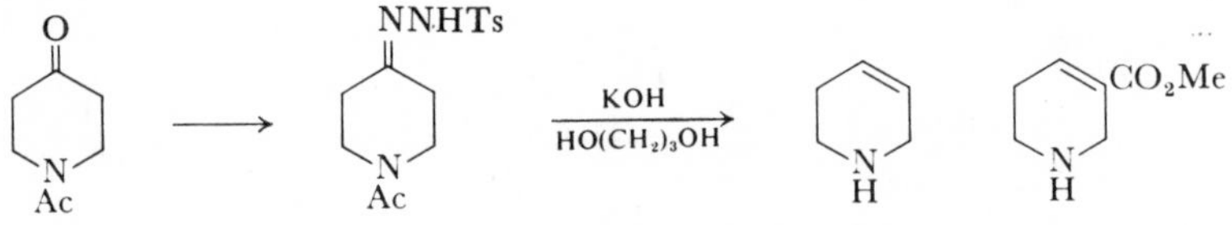

O NNHTs KOH $HO(CH_2)_3OH$ CO_2Me N Ac N Ac N H N H

57 **58** **59**

(e.g., **61**) are in equilibrium with the corresponding carbinolamines (**60**), and benzoylation gives the *N*-benzoylaminoketone. Although derivatives of

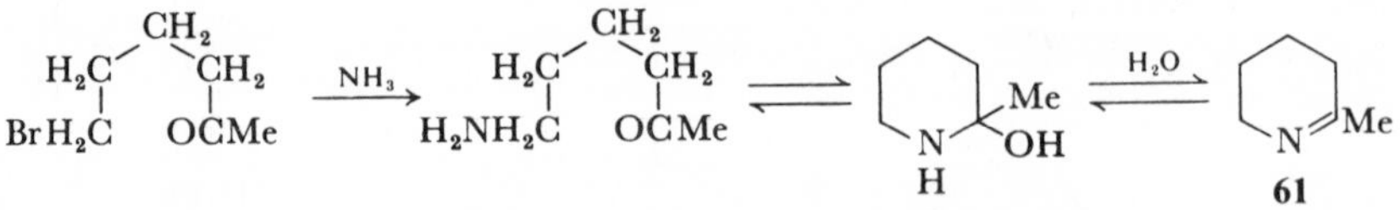

CH_2 H_2C CH_2 BrH_2C OCMe NH_3 CH_2 H_2C CH_2 H_2NH_2C OCMe Me OH N H H_2O N Me

60 **61**

1,2,3,4-tetrahydropyridine (**62**) unsubstituted on the nitrogen atom have been reported, tautomerism to the corresponding 2,3,4,5-tetrahydro compounds (cf. **61**) very probably took place, as is the case with corresponding pyrrolines.

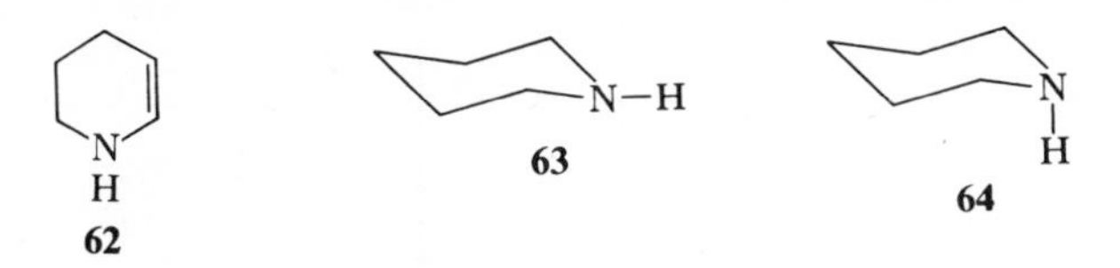

Piperidine or hexahydropyridine (**63** ⇌ **64**) is obtained commercially by the hydrogenation of pyridine over nickel at about 200°C, or in acetic acid solution over platinum in the laboratory at ambient conditions. Dehydrogenation of piperidines to pyridines can be effected by palladium on charcoal, hot nitrobenzene, and other reagents. Piperidine is a colourless liquid, b.p. 106°C, dipole moment 1·17 ± 0.01 D, which is miscible with water and has an amine-like smell. It is a slightly stronger base (pK_a 11·28) than diethylamine (pK_a 11·04). Piperidine fumes in air with the formation of the carbonate. *N*-Alkyl- and *N*-arylpiperidines can be obtained by the reduction or hydrogenation of the *N*-substituted pyridinium salts and by the cyclisation of 1,5-dichloropentanes with the appropriate primary amine.

Like cyclohexane, piperidine and its derivatives exist in the chair form; very few derivatives are non-chair.[39] Two conformations are possible for the *N*-hydrogen atom (**63** and **64**), and for both piperidine[40] and its 1-methyl derivative[41] the equatorial arrangement (e.g., **63**) is the more stable in non-interacting solvents.

Quaternisation of piperidines takes place by both axial and equatorial approach, the ratio of products depending on the alkylating agent employed. Methyl iodide and benzyl chloride alkylate mainly by axial and equatorial attack, respectively.[42]

Piperidine has many of the properties of a typical aliphatic secondary amine. It forms salts, nitroso, acyl, alkyl, and aryl derivatives and reacts with aryl isocyanates to form ureas. Piperidine and sodium amide, which behaves as the *N*-sodio derivative of piperidine, react with aryl halides via benzyne intermediates yielding *N*-arylpiperidines, and with aryl methyl ethers to give phenols and *N*-methylpiperidine.

OMe OH
N + $\xrightarrow{NaNH_2}$ N +
H Me

The *N*-oxidation of piperidine by hydrogen peroxide or benzoyl peroxide takes place by preferential axial attack[43] to give **66** and its *O*-benzoyl derivative, respectively. The hydroxy compound may tautomerise, and derivatives of both forms are known. Oxidation of **66** (2,2′,6,6′-tetramethyl) by hydrogen peroxide and a catalyst such as sodium tungstate gives the stable free radical **65**, derivatives of which are used as 'spin labels' in investigating biological systems.[44]

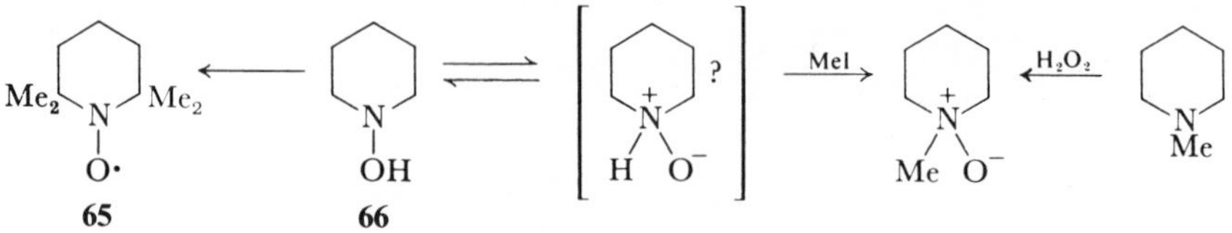

Piperidine is widely used to make enamines, which often undergo condensation reactions with a complementary orientation to base-catalysed aldol reactions. Aldehydes form enamines in preference to ketones. This can be of immense value in synthesis, and an example is shown below[45]:

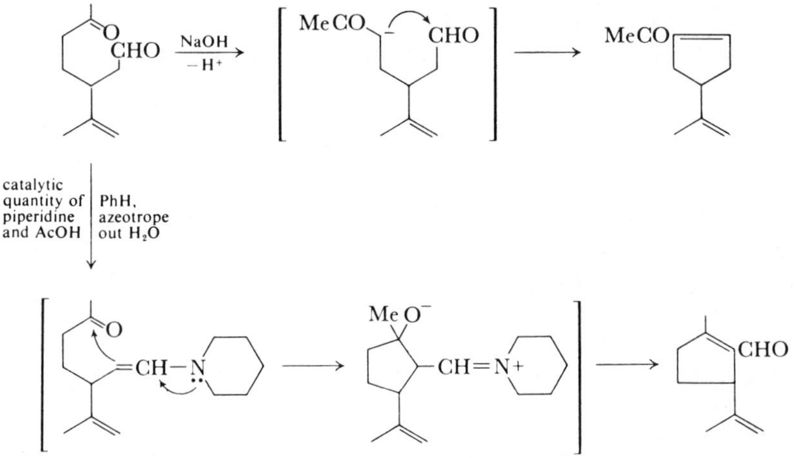

The piperidine, and similar rings, occur widely in natural products. The 'Hofmann exhaustive methylation' procedure has been very much used with such naturally occurring bases to provide structural information, and in the case of piperidine it opens the ring. The aromatic pyridine ring is not opened by this method. The base is first fully methylated to the quaternary iodide (e.g., **67**), which is then converted to the hydroxide **68**. Dry distillation then causes loss of

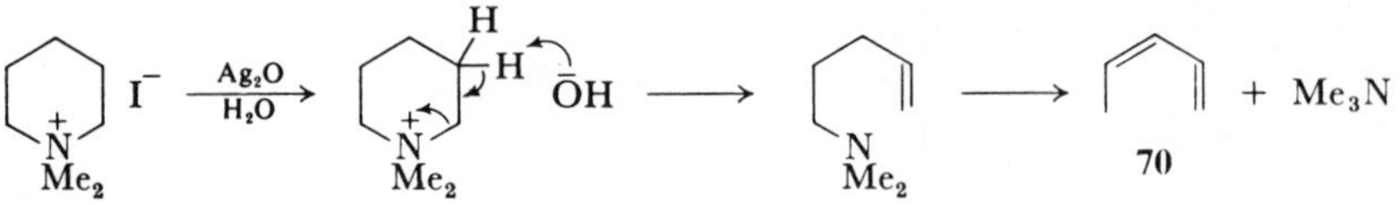

a β-proton, to the nitrogen atom, and breaking of one C–N bond to give **69**. Repetition of the sequence breaks a further C–N bond to give trimethylamine and penta-1,3-diene (**70**). It is not known if penta-1,4-diene is an intermediate in the formation of the last compound. Three applications of the reaction are needed to remove the nitrogen atom from quinolizidine (p. 331).

Two other methods of opening the piperidine ring were developed by von Braun. In the first of these, piperidine is benzoylated and the product (**71**) is treated with phosphorus pentachloride. This is a convenient method of making

$$\textbf{71}\ (\text{N-COPh piperidine}) \xrightarrow{PCl_5} CH_2{=}\ldots \; PhClC{=}N(CH_2)_5Cl \begin{cases} \xrightarrow{PCl_5,\ 200^\circ} Cl(CH_2)_5Cl + PhCN \\ \xrightarrow{H_2O} PhCONH(CH_2)_5Cl + HCl \end{cases}$$

1,5-dichloropentane. In the second method,[46] *N*-alkyl- or *N*-arylpiperidines are quaternized with cyanogen bromide and the quaternary salt (**72**) is hydrolysed by 48% aqueous hydrobromic acid.

$$\text{N-R piperidine} \xrightarrow{BrCN} \textbf{72}\ (\overset{+}{N}(R)(CN)\ Br^-) \longrightarrow RN(CN)(CH_2)_5Br \longrightarrow R\overset{+}{N}H(CO_2H)(CH_2)_5Br\ Br^- \xrightarrow{-CO_2} R\overset{+}{N}H_2(CH_2)_5Br\ Br^-$$

2. *Alkylpyridines*

Many of these compounds, like pyridine itself, are obtained from coal tar. The monomethylpyridines are sometimes called picolines, dimethylpyridines lutidines, and trimethylpyridines collidines. The collidines are useful chromatographic solvents.

2-or α- 3-or β- 4-or γ- Picoline; 2,4- Lutidine; 2,4,6- Collidine

The chemical properties of the three monomethylpyridines as representative examples of the alkylpyridines are considered here in some detail. Although the electron-donating characteristic of the methyl group does activate the pyridine ring appreciably towards both electrophilic and, surprisingly, nucleophilic attack,[47] the electron-attracting property of the pyridine ring on the methyl groups is much more striking. Methyl groups at positions 2 and 4 of the pyridine ring are denuded of electrons by the inductive and resonance effect of the nitrogen atom (p. 230). A proton is easily lost from these activated methyl groups under suitable conditions, and treatment of 2-methylpyridine with phenyl- or butyllithium, or sodamide in liquid ammonia, convert it to the corresponding metal derivatives, which give the typical reactions of such compounds (e.g., ref. 52). It is very interesting that the anion **74** in liquid ammonia does not react with chlorobenzene unless stimulated by potassium metal or ultraviolet light, in which case 2-benzylpyridine is formed.[48] 2,4-Dimethylpyridine loses a proton from the 4-methyl group with sodamide in liquid ammonia or lithium diisopropylamide in ether–hexane.[49] When the metal can coordinate with the ring nitrogen, as is the case using butyl-[49] or phenyllithium[50] in suitable solvents, exclusive metalation of the 2-methyl group occurs. Tautomerism of 2-methylpyridine to **73** does not take place to a

N Me
PhLi
PhH
N CH_2^- Li^+
$(EtO)_2CO$ (ref. 52)
N CH_2CO_2Et
$NaNH_2/NH_3$
N H CH_2
73
N CH_2^- Na^+
74
MeI
N Et

significant extent,[51] although stable compounds of this type, bearing an *N*-substituent, can be formed from the corresponding pyridinium salts with bases.[53] The rate of proton exchange for deuterium in the methyl groups of 2-, 3-, and 4-methylpyridine caused by sodium dissolved in MeOD is in the ratio 130 : 1 : 1810 and must proceed through species such as **74**. Exchange occurs even more readily under acid conditions,[54] for the reasons discussed below.

The activity of the methyl groups of 2- and 4-methylpyridine is increased by converting the compounds into quaternary salts, for example, with methyl iodide or by conversion to the 1-oxides (p. 249), and carrying out reactions in the presence of Lewis acids, for example, zinc chloride or acetic anhydride (below), can have the same effect.

Alkyl substituents at the 2- and 6-positions can sterically protect the nitrogen atom from electrophilic attack, the ratio of the rates of methylation of pyridine and its 2-methyl and 2,6-dimethyl derivatives by methyl iodide in dimethyl

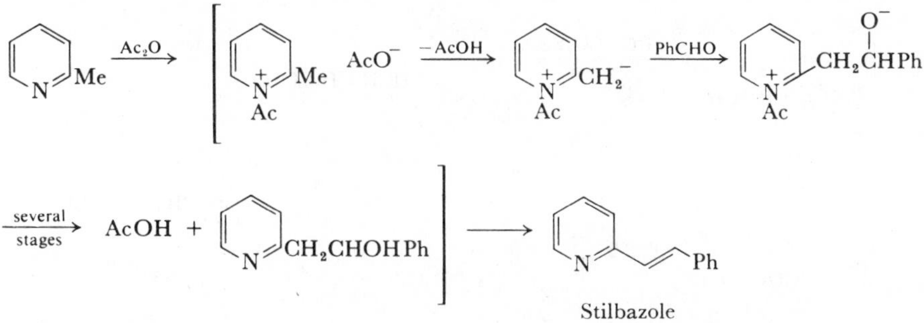

Stilbazole

sulphoxide being 1 : 0·39 : 0·023, respectively.[55] 2,6-Di-*t*-butylpyridine does not react with methyl iodide at all under normal conditions, because of the steric effect of the *t*-butyl groups, but at 5000 atm gives[56] a very stable methiodide which sublimes at 250°C at 20 torr. The *t*-butyl groups are thought to cover the *N*-methyl substituent, thus preventing the usual reversion to the original pyridine and alkyl halide under such conditions.

2,6-Dimethylpyridine (sometimes called 2,6-lutidine) is a useful dehydrohalogenation reagent, being a great improvement on pyridine where quaternary salt formation is not sterically inhibited. Like pyridine, 2,6-dimethylpyridine forms addition compounds with sulphur trioxide and boron trifluoride (**75**), but 2,6-di-*t*-butylpyridine does not. Sulphur trioxide does, however, attack 2,6-di-*t*-butylpyridine to give the 3-sulphonic acid (**76**). Di-*t*-butylpyridine (pK_a 3·58) forms salts with protonic acids; it is a much weaker base than pyridine itself (pK_a 4·38, also in 50% EtOH).

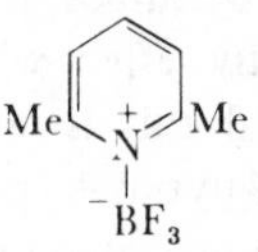

75 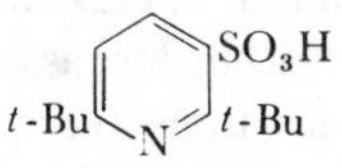 **76**

The chlorination of 2-methylpyridine gives first 2-trichloromethylpyridine. The ring is then attacked successively at the 5- and 3-positions, and the halogenation of 4-methylpyridine gives tars containing ionic halogen.

2-Vinylpyridine is made commercially as shown below. It can be polymerized

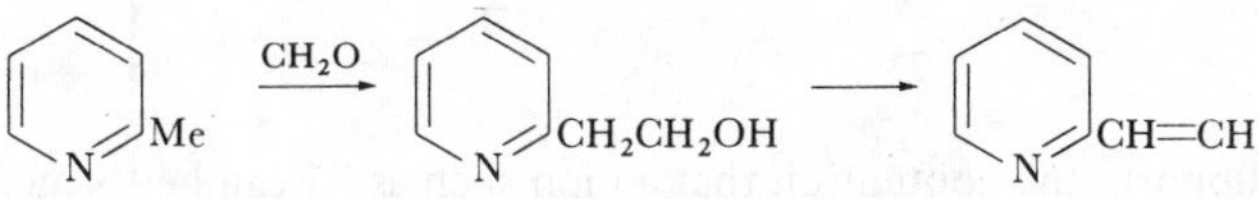

like styrene, and its methylene group is sufficiently electrophilic to react with nucleophilic reagents as indicated. 4-Vinylpyridine is similar in preparation and properties. These additions may be compared with the Michael reaction, one example of which is given.

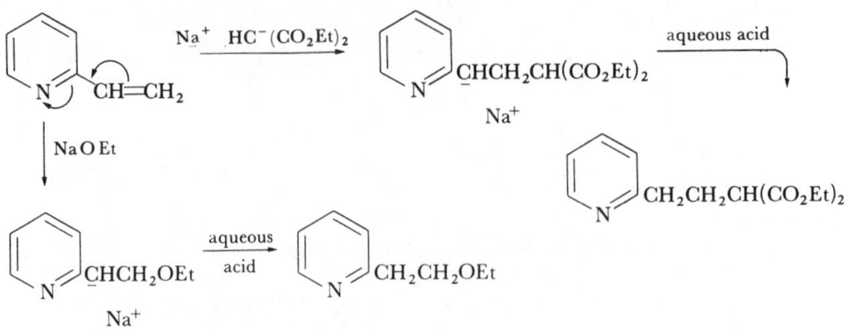

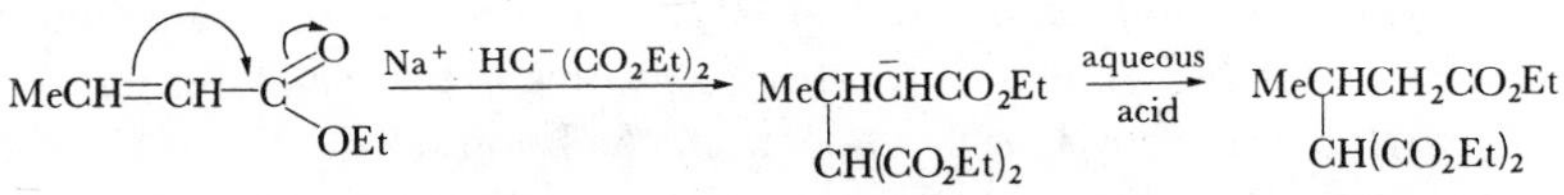

Until recently the methyl group of 3-methylpyridine was thought to possess none of the activity well known in its isomers. It is in fact comparatively unreactive and does not react with benzaldehyde in acetic acid anhydride or with selenium dioxide, which oxidises the 2- and 4-isomers to the corresponding acids. It is also unreactive to phenyllithium, which gives 3-methyl-2-phenylpyridine. These results are expected, as the methyl group is *meta* to the electron-attracting centre, the nitrogen atom, and activation through resonance, which is so important in the 2- and 4-methylpyridines, is therefore precluded. However, the nitrogen atom does assert an inductive effect which is not dependent on resonance considerations (cf. p. 230), and this is shown by the greater basicity of 3-methylpyridine ($pK_a = 5{\cdot}68$) over pyridine ($pK_a = 5{\cdot}17$). 3-Methylpyridine does in fact react with sodamide in liquid ammonia to give the ion **77**, which with methyl chloride or iodide yields 3-ethylpyridine.[57] The racemization of the optically active **78** by boiling with potassium dissolved in triethylmethanol supports the contention that an ion such as **77** can be formed. Further treatment of 3-ethylpyridine gives successively 3-isopropylpyridine and 3-*t*-butylpyridine. It appears that the order of reactivity of the picoline methyl groups in this reaction is 2 > 3 and 4 > 2.[58]

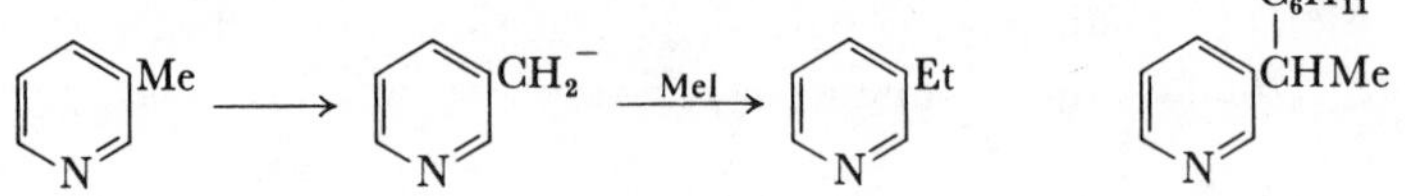

The bromination of 3-methylpyridine hydrochloride gives 3-bromomethylpyridine. The methylpyridine 1-oxides are discussed with pyridine 1-oxide (below).

3. *Arylpyridines*

The phenylpyridines are of particular interest in connexion with their oxidation and reduction, as this gives information about the comparative reactivity, but not necessarily about the aromaticity, of the two rings. 2-Phenylpyridine is easily obtained from pyridine and phenyllithium, and 4-phenylpyridine is obtained by the synthesis outlined. The hydrogenation of 2-

Ph Me C CH_2 O CH_2 CH_2O $\overset{+}{N}H_4$ Cl^- → Ph Me O N H —HCl→ Ph N H —Pd/C, $PhNO_2$→ Ph N

and 3-phenylpyridines over platinum oxide (Adams) catalyst gives the corresponding phenylpiperidines. In general, the hydrogenation of a pyridine proceeds more easily than that of the corresponding benzene. The phenyl-

N CO_2H ← N Ph → N H Ph

pyridines all give pyridinecarboxylic acids on oxidation with potassium permanganate, indicating the greater stability of the pyridine ring under these conditions.

4. *Pyridine* 1-*oxide, or pyridine* N-*oxide*

This compound (**79**) is easily prepared by the oxidation of pyridine in glacial acetic acid by 30% aqueous hydrogen peroxide at 70–80°C. It forms water-soluble colourless crystals, m.p. 66°C, and has attracted much interest in recent years owing to the diverse reactions it can undergo.[59] It is a much weaker base than pyridine but does form some salts.

N —H_2O_2, HOAc→ $\overset{+}{N}$–O^-

79

As the dipole moment (4·24 D) of pyridine 1-oxide is lower than that calculated from the moments of pyridine and the trimethylamine oxide, Ochiai concluded that the resonance structures **79a**–**79c** must be important, and he predicted that nitration should take place at position 2 or 4. Later dipole moment studies with a series of pyridine 1-oxides substituted at position 4 have shown that the oxygen atom can either release or accept electrons according to the substituent. The resonance structures **79d**, **79e**, and **79f** must therefore make appropriate contributions to the resonance hybrid, and it can therefore be understood why the oxide undergoes many types of reaction. Experiment showed that Ochiai's prediction was correct and that in nitric–sulphuric acids at

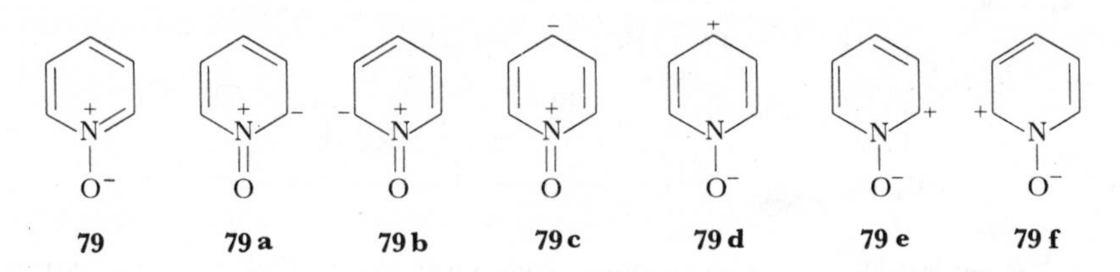

100°C 4-nitropyridine 1-oxide (**80**) is by far the major product; at 150°C some 2-nitropyridine (7·6%) is also obtained, presumably through deoxygenation of the corresponding *N*-oxide which would be formed first. Nitration occurs through attack on the free base, not on the conjugated acid, although acid-catalysed exchange of the ring hydrogen atoms does take place via the conjugate acid. The nitration reaction is of particular interest, as it opens a new route to the synthesis of 4-substituted pyridines from pyridines unsubstituted at this position. Pyridine 1-oxide rather surprisingly does not sulphonate at all readily. The mildest condition effecting reaction, sulphuric acid containing 20% of sulphur trioxide and mercuric sulphate at 220–240°C, gives 40–45% of the

80 **81** **82**

3-sulphonic acid (**82**) and ca. 1% of the 2- and ca. 2% of the 4-isomers; ca. 40% of the 1-oxide is unchanged. The remarkable change in orientation from the nitration reaction, which is also carried out in sulphuric acid solution, is probably due to the formation of **81**, in which the electron-donating property of the oxygen atom is suppressed. Bromine and silver sulphate, a source of bromonium ions, in 90% sulphuric acid give a mixture of the 2- and

4-bromopyridine 1-oxides in a 1 : 2 ratio and may be compared with the nitration reaction above. However, bromination in fuming sulphuric acid, when the species being attacked is probably **81**, gives 3-bromopyridine 1-oxide as the main product.[60] With mercuric acetate in acetic acid the 1-oxide mercurates[61] at position 2 and then 6, but some attack also occurs at position 3. Phenyl radicals, obtained from diazoaminobenzene at 181°C, phenylate pyridine 1-oxide at positions 2-, 3-, and 4- in a 9·9 : 1 : 3·9 ratio.[62] This ratio is similar to that obtained in the phenylation of nitrobenzene (p. 238).

Carbanion formation from deuteriated pyridine 1-oxide is very easy and much more facile than that from pyridine, the relative rates for proton exchange for deuterium at the 2,6-, 3,5-, and 4-positions being 1500 : 10 : 1,[63] respectively, using sodium methoxide in methanol. The 2,6-exchange occurs rapidly at 75°C.

With butyllithium at –60°C pyridine 1-oxide gives the 2-lithium derivative, which has normal properties.[64] In contrast, phenylmagnesium bromide adds to pyridine 1-oxide at –50°C to give **83**, which is then converted by acid to **84**, this last undergoes an electrocyclic ring opening at 0°C to **85**, the position of equilibrium between the open chain and cyclic isomers of this type being greatly dependent on the substituents present.[65] On heating with acetic anhydride or to 200°C dehydration of **84**–**85** occurs, giving 2-phenylpyridine.

PhMgBr, –50 °C; H^+, –50 °C

OMgBr **83**   OH **84**   OH **85**

The reduction of pyridine 1-oxides is more difficult than that of most tertiary amine oxides, probably because of resonance interaction. They are only reduced by sulphur dioxide at[66] ca. 100°C, but hydrogenation or reaction with phosphorus trichloride is generally effective.

NH_2 ← H_2, Pd, HCl — NH_2 ← H_2, Pd — NO_2 → PCl_3 in $CHCl_3$, $-POCl_3$ → NO_2

Pyridine 1-oxide can be alkylated on the oxygen by methyl iodide. Hydrolysis of the product (**86**) with alkali yields pyridine and formaldehyde, and aromatic aldehydes can be obtained in excellent yields by this method.[67]

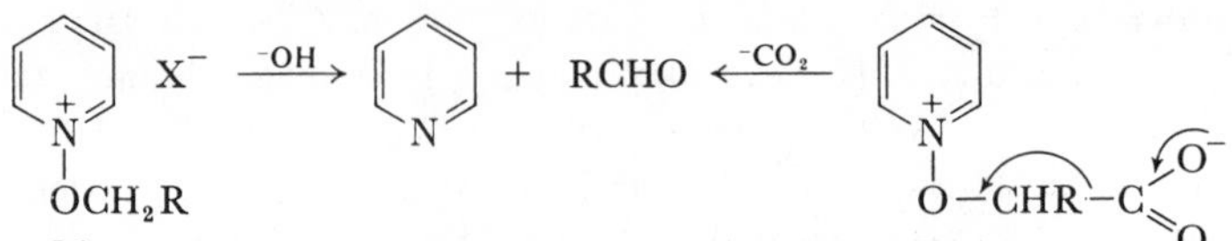

X^- $\xrightarrow{^-OH}$ + RCHO $\xleftarrow{-CO_2}$

OCH_2R O–CHR–C

86 87

Corresponding derivatives of α-halogenated acids (e.g., **87**) also give aldehydes or ketones.[68] Cyclic quaternary salts such as **88** decompose similarly.[69]

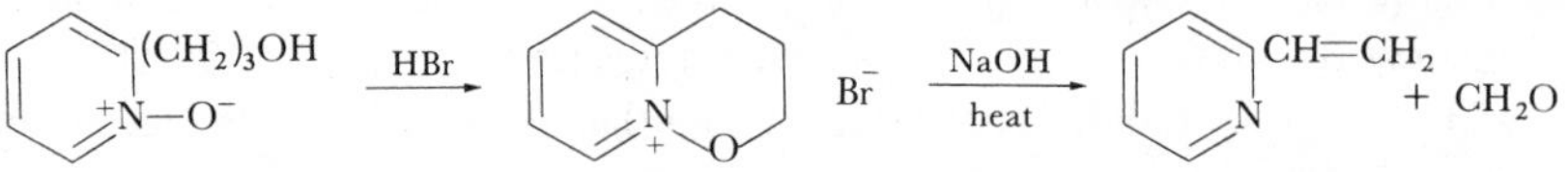

$(CH_2)_3OH$ HBr Br^- NaOH heat CH=CH$_2$ + CH_2O

88

1-Methoxypyridinium salts (cf. **86**) react with nucleophiles, such as the cyanide anion, mainly at the 2-position to give the 2- and 4-cyanopyridines.[70] This type of reaction has valuable potentialities and is probably similar in mechanism to the much better known conversion of pyridine 1-oxide into 2-pyridone by acetic anhydride. The first stage of the latter reaction is almost certainly the formation of **89**, as the corresponding perchlorate is formed from the same reactants in the presence of perchloric acid. One or two other intermediates, **91**, which has been isolated,[71] and possibly also **90**, are formed as hydrogenation of the reaction mixture yields **92**, while 2-pyridone is not reduced under these conditions. Pyridine 1-oxide with tosyl (4-toluene-

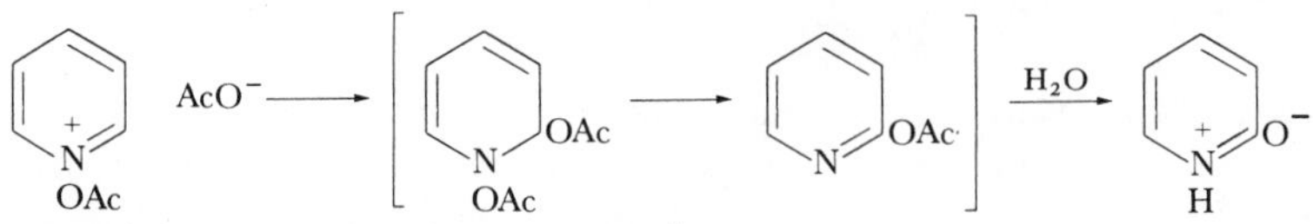

AcO^- OAc OAc OAc OAc H_2O O^- H

89 90 91

$OSO_2C_6H_4Me$

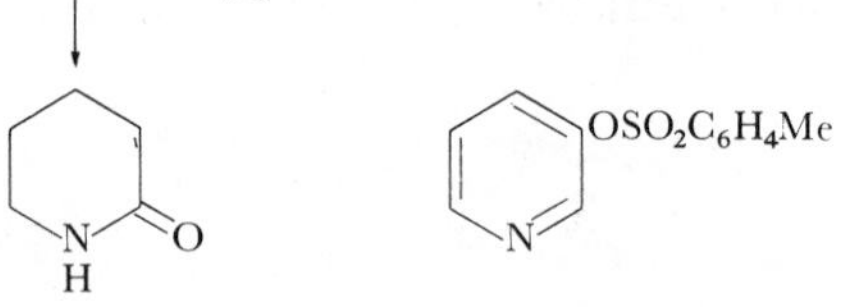

92 93

sulphonyl) chloride gives largely the 3-derivative (**93**), and with thionyl chloride a mixture of 2- and 4-chloropyridines is formed. The oxide hydrochloride with phosphorus pentachloride yields 4-chloropyridine. In the last reaction a chloride anion probably attacks the 4-position.

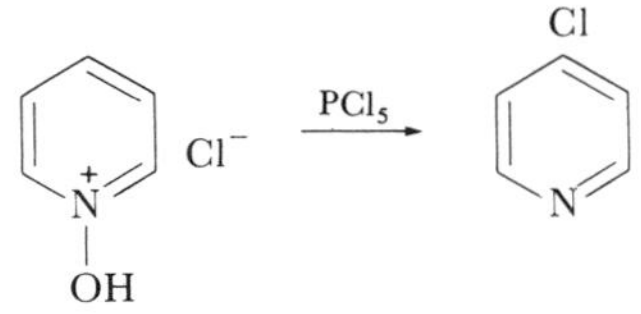

2-Methylpyridine forms a 1-oxide (**94**) in the usual way (p. 249), and on treatment with acetic anhydride a mixture of **100** with 3- and 5-acetoxy-2-methylpyridine (67 : 15 : 18 ratio, respectively), containing a small amount of the pyridone (**99**), is formed.[72] The reaction appears to proceed through the formation of **95**, which gives[73] the anhydro base **96**. Tracer studies show that there is 50% incorporation of oxygen in **100** when ^{18}O-labelled acetic anhydride is used.[74] Partial dissociation of **96** to a radical[74] (**97**) or ion pair[75] followed by recombination can account for the products formed. 4-Methylpyridine 1-oxide

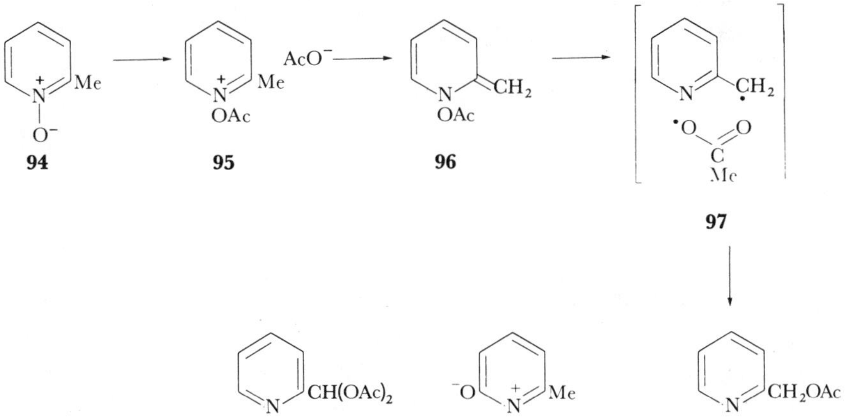

94 **95** **96** **97** **98** **99** **100**

in a similar way with acetic anhydride yields 4-acetoxymethylpyridine and some 3-acetoxy-4-methylpyridine. Tracer studies[76] show that the reaction does not proceed in exactly the same way as for 2-methylpyridine 1-oxide, but that the acetoxy anion appears to attack the anhydro base analogous to **96**.

Conversion of 2-acetoxymethylpyridine into the 1-oxide and treatment with acetic anhydride gives a useful preparation of pyridine-2-carbaldehyde diacetate (**98**). The methyl groups of 2- and 4-methylpyridine 1-oxide are activated (p. 246).

The photolysis of pyridine 1-oxide is solvent dependant, and a number of derivatives give various compounds, the exact mode of formation of which has not been established in most cases.[77] The following scheme nevertheless accounts for the formation of the products. Ionic intermediates are shown, although radicals would be equally acceptable in some of the reactions. Pyridine 1-oxide itself in methanol[78] gives a mixture of **104** (R = H), and its 'dimethyl ketal,' and **103** (R = H), along with pyridine and formaldehyde, while 2,6-dicyanopyridine 1-oxide[79] in methylene dichloride yields 35% of **102** (R = CN) and 20% of **103** (R = CN). When the pyridine ring is unsymmetrically substituted the initial three-membered ring can be formed in both possible positions,[80] and 3-hydroxypyridines are also obtainable, possibly by a suitable opening of the oxaziridine ring of **101**.

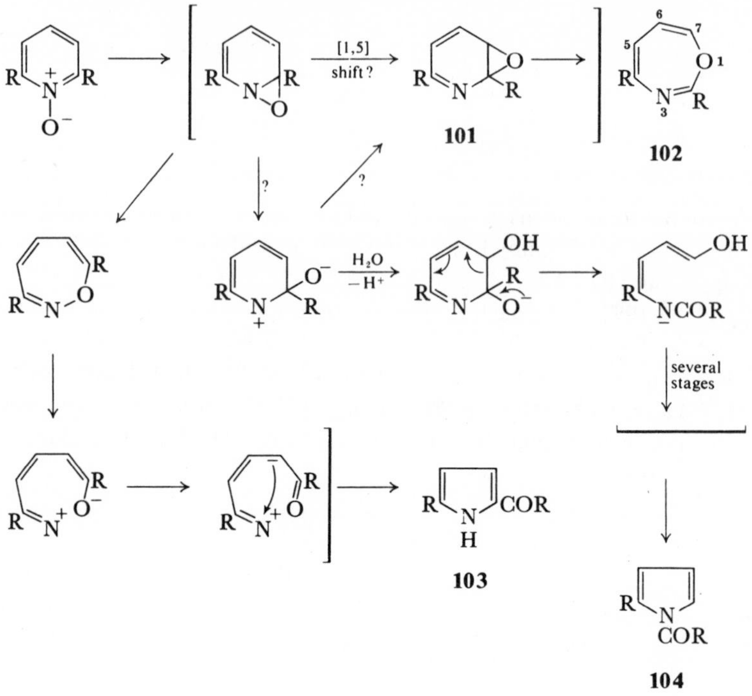

5. *Halogenated pyridines*

2- and 4-Chloropyridine can be obtained in the laboratory by heating the corresponding pyridone with phosphorus oxychloride or pentachloride, a reaction reminiscent of the conversion of amides into iminochlorides (**107**) by phosphorus pentachloride. This method is the most convenient for 4-chloro- and 4-bromopyridines, while the 2-substituted isomers are often more readily available from 2-aminopyridine (p. 260). 1-Methyl-2-pyridone (**105**) loses its methyl group with phosphorus pentachloride, and some 2,5-dichloropyridine is also formed. 3-Hydroxypyridine does not react with any of the phosphorus halides

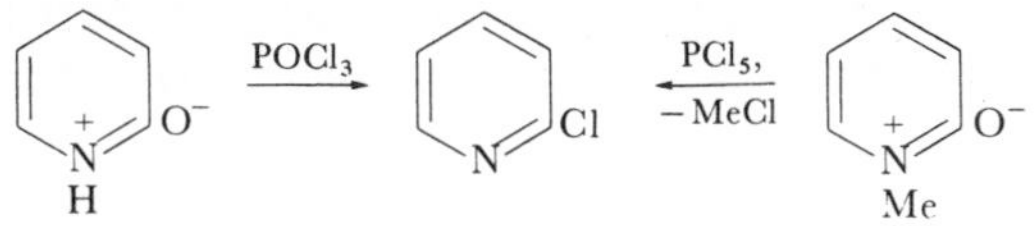

105

and has not been converted directly into a 3-halogenated pyridine. 3-Bromo- and 3-chloropyridine can be prepared from 3-aminopyridine by ordinary diazotization sequences, (and by other methods, p. 238 and 100, respectively).

The chemical properties of the 2- and 4-halogenated pyridines differ very widely from those of 3-halogenated pyridines and so are considered separately. The halogen atoms of 2- and 4-chloro- or 2- and 4-bromopyridines are very easily displaced by nucleophilic reagents (cf. 2- and 4-bromonitrobenzene), a 4-halogen being more easily displaced than the corresponding one at position 2.[81] Attack of an anion at the 2- and 4-positions is particularly facilitated by the electron-attracting properties of the nitrogen atom, which are transmitted by resonance. The electron movements are illustrated by **106** and **108**. 2-Halogenated pyridines are similar, structurally, to the very reactive imino halides (e.g., **107**), which are the nitrogen analogues of acid chlorides obtained from phosphorus pentahalides and amides. 4-Halogenated pyridines may be considered as 'vinylogous' iminohalides, the nitrogen atom interacting with the

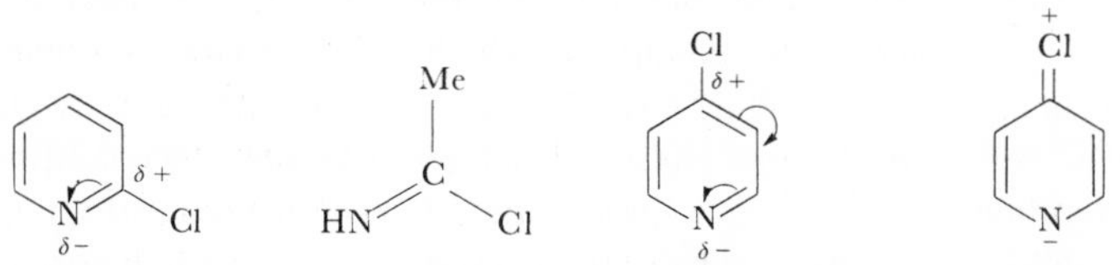

106 **107** **108** **109**

halogen through an additional double bond. In the case of 4-chloropyridine, dipole-moment studies have shown that structure **109** is an important contributor to the resonance hybrid. Examples of (nucleophilic) attack by anions are the conversion of 2-bromo- or 2-chloropyridine into **110** with primary amines, into **111** with sodium ethoxide, into **112** with potassium hydrogen sulphide, and into 2-pyridone (**113**) with aqueous mineral acid. The 4-halogenated pyridines behave similarly.

110 **111** **112** **113**

4-Bromopyridine, but not 2-bromopyridine, which is more stable, reacts slowly with itself in the cold to give **114** and more complex products.

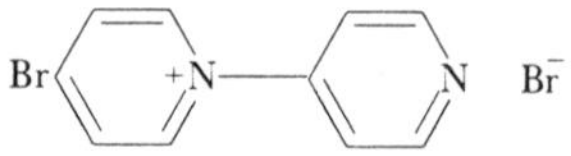

114

The conversion of 2- or 4-halogenated pyridines to the 1-methylpyridinium chlorides increases the ease with which anions displace the halogen atoms. 2-Bromopyridine does not form a Grignard reagent under ordinary conditions, but it will react with magnesium in the presence of reacting ethyl bromide. This is the 'entrainment' method of preparing Grignard reagents. 3-Bromopyridine forms a Grignard reagent normally, but all the bromopyridines give the corresponding pyridyl lithiums with butyllithium at low temperatures. The lithium derivatives are particularly convenient for synthetic purposes and give the usual reactions for such compounds. Halogen atoms at positions 2 and 4 of a pyridine ring are easily displaced by hydrogen on hydrogenation over palladium on strontium carbonate or Raney nickel in the presence of alkali. Chemical methods of reduction have also been used but are not so convenient.

3-Bromopyridine is quite unlike its 2- and 4-isomers in chemical reactivity and is much more like bromobenzene in this respect. The bromine atom is not replaced by hydrogen under conditions which dehalogenate its isomers. Like bromobenzene, 3-bromopyridine gives 3-cyanopyridine with cuprous cyanide at 165–170°C, but unlike bromobenzene, which is inert under the conditions used, 3-bromopyridine reacts with sodium methoxide in methanol (150°C), and with ammonia and copper sulphate (150°C). This suggests that the inductive effect of

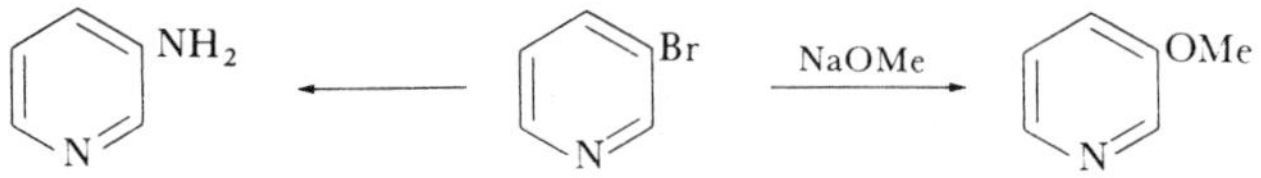

NaOMe

the nitrogen atom is sufficient to activate the 3-position of the ring somewhat towards nucleophilic reagents and confirms observations made in this respect regarding 3-methylpyridine (p. 248). In agreement with this, the activation energy for the reaction of 3-bromopyridine with piperidine is less than that of the corresponding reaction with bromobenzene.

A number of reactions involving halogenated pyridines with alkali metal amides in liquid ammonia proceed via proton abstraction and pyridyne (e.g., **115**) formation. For example, 3- and 4-chloropyridine give the same mixture of aminopyridines.

$$\text{3-chloropyridine} \xrightarrow[NH_3]{NaNH_2} [\text{3-chloropyridyl anion}] \longrightarrow [\text{3,4-pyridyne } \mathbf{115}] \longrightarrow \text{4-aminopyridine (44\%)} + \text{3-aminopyridine (22\%)}$$

6. *Nitropyridines*

3-Nitropyridine (p. 237) is of little interest, as 3-aminopyridine, which can be obtained on reduction, is more easily prepared from nicotinamide (below).

2- and 4-Nitropyridines have been obtained by the hydrogen peroxide–sulphuric acid oxidation of the corresponding amines, but 4-nitropyridine is now much more readily available from pyridine 1-oxide. These nitropyridines have been little investigated. 4-Nitropyridine is less stable than 4-bromopyridine and rapidly polymerizes. 2-Nitropyridine 1-oxide cannot be prepared by the

$$\text{2-aminopyridine} \longrightarrow \text{2-aminopyridine 1-oxide} \xrightarrow[H_2SO_4]{H_2O_2,\ \text{fuming}} \text{2-nitropyridine 1-oxide} \not\longleftarrow \text{2-nitropyridine}$$

oxidation of 2-nitropyridine, but it has been obtained indirectly. Both 2- and 4-nitropyridine 1-oxides with acetyl chloride at 70°C give the corresponding chloropyridine 1-oxides, with the elimination of the nitro group, and 4-nitropyridine 1-oxide also reacts with sodium ethoxide or benzyloxide. This last displacement is similar in mechanism to the conversion of 1,2-dinitrobenzene to 2-nitrophenol by sodium hydroxide.

$$\text{4-nitropyridine 1-oxide} \xrightarrow{PhCH_2O^-\ Na^+} \text{4-}OCH_2Ph\text{-pyridine 1-oxide} \xrightarrow{H_2,\ Pd} \text{4-}OH\text{-pyridine 1-oxide} + PhMe$$

7. *Aminopyridines*

All the monoaminopyridines can be obtained from the appropriate halogenated pyridines and ammonia (pp. 254 and 255), or from the corresponding acid amides via the Hofmann reaction. The most convenient methods of making 2-, 3-, and 4-aminopyridines, respectively, are from pyridine and sodamide in dimethylaniline (p. 239), from nicotinamide by Hofmann's reaction, and by reduction of 4-nitropyridine 1-oxide (p. 251).

3-Aminopyridine behaves as a typical aromatic amine. It can be diazotized normally, and 3-pyridyldiazonium salts undergo the same reactions as benzenediazonium salts.

The structures of 2- and 4-aminopyridine have been subject to much discussion,[82] as tautomerism to the corresponding imino compounds is possible and the reactions of the compounds differ from those of 3-aminopyridine, where such tautomerism cannot take place. The structure of 2-aminopyridine (**116** ⇌ **117**) has been investigated by a comparison of its ultraviolet absorption spectrum with those of the *N*-methyl compounds **118** and **119**. As it has been shown that the replacement of a hydrogen atom by a methyl group in such circumstances usually causes comparatively little change in the spectrum, compounds **118** and **119** serve as 'spectral models' for the tautomers **116** and **117**, respectively. Because the absorption spectrum of 2-aminopyridine is quite dissimilar to that of **119** but virtually identical with that of **118**, it appears that 2-aminopyridine is best represented as the amino tautomer. Similar results have been obtained in the 4-aminopyridine series, and it has been estimated that the amount of imine tautomer (e.g., **117**) in these pyridines is less than 0·1% under the conditions of examination.

Another approach to the amine–imine problem can be made from a study of dissociation constants on the assumption that the replacement of a hydrogen atom attached to a nitrogen atom by a methyl group does not affect the basicity of an amine very much. This assumption is thought to be sufficiently valid in the cases under consideration. The addition of a proton to the amine (**120**) or imine (**122**) tautomer gives the same mesomeric cation (**121**).

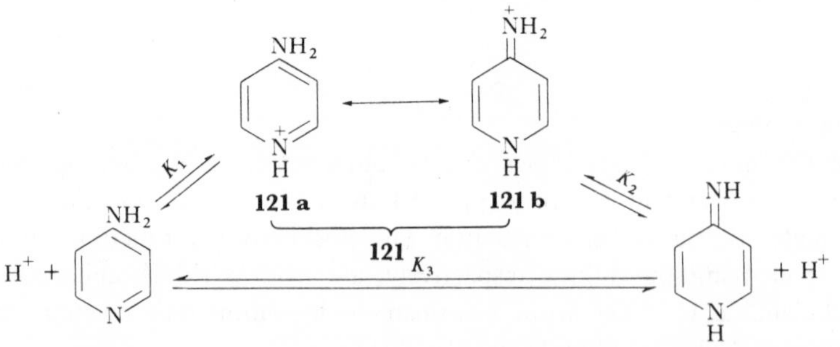

Three equilibria therefore exist, and approximations to the equilibrium constants K_1 and K_2 may be found experimentally by potentiometric titrations

$$K_1 = \frac{[120]\,[H^+]}{[121]} \qquad K_2 = \frac{[122]\,[H^+]}{[121]}$$

$$K_3 = \frac{[\text{amino tautomer}]}{[\text{imino tautomer}]} = \frac{K_1}{K_2} = 2 \times 10^3$$

of the appropriate methyl derivatives of **120** and **122**. These would be **123** and **124**, respectively. In this particular instance titrations of **124** and 4-amino-

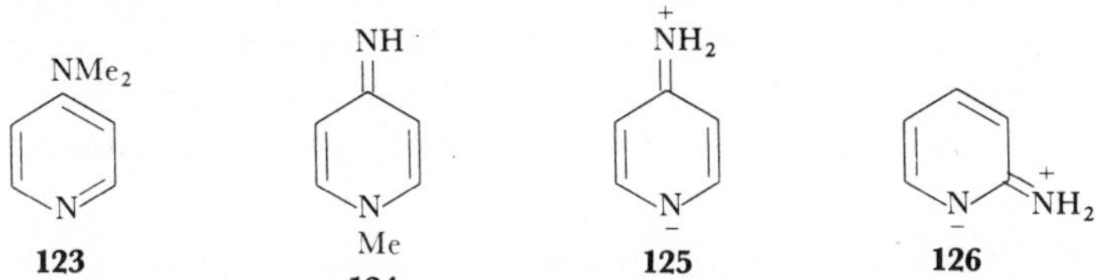

123 **124** **125** **126**

pyridine itself were carried out. As the results show that the amino structure is very much more stable than the imino form, the substitution of the dissociation constant of 4-aminopyridine for that of 4-dimethylaminopyridine (**123**) does not introduce much error. Very similar results were obtained in the 2-aminopyridine series, thus confirming the deductions made from absorption spectrum measurements. The dipole moment of 4-aminopyridine (4·36 D) is appreciably larger than that calculated by combining the dipole moments of aniline and pyridine. This shows that there is an additional resonance interaction in this pyridine that is due to a contribution from **125** to the resonance hybrid. A similar charged contributor **126** can be written in the case of 2-aminopyridine but not for 3-aminopyridine.

The addition of a proton to the ring nitrogen atom of 4-aminopyridine (**120**), but not the amino nitrogen atom, gives an ion (**121a**) for which an additional resonance structure (**121b**), above those usually associated with the pyridine ring, can be written. As this monocation has more resonance energy ('additional ionic resonance') than 4-aminopyridine itself, it has a great tendency to form, and 4-aminopyridine is therefore a strong base (pK_a 9·1). 2-Aminopyridine is a much weaker base (pK_a 7·2) and 3-aminopyridine is weaker still (pK_a 6·6), as

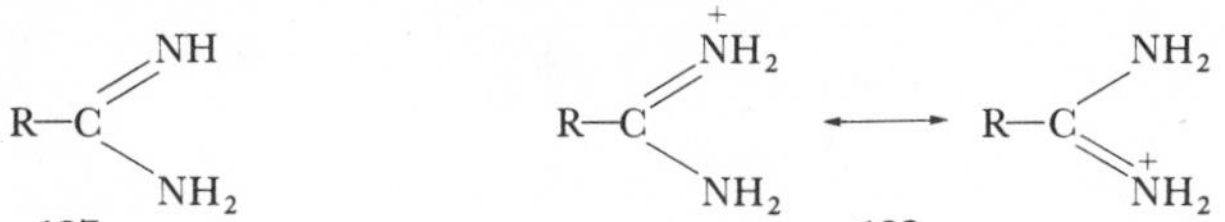

127 **128**

the mono cation has no additional resonance possibilities. 2-Amino- and 4-aminopyridines may be compared with the strongly basic amidine **127**, which accepts a proton to give the highly resonance stabilised cation **128**. It is almost impossible to add a second proton to the monocation (**121**) of 4-aminopyridine because if this occurred the 'additional ionic resonance' would necessarily be lost.

Attempted diazotization of 2- and 4-aminopyridines in aqueous acid gives the corresponding pyridone; no trace of diazonium salt can be detected by attempted coupling with alkaline 2-naphthol. However, 2-aminopyridine with boiling sodium ethoxide and amyl nitrite gives the sodium diazotate (**129**), which couples with 2-naphthol in the normal manner. It therefore appears that the failure to obtain a diazonium salt under ordinary conditions is due to its reactivity and not to its inherent instability. This reactivity is almost certainly due to the pyridine ring of the diazonium cation (**130**) being protonated in the mineral acid solution. The positive charge on the ring greatly facilitates the attack

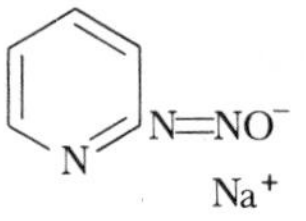

N=NO⁻ Na⁺ OH₂ N₂⁺ $-N_2\uparrow$ $-H^+$ OH O⁻ + H⁺

129 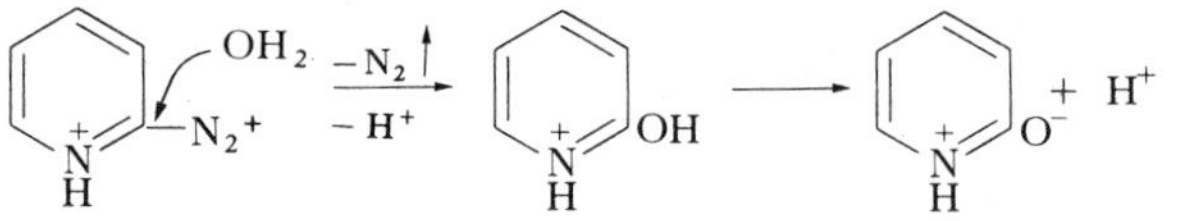 **130**

of nucleophilic reagents, including water and hydroxyl ions, at position 2, with the elimination of nitrogen. The situation is similar to the attempted diazotization of 2,4,6-trinitroaniline under ordinary conditions, when the product is 2,4,6-trinitrophenol. 2-Bromopyridine can be obtained in excellent yield from 2-aminopyridine hydrobromide perbromide in 48% hydrobromic acid and sodium nitrite. Here the bromide anion successfully competes with the water molecule for the pyridinium–diazonium dication (**130**). 4-Fluoro-[83] and 4-chloropyridine have been obtained similarly from 4-aminopyridine using conditions where the tetrafluoroborate or chlorine anions successfully compete with solvolysis.

2-Aminopyridine on irradiation in dilute hydrochloric acid gives a dimer[84] which is structurally analogous to that from 2-pyridone (p. 264), and with excess 4-toluenesulphonyl chloride in pyridine at 20°C very surprisingly gives the bis-derivative **131**.[85] The mono derivative **132** is obtained, with some of **131**,

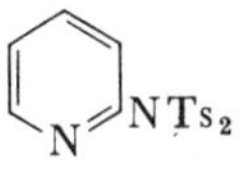

NTs_2 NHTs NH_2 NH OH O⁻

131 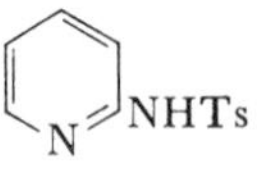 **132** 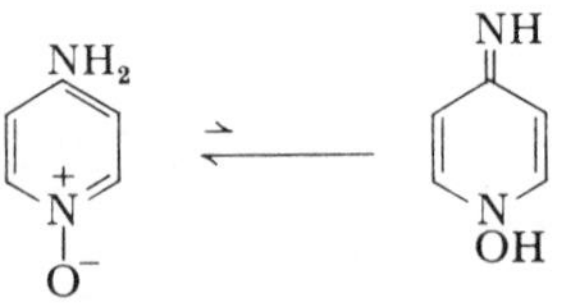 **133**

when 1 mole of the reagent is employed and hydrolysis of **131** to **132** by aqueous sodium hydroxide is rapid.

4-Aminopyridine 1-oxide is easily prepared from pyridine 1-oxide by nitration and subsequent reduction with ammonium hydrogen sulphide or hydrogen over palladium. It can tautomerize but, as in the case of 4-aminopyridine, ultraviolet absorption studies have shown that the amino form (**133**) preponderates. This 1-oxide is of particular interest, as it can be diazotized in the usual way, and the diazonium salt undergoes the common reactions of aromatic diazonium compounds. As 1-oxides are easily reduced to the corresponding pyridines, this is a convenient route to 4-substituted derivatives. 4-Aminopyridine 1-oxide gives 4-aminopyridine with sodium hydrosulphite or with hydrogen and palladium in the presence of hydrochloric acid.

2-Aminopyridine 1-oxide is obtained directly from 2-aminopyridine with peracetic acid. It also exists largely in the amino form and can also be diazotized and coupled with alkaline 2-naphthol. The greater stability of these diazonium salts of the 1-oxides, in comparison with those derived from pyridine itself, is probably due both to an electron release from the oxygen atom, which discourages attack by water molecules and anions on the ring, and to a resonance stabilization of the cation.

8. *Hydroxypyridines*

3-Hydroxypyridine (**134**) is best obtained by fusing pyridine-3-sulphonic acid with potassium hydroxide but has been prepared by decomposing pyridine-3-diazonium salts with water. 3-Hydroxypyridine does tautomerize partially to a

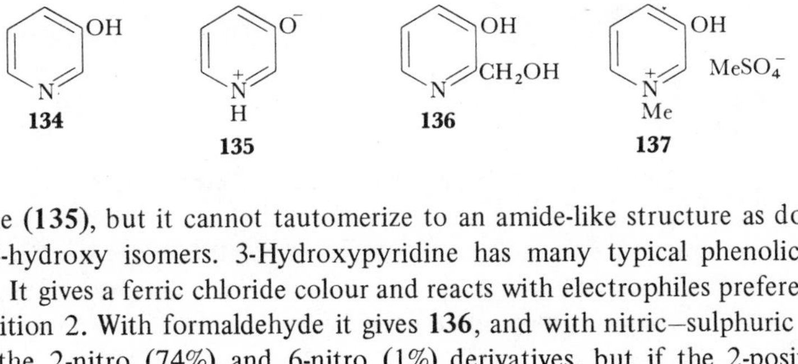

OH N **134** O^- $\overset{+}{N}$ H **135** OH CH_2OH N **136** OH $\overset{+}{N}$ Me $MeSO_4^-$ **137**

betaine (**135**), but it cannot tautomerize to an amide-like structure as do its 2- and 4-hydroxy isomers. 3-Hydroxypyridine has many typical phenolic properties. It gives a ferric chloride colour and reacts with electrophiles preferentially at position 2. With formaldehyde it gives **136**, and with nitric–sulphuric acid it gives the 2-nitro (74%) and 6-nitro (1%) derivatives, but if the 2-position is blocked by a methyl group then the 4-position is most easily nitrated.[86] Both diazomethane and methyl sulphate with alkali attack the ring nitrogen in preference to the hydroxyl group, and with the second reagent **137** is the product. However, with diazomethane in *t*-butyl alcohol at −15°C it (**134**) gives 3-methoxypyridine (70% yield).

2-Pyridone, which is sometimes written as the tautomeric 2-hydroxypyridine, is obtained most readily from 2-aminopyridine and nitrous acid or from pyridine

1-oxide and acetic anhydride. It can also be prepared from pyridine and potassium hydroxide at about 300°C or from 2-chloropyridine with hot dilute

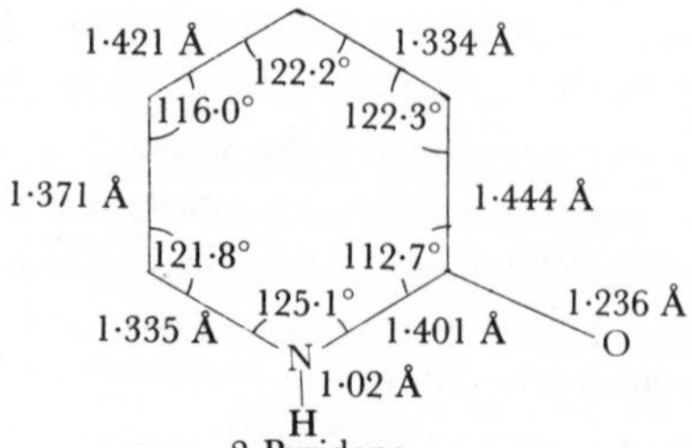

2-Pyridone

mineral acid. It is a colourless solid, m.p. 107°C, soluble in alcohol and water. X-Ray diffraction studies of the solid show that the keto tautomer is present and that the lattice is hydrogen bonded. In methanol the ultraviolet absorption spectrum of the compound is very similar to that of 1-methyl-2-pyridone (**138**), but it differs markedly from that of 2-ethoxypyridine (**141**). The infrared absorption spectra both of 1-methyl-2-pyridone and of the unmethylated compound show a similar absorption band in the carbonyl region, which is in the amide range and is absent in the spectrum of 2-methoxypyridine. It therefore appears that the equilibrium between the 2-hydroxypyridine (**140**) and 2-pyridone (**139**) structures also greatly favours the latter under these

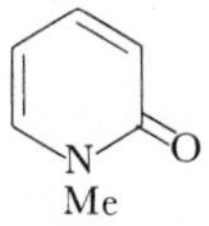

138 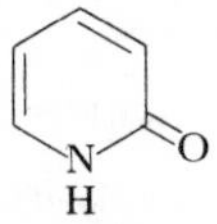 **139** 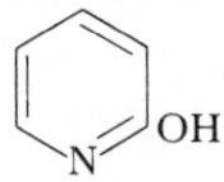 **140** 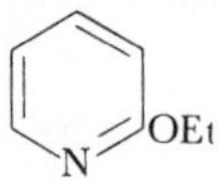 **141**

conditions. As 2-pyridone is very weakly basic, has no normal carbonyl properties, and has a strongly aromatic type of ultraviolet absorption spectrum, it cannot be represented accurately by structure **139**. It is better written as **142** to emphasize its aromatic character and is better considered as a resonance hybrid to which contribute both the uncharged structure **139** and charged

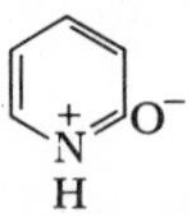

142 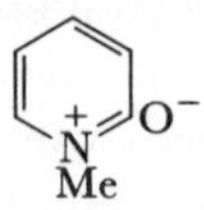 **143** 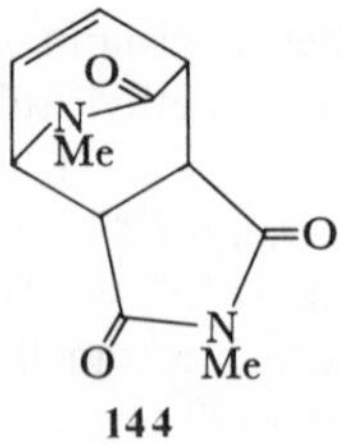 **144**

structures such as **142**, where the ring is benzenoid and has its usual various Kekulé forms. The resonance energy is ca. 105 kJ (25·6 kcal)/mole,[87] a little less than that of pyridine. 1-Methyl-2-pyridone (**138** and **143**) is similar and, unlike pyridine, shows sufficient diene-like reactivity to undergo Diels–Alder reactions, for example, with *N*-methylmaleic imide **144** is formed.[88]

2-Pyridone forms salts with sodium ethoxide and with strong acids, such as hydrogen chloride in ether or concentrated sulphuric acid. These salts are derived from the resonance-stabilized ions **145** and **146**, respectively. 2-Pyridone can behave as such or as 2-hydroxypyridine towards chemical reagents, and

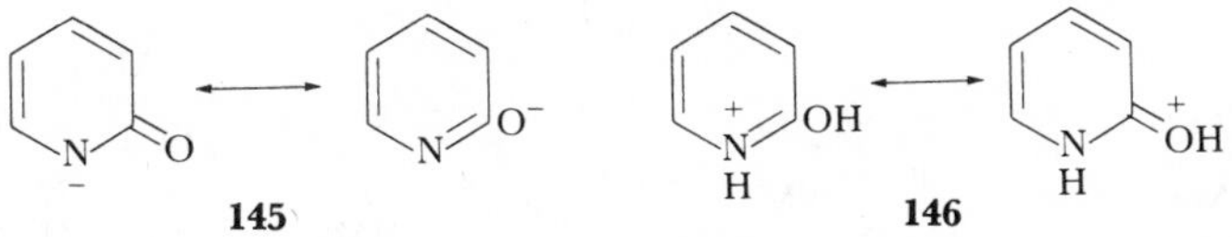

N O⁻ O OH H OH H

145 **146**

reactions at nitrogen and oxygen are in competition with each other. For example, 2-pyridone with acetic anhydride gives a 9 : 1 mixture of 2-acetoxypyridine and 1-acetyl-2-pyridone. With methyl iodide it gives 1-methyl-2-pyridone, while with diazomethane a mixture of this and 2-methoxypyridine is formed.[89] The ratio of the products formed in these types of reactions depends

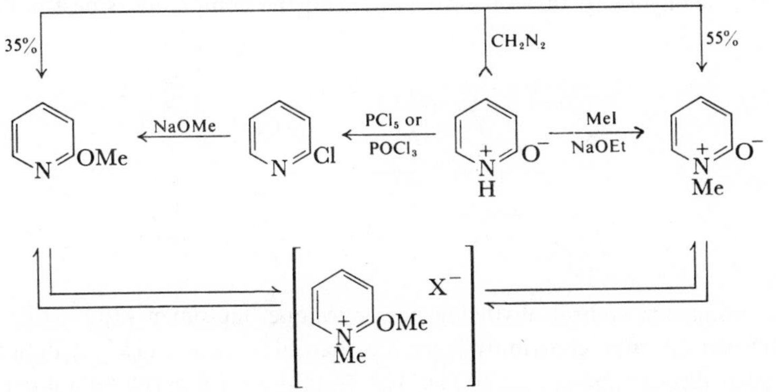

35% CH_2N_2 55%

N OMe NaOMe N Cl PCl_5 or $POCl_3$ N H O⁻ MeI NaOEt N Me O⁻

X⁻ N Me OMe

on the precise conditions. 2-Methoxypyridine and 1-methyl-2-pyridone give an equilibrium mixture when heated with a trace of methyl iodide, or what must be the intermediate in such an equilibration, 1-methyl-2-methoxypyridinium iodide. 2-Methoxypyridine is hydrolysed by aqueous mineral acid to methanol and 2-pyridone. This reaction is similar to the hydrolysis of imino ethers (e.g., **147**) to amides by water. 2-Allyloxypyridine undergoes a normal Claisen rearrangement, giving a mixture of equal quantities of 1- and 3-allyl-2-pyridone.

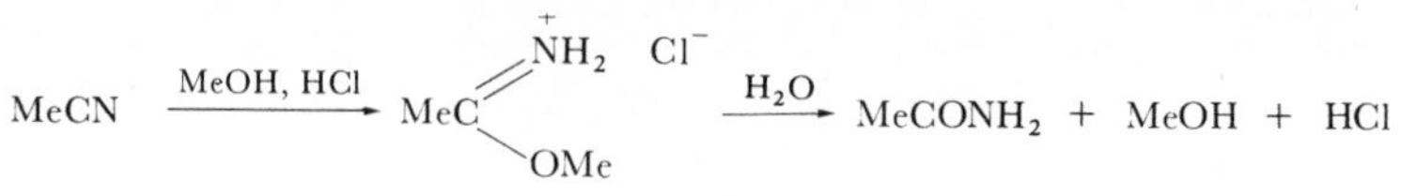

$$MeCN \xrightarrow{MeOH,\ HCl} MeC(=\overset{+}{N}H_2)(OMe)\ Cl^- \xrightarrow{H_2O} MeCONH_2 + MeOH + HCl$$

147

2-Pyridone is easily halogenated to first the 5- and then the 3,5-disubstituted derivative, and nitration also takes place preferentially at position 5.

1-Methyl-2-pyridone on photolysis[90] in dilute ether solution yields **149**, presumably via the diradical **148**, which dimerises to **150** if formed in concentrated solution.[91]

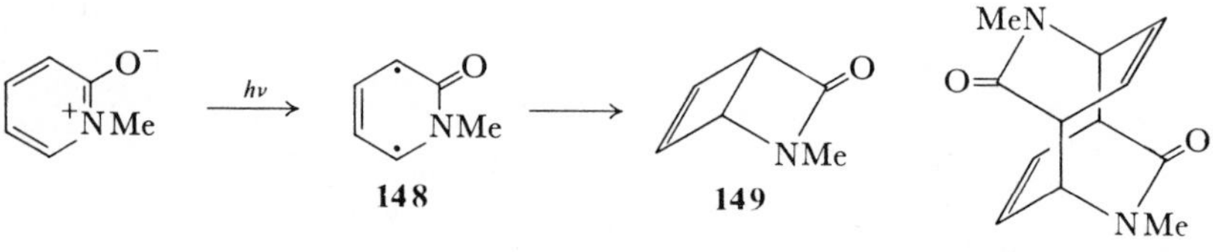

148 **149** **150**

4-Pyridone is readily obtainable by the route indicated; 4-chloropyridine is presumably first formed and reacts with pyridine to give the salt (**151**). The structure of 4-pyridone according to ultraviolet absorption measurements is exactly analogous to that of 2-pyridone, and the compound is best represented

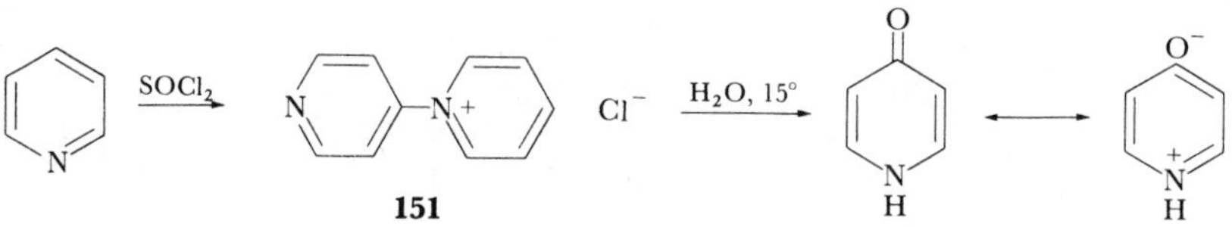

$SOCl_2$ Cl^- H_2O, 15°

151 **152** **153**

as a resonance hybrid involving the pyridone tautomer (**152**–**153**) almost exclusively. A large contribution by the charged structure (**153**) is indicated by the high dipole moment of 6·0 D. The reactions of 4-pyridone are parallel to those of 2-pyridone and under the appropriate conditions it gives *N*- or *O*-derivatives. The pyridone itself undergoes acid-catalysed proton exchange at positions 3 and 5 at low acidities, reminiscent of enamines, but at high acidities the oxygen atom accepts a proton to form the 1-hydroxypyridinium cation and then exchange occurs.[92] Electrophilic substitution, such as bromination, nitration, and attack of carbon dioxide on the sodium salt (cf. Kolbe's reaction with sodium phenate), gives 3-, and then 3,5-disubstitution. Some 4-pyridones with very strongly electronegative substituents, such as the tetrafluoro derivative, exist 'entirely' as the 4-hydroxypyridine tautomers.

'2,4-Dihydroxypyridine' (**154**) is of interest, as three tautomeric structures are possible. From an absorption spectra study it has been shown that **155–156** is predominant in 50% aqueous ethanol. This is expected, as the charge separation in the ionic resonance form (**156**) of this structure is less than that in the 4-pyridone tautomer (**157**).

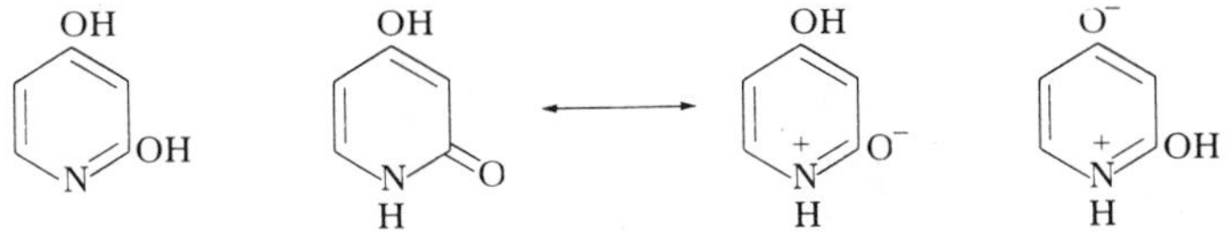

154 **155** **156** **157**

2-Pyridinethiol (**158**) is in equilibrium with 2(1*H*)-pyridinethione (**159**), which is the predominant tautomer. This compound can be obtained from 2-chloropyridine and sodium hydrogen sulphide (p. 255) and is of special interest in peptide synthesis.[93] With protected amino acids, such as **160**, and

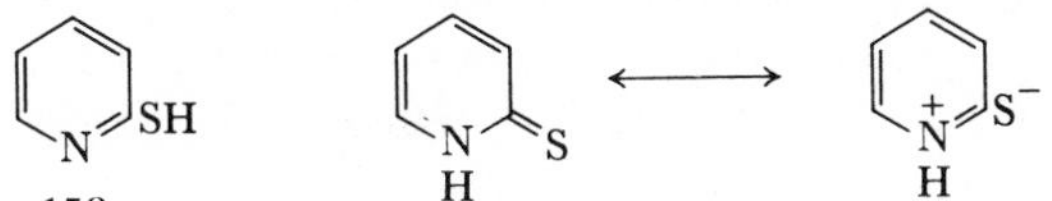

158 **159**

dicyclohexylcarbodimide (DCC), the corresponding ester (**161**) is obtained. These esters are particularly valuable because they condense rapidly and completely with amino acid esters (**162**) to form a new peptide (**163**), even when much steric hindrance is present. No detectable racemisation occurs. The coupling reaction must be facilitated by anchimeric assistance, as indicated, and has great promise for peptide synthesis. Compounds like **161**, where R is unreactive to the reagent employed, give excellent yields of ketones with Grignard reagents.[94] Presumably a similar cyclic intermediate is involved, and this eliminates tertiary alcohol formation.

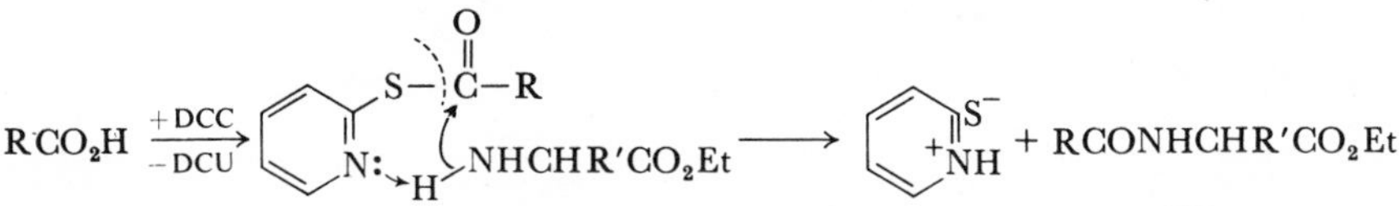

RCO_2H $\xrightarrow[-DCU]{+DCC}$ (161, S–C(=O)–R) + $NH_2CHR'CO_2Et$ ⟶ pyridinethione + $RCONHCHR'CO_2Et$

160 **161** **162** **163**

R = $PhCH_2OCONHCH_2$–, for example

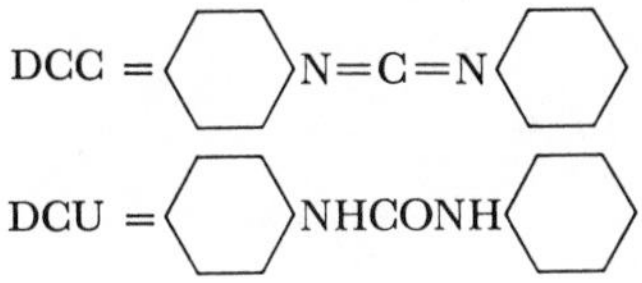

DCC = C_6H_{11}N=C=NC_6H_{11}

DCU = C_6H_{11}NHCONHC_6H_{11}

9. *Pyridine alcohols, aldehydes, carboxylic acids and derivatives*

Reduction of the corresponding aldehydes and ketones yields pyridinemethanols, such as **164**, and 2- and 4-(2-hydroxyethyl)pyridine (e.g., **165**) are obtained from the corresponding methylpyridines with formaldehyde. They dehydrate readily to the vinylpyridines (p. 247).

CHROH **164** CH_2CH_2OH **165**

4-Picolyl ethers and esters have recently been much used in polypeptide chemistry.[95] 4-Pyridinemethanol (4-picolyl alcohol), for example, combines with a benzyloxycarbonylamino acid (e.g., **166**) in the presence of dicyclohexylcarbodiimide to give the ester **167**. These esters undergo normal basic hydrolysis and yield hydrazides with hydrazine. However, they are stable to trifluoroacetic acid or hydrogen bromide, perhaps because the pyridine ring becomes positively charged by accepting a proton and thereby discourages the further proton attack at the nearby ester group essential for hydrolysis of this group. They are reduced to the corresponding acids (**168**) and 4-methylpyridine (**169**) by sodium and liquid ammonia or by hydrogen over palladised charcoal in the same way as benzyl esters, but in contrast to benzyl esters, the picolyl group is also reduced off electrolytically at a mercury cathode or by dithionite in acetic acid.

The 4-picolyl group is valuable for protecting thiol[96] or phenolic hydroxyl[96] and hydrazino groups[97] in peptide synthesis. Adding 4-chloromethylpyridine

$PhCH_2OCONHCHRCO_2H$ + 4-MePy

dithionite/HOAc, or electrolytic reduction

$PhCH_2OCONHCHRCO_2H$ (**166**) → (DCC + $HOCH_2Py$ or TMG salt + $ClCH_2Py$) → $PhCH_2OCONHCHRCOOCH_2$-(4-pyridyl) (**167**)

Na/NH_3 or H_2, Pd/C

PhMe CO_2 | HBr or CF_3CO_2H

NH_2CHRCO_2H + Me-(4-pyridyl) (**168**, **169**) ← (Electrolytic reduction or $Na_2S_2O_4/HOAc$) ← $NH_2CHRCOOCH_2$-(4-pyridinium $\overset{+}{N}H$) X^-

DCC = $C_6H_{11}N{=}C{=}NC_6H_{11}$,
TMG = $(Me_2N)_2C{=}NH$

(4-picolyl chloride) to the sodium salt of cysteine in liquid ammonia and to the nickel complex of tyrosine gives **170** and **171**, respectively, and in both cases after conversion to more complicated peptides, the 4-picolyl group can be removed by electrolytic reduction leaving the original SH and OH groups intact and without affecting the new peptide links.

$HO_2CCH(NH_2)CH_2SCH_2$–(4-pyridyl) **170**

$HO_2CCH(NH_2)CH_2$–(C_6H_4)–OCH_2–(4-pyridyl) **171**

4-Picolyloxycarbonylhydrazide (**172**) is easily obtained and can be coupled with *t*-butoxycarbonyl-α-amino acids using the 1-hydroxybenzotriazole method (p. 438). The butoxy group, but not the picolyl group, is split off with trifluoroacetic acid and the peptide chain is then extended by coupling with further amino acid derivatives. Finally, the picolyl group is removed by catalytic hydrogenation to give the hydrazide **173**. This with nitrous acid yields the corresponding azide, ready for coupling with a carboxyl-protected amino acid to build a bigger polypeptide.[97]

$(4\text{-}NO_2C_6H_4O)_2CO \xrightarrow[\text{alcohol}]{\text{4-picolyl}} 4\text{-}NO_2C_6H_4OCOOCH_2$–(4-pyridinium, $\overset{+}{N}H$) $\xrightarrow{NH_2NH_2}$

$4\text{-}NO_2C_6H_4O^-$ + $NH_2NHCOOCH_2$–(4-pyridyl) **172** $\longrightarrow$ $t\text{-}BuOCONHCHRCONHNHCOOCH_2$–(4-pyridyl)

$\xrightarrow[\text{(b) } H_2,\ Pd/C]{\text{(a) Build polypeptide chain}}$ $t\text{-}BuOCO(NH\cdots)_nCONHNH_2$ + CO_2 + Me–(4-pyridyl)

173

All the pyridinecarbaldehydes can be obtained from the corresponding carboxylic esters by reduction to the alcohols by lithium aluminium hydride and subsequent selenium dioxide oxidation. Pyridine-2-carbaldehyde is readily available by a convenient laboratory procedure involving hydrolysis of a diacetate (**98**, p. 253) obtained from 2-methylpyridine 1-oxide. Interesting new routes are outlined for both the 2-[98] (**174**) and 3-carbaldehydes (**175**).[99] In the latter, crotonaldehyde is converted to the enamine, which undergoes electrophilic attack with Vilsmeier's reagent in typical enamine fashion at carbon atoms 2 and 4. Hydrolysis of the product by aqueous ammonium chloride gives the pyridine (**175**).

N Br

NaH | $MeSOCH_2SMe$

O^- SMe CH SMe N $\xrightarrow[200\ ^\circ C]{\Delta}$ [N CH OSMe SMe ?] $\longrightarrow$ N CHO + MeSSMe

174, 100%

$MeCH{=}CHCHO \xrightarrow{MeNH_2}$ NMe_2 $\xrightarrow[COCl_2]{Me_2NCHO}$

[$Me_2\overset{+}{N}$ $\overset{+}{N}Me_2$ NMe_2 $\xrightarrow[aq.]{NH_4Cl}$ O CHO OH + NH_3] $\longrightarrow$ N CHO

175, 40% yield

The pyridinecarbaldehydes are all water-soluble colourless liquids, which have many of the usual reactions of aromatic aldehydes. Pyridine-2-carbaldehyde with aqueous potassium cyanide undergoes a normal condensation with the precipitation of 2-pyridoin, which actually exists as the tautomer (**176**) in neutral or weakly acid solution.

O H N N H O — **176**

CO_2H N — **177**

CH_2OH N — **178**

Pyridine-3-carbaldehyde gives no visible reaction under the same conditions, while pyridine-4-carbaldehyde gives a mixture of the acid (**177**) and the alcohol (**178**). These are the products expected of a Cannizzaro reaction which is perhaps initiated by reaction with potassium hydroxide formed by the hydrolysis of the potassium cyanide.

Pyridinecarboxylic acids are obtainable from many pyridine syntheses and can be prepared by general methods, including the oxidation of alkylpyridines with alkaline potassium permanganate. Many of the acids were first obtained by the oxidation of natural products at early dates and have been given trivial

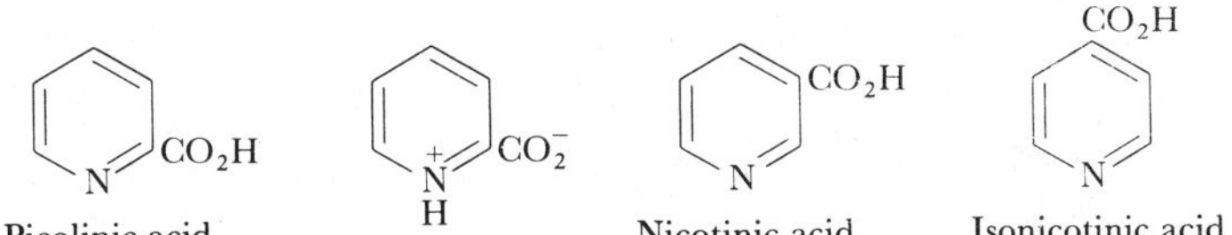

Picolinic acid, m.p. 137°

179

Nicotinic acid, m.p. 229°

Isonicotinic acid m.p. 299°

names which are still widely used. The oxidation of the alkaloid nicotine gives nicotinic acid. This acid is of particular importance, as it is a vitamin for man, and it is obtained commercially by the oxidation of quinoline. The first product, quinolinic (pyridine-2,3-dicarboxylic) acid, can be isolated and subsequently selectively decarboxylated, or both reactions can be carried out together. A similar oxidation of isoquinoline gives pyridine-3,4-dicarboxylic acid, which on decarboxylation gives pyridine-3-carboxylic (nicotinic) acid along with some pyridine-4-carboxylic (isonicotinic) acid. If the structures of quinoline and isoquinoline are assumed, these relationships prove the structures of the pyridinemonocarboxylic acids involved, and the remaining one must be picolinic acid. In the decarboxylation of pyridine acids carboxyl groups at position 2 are

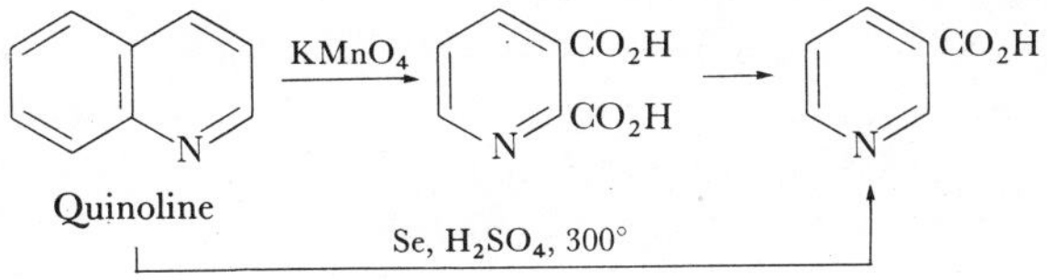

Quinoline

$KMnO_4$

Se, H_2SO_4, 300°

always lost preferentially and then usually carboxyl groups at position 4. The decarboxylation of pyridine-2-carboxylic acid proceeds[100] via the intermediate **179a** and ylid **180**, which can be trapped as **181** by benzaldehyde. In contrast, substituted pyridine-2-carboxylic acids[100] and pyridine-2-acetic acid[101] decarboxylate via the corresponding zwitterions (cf. **179** and **182**, respectively).

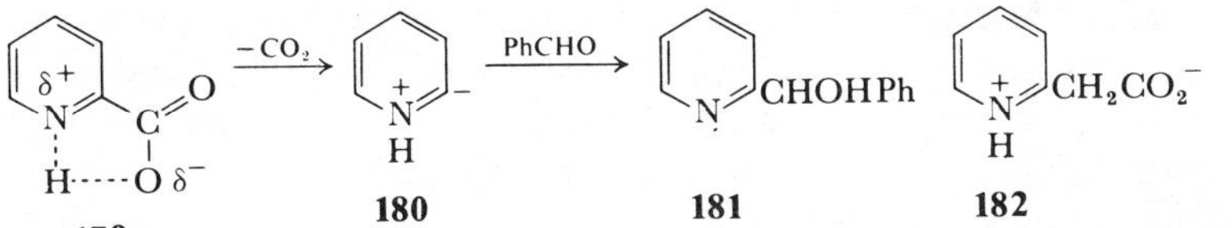

179a —($-CO_2$)→ **180** —(PhCHO)→ **181**; **182**

Pyridinecarboxylic acids in general have properties normal to the pyridine ring and aromatic carboxyl groups. They are betaines (e.g., **179**), at least to some extent, and it is interesting that picolinic acid melts (137°C) at a much lower temperature than its isomers. This is because interaction between the charged centres of the zwitterion (**179**) has more intramolecular character than is possible with its isomers. Pyridine-2-carboxylic acids give a yellow-brown colour with ferrous salts, Skraup's test.[102] The constitution of the coloured complex from picolinic acid is not known with certainty, but may be **183**. Pyridine-3- and pyridine-4-carboxylic acids do not give a colour under these conditions.

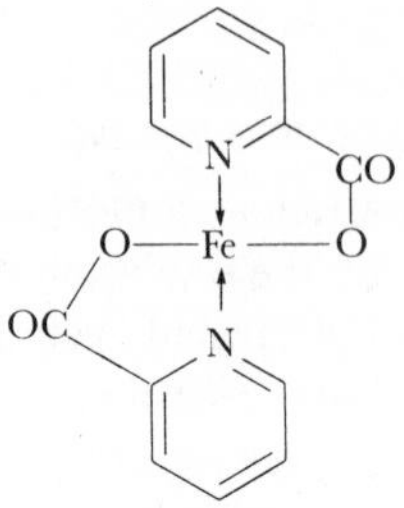

183

E. Synthetic methods

There are many ways in which the pyridine ring can be synthesized, but only syntheses of particular theoretical or practical interest are considered here. The first, due to Hantzsch, was published in 1881.

(1) Pyridine has been synthesized from glutaconic aldehyde (**184**) and from pentamethylenediamine (p. 229). These methods have little practical interest, but useful vapour-phase syntheses include passing tetrahydrofurfuryl alcohol and ammonia over alumina[103] at 500°C and that shown below.

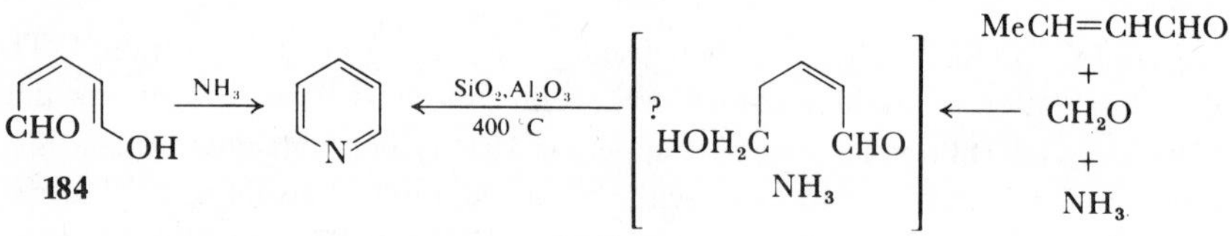

$MeCH{=}CHCHO + CH_2O + NH_3 \longrightarrow [HOH_2C\text{–}CH{=}CH\text{–}CH_2\text{–}CHO,\ NH_3] \xrightarrow[400^\circ C]{SiO_2, Al_2O_3}$ pyridine $\xleftarrow{NH_3}$ **184** (CHO, OH)

(2) Many aldehydes and ketones can react with ammonia to give pyridines under the correct conditions. Owing to the availability of the reactants, this type of synthesis has attracted much interest, but mixtures are often formed, even if yields are good. 5-Ethyl-2-methylpyridine (**185**) can be obtained from ammonia and acetaldehyde. In the laboratory the reaction is best carried out by heating paraldehyde, ammonia, and ammonium acetate to 230°C for an hour in a steel autoclave, when 50–53% yields can be obtained along with some 2-methyl-pyridine. Yields of up to 70% of the pyridine (**185**) have been reported occasionally. The mechanism of the reaction is not clear, although it almost certainly proceeds via aldol condensations. These may involve acetaldehyde or the corresponding imine, acetaldimine, which is present at least to some extent in the mixture. For simplicity, only one of the possible reaction sequences is

CHO * Me Me OCHMe MeCHO CHO NH_3 — CHO CH_2 $CH_2CHOHMe$ MeCHOH CHO NH_3 → 185

formulated, and it must be emphasized that other equally plausible mechanistic routes are possible and that direct experimental evidence concerning any mechanism is entirely lacking. It is easily seen that if the acetaldehyde molecule marked with an asterisk is missing, the formation of 2-methylpyridine is accounted for, providing that a dehydrogenation can take place.

(3) The Hantzsch synthesis of pyridines, if variations are included, is the most flexible and widely used synthesis of pyridine derivatives. It consists essentially of treating a β-keto ester with an aldehyde in the presence of ammonia. A 1,4-dihydropyridine (**187**) is first formed, and this is oxidized *in situ*, often with nitric or chromic acid, to the pyridine (**188**). Examples are the condensations of

R CHO EtO_2CCH_2 CH_2CO_2Et MeCO OCMe NH_3 —A→ R HC EtO_2CCH CCO_2Et MeC COMe NH_2 ↓A ↓B R CH EtO_2CHC $CHCO_2Et$ MeCO OCMe 186 —B, NH_3→ 187 —HNO_3→ 188 (NH_2OH: 186 → 188)

acetaldehyde and of benzaldehyde with ethyl acetoacetate. Two routes by which the reactions could proceed are outlined, and as excellent yields of pyridines are often obtained when the suggested intermediates of either scheme are used, it may be concluded that equilibrium between the intermediates of both schemes is rapidly attained or, perhaps, that both sequences give the pyridine with comparable facility. The production of undesirable isomers can sometimes be limited by preparing the intermediates of scheme A (or B) separately and then allowing them to react together. Reaction of the 1,5-diketone (**186**) of route B with hydroxylamine instead of ammonia gives the pyridine (**188**) directly, thereby avoiding the oxidation stage.

Another variation of Hantzsch's synthesis is the condensation of 2 moles of aldehyde (e.g., acetaldehyde) with 1 mole of the β-keto ester (e.g., ethyl acetoacetate) in the presence of ammonia. The yields are only 20–30%, but the reaction has been widely used. A more satisfactory procedure is to use the self-condensation product of the aldehyde (e.g., crotonaldehyde) and the ethyl β-aminocrotonate, as this reduces the possibility of unwanted products. These reactants, in pyridine, give **189**, **190**, and **191**. On distillation *in vacuo* the

HNO_3

189 57% **190** 15% **189** **191** 28%

$LiAlH_4$

dihydropyridine **189** gives some **190**, doubtless through disproportionation; a similar disproportionation of **189** to **190** and **191** takes place on shaking with a palladium-on-charcoal hydrogenation catalyst. Occasionally in the Hantzsch synthesis the substituent, which should appear at position 4 of the final pyridine, is lost. This is particularly the case when it is big and is an electron-releasing group.[104] The structures of dihydropyridines such as **189** are usually deduced from their nuclear magnetic resonance spectra.

Guareschi and Thorpe have used β-keto esters instead of the two acetaldehyde molecules in the above variation, and in this case it is necessary to use cyanacetic ester or amide as the second component. As the oxidation state of the reactants is higher than that employed in the original Hantzsch synthesis, a pyridine is formed directly. There are many variations of this condensation. The β-keto ester can be replaced by a 1,3-diketone, cyanoacetic acid is an alternative to cyanoacetic ester, and cyanoacetamide can be used instead of both the cyanoacetic ester and the ammonia. An example of the last reaction is the synthesis of **192**, which is used for the preparation of vitamin B_6 (p. 278).

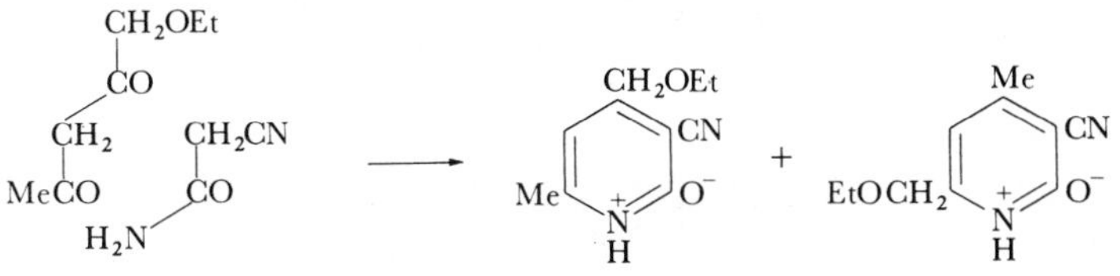

CH_2OEt
CO
CH_2 CH_2CN
MeCO CO
H_2N
→
CH_2OEt
CN
Me $\overset{+}{N}$ O^-
H
+
Me
CN
$EtOCH_2$ $\overset{+}{N}$ O^-
H

192 (75% yield)

A variation of this synthesis is to condense a 1,3-dicarbonyl compound, or a

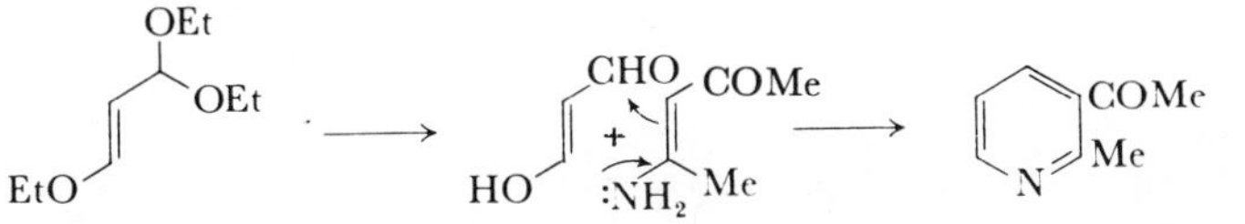

OEt
OEt
EtO
→
CHO COMe
+
HO :NH_2 Me
→
COMe
Me
N

precursor, with an enamine. Addition of an enamine to an acetylenic ketone also provides a general synthesis of pyridines.

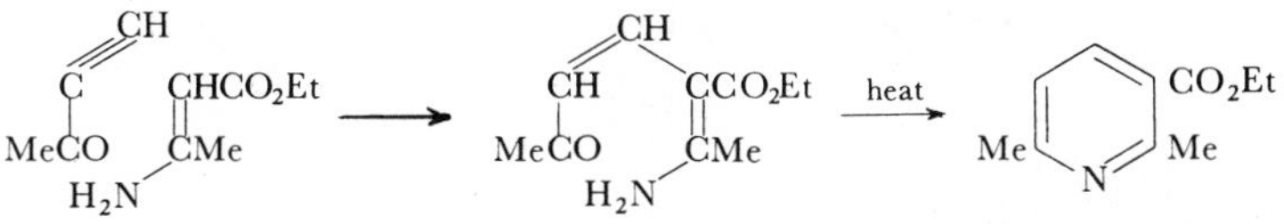

CH
C $CHCO_2Et$
MeCO CMe
H_2N
→
CH
CH CCO_2Et
MeCO CMe
H_2N
→ (heat)
CO_2Et
Me N Me

(4) Two adaptable and useful syntheses of pyridines from furans have been developed.[105] Many 2-furyl ketones **(193)** on heating with ammonia and ammonium chloride give 3-hydroxypyridines **(194)**. Another route, starting from 2-acetylfuran, gives the pyridone 1-oxide **(195)**. A third synthesis starts

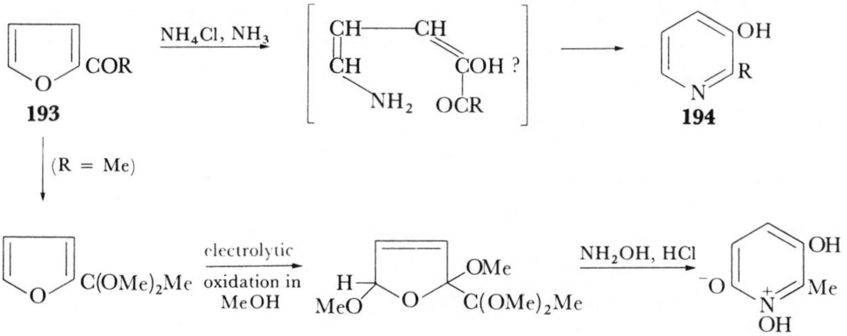

O COR
193
→ (NH_4Cl, NH_3)
[CH—CH
CH COH ?
NH_2 OCR]
→
OH
R
N
194

(R = Me)
↓

O $C(OMe)_2Me$
→ (electrolytic oxidation in MeOH)
H
MeO O OMe
$C(OMe)_2Me$
→ (NH_2OH, HCl)
OH
^-O $\overset{+}{N}$ Me
OH
195

from **196**, which can be prepared by the reductive acetylation (hydrogen, Raney nickel, and acetic anhydride) of 2-acetylfuran oxime.

CHMeNHAc
196
Electrolytic oxidation in MeOH
MeO OMe H O CHMeNHAc
NaOH aq.
MeO OMe H O $CHMeNH_2$
H^+ aq.
[CH=CH CHO CO ? H_2NCHMe]
OH Me N

(5) A new synthesis[106] of 1,4-dihydropyridines, easily oxidised to pyridines, starts with isoxazoles and appears to be a general method. The key stage is the hydrogenolytic splitting of the N–O bond of the isoxazole.

Me CH_2Cl O N Me
$AcCH_2CO_2Et$ base
Me $CH_2CHAcCO_2Et$ O N Me
H_2, Pd/C Et_3N
[MeCO CO_2Et ? Me N H_2 O Me]
Ac CO_2Et Me N H Me

(6) The Diels–Alder reaction has been used[107] to build up the pyridine ring, and the combination of butadiene and cyanogen at 400°C provides a synthesis of 2-cyanopyridine.[108]

Me R + $CHCO_2C_4H_9$ N $SO_2C_6H_4Me$
Me R N $CO_2C_4H_9$ $SO_2C_6H_4Me$
$\bar{O}H$
Me R N CO_2H

Oxazoles such as **196a** with dienophiles yield pyridines, and this reaction has been employed to synthesise vitamin B_6 (p. 278) and analogues.[109]

CHO CO_2Et HN—CHMe
P_2O_5
O OEt N Me
196a
O
[O O OEt N Me]
O OH N Me

(7) Pyridines can be obtained from pyrylium salts and ammonia (p. 282) and 2- (e.g., **197**) and 4-pyridones (e.g., **199**) similarly can be prepared from the corresponding pyrones. The usefulness of this type of synthesis depends on the availability of the starting materials. The reverse reaction, the conversion of **199** to **198**, occurs in cold dilute aqueous hydrochloric acid.[110]

Methyl coumalate $\xrightarrow[20^\circ]{NH_3}$ **197**   **198** $\xrightarrow[EtOH]{NH_3}$ **199**

F. Natural occurrence and compounds of special interest

Pyridine itself does not occur free in nature, although piperidine is present as a salt in black pepper. Numerous derivatives of both these compounds occur in the plant kingdom.

1. *Nicotinamide adenine dinucleotides*

The most widely distributed derivative of pyridine is probably nicotinamide (**200**), which is an essential part of these coenzymes. Nicotinamide was

200 ($CONH_2$)   **201** (CO_2H)

identified in 1937 as the substance necessary for the prevention of the deficiency disease, pellagra, in man. The free acid, nicotinic acid (**201**), is equally effective. Nicotinic acid is formed *in vivo* from the essential amino acid tryptophan in animals and is built up from glycerol and aspartic acid by *M. tuberculosis.*[111] It is stored in the liver and is excreted in human urine as a glycine conjugate, nicotinuric acid (**202**), as the betaine, trigonelline (**203**), and as the pyridone (**204**).

202 ($CONHCH_2CO_2H$)   **203** (CO_2^-, N-Me)   **204** (CO_2H, N-Me)

Nicotinamide adenine dinucleotide (NAD, **205**) and its phosphate (NADP, **206**), also known as di- (DPN) and triphosphopyridine nucleotide (TPN), respectively, are very widely distributed. They are involved in oxidation–reduction processes in living organisms. It is very probable that during the reduction a proton and two electrons are added.[112] The system acts as

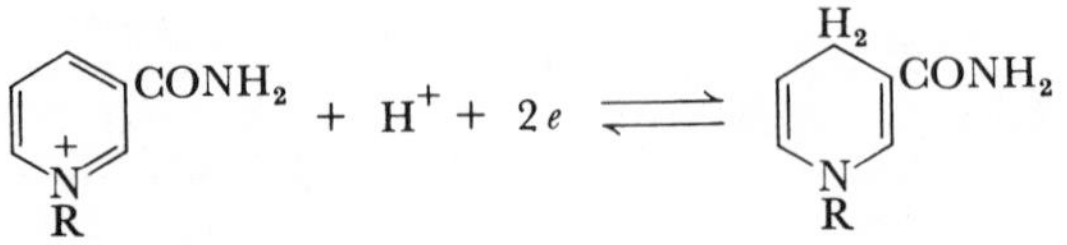

effectively as a source and acceptor of hydride ions. Although the ring is essentially planar the molecule is chiral because of the 1-substituent, and the two hydrogen atoms at position 4 are therefore not identical in enzymic reactions.

The structure of NAD (**205**) has been clearly deduced from the results of degradative studies.[113] Acid hydrolysis of the nucleotide (**205**) gives adenine

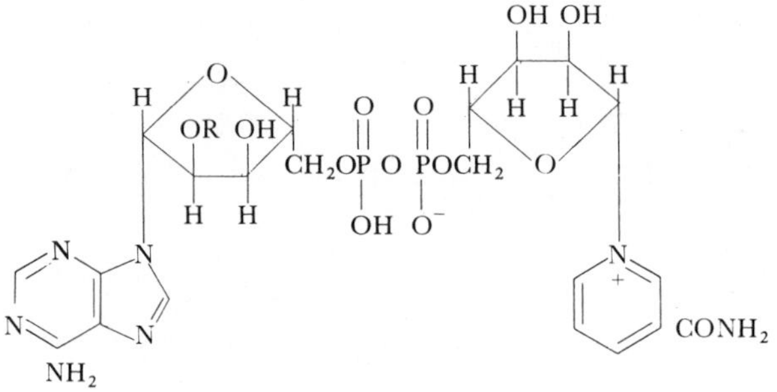

205, R = H. Nicotinamide adenine nucleotide (NAD)
206, R = PO_3H_2. Nicotinamide adenine nucleotide phosphate (NADP)

(**207**), ribose-5-phosphate (**208**), and nicotinamide (**209**). Dilute cold alkali splits off the nicotinamide and leaves the rest of the molecule intact, while if this

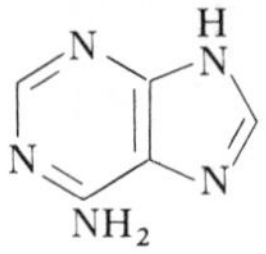

207 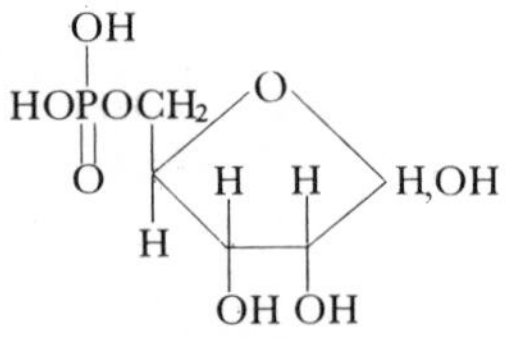 **208**  **209**

hydrolysis is carried out at higher temperatures, adenosine-5′-phosphate (**215**) is produced. Enzymic hydrolysis with phosphomono- and phosphodiesterases give nicotinamide-riboside (**213**), adenosine, and phosphate. NAD (**205**) was also shown to be monoacidic and so its structure was on a firm basis before being confirmed by synthesis.[114]

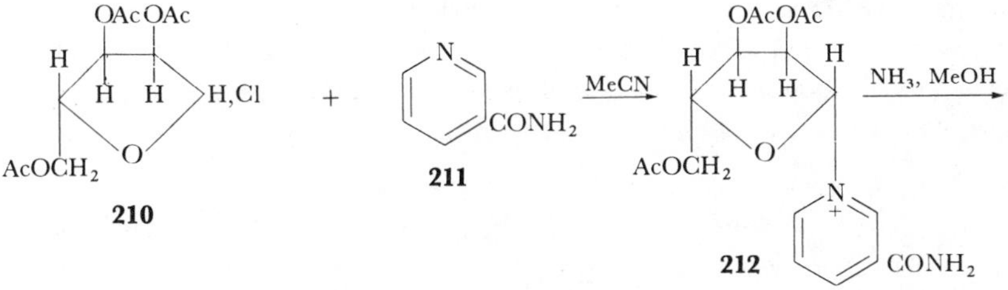

OAc OAc H H H H,Cl AcOCH$_2$ O **210** + N CONH$_2$ **211** $\xrightarrow{\text{MeCN}}$ OAcOAc H H H H AcOCH$_2$ O N$^+$ CONH$_2$ **212** $\xrightarrow{\text{NH}_3\text{, MeOH}}$

OH OH H H H H HOCH$_2$ O N$^+$ CONH$_2$ **213** $\xrightarrow{\text{POCl}_3\text{, MeNO}_2}$ OH OH H H H H HOPOCH$_2$ O O$^-$ N$^+$ CONH$_2$ **214**

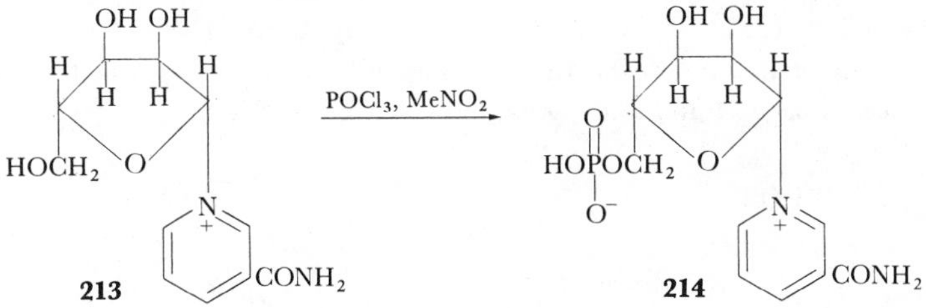

H O H OH OH H H CH$_2$OPOH OH N N N N NH$_2$ **215**

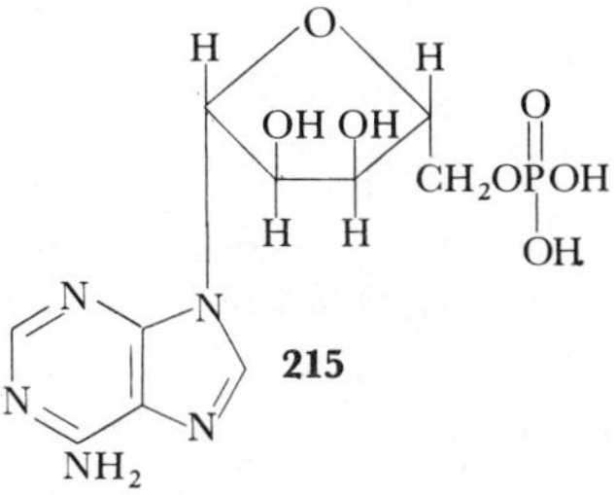

$C_6H_{11}N{=}C{=}NC_6H_{11}$

216

Acetylchlororibofuranose (**210**), prepared from ribose by acetylation and subsequent treatment with ethereal hydrogen chloride, reacts with nicotinamide (**211**) in acetonitrile as solvent giving **212**. On hydrolysis with ammonia,

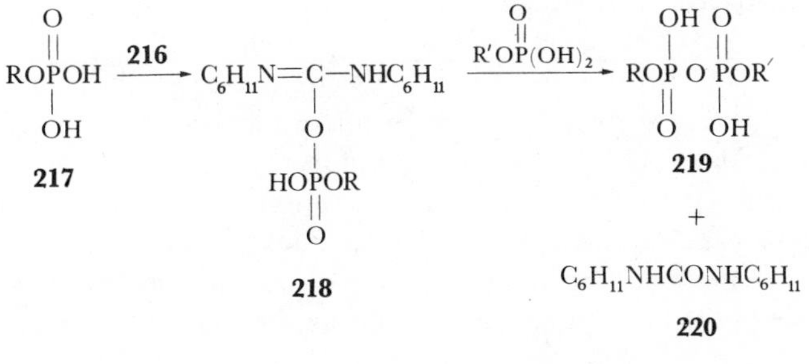

ROPOH (O, OH) **217** $\xrightarrow{\textbf{216}}$ $C_6H_{11}N{=}C{-}NHC_6H_{11}$ O HOPOR O **218** $\xrightarrow{\text{R}'\text{OP(O)(OH)}_2}$ ROP O POR′ (OH, O, O, OH) **219** + $C_6H_{11}NHCONHC_6H_{11}$ **220**

nicotinamide nucleoside (**213**) is formed and is phosphorylated as shown to nicotinamide nucleotide (**214**). This reacts with adenosine-5′-phosphate (**215**) in the presence of dicyclohexylcarbodiimide (**216**) to give nicotinamide adenine nucleotide (**205**). The diimide probably reacts with one or other of the phosphates (**217**) to give an ester (**218**), which then combines with the other phosphate. The pyrophosphate (**219**) and the urea (**220**) are the products. Dicyclohexylcarbodiimide is also used as a dehydration reagent in the synthesis of penicillin (p. 66) and polypeptides (p. 438).

2. *Pyridoxine*

Pyridoxine, or vitamin B_6, is the factor required to prevent the development of a dermatitis in rats, and through the use of this property for assay it was isolated in several laboratories in 1938. Chemical studies soon disclosed the structure (**221**) of the vitamin. It was amphoteric, gave a positive ferric chloride colour test, and showed the presence of three active hydrogens by reaction with methylmagnesium iodide (Zerewitinoff's method).

CH_2OH HO CH_2OH Me 2 6 N — CH_2N_2 → CH_2OH MeO CH_2OH Me N → CO_2H MeO CO_2H Me N — soda-lime → MeO Me N

221 **222** **223** **224**

The absorption spectrum of pyridoxine closely resembled that of 3-hydroxypyridine, and with diazomethane the ether (**222**) was obtained. The acid (**223**), obtained on oxidation of the ether, gave a phthalein with resorcinol. This indicated a 1,2-arrangement of the carboxyl groups. The acid gave no colour with ferrous sulphate, which showed that no carboxyl group was at position 2 of the pyridine ring (p. 270). These data leave only two possible structures for pyridoxine, and the formation of **224** through decarboxylation of the acid confirms one (**221**) and excludes the other, which has the methyl group at position 6.

The commercial synthesis of Harris and Folkers is outlined on the next page; other syntheses have been carried out subsequently.

Pyridoxine is converted into pyridoxal-5-phosphate (**225**) *in vivo*, and this phosphate is required in a great variety of enzymic reactions involving amino acids. Snell[115] has succeeded in reproducing a number of these reactions by incubating pyridoxal (**226**) and a metal (copper, cobalt, or iron) salt with the amino acid and has suggested a general mechanism. This requires the formation of the complex **227** from the metal ion, pyridoxal, and the amino acid. The succeeding reactions all depend on the ability of the electron-attracting groups in the molecule (**227**) to withdraw electrons from the region of the α-carbon atom

CH_2OEt / CO_2Et + Me / MeCO $\xrightarrow{NaOEt}$ CH_2OEt–CO–CH_2–MeCO + CH_2CN–CO–H_2N $\xrightarrow{\text{piperidine}}$ CH_2OEt, CN, O^-, Me, $\overset{+}{N}$H $\xrightarrow{HNO_3}$

O_2N, CH_2OEt, CN, O^-, Me, $\overset{+}{N}$H $\xrightarrow{PCl_5}$ O_2N, CH_2OEt, CN, Cl, Me, N $\xrightarrow[Pt]{H_2}$ H_2N, CH_2OEt, CN, Me, N $\xrightarrow{H_2,\ Pd}$

H_2N, CH_2OEt, CH_2NH_2, Me, N $\xrightarrow[175°]{2\cdot5\ M\ HCl}$ H_2N, CH_2OH, CH_2NH_2, Me, N $\xrightarrow{HNO_2}$ HO, CH_2OH, CH_2OH, Me, N

221

CHO, HO, $CH_2OP(O)(OH)OH$, Me, N

225

$RC(H)CO_2^-$ / $^+NH_3$ + M^{3+} + CHO, $HOCH_2$, O^-, Me, $\overset{+}{N}$H (**226**) $\underset{H^+}{\overset{-H^+}{\rightleftharpoons}}$ $H_3\overset{+}{O}$ + **227** $\underset{H^+}{\overset{-H^+}{\rightleftharpoons}}$ **228**

227: H (a) (b), R–C–CO, O, $\overset{+}{N}$, M, O, $HOCH_2$, Me, $\overset{+}{N}$H

228: R–C–CO, (c), O, $\overset{+}{N}$, M, O, $HOCH_2$, Me, NH

228 ⟷ **229**

$RCOCO_2^-$ + M^{3+} + CH_2NH_2, $HOCH_2$, OH, Me, N (**230**) $\underset{-H_3O^+}{\overset{H_3O^+}{\rightleftharpoons}}$ **229**

229: O^-, R–C=C, O, $\overset{+}{N}$, M, O, $HOCH_2$, Me, $\overset{+}{N}$H of the amino acid moiety. Ionization of a hydrogen atom by the breaking of bond (a) to give **228** followed by recombination to **227** and hydrolysis gives the original amino acid, racemized. Structure **228**, which requires the loss of the resonance energy of the pyridine ring, is perhaps unlikely for the ionized intermediate involved in the racemization. This intermediate is certainly better considered as a resonance hybrid to which **228** and **229** contribute. Hydrolysis of this intermediate at (c) gives pyridoxamine (**230**) and a keto acid. This mechanism for the reversible conversion of an amino acid to the corresponding keto acid, a process called 'transamination,' explains how the nitrogen atom of the amino group of one amino acid can be interchanged with the corresponding nitrogen atom of the same, or a different, amino acid *in vivo*. These ideas account for the identical *in vivo* vitamin activity of pyridoxal (**226**) and pyridoxamine (**230**); pyridoxine (**221**) is equally potent.

Some enzymic decarboxylations of α-amino acids to carbon dioxide and the amine corresponding to the amino acid require pyridoxal as a coenzyme. This can be accounted for by the breakage of bond (b) of **227** to give **231**, which then protonates and hydrolyses as shown.

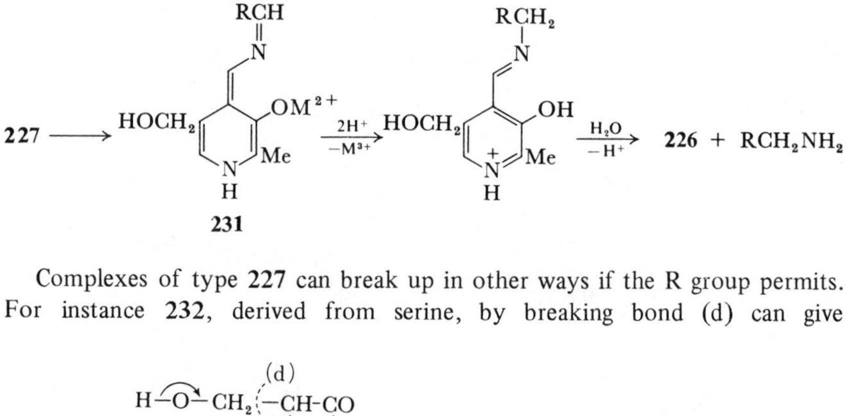

$$\mathbf{227} \longrightarrow \underset{\mathbf{231}}{\text{RCH=N, HOCH}_2\text{, OM}^{2+}\text{, Me, NH}} \xrightarrow[-M^{3+}]{2H^+} \text{RCH}_2\text{–N=, HOCH}_2\text{, OH, Me, }\overset{+}{\text{N}}\text{H} \xrightarrow[-H^+]{H_2O} \mathbf{226} + RCH_2NH_2$$

Complexes of type **227** can break up in other ways if the R group permits. For instance **232**, derived from serine, by breaking bond (d) can give

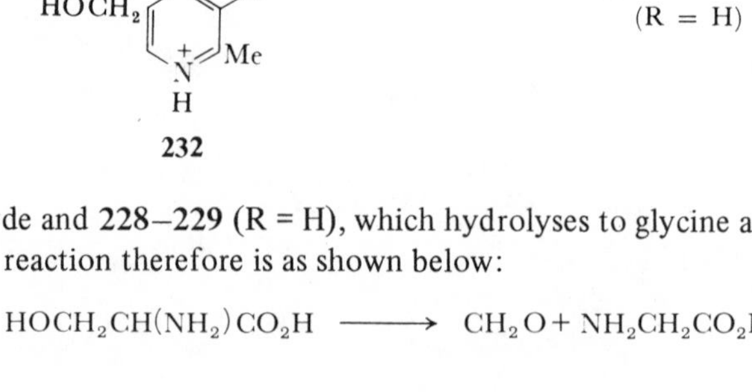

$$\underset{\mathbf{232}}{\text{H–O–CH}_2\overset{(d)}{\vdots}\text{–CH–CO–O–M–O, }\overset{+}{\text{N}}\text{, HOCH}_2\text{, Me, }\overset{+}{\text{N}}\text{H}} \longrightarrow H^+ + O{=}CH_2 + \mathbf{228\text{–}229}\ (R = H)$$

formaldehyde and **228–229** (R = H), which hydrolyses to glycine and pyridoxal. The overall reaction therefore is as shown below:

$$HOCH_2CH(NH_2)CO_2H \longrightarrow CH_2O + NH_2CH_2CO_2H$$

Even the β-hydrogen atoms of α-amino acids can be exchanged, as established[116] by experiments in deuterium oxide.

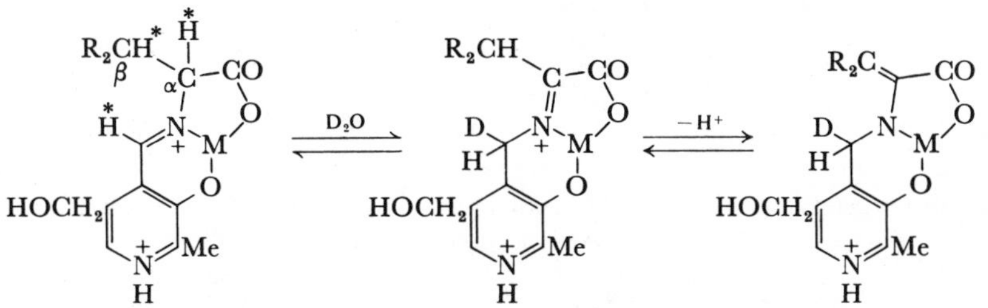

* = H exchanged

The success of the model system in reproducing the above, and other reactions of amino acids that occur *in vivo*, does not prove that the *in vivo* and model reactions take place by the same mechanism. The correspondence is, however, so great that it is unlikely that there are major differences in mechanism in the two cases.

3. *Pharmaceuticals*

There are many medicinal agents derived from pyridines, but only three simple ones are mentioned here. Iproniazide (**233**) is a valuable antitubercular drug and amine oxidase inhibitor, and Coramine (**234**) is a widely used heart stimulant. Sulphapyridine (**235**), one of the earlier sulphonamides, was once used as an antibacterial agent but has severe side reactions.

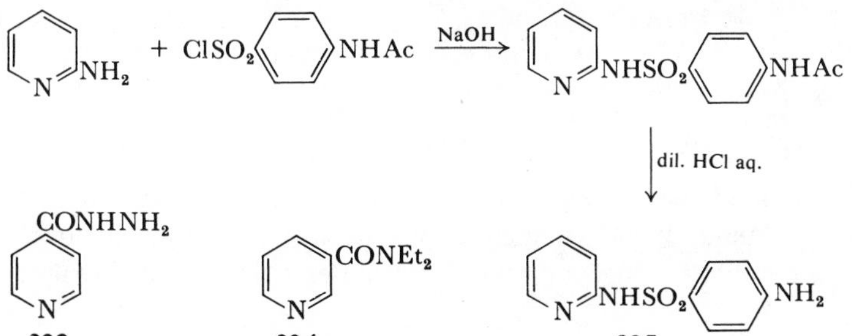

233 **234** **235**

2. PYRYLIUM SALTS AND PYRANS

A. Introduction

Few simple derivatives of the aromatic pyrylium cation (**1**) are known, although benzopyrylium salts (p. 337) are widely distributed as flower petal colouring matters. It appears that 2-pyran (**2**) may be in equilibrium with an open-chain form (**3**), and the reactive 4-pyran (**4**) has been synthesised. The pyrones **5** and **6** are well known. 2,3-Dihydro-4-pyran (**7**) is readily available, and tetrahydropyran (**8**) is used as a solvent and as a synthetic intermediate. The sulphur analogues of these compounds received little attention until very recently.

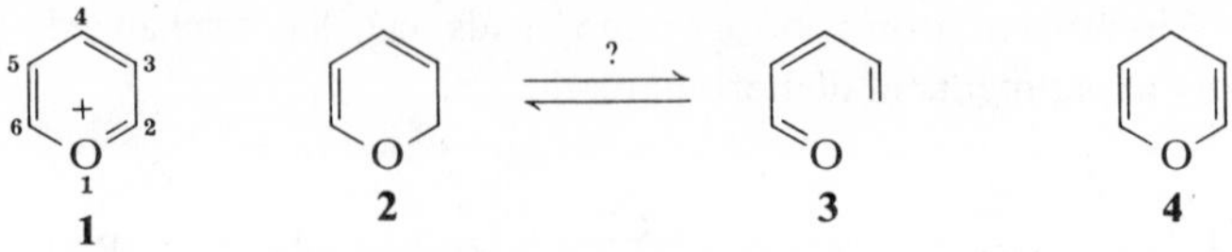

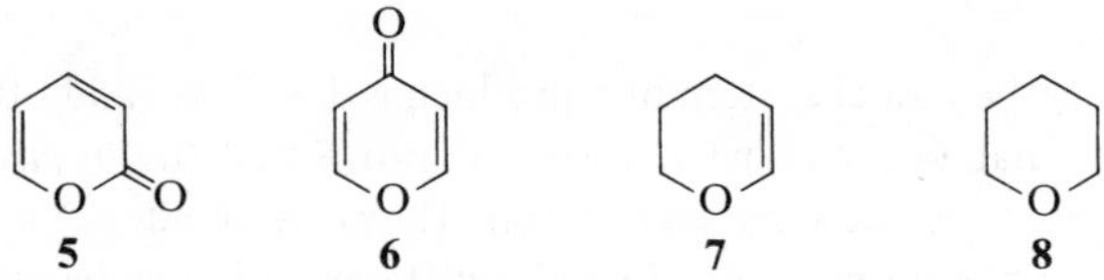

B. Pyrylium salts

The sodium salt of glutaconic aldehyde (p. 234) with perchloric acid at –20°C gives a red[117] oxonium salt, which on standing at 0°C cyclizes to the colourless pyrylium perchlorate (**9**). This perchlorate is converted to pyridine by ammonia. 2,4,6-Triphenylpyrylium ferrichloride (**10**) can be prepared quite readily, and a number of related compounds have been obtained by essentially the same method. It is extremely stable in acid solution and nitrates with difficulty only in the phenyl groups. 2,4,6-Trimethylpyrylium perchlorate (**11**) is easily available from 2,6-dimethyl-4-pyrone (**44**) by the quite general procedure shown. These salts are very easily attacked by nucleophiles, usually at position 2, and in suitable cases sequential ring opening to a ketone (e.g., **12**)

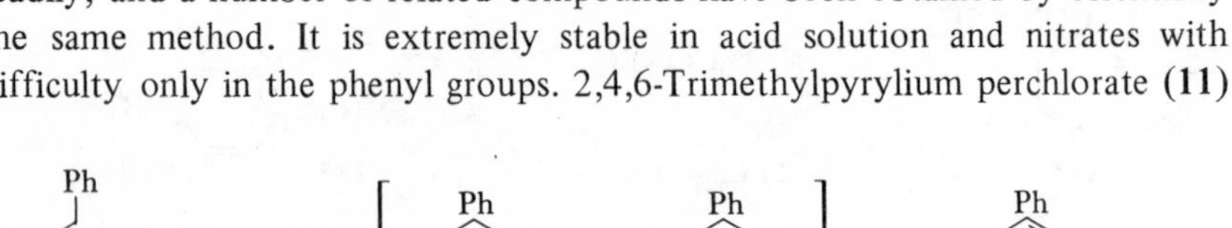

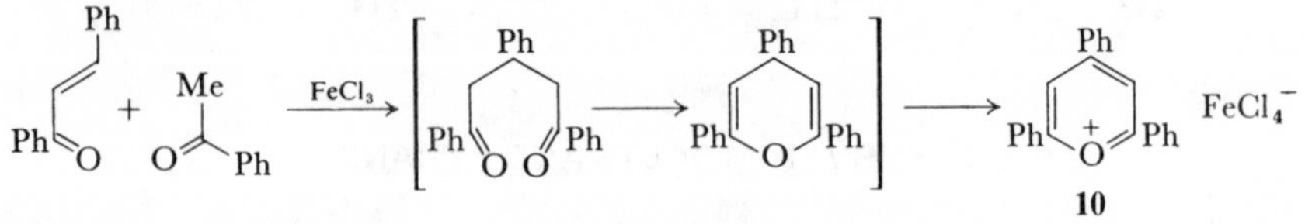

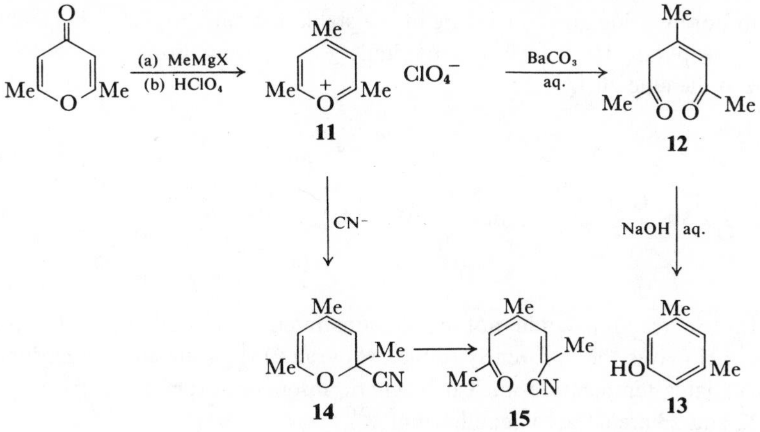

O
Me O Me
(a) MeMgX
(b) $HClO_4$
Me
Me O Me
+
ClO_4^-
11
$BaCO_3$
aq.
Me
Me O O Me
12
CN^-
NaOH aq.
Me
Me O Me CN
14
Me
Me Me O CN
15
Me
HO Me
13

and cyclisation to a benzene derivative (e.g., **13**) can be achieved. With aqueous sodium cyanide the product is **15**. Presumably addition of the anion gives **14**, which undergoes an electrocyclic ring opening. Photolysis[118] of the perchlorate (**11**) in water yields the unsaturated aldehyde **16**, but it has not been established how the reaction proceeds.[76]

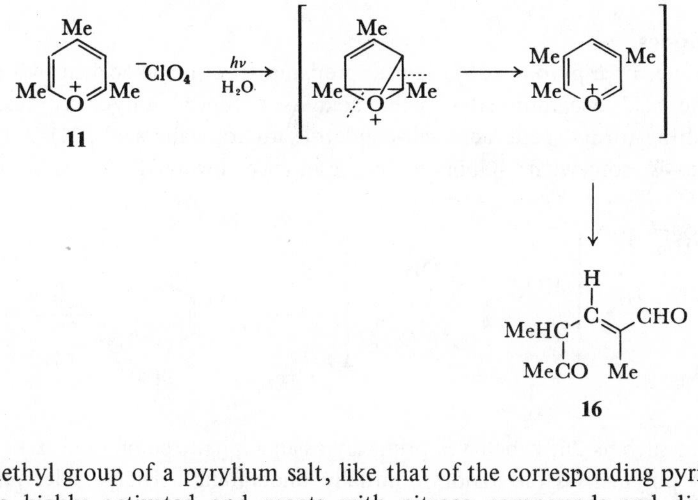

Me
Me O Me
+
$^-ClO_4$
11
$h\nu$
H_2O
Me
Me O Me
+
Me Me
Me O
+
H
MeHC C CHO
MeCO Me
16

A 2-methyl group of a pyrylium salt, like that of the corresponding pyridinium salt, is highly activated and reacts with nitroso compounds and diazotised aromatic anions.

C. 2*H*-Pyrans

2*H*-Pyran (**2**) has not yet been obtained, although some substituted derivatives have been prepared. 2,4,6-Trimethylpyrylium perchlorate (**17**) with sodium borohydride gives a mixture of the stable 1,4-dihydropyran (**18**) and the 1,2-dihydropyran **19**, which rapidly undergoes valency isomerisation to the *cis–trans*-dienone **20**:[119]

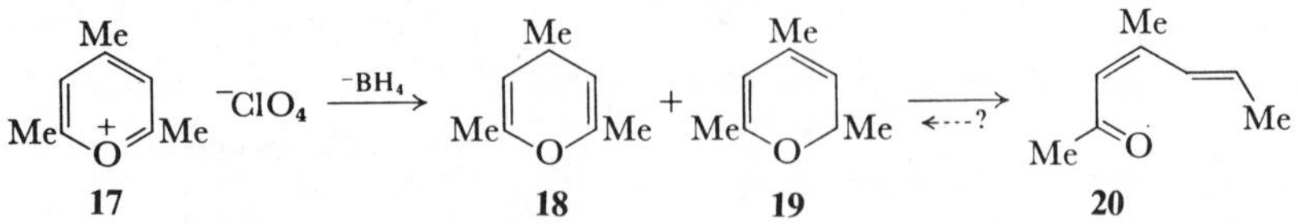

The reverse reaction could not be demonstrated, but irradiation of *trans*-β-ionone (**21**) gives the apparently stable 2*H*-pyran (**23**) presumably via *cis*-ionone (**22**). Variable temperature nuclear magnetic resonance experiments have shown that **22** and **23** are in fact in equilibrium.[120]

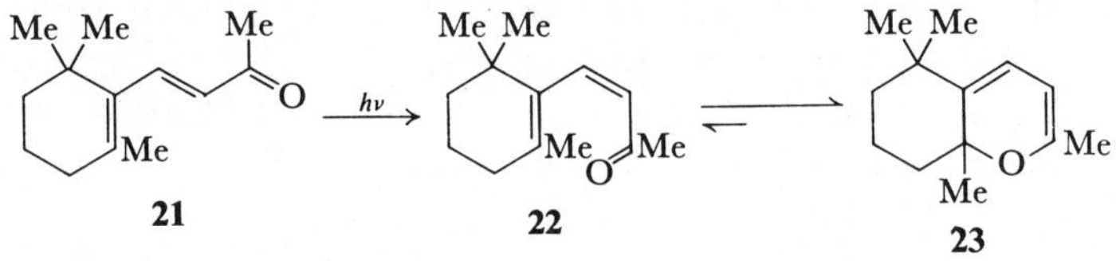

D. 2-Pyrones

2-Pyrone, or α-pyrone (**25**), can be prepared from malic acid in two stages. Sulphuric acid decarbonylates malic acid as a typical α-hydroxy acid, and the resulting formylacetic acid self-condenses to coumalic acid (**24**). Although 2-pyrone is somewhat stabilised by resonance involving **25** and charged structures such as **26**, it behaves primarily as an aliphatic compound. Like other dienes it polymerises on standing, initially undergoing a Diels–Alder reaction with itself.[121] The most generally useful type of Diels–Alder reaction with α-pyrones involves hot dimethyl acetylenedicarboxylate, when the initial adducts (e.g., **27**) lose carbon dioxide to give derivatives of dimethyl phthalate. Bromination occurs by an addition–elimination sequence to the 4,5-double bond at low temperatures, giving 5-bromo-2-pyrone,[122] and trimethyloxonium tetrafluoroborate methylates the carbonyl group, giving the 2-methoxypyrylium

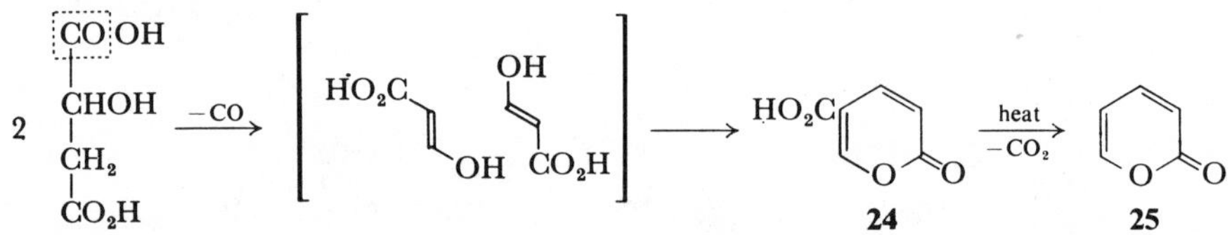

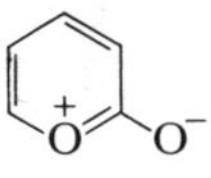

HO_2C / CHO CO_2H / $HO_2CCH_2CH{=}CHCO_2H$ / CO_2Me / CO_2Me

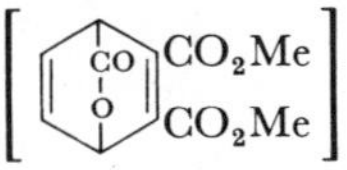

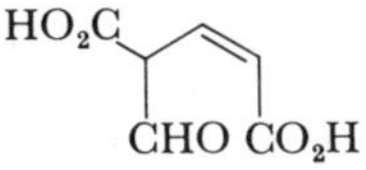

26 27 28 29

salt. 2-Pyrone is easily hydrolysed by dilute alkali; coumalic acid is similar and gives formic and glutaconic acids (**29**), possibly through the intermediary of formylglutaconic acid (**28**). Both 2-pyrone and methyl coumalate give pyridines with ammonia (p. 275), and coumalic acid can be converted into furans (p. 150).

Photolysis of 2-pyrone (**25**) sets up an equilibrium with the aldehydoketene (**30**), which if methanol is present forms the ester **31**. In the absence of such a solvent the bicyclic compound **32** is formed, and this under further irradiation gives cyclobutadiene and carbon dioxide.[123]

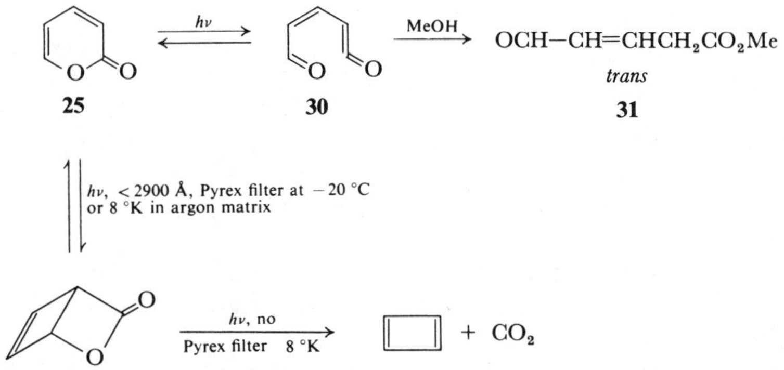

$h\nu$ — MeOH → $OCH{-}CH{=}CHCH_2CO_2Me$ *trans*

25 30 31

$h\nu$, < 2900 Å, Pyrex filter at −20 °C or 8 °K in argon matrix

$h\nu$, no Pyrex filter 8 °K → + CO_2

32

E. 4*H*-Pyrans and 4-Pyrones

4*H*-Pyran (**35**), the structure of which has been established[124] from its nuclear magnetic resonance spectrum, has been obtained from the pyrolysis of the acetate (**33**) and by the dehydrohalogenation of the dichloro compound (**34**) with dimethylaniline.[125] It is a colourless liquid, b.p. 84°C with slight decomposition which rapidly decomposes at room temperature in air. Hydrogenation yields tetrahydropyran, and 2,4-dinitrophenylhydrazine gives the corresponding bis derivative of glutardialdehyde.

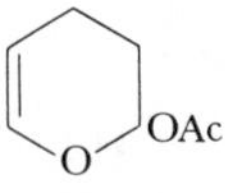

OAc Cl Cl

33 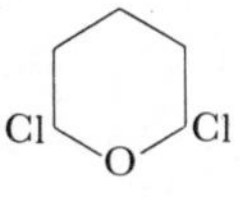 34 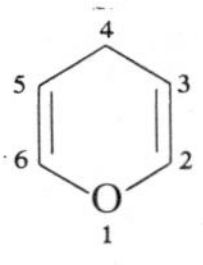 35

The cyclisation of α,α'-diketopimelic acid (**36**) gives 1,4-pyran-2,6-dicarboxylic acid (**37**), but decarboxylation to 4*H*-pyran has not been achieved. The 4*H*-pyran **38** can be prepared from ethyl acetoacetate, formaldehyde, and zinc chloride, an example of a general synthesis[126] of these compounds which closely resembles the Hantzsch synthesis of pyridines (p. 271).

HO_2CCO $COCO_2H$ **36** $\xrightarrow{H_2SO_4}$ HO_2C CO_2H **37**; EtO_2C CO_2Et Me Me **38**

4- (or γ-) Pyrone (**41**) can be obtained from ethyl acetonedioxalate (**39**). Treatment with hydrochloric acid gives chelidonic acid (**40**), which on decarboxylation over copper gives the pyrone.

EtO_2CCO CH_2 C O CH_2 $OCCO_2Et$ **39** $\longrightarrow$ HO_2C CO_2H **40** $\longrightarrow$ **41**

Dehydroacetic acid (**42**) is one of the products obtained by pyrolysis of ethyl acetoacetate but is best prepared from the ester by self-condensation in the presence of sodium bicarbonate. The acidic properties are due to the ionization of the hydroxyl hydrogen atom at position 4 under the influence of the carbonyl groups. It undergoes an interesting rearrangement with sulphuric acid to give 2,6-dimethyl-4-pyrone-3-carboxylic acid (**43**). The ring clearly opens and

CO_2Et H_2C CH_2COMe MeCO CO EtO $\xrightarrow{NaHCO_3}$ **42** (OH, COMe, Me, O) $\xrightarrow{H_2SO_4}$ **43** (CO_2H, Me, Me) $\xrightarrow{-CO_2}$ **44** (Me, Me)

$MeCO_2H$ or Ac_2O $\xrightarrow{\Delta,\ polyphosphoric\ acid}$ **44**

42 $\xrightarrow{Boil\ conc.\ HCl\ aq.}$ **44**

closes in the alternative fashion. The structure of the acid (**43**) has been proved by its conversion with ammonia into 2,6-dimethyl-4-pyridone-3-carboxylic acid, which was synthesized by an independent route. Decarboxylation of the carboxylic acid gives 2,6-dimethyl-4-pyrone (**44**), which is also obtainable in a remarkable 70% yield by heating acetic acid or anhydride in polyphosphoric acid; the mechanism of this conversion is not known.[127]

In contrast to the less readily available 4-pyrone, the reactions of its 2,6-dimethyl derivative (**44**) have been quite fully investigated. On treatment with alkali, crystalline diacetylacetone (**45**) is formed, and this gives the

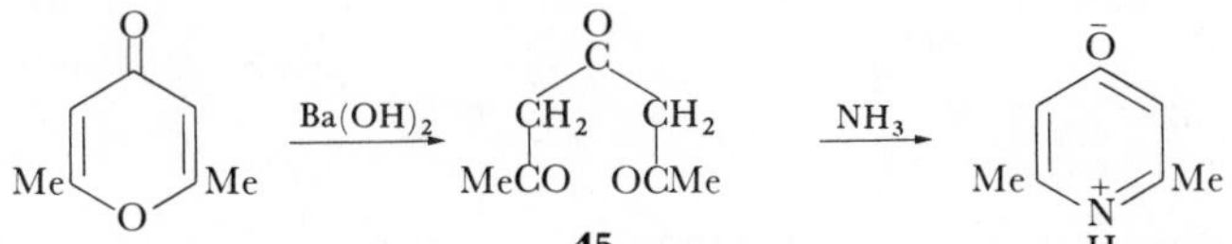

$Ba(OH)_2$ NH_3

44 **45**

corresponding pyridone with ammonia. The 4-pyrone (**44**) is neutral in solution, forms ionized salts (e.g., **46**) with a wide variety of acids, is not reduced by zinc and acetic acid, and does not form a phenylhydrazone. It reacts with methyl iodide to form an addition compound, which must be **47**, as after conversion to the perchlorate and heating with ammonium carbonate 4-methoxy-2,6-dimethyl-pyridine is formed.

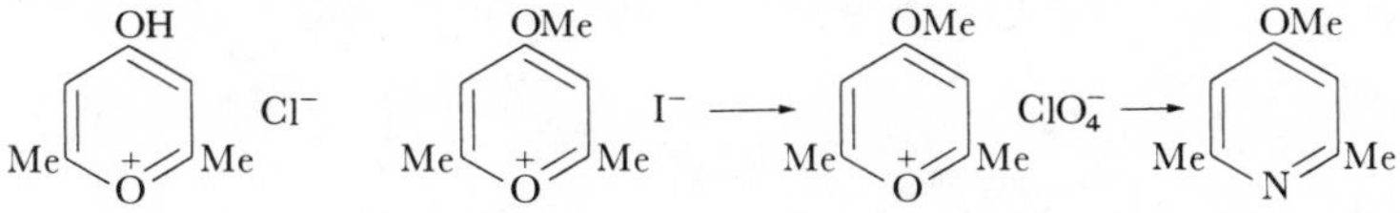

46 **47**

The pyrone cannot therefore be correctly represented as **44**, but it must be considered as a resonance hybrid to which **44** and charged structures such as **48**–**51** contribute. This is in agreement with the observed dipole moment

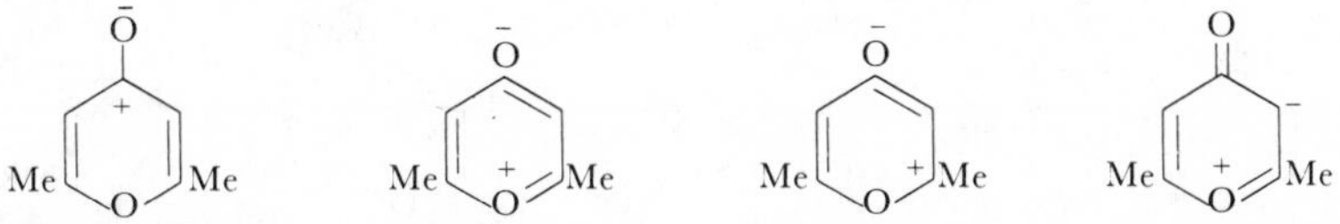

48 **49** **50** **51**

(4·05 D) of the compound, which is between those calculated for structures **44** (1·75 D) and **49** (22 D). Electrophilic substitution and acid-catalysed proton exchange occur at positions 3 and 5.

Irradiation of 2,6-dimethylpyrone in the solid state, or solution, yields the dimer **52**, but if the reaction is carried out at great dilution the furan **53** is obtained.[128]

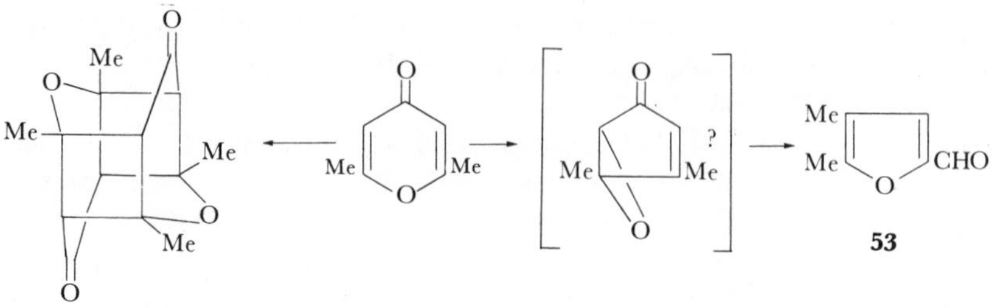

52 **53**

The 4-pyrone antibiotic LL-Z1220, isolated from a fungus, possesses structure **54**. It is the first naturally occurring compound discovered with a benzene dioxide system, and it undergoes a most interesting reversible valency tautomerism to the 1,4-dioxocin **55**.[129]

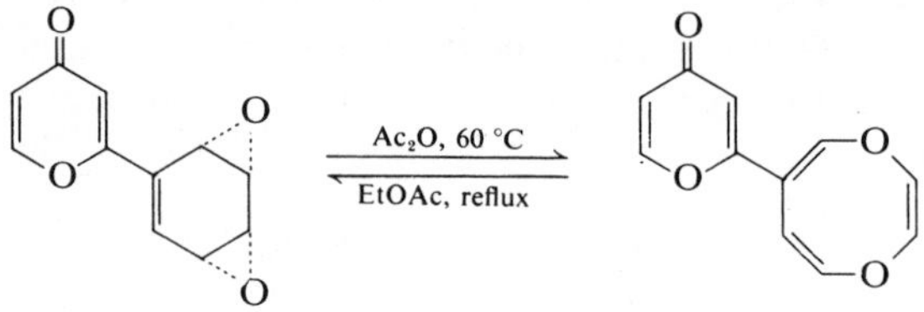

Ac_2O, 60 °C
EtOAc, reflux

54 **55**

F. Reduced pyrans

2,3-Dihydro-4-pyran (**56**) is easily obtained from tetrahydrofurfuryl alcohol (p. 145), and derivatives have been obtained from Diels–Alder reactions between butadienes and aldehydes or ketones.[130] This dihydropyran (**56**) is of particular interest owing to the reactivity of its double bond. It reacts vigorously with water forming **57** and **58**, and with other hydroxylic compounds, such as

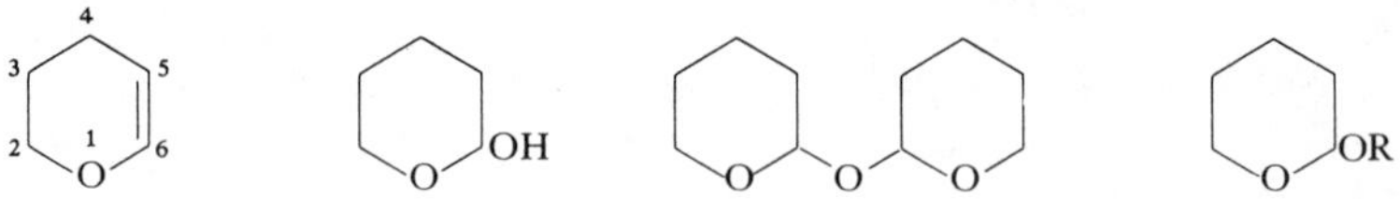

56 **57** **58** **59**

alcohols, phenols, and acids, to give ethers or esters (**59**). As these last compounds (**59**) are stable to alkali but are easily split to 2-hydroxytetrahydropyran (**57**) and the parent hydroxylic compound by dilute acids, a very useful method is available for the protection of alcoholic groups in the presence of alkalis. 2-Hydroxytetrahydropyran is the semiacetal of the tautomeric 5-hydroxypentanal (**60**). The equilibrium is in favour of the cyclic form.

OH

57

CH_2 CH_2 CH_2 CH_2OH CHO

60

Halogens and hydrogen halides add to 2,3-dihydro-4-pyran. In the products the halogen at position 3 is relatively inert, but that at position 2 is reactive as in typical α-chloro ethers. It can be displaced by methylmagnesium iodide and

HCl Cl MeMgI Me OMe

61 **62**

sodium methoxide giving **61** and **62**, respectively. A useful synthesis of acetylenic alcohols is available from 2,3-dichlorotetrahydropyrans.

Cl_2 Cl Cl distil $-HCl$ Cl BuLi $HOCH_2CH_2CH_2C{\equiv}CH$

2,3-Dihydro-6-pyran (**63**) has normal ethylenic properties and on hydration, for example, gives 3-hydroxytetrahydropyran.

Br KOH glycol H_2SO_4 OH

63

Tetrahydropyran (**64**) is readily available from the hydrogenation of 2,3-dihydro-4-pyran and has been obtained similarly from **63**. It is a typical strainless cyclic ether.

CH_2 CH_2 CH_2 CH_2Cl CH_2OAc AcCl, $ZnCl_2$

64

HNO_3 → $HO_2C(CH_2)_3CO_2H$ Glutaric acid

CO, steam, catalyst → $HO_2C(CH_2)_5CO_2H$ Pimelic acid

CO, air → δ-Valerolactone

G. Sulphur-containing analogues

2*H*-Thiopyran (**65**) has been obtained[131] as outlined and is a liquid b.p. 32–34°C at 12 torr. The 1,1-dioxide (**67**) is obtained,[132] surprisingly, from the dehydrobromination of **66** and undergoes a normal Diels–Alder addition with dimethyl acetylenedicarboxylate.[133] These dioxides readily lose a proton to base.[134]

SLi Br → NH_3 → S → pyridine $(Me_2N)_3PO$ → [S] → S **65**

66 → **67** → $MeO_2CC{\equiv}CCO_2Me$ → O_2S CO_2Me CO_2Me

4*H*-Thiopyran (**68**) has been obtained in a similar way[135] to its oxygen analogue (p. 285) and has a b.p. 30°C at 12 torr. It is readily oxidised[136] by chlorine to thiapyrylium (**69**) chloride, and an alternative way of making this class of compound is outlined.[137]

Cl S Cl → $PhNMe_2$ → **68** → Cl_2, −40° → **69** ← $HClO_4$ ← OH S ← O S

Thia(IV)benzene itself, sometimes called 1*H*-thiopyran is not known, but a number of derivatives have been obtained[138] from thiopyrylium salts (e.g., **70**) and aryllithiums, or better by hydride abstraction from sulphonium salts such as

70 → PhLi → **71** ↔ **72** → **73**

74. These compounds are deep red or purple, are extremely reactive, and are usually obtainable only in solution. The deep purple 1,2,4,6-tetraphenylthia(IV)benzene (**71**) rearranges rapidly to the colourless 4*H*-thiopyran (**73**), and in less highly substituted cases polymerisation occurs. The most stable compound of this class obtained is **75**, which does not decompose at 40°C. The thia(IV)naphthalene **76** shows two methyl signals in its n.m.r. spectrum, and these do not become equivalent at 60°C, at which temperature the compound decomposes rapidly. It is therefore clear that the sulphur atom is pyramidal in

S, C_6F_5 —MeI, $AgBF_4$→ SMe BF_4^-, C_6F_5 **74** —$MeSOCH_2^-$, Na^+→ SMe, C_6F_5 **75**

SPh, $CHMe_2$ **76**

these compounds and that the *S*-substituent is not coplanar with the ring (cf. 1-alkylthiophenium salts, p. 155). The pyramidal inversion barrier for **76** is over 99 kJ (23·7 kcal)/mole. Thia(IV)benzenes are therefore not aromatic and are best considered as sulphonium ylides (**71**–**72** and other charged resonance forms). The electronic hybridisation of the sulphur atom is not known.

1-Methyl-3,5-diphenylthia(IV)benzene 1-oxide (**77**) has been obtained as indicated below. It has m.p. 148°C and can be sublimed near this temperature at 0·05 mm pressure. The compound is therefore very much more stable than thia(IV)benzenes such as **71**. Acetic acid catalyses the exchange of the hydrogen atoms at positions 2, 4, and 6, and sodium methoxide catalyses those of the methyl group, for deuterium when deuterium oxide is present.[139]

Ph, Ph, O + H_2C, Me, $\overset{+}{S}$, ^-O, Me → Ph, Ph, $\overset{+}{S}$, ^-O, Me **77**

H. Phosphabenzenes

2,4,6-Triphenylpyrylium tetrafluoroborate (**78**) with tris-trimethylsilylphosphine (**79**) in boiling acetonitrile yields 2,4,6-triphenylphosphabenzene (**80**) in 62% yield.[140] An X-ray study[141] of 2,6-dimethyl-4-phenylphosphabenzene has shown that the heterocyclic ring is flat and that the C–C bonds average 1·396 Å in length with no significant deviations. The phosphabenzene system must therefore possess considerable aromaticity, and this agrees with the observations that **80** does not react with most dienophiles but does so[142] with hexafluoro-2-butyne to give the phosphabarrelene **81**.

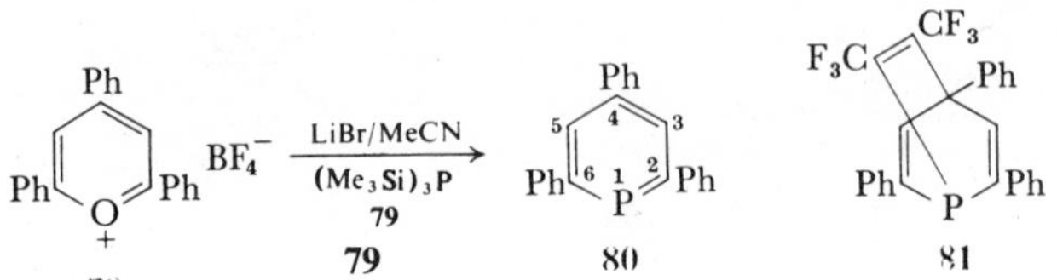

Ph Ph Ph BF_4^- **78** LiBr/MeCN $(Me_3Si)_3P$ **79** **79** Ph Ph Ph P **80** F_3C CF_3 Ph Ph Ph P **81**

1,1-Diphenylphosphabenzene (**84**) has been obtained[143] as a yellow powder by the route shown below. In solution it is easily oxidized, and with acids it

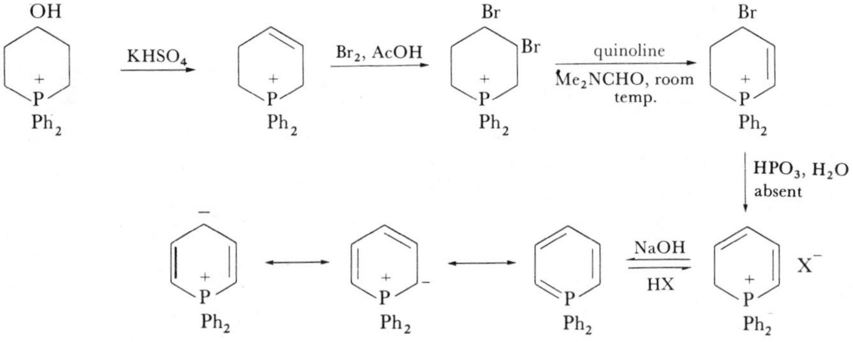

OH $\overset{+}{P}Ph_2$ $KHSO_4$ $\overset{+}{P}Ph_2$ Br_2, AcOH Br Br $\overset{+}{P}Ph_2$ quinoline Me_2NCHO, room temp. Br $\overset{+}{P}Ph_2$ HPO_3, H_2O absent

82 ⟷ **83** ⟷ **84** NaOH / HX **85** X^-

accepts a proton to form stable salts (**85**). It is perhaps best considered as a resonance-stabilized ylid (**82–83**) and not as a benzene derivative at all.

GENERAL BIBLIOGRAPHY

PYRIDINES

E. Klingsberg (ed.), *Pyridine and its Derivatives*, Parts 1–4, 1960–1964, and *Supplements* Parts 1–4, R. A. Abramovitch (ed.), 1974–1975, Interscience, Wiley, New York.

H. D. Springall and I. T. Millar, *Sidgwick's Organic Chemistry of Nitrogen*, Oxford University Press, London, 1966.

K. Schofield, *Heteroaromatic Nitrogen Compounds: Pyrrole and Pyridine*, Butterworths, London, 1967.

R. A. Abramovitch and J. G. Saha, 'Substitution in the Pyridine Series: Effect of Substituents,' *Adv. Heterocycl. Chem.*, **6**, 229 (1966).

E. H. Rodd (ed.), *Chemistry of Carbon Compounds*, Vol. 4A, Elsevier, New York, 1957.

PYRYLIUM SALTS, PYRANS AND PYRONES, AND SULPHUR ANALOGUES

F. M. Dean, *Naturally Occurring Oxygen Ring Compounds*, Butterworths, London, 1963.

A. T. Balaban, W. Schroth, and G. Fischer, 'Pyrylium Salts. Part I. Synthesis,' *Adv. Heterocycl. Chem.*, **10**, 241 (1969).

J. Fried, in *Heterocyclic Compounds*, ed. R. C. Elderfield, Vol. I, Wiley, New York, 1950.

REFERENCES

1. A. F. Bedford, A. E. Beezer, and C. T. Mortimer, *J. Chem. Soc.*, 2039 (1963).
2. B. Bak, L. Hansen, and J. Rastrup-Andersen, *J. Chem. Phys.*, **22**, 2013 (1954).
3. R. D. Brown and M. C. Heffernan, *Australian J. Chem.*, **9**, 83 (1956).
4. S. K. Chakrabartty and H. O. Kretschmer, *J.C.S. Perkin I*, 222 (1974).
5. A. R. Fersht and W. P. Jencks, *J. Amer. Chem. Soc.*, **92**, 5432 (1970).
6. D. Audman and L. W. Deady, *Chem. Comm.*, 32 (1973).
7. P. A. Claret and G. H. Williams, *J. Chem. Soc. (C)*, 146 (1969).
8. P. Hemmes, J. Costanzo, and F. Jordan, *Chem. Comm.*, 696 (1973).
9. J. A. Zoltewicz and R. E. Cross, *J.C.S. Perkin II*, 1363, 1368 (1974).
10. D. L. Glusker, H. W. Thompson, and R. S. Mulliken, *J. Chem. Phys.*, **21**, 1407 (1953).
11. E. N. Marvell and I. Shahidi, *J. Amer. Chem. Soc.*, **92**, 5646 (1970) and earlier papers.
12. H. von Dobeneck and W. Goltzsche, *Chem. Ber.*, **95**, 1484 (1962).
13. F. T. Boyle and R. Hull, *J.C.S. Perkin I*, 1541 (1974).
14. K. E. Wilzbach and D. J. Rausch, *J. Amer. Chem. Soc.*, **92**, 2178 (1970).
15. O. L. Chapman, C. L. McIntosh, and J. Pacansky, *J. Amer. Chem. Soc.*, **95**, 616 (1973).

15a. R. D. Chambers, R. Middleton, and R. P. Corbally, *Chem. Comm.*, 731 (1975).

16. D. L. Currell and L. Zechmeister, *J. Amer. Chem. Soc.*, **80**, 205 (1958).
17. R. Taylor, *J. Chem. Soc.*, 4881 (1962).
18. D. E. Pearson, W. W. Hargrove, J. K. T. Chow, and B. R. Southers, *J. Org. Chem.*, **26**, 789 (1961).
19. E. E. Garcia, C. V. Greco, and I. M. Hunsberger, *J. Amer. Chem. Soc.*, **82**, 4430 (1960).
20. H. J. den Hertog, L. van der Does, and C. A. Landheer, *Rec. Trav. Chim.*, **81**, 864 (1962).
21. R. A. Abramovitch and K. Kenaschuk, *Can. J. Chem.*, **45**, 509 (1967).
22. H. J. M. Dou and B. M. Lynch, *Bull Soc. Chim. Fr.*, 3815, 3820 (1966); K. C. Bass and P. Nababsing, *J. Chem. Soc. (C)*, 338 (1969).
23. C. S. Giam, E. E. Knaus, and F. M. Pasutto, *J. Org. Chem.*, **39**, 3565 (1974).
24. F. J. Scalzi and N. F. Golob, *J. Org. Chem.*, **36**, 2541 (1971); R. J. Francis, J. T. Wiesner, and J. M. Paul, *Chem. Comm.*, 1420 (1971).
25. A. T. Nielsen, D. W. Moore, G. M. Muha, and K. H. Berry, *J. Org. Chem.*, **29**, 2175 (1964).
26. G. M. Badger and W. H. F. Sasse, *Adv. Heterocycl. Chem.*, **2**, 179 (1963).

27. U. Eisner and J. Kuthan, *Chem. Rev.*, **72**, 1 (1972).
28. N. C. Cook and J. E. Lyon, *J. Amer. Chem. Soc.*, 88, 3396 (1966).
28a. A. J. Birch and E. A. Karakhanov, *Chem. Comm.*, 480 (1975).
29. D. C. Horwell and J. A. Dreyup, *Chem. Comm.*, 485 (1972); see also A. I. Meyers, P. M. Stout, and T. Takaya, *Chem. Comm.*, 1260 (1972).
30. W. S. Caughey and K. A. Schellenberg, *J. Org. Chem.*, **31**, 1978 (1966).
31. R. E. Lyle and P. S. Anderson, *Adv. Heterocycl. Chem.*, **6**, 45 (1966).
32. R. E. Lyle and C. B. Boyce, *J. Org. Chem.*, **39**, 3708 (1974).
33. K. Bláha and O. Červinka, *Adv. Heterocycl. Chem.*, **6**, 147 (1966).
34. A. C. Anderson and G. Berkelhammer, *J. Amer. Chem. Soc.*, **80**, 992 (1958).
35. R. M. Acheson, N. D. Wright and P. A. Tasker, *J.C.S. Perkin I*, 2918 (1972).
36. R. E. Lyle and G. J. Gauthier, *Tetrahedron Lett.*, 4615 (1965).
37. E. M. Kosower and E. J. Poziomek, *J. Amer. Chem. Soc.*, **86**, 5515 (1964); cf. M. Itoh and E. M. Kosower, *ibid.*, **90**, 1843 (1968).
38. N. L. Weinberg and E. A. Brown, *J. Org. Chem.*, **31**, 4054 (1966).
38a. M. Ferles and J. Plime, in *Adv. Heterocycl. Chem.*, **12**, 43 (1970).
39. G. M. Kellie and F. G. Riddell, *Topics in Stereochemistry*, Vol. 8, eds. E. L. Eliel and N. L. Alliger, Wiley-Interscience, New York, 1974.
40. R. A. Y. Jones, A. R. Katritzky, A. C. Richards, R. J. Wyatt, R. J. Bishop, and L. E. Sutton, *J. Chem. Soc. (B)*, 127 (1970).
41. P. J. Crowley, M. J. T. Robinson, and M. G. Ward, *Chem. Comm.*, 825 (1974).
42. V. J. Baker, I. D. Blackburne, and A. R. Katritzky, *J.C.S. Perkin II*, 1557 (1974) and earlier papers.
43. R. P. Duke, R. A. Y. Jones, and A. R. Katritzky, *J.C.S. Perkin II*, 1553 (1973).
44. H. M. Swartz, J. R. Bolton, and D. C. Borg, *Biological Applications of Electron Spin Resonance*, Wiley, New York 1972; C. L. Hamilton and H. M. McConnell in *Structural Chemistry and Molecular Biology*, ed. A. Rich and N. Davidson, Freeman and Co, San Francisco, 1968; E. G. Rozantsev, *Free Nitroxyl Radicals*, Plenum Press, New York, 1970.
45. M. E. Kuehne, *Synthesis*, 510 (1970).
46. H. A. Hageman, *Org. React.*, **7**, 198 (1963).
47. R. A. Abramovitch, F. Helmer, and M. Liveris, *J. Org. Chem.*, **34**, 1730 (1969).
48. J. F. Bunnett and B. F. Gloor, *J. Org. Chem.*, **39**, 382 (1974).
49. E. M. Kaiser, G. J. Bartling, W. R. Thomas, S. B. Nichols, and D. R. Nash, *J. Org. Chem.*, **38**, 71 (1973).
50. R. Levine, D. A. Dimmig, and W. M. Kadunce, *J. Org. Chem.*, **39**, 3834 (1974).
51. C. Swee-Ong, M. J. Cooke, and A. R. Katritzky, *J.C.S. Perkin II*, 2111 (1973).
52. G. R. Newkome, J. M. Robinson, and N. S. Bhacca, *J. Org. Chem.*, **38**, 2234 (1973).
53. G. V. Boyd and A. D. Ezekiel, *J. Chem. Soc. (C)*, 1866 (1967).
54. W. E. Parkam and P. E. Olsen, *Tetrahedron Lett.*, 4783 (1973).
55. L. W. Deady and J. A. Zoltewicz, *J. Org. Chem.*, **37**, 603 (1972).
56. Y. Okamoto and Y. Shimasawa, *Tetrahedron Lett.*, 317 (1966).
57. H. C. Brown and W. A. Murphey, *J. Amer. Chem. Soc.*, **73**, 3308 (1951).
58. H. L. Lochte and T. H. Cheavens, *J. Amer. Chem. Soc.*, **79**, 1667 (1957).
59. A. R. Katritzky and J. M. Lagowski, *Heterocyclic* N-*oxides*, Methuen, London, 1971; E. C. Taylor in *Topics in Heterocyclic Chemistry*, ed. R. N. Castle, Interscience, Wiley, New York, 1969.
60. M. van Ammers, H. J. den Hertog, and B. Haase, *Tetrahedron*, **18**, 227 (1962).
61. M. van Ammers and H. J. den Hertog, *Rec. Trav. Chim.*, **81**, 124 (1962).

62. L. K. Dyall and K. H. Pausacker, *J. Chem. Soc.*, **18**, (1961).
63. J. A. Zoltewicz and G. M. Kauffman, *J. Org. Chem.*, **34**, 1405 (1969).
64. R. A. Abramovitch, R. T. Coutts, and E. M. Smith, *J. Org. Chem.*, **37**, 3584 (1972) and earlier papers.
65. P. Schiess and P. Ringele, *Tetrahedron Lett.*, 311 (1972); T. J. Van Bergen and R. M. Kellogg, *J. Org. Chem.*, **36**, 1705 (1971).
66. F. A. Daniher and B. E. Hackley, *J. Org. Chem.*, **31**, 4267 (1966).
67. W. Feely, W. L. Lehn, and V. Boekelheide, *J. Org. Chem.*, **22**, 1135 (1957).
68. T. Cohen and I. H. Song, *J. Org. Chem.*, **31**, 3058 (1966).
69. V. Boekelheide and W. Feely, *J. Amer. Chem. Soc.*, **80**, 2217 (1958).
70. W. E. Feely, G. Evanega, and E. M. Beavers, *Org. Synth.*, **42**, 30 (1962).
71. T. Cohen and G. L. Deets, *J. Org. Chem.*, **37**, 55 (1972).
72. P. W. Ford and J. M. Swan, *Australian J. Chem.*, **18**, 867 (1965).
73. V. J. Traynelis and P. L. Pacini, *J. Amer. Chem. Soc.*, **86**, 4917 (1964).
74. S. Oae, T. Kitao, and Y. Kitaoka, *J. Amer. Chem. Soc.*, **84**, 3359 (1962).
75. R. Bodalski and A. R. Katritzky, *J. Chem. Soc. (B)*, 831 (1968).
76. S. Oae, T. Kitao, and Y. Kitaoka, *J. Amer. Chem. Soc.*, **84**, 3362 (1962).
77. A. C. Day and J. A. Barltrop, *Chem. Comm.*, 177 (1975).
78. A. Alkaitis and M. Calvin, *Chem. Comm.*, 292 (1968).
79. M. Ishikawa, C. Kaneko, I. Yokoe, and S. Yamada, *Tetrahedron*, **25**, 295 (1969).
80. O. Burchart, C. L. Pedersen, and N. Harrit, *J. Org. Chem.*, **37**, 3592 (1972).
81. S. Oae, T. Kitao and Y. Kitaoka, *J. Chem. Soc.*, 3736 (1964).
82. A. R. Katritzky and J. M. Lagowski, *Adv. Heterocycl. Chem.*, **1**, 341 (1963).
83. P. B. Desai, *J.C.S. Perkin I*, 1865 (1973).
84. E. C. Taylor and R. O. Kan, *J. Amer. Chem. Soc.*, **85**, 776 (1963).
85. P. A. Lyon and C. B. Reese, *J.C.S. Perkin I*, 2645 (1974).
86. R. C. de Selms, *J. Org. Chem.*, **33**, 478 (1968).
87. M. J. Cook, A. R. Katritzky, P. Linda, and R. D. Tack, *J.C.S. Perkin II*, 1080 (1973).
88. H. Tomisawa and H. Hongo, *Tetrahedron Lett.*, 2465 (1969), see also R. M. Acheson and P. A. Tasker, *J. Chem. Soc. (C)*, 1542 (1967); L. A. Paquette, *J. Org. Chem.*, **30**, 2107 (1965).
89. N. Kornblum and G. P. Coffey, *J. Org. Chem.*, **31**, 3449 (1966).
90. E. J. Corey and J. Streich, *J. Amer. Chem. Soc.*, **86**, 950 (1964).
91. W. A. Ayer, R. Hayatsu, P. de Mayo, S. T. Reid, and J. B. Stothers, *Tetrahedron Lett.*, 648 (1961).
92. P. Bellingham, C. D. Johnson, and A. R. Katritzky, *Chem. Comm.*, 1047 (1967), and earlier papers.
93. K. Lloyd and G. T. Young, *J. Chem. Soc. (C)*, 2890 (1971).
94. T. Mukaiyama, M. Araki, and H. Takei, *J. Amer. Chem. Soc.*, **95**, 4763 (1973).
95. G. T. Young, in *Chemistry of Polypeptides*, ed. P. G. Katsoyannis, Plenum, New York, 1973; R. Camble, R. Garner, and G. T. Young, *J. Chem. Soc. (C)*, 1911 (1969).
96. A. Gosden, D. Stevenson, and G. T. Young, *Chem. Comm.*, 1123 (1972).
97. R. Macrae and G. T. Young, *Chem. Comm.*, 446 (1974).
98. G. T. Newkome, J. M. Robinson, and J. D. Sauer, *Chem. Comm.*, 410 (1974).
99. Z. Arnold, *Experientia*, **15**, 415 (1959).
100. R. J. Mosher and E. U. Brown, *J. Org. Chem.*, **37**, 3938 (1972), and papers cited.
101. R. G. Button and P. J. Taylor, *J.C.S. Perkin II*, 557 (1973).
102. R. M. Acheson and G. A. Taylor, *J. Chem. Soc.*, 4141 (1959).

103. J. D. Butler, J. H. Dodsworth, and A. T. Groom, *Chem. Comm.*, 54 (1966).
104. B. Leov and K. M. Snader, *J. Org. Chem.*, **30**, 1914 (1965).
105. P. Bosshard and C. H. Eugster, *Adv. Heterocycl. Chem.*, **7**, 377 (1966).
106. G. Stork, M. Ohashi, H. Kamashi, and H. Kakisawa, *J. Org. Chem.*, **36**, 2784 (1971).
107. R. Albrecht and G. Kresze, *Chem. Ber.*, **98**, 1431 (1965).
108. G. J. Janz and A. R. Monahan, *J. Org. Chem.*, **29**, 569 (1964).
109. N. D. Doktorova, L. V. Ionova, M. Ya. Karpeisky, N. Sh. Padyukova, K. F. Turchin and V. L. Florentiev, *Tetrahedron*, **25**, 3527 (1969); R. A. Firestone, E. E. Harris, and W. Reuter, *Tetrahedron*, **23**, 943 (1967).
110. T. B. Windholz, L. H. Peterson, and G. J. Kent, *J. Org. Chem.*, **28**, 1443 (1963).
111. D. Gross, H. R. Schütte, G. Hübner, and K. Mothes, *Tetrahedron Lett.*, 541 (1963).
112. W. L. Meyer, H. R. Mahler, and R. H. Baker, *Biochem. Biophys. Acta*, **64**, 353 (1962).
113. F. Schlenk, in *The Enzymes*, eds. J. B. Sumner and K. Myrbäck, Vol. II, Part I, Academic Press, New York, 1951, p. 250.
114. N. A. Hughes, G. W. Kenner, and A. Todd, *J. Chem. Soc.*, 3733 (1957).
115. E. E. Snell, A. E. Braunstein, E. S. Severin and Yu. M. Torchinsky, *Pyridoxal Catalysis: Enzymes and Model Systems*, Wiley, New York, 1968; E. E. Snell and S. J. Di Mari in *The Enzymes*, ed. P. O. Boyer, Vol. II, 3rd ed., Academic Press, New York, 1970, p. 335.
116. E. H. Abbott and A. E. Martell, *Chem. Comm.*, 1501 (1968).
117. D. E. Games, *Aromat. Heteroaromatic Chem.*, **1**, 369 (1973).
118. J. A. Barltrop, K. Dawes, A. C. Day, and A. J. H. Summers, *Chem. Comm.*, 1240 (1972); cf. J. A. Barltrop, R. Carder, A. C. Day, J. R. Harding, and C. Samuel, *ibid.*, 729 (1975).
119. E. N. Marvell and T. Gosink, *J. Org. Chem.*, **37**, 3036 (1972).
120. E. N. Marvell, T. Chadwick, G. Caple, T. Gosink, and G. Zimmer, *J. Org. Chem.*, **37**, 2992 (1972).
121. D. C. White and D. Seyferth, *J. Org. Chem.*, **37**, 3545 (1972).
122. W. H. Pirkle and M. Dines, *J. Org. Chem.*, **34**, 2239 (1969).
123. O. L. Chapman, C. L. McIntosh, and J. Pacansky, *J. Amer. Chem. Soc.*, **95**, 614 (1973), and papers cited.
124. S. Masamune and N. T. Castellucci, *J. Amer. Chem. Soc.*, **84**, 2452 (1962).
125. J. Strating, J. H. Keijer, E. Molenaar, and L. Brandsma, *Angew. Chem.*, **74**, 465 (1962).
126. J. Wolinsky and H. S. Hauer, *J. Org. Chem.*, **34**, 3169 (1969).
127. E. B. Mullock and H. Suschitzky, *J. Chem. Soc. (C)*, 828 (1967).
128. P. Yates and M. J. Jorgensen, *J. Amer. Chem. Soc.*, **85**, 2956 (1963) and earlier papers.
129. D. B. Borders and J. E. Lancaster, *J. Org. Chem.*, **39**, 435 (1974).
130. D. G. Kubler, *J. Org. Chem.*, **27**, 1435 (1962); O. Achmatowicz and A. Zamojski, *Rocz. Chem.*, **35**, 1251 (1961).
131. L. Brandsma and P. J. W. Schuijl, *Rec. Trav. chim.*, **88**, 30 (1969).
132. E. Molenaar and J. Strating, *Rec. Trav. chim.*, **86**, 1047 (1967).
133. N. H. Fischer and H-N. Lin, *J. Org. Chem.*, **38**, 3073 (1973).
134. G. Gaviraghi and G. Pagani, *J.C.S. Perkin II*, **50** (1973).
135. J. Strating, J. H. Keijer, E. Molenaar, and L. Brandsma, *Angew. Chem.*, **74**, 405 (1962).
136. E. Molenaar and J. Strating, *Tetrahedron Lett.*, 2941 (1965).
137. I. Degani, R. Fochi, and C. Vincenzi, *Tetrahedron Lett.*, 1167 (1963).

138. B. E. Marganoff, J. Stackhouse, G. H. Senkler, and K. Mislow, *J. Amer. Chem. Soc.*, **97**, 2718 (1975).
139. A. G. Hortmann, *J. Amer. Chem. Soc.*, **87**, 4972 (1965).
140. G. Märkl, F. Lieb, and A. Merz, *Angew. Chem. Int. Ed.*, **6**, 458 944 (1967).
141. J. C. J. Bart and J. J. Daly, *Angew. Chem. Int. Ed.*, **7**, 811 (1968).
142. F. Märkl, and F. Lieb, *Angew. Chem. Int. Ed.*, **7**, 773 (1968).
143. F. Märkl *Angew. Chem.*, **75**, 669 (1963); **77**, 1109 (1965).

CHAPTER VI

HETEROCYCLIC ANALOGUES OF NAPHTHALENE WITH ONE HETEROATOM

The most investigated compounds of this group contain a nitrogen atom and theoretically can be derived from naphthalene by the substitution of a carbon, or carbon and hydrogen atoms, by a nitrogen atom. The same results can be obtained by fusing a benzene ring onto that of pyridine, and three structures result. Derivatives of all three ring systems, of quinoline (**1**), of isoquinoline (**2**),

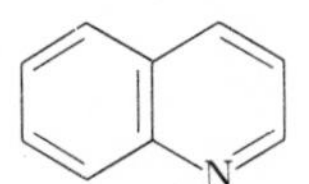
1, Quinoline

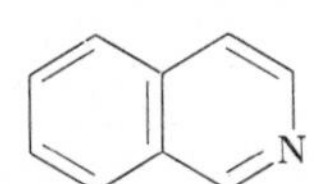
2, Isoquinoline

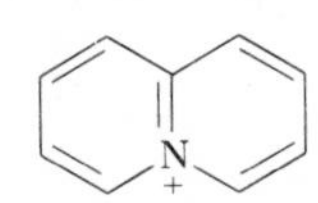
3, Quinolizinium

and of the quinolizinium (or pyridocolinium) cation (**3**) occur naturally in the form of alkaloids, although those derived from the last-named system are comparatively rare. Quinoline and isoquinoline have received much attention from research workers for many years, but interest in simple quinolizinium salts is comparatively recent. These three types of compounds are essentially aromatic in character and to varying degrees combine the properties of pyridine and naphthalene. In this chapter emphasis is placed on properties that differ from those which might be anticipated by analogy to behaviour in the pyridine series.

The oxygen analogues of naphthalene which are considered here are those derived from the benzopyrylium cation (**4**) and the benzopyrans (**5** and **6**), in which the oxygen atom is adjacent to the carbocyclic ring. Derivatives of these systems occur widely in plants and the anthocyanins, substituted benzopyrylium salts, are responsible for most red and blue flower colours. Very few sulphur analogues of these oxygen-containing compounds are known, and at present they are not of general interest.

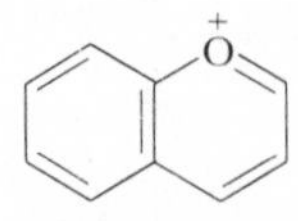
4, Benzopyrylium

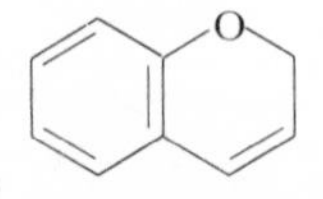
5, 2*H*-Benzopyran

6, 4*H*-Benzopyran

1. QUINOLINE

A. Introduction

In 1834 Runge isolated crude quinoline from coal tar, and in 1842 the same compound was obtained by heating the alkaloid cinchonine with alkali. Another 40 years elapsed before the identity of the materials from these two sources was established.

The positions of the quinoline ring are usually numbered (**1**), and the use of letters (**2**) has been virtually discontinued. A great many quinolines are known,

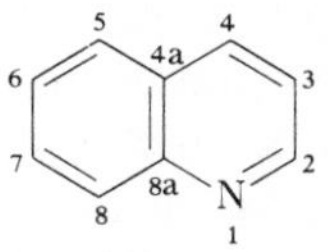

1 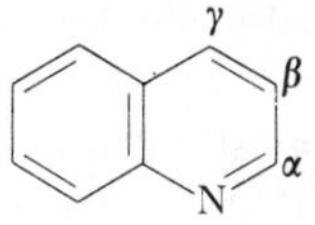 **2**

and most have been obtained by direct synthesis involving the building of the nitrogen-containing ring or by the transformation of existing substituents. The direct introduction of substituents into the nucleus, and particularly into the carbocyclic ring, is easier than in the pyridine series.

B. Physical properties and structure

Quinoline is a colourless liquid, which turns yellow on standing (cf. pyridine, p. 235) and which possesses a characteristic odour resembling that of pyridine. It has f.p. – 15·6°C, b.p. 237°C, and d_4^{15} 1·097. It is miscible with most organic solvents, dissolves in water to about 0·7% at room temperature, and is a useful high-boiling solvent for many types of organic compounds.

The structure now accepted for quinoline was first proposed by Körner and was supported at an early stage by the oxidation to quinolinic acid (**4**) and by several syntheses, of which Friedlander's is shown.

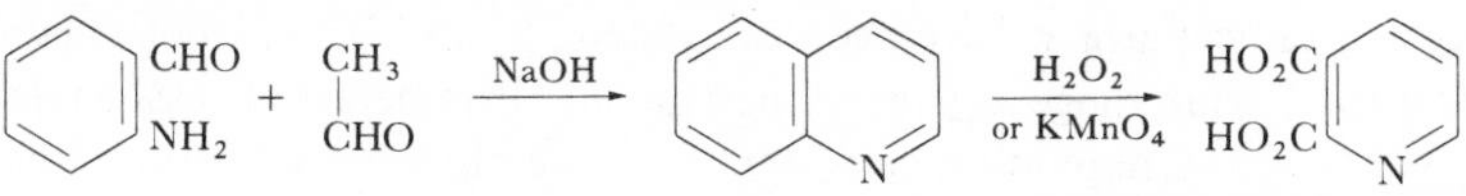

CHO, NH_2 + CH_3–CHO —NaOH→ **3** —H_2O_2 or $KMnO_4$→ HO_2C, HO_2C **4**

The resonance energy of quinoline [198 kJ (47·3 kcal)/mole], calculated by Klages' method[1] from heat of combustion data, is substantially less than that of naphthalene [255 kJ (61·0 kcal)/mole]; this method of calculation also gives an unexpectedly low value for the resonance energy of pyridine. The interatomic distances in quinoline have not yet been determined but are probably between those of naphthalene[2] and those of pyridine.

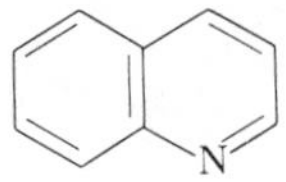

3 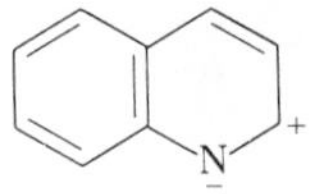 **3a** 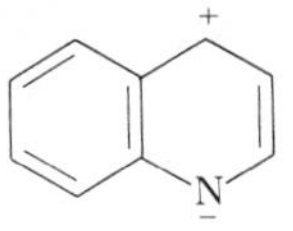 **3b**

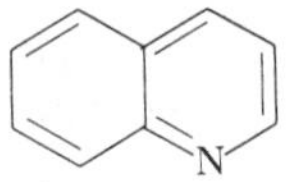

3c 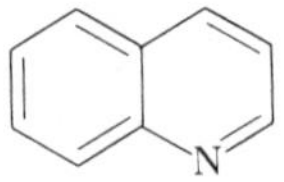 **3d**

Quinoline is best considered as a resonance hybrid that corresponds to that of naphthalene in regard to the uncharged contributing structures (**3**, **3c**, and **3d**). As in the case of naphthalene, where bond-length determinations[2] have confirmed the supposition that some 'bond fixation' occurs, chemical evidence (p. 309) suggests that structure **3** with the symmetrical double-bond arrangement in the rings is probably the most important. The dipole moment (2·10 D)

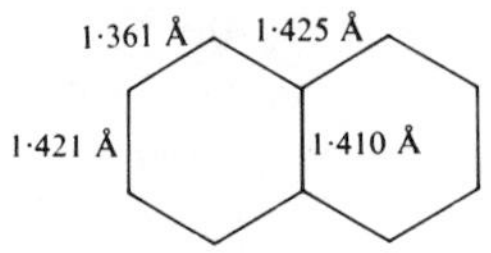

1·361 Å 1·425 Å 1·421 Å 1·410 Å

Interatomic distances in naphthalene

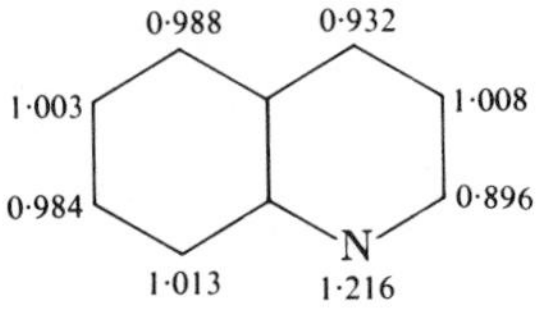

0·988 0·932 1·003 1·008 0·984 0·896 1·013 1·216

π-Electron densities in quinoline

and the facile nucleophilic substitutions that take place at positions 2 and 4 (p. 306) of quinoline indicate that charged structures, such as **3a**, **3b** and similar structures with different bond arrangements in the carbocyclic ring, must also contribute significantly to the resonance hybrid. All this is in agreement with the general, but not detailed, similarity of the ultraviolet absorption spectra of quinoline and naphthalene.

The π-electron densities and localisation energies for electrophilic, nucleophilic, and radical attack have been calculated[3] by the Hückel molecular orbital method for a wide range of Hückel parameters. Parameters of reasonable values can be chosen to give pictures consistent with the observed substitution patterns, and the π-electron densities obtained from one set of parameters is shown above.

C. Chemical properties

The chemical properties of quinoline are, in general, those that might be anticipated from an amalgamation of those of pyridine and naphthalene. The effect of the nitrogen atom is largely confined to its own ring, which possesses most of the chemical properties of the pyridine system.

1. *Addition and ring-opening reactions*

With a few exceptions, most of these concern the heterocyclic ring. Quinoline is a slightly weaker base (pK_a = 4·94) than pyridine (pK_a 5·23), and forms many salts which are sparingly soluble in water. It readily forms a 1-oxide (p. 307), and quaternary salts with reagents such as methyl iodide, methyl sulphate, and benzoyl chloride. 1-Alkylquinolinium salts, such as **5**, with alkali give the corresponding hydroxides (**6**). These are certainly in equilibrium with the corresponding pseudobases (e.g., **7**), but the position of equilibrium in water is not known with any certainty. Analogy with naphthalene and benzene, where conversion into the 'corresponding' dihydro derivatives (styrene and buta-1,3-diene) involves the loss of 88 kJ (21 kcal)/mole and 136 kJ (33 kcal)/mole of resonance energy, respectively, suggests that the position of equilibrium is further towards the pseudohydroxide than in the pyridine series. The pseudohydroxide could also tautomerize to an open-chain form (**8**). The oxidation of **6** with alkaline potassium ferricyanide gives 1-methyl-2-quinolone

I⁻ NaOH / HI ⁻OH

5 **6**

7 **8** **9**

(**9**, compare 1-methyl-2-pyridone, p. 262). Crystallization of the hydroxide (**6**~**7**) from ethanol gives the corresponding ethoxide, which can have either an ionic (**10**) or covalent (**11**) structure.

10 **11**

Treatment of 1-methylquinolinium iodide (**5**) with potassium cyanide gives a compound (**12**) corresponding to the pseudobase (**7**), but where the cyanide group has unexpectedly entered position 4. This is proved by successive oxidation and thermal decomposition, when 4-cyanoquinoline is formed. The whole sequence is a useful synthetic procedure, and it is noteworthy that the decomposition of 1-alkylquinolinium halides (e.g., **5** and **13**) is quite general and may be carried out *in vacuo* or in a high-boiling solvent such as ethyl benzoate.

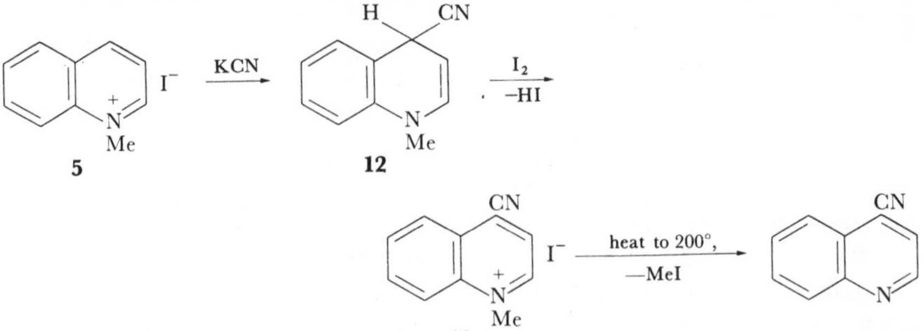

1-Benzoylquinolinium chloride (**14**) behaves differently with potassium cyanide and, perhaps more understandably, gives **15**, where the double bond of the 'pyridine' ring is still conjugated with the benzene system. Compounds such as **15**, which are easily made,[4] are sometimes known as Reissert compounds after their discoverer. They have special interest, as on treatment with concentrated hydrochloric acid they break up as shown to give quinoline-2-carboxylic acids. The mechanism proposed[4] is consistent with the facts that 2-cyanoquinoline is not an intermediate and that the reaction proceeds similarly with isoquinoline and phenanthridine but not with acridine (p. 332), where no cyclic intermediate is possible, with the isolation of some quinoline-2-carboxyamide as a minor product, and with labelling studies which show that the asterisked hydrogen atom of the benzaldehyde **16** formed comes from the solvent.

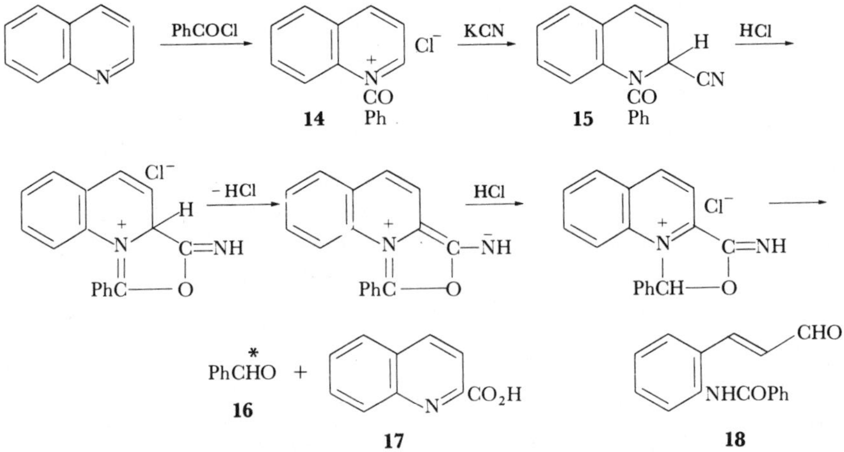

1-Benzoylquinolinium chloride (**14**) with sodium hydroxide may give initially the hydroxide corresponding to structure **7**, but the product actually isolated is the *trans* aldehyde (**18**). In contrast, the rings of 1-cyanoquinolinium salts do not open with alkali.

The reduction of quinoline, 2-quinolone, or of 1-methylquinolinium iodide by lithium aluminium hydride yields the corresponding 1,2-dihydroquinoline.[4a] 1,2-Dihydroquinoline (**23**) is reported[5] to be obtained in good yields from the sodium–liquid ammonia reduction of quinoline, but it disproportionates with acid and is easily oxidised. However, more recent work[6] using lithium and liquid ammonia has shown that either the carbocylic or heterocyclic ring can be reduced, depending on the proton source used, and conditions have been found[7] where the intermediate anion (**19**) can be trapped by alkylating agents to give the 1-alkyl-1,4-dihydroquinolines (e.g., **20**) exclusively.

2 Li
2 stages
N
2Li$^+$
NH_3
$-NH_2^-$
N
19
Me_2SO_4
N
Me
20

Quinolines are attacked by organolithiums[8] at position 2, to give salts (**21**). These yield stable 1-ethoxycarbonyl-1,2-dihydroquinolines (e.g., **22**) with ethyl chloroformate or can be protonated to 1,2-dihydroquinolines, which are easily oxidised by air or nitrobenzene to the corresponding quinolines.

N
RLi
N
R
Li$^+$
21
$ClCO_2Et$
N
R
CO_2Et
22

Quinoline is easily reduced to tetrahydroquinoline (**24**) by tin and hydrochloric acid, or by hydrogenation. This preferential reduction of the 'pyridine' ring can be altered by substitution. 1,2,3,4-Tetrahydroquinoline is a stronger base than quinoline, it has most of the properties of a secondary aromatic amine, and it is dehydrogenated to quinoline by many oxidants, including even iodine.

H$_2$, Ni, or Sn/HCl

oxidation

LiAlH$_4$ / O$_2$ — H$_2$, Ni

23 — 24

H$^+$

It gives a quaternary salt with excess methyl iodide. The corresponding hydroxide (**26**), contrary to earlier ideas, does[9] undergo the Hofmann degradation to 2-allyldimethylaniline (**25**) which must be distilled out *in vacuo*. As at ordinary pressures the degradation leads to methanol and 1-methyl-1,2,3,4-tetrahydroquinoline (cf. **24**) it appears that the formation of **25** from the quaternary hydroxide (**26**) is reversible. Reduction of the hydroxide with sodium amalgam (Emde's method, see also p. 323) opens the heterocyclic ring in the alternative mode, giving **27**, and is a useful complementary procedure. 1-Benzoyl-1,2,3,4-tetrahydroquinoline undergoes the normal von Braun ring

160° at 15 mm. — $\bar{O}H$ — Na/Hg

NMe$_2$ — $\overset{+}{N}$ Me$_2$ — N Me$_2$

25 — 26 — 27

opening (p. 245). Powerful hydrogenation of quinoline in acetic acid over platinum gives a mixture of *cis*- and *trans*-decahydroquinoline.

Oxidation of quinoline can open the carbocyclic ring (p. 321), and with ozone, quinoline gives mainly glyoxal and pyridine-2,3-carbaldehyde in conformity with the idea of some bond fixation (p. 309). In other cases, such as 2-phenylquinoline (**28**) and pseudohydroxides (**30**), the nitrogen containing ring is split to give the corresponding acylanthranilic acid (**29**).

KMnO$_4$ — KMnO$_4$

Ph — CO$_2$H NRCOR′ — R′ OH N R

28 — 29 — 30

2. *Substitution reactions*

In concentrated sulphuric acid, quinoline is converted almost entirely into the cation. Bromination[10] and both chlorination and iodination[11] with silver sulphate in this solvent yield a mixture of the 5- and 8-halogenated quinolines.

Sulphonation at 220°C gives largely the 8-sulphonic acid, which rearranges to the 6-sulphonic acid at 300°C, and fuming sulphuric acid gives a mixture of the 5-, 7-, and 8-sulphonic acids. Nitration in sulphuric acid at 0°C is rapid and also gives[12] a mixture of 5- (52·3%) and 8-nitroquinolines (47·7%); no other isomers were detected by methods that would easily have estimated quantities of the order of 1%. The reaction kinetics[13] show that the quinolinium cation is the species actually nitrated, and that the proton is present in the transition state. These results are in agreement with molecular-orbital calculations,[3] which show that the localization energies for electrophilic attack on the quinolinium cation are lowest at positions 5 and 8.

Nitration in acetic anhydride[14] however, gives 3-nitroquinoline (up to 6·6%) along with a little (0·9%) of the mixed 6- and 8-nitro isomers; the 3-nitro compound is not obtained if the reaction conditions preclude the formation of nitrous fumes. The same products are also obtained on nitration with dinitrogen

31

$X = NO_2$ or AcO

tetroxide. The positions chosen by the entering groups do not suggest a free-radical attack. A possible course of reaction is through an initial addition, giving compound **31** followed by electrophilic attack on position 3 and subsequent eliminations. This appears to be the reaction scheme that gives the highest electron density at the 3-position of the intermediate (e.g., **31**) which must be substituted.[15]

Bromine combines with quinoline to give initially a π-donor complex (**32**).[16] This in hot carbon tetrachloride in the dark gives[17] 3-bromoquinoline (**33**, 43%), but a much higher yield (82%) can be obtained along with a little 3,6-dibromoquinoline if pyridine is present.[17] 3-Bromoquinoline (25% yield) can also be formed by a vapour-phase bromination over pumice at 300°C, by bromination in the presence of sulphur, sulphur chloride, and by heating quinoline in hydrochloric acid with bromine. The mechanism is probably similar to that of nitration at position 3. Heating 3-bromoquinolinium bromide to 300°C gives[18] mainly quinoline (32%), bromine, and 4-bromoquinoline. This

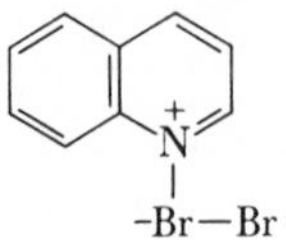

$\overset{+}{N}$–Br^-—Br

32

Br

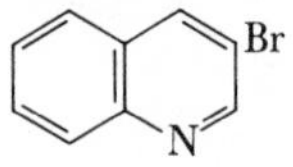

33

remarkable debromination process may proceed by the reverse of the nitration mechanism (above) leading to 3-nitroquinoline. Vapour-phase bromination of quinoline at 500°C gives 2-bromoquinoline (60%), probably via a radical attack.

Phenyl radicals, obtained through the decomposition of benzoyl peroxide, attack quinoline, giving a mixture of all the phenylquinolines. 8-Phenyl- (30% yield) and 4-phenylquinoline (20%) are the major products. Benzyl radicals substitute mainly at positions 2 and 4 for both quinoline and its cation.[19] The acetyl radical ($CH_3\dot{C}O$), from acetaldehyde and *t*-butylhydroperoxide, attacks the quinolinium cation in sulphuric acid to give about 60% of 2,4-diacetylquinoline.[20]

Nucleophilic reagents, such as sodamide (p. 239), potassium hydroxide (p. 262), Grignard reagents, and alkyl- or aryl-lithiums (p. 238), attack quinoline preferentially at position 2, and on occasion at position 4, as in the pyridine series.

D. Derivatives of quinoline

Substituted quinolines can be divided into two classes, depending on whether the substituent under consideration is attached to the carbocyclic (**34**, ring A) or heterocyclic (**34**, ring B) ring. In general, substituents of these rings have properties corresponding to those of the analogous naphthalene or pyridine, respectively.

2-Methyl- (quinaldine, **35**) and 4-methylquinoline (lepidine, **37**) occur to a small extent in coal tar and have methyl groups that are very reactive in an analogous way to those of 2- and 4-methylpyridine (p. 246), and they show the same preferences to reagents causing metallation (p. 246). The methyl group of

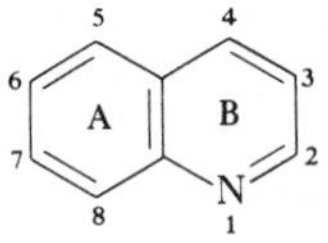

5 4 6 3 A B 7 2 N 8 1

34

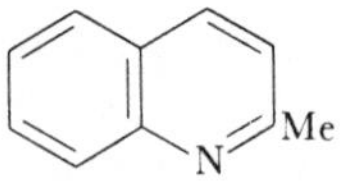

Me

35

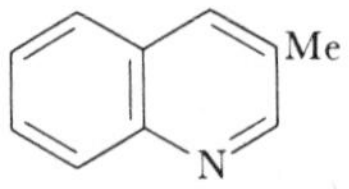

Me

36

Me

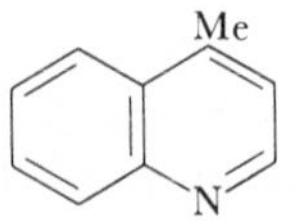

37

3-methylquinoline (**36**) is similar to that of 3-methylpyridine (p. 248) and is comparatively unreactive.

Quinoline 1-oxide is interesting, as the products obtained on nitration depend on the temperature.[21] Between 0 and 20°C in sulphuric acid solution a mixture

H_2O_2, AcOH; HNO_3, H_2SO_4 65–70°; NO_2; 5; 8; $\overset{+}{N}$; O^-

of 8-(major) and 5-nitroquinoline 1-oxides is obtained, while at 65–70°C 4-nitroquinoline 1-oxide is the main product and corresponds to results in the pyridine series (p. 250). Quinoline 1-oxide with benzoyl chloride and potassium cyanide gives benzoic acid and 2-cyanoquinoline quantitatively. The reaction probably proceeds through the formation of 1-benzoyloxyquinolinium chloride. Very different results are obtained if the 1-oxide is replaced by quinoline itself (p. 302).

PhCOCl; Cl^-; KCN; PhCOO; H; CN; ?; + $PhCO_2H$

The products of photolysis of quinoline 1-oxides[22] depend on the solvent employed and are probably formed via an oxaziridine (**38**) intermediate, as in the case of pyridine 1-oxide (p. 254). Ring opening to a charge species (**38a**) is thought[22] more probable than to the corresponding diradical, and tautomerism then gives 2-quinolone (**38b**), which is the main product using aqueous solvents. With acetone, 2-cyano- and 2-phenylquinoline 1-oxides give stable oxazepines (**40**). In other cases the oxazepines are not isolable, and if some water or ethanol is present, hydrolysis occurs to give equilibrium mixtures (**39** and **39a**) of compounds that can dehydrate to corresponding 1-acylindoles.

2- and 4-Chloro- or bromoquinolines can be obtained from the quinolones (e.g., **44**) by treatment with phosphorus halides, and have very reactive halogen atoms for the same reasons as do the corresponding pyridines (p. 254); 3-bromoquinolines are unreactive. The rate of formation of bromide ion in the reaction of 5-, 6-, 7-, and 8-bromoquinoline with piperidine has been studied

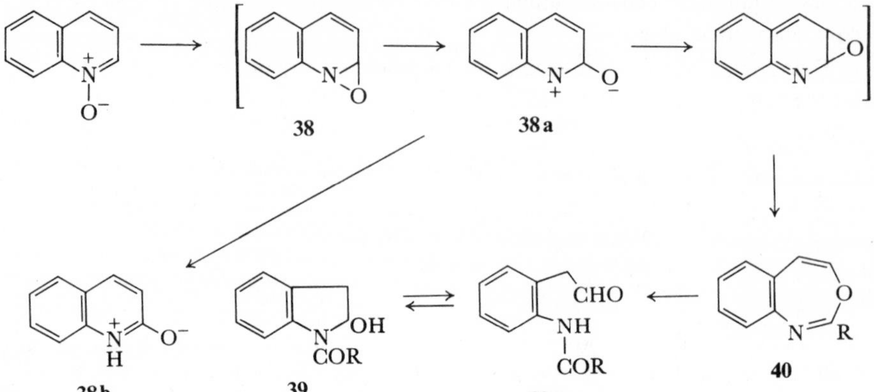

38 38a

38b 39 39a 40

kinetically.[23] Although the activation energies for the four reactions are all very similar, the rates of reaction increased steadily for the compounds in the order given. Hence structures **41** and **42**, which can be written for quinoline and which would facilitate nucleophilic attack at positions 5 and 7, do not contribute greatly to the resonance hybrid. It can, however, be concluded that **42** makes a

41 42

larger contribution than **41**, as 7-bromoquinoline reacts faster than 5-bromoquinoline in the above reaction and because 7-aminoquinoline is a stronger base than 5-aminoquinoline (see p. 309).

Nitroquinolines can be obtained by direct nitration (p. 305) or by syntheses outlined later. 6-Nitroquinoline nitrates at position 8, and nucleophilic attack takes place at position 5, which corresponds to tne 'active' α-position of naphthalene. In the example shown the initial nucleophilic attack is by the

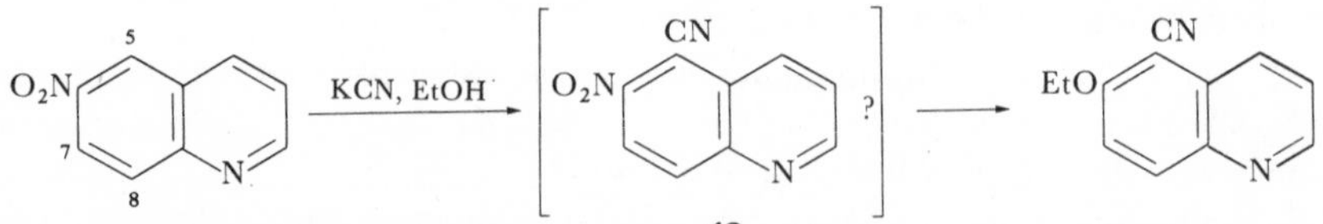

O_2N KCN, EtOH O_2N CN ? EtO CN

43

cyanide ion, to give **43**, and is followed by the displacement of the nitro group by an ethoxide ion; no attack appears to occur at position 7.

5-, 6-, 7-, and 8-Aminoquinoline can be obtained by the reduction of the corresponding nitro compounds or by one of the synthetic methods; 2- and 4-aminoquinoline are obtained by methods analogous to those of the pyridine series (p. 257), while 3-aminoquinoline is prepared from 3-bromoquinoline and ammonia in the presence of copper (cf. 3-bromopyridine, p. 256). The chemical properties of the aminoquinolines, with the exception of the 2- and 4-derivatives, which structurally and chemically resemble 2- and 4-aminopyridine, are comparable to those of aniline. 6- and 7-Aminoquinoline couple with aromatic diazonium compounds at the 5- and 8-positions, respectively, in conformity with the 'bond fixation' expected of a naphthalene. 2- and 4-Aminoquinolines are relatively strong bases because of 'additional ionic resonance,' which is discussed in connexion with the corresponding pyridines (p. 259). The fact that 7-aminoquinoline also shows enhanced basicity indicates that some 'additional ionic resonance' involving both rings is possible (cf. **42**), and the low basicity of 8-aminoquinoline is probably due to interaction of the amino group with the nitrogen atom in the adjacent ring (Table 1).

TABLE 1. The first dissociation constants (pK_a) of the aminoquinolines in water at 20°C

Substituent	pK_a	Substituent	pK_a
Unsubstituted	4·94	5-Amino-	5·51
2-Amino-	7·34	6-Amino-	5·62
3-Amino-	4·95	7-Amino-	6·65
4-Amino-	9·17	8-Amino-	3·93

Carbostyril or 2-quinolone (**44**), a colourless material of m.p. 199°C, is very similar to 2-pyridone in chemical properties (p. 261), and likewise does not tautomerize appreciably to 2-hydroxyquinoline (**45**); the lactam structure is supported by ultraviolet and infrared absorption spectrum measurements. The hydrolysis of 2-chloroquinolines to 2-quinolones in dimethyl sulphoxide[24] is much easier than by acid alone. 2-Quinolone is oxidised by potassium permanganate to give isatin (**47**), through the intermediacy of isatinic acid (**46**), and it is nitrated, via attack on the neutral molecule, at position 6.

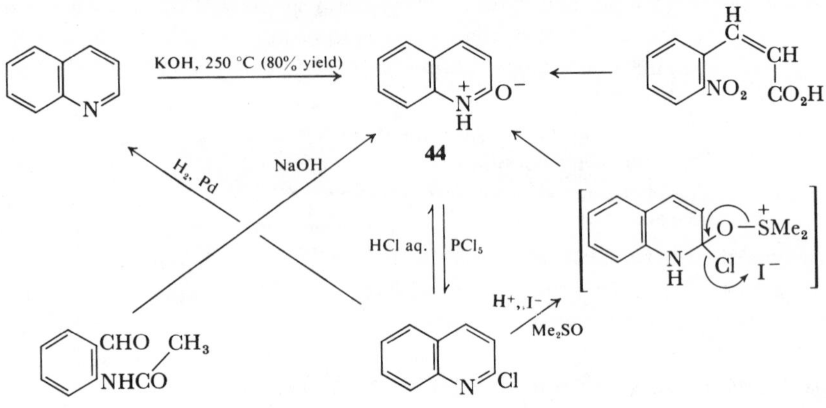

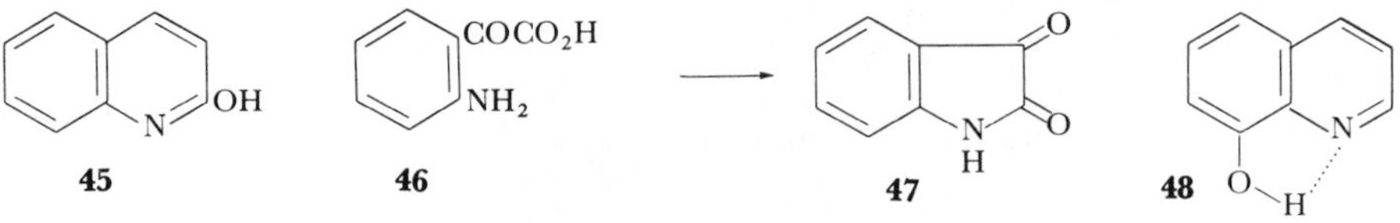

4-Quinolone[25] is similar in constitution and properties to 4-pyridone (p. 264). It behaves as an enamine towards nitric acid at 100°C, when the neutral molecule is nitrated to give 3-nitro-4-quinolone, but in concentrated sulphuric acid at 0°C the protonated molecule is attacked at position 6.[26] In contrast, all the other hydroxyquinolines, obtainable by general methods, are phenolic in nature. The most important of these is 8-hydroxyquinoline (**48**) or oxine,[27] which can be obtained in excellent yields either from quinoline-8-sulphonic acid by fusion with alkali or from 2-amino- or 2-nitrophenol by the Skraup synthesis (p. 311) under special conditions. The volatility of oxine in steam and its low melting point (75–76°C) suggest that it is hydrogen bonded as indicated. Its importance lies in the fact that with many metals it forms very sparingly soluble chelated complexes of suitable characteristics for gravimetric determinations. Volumetric determinations are also possible, as the complexes, after filtration and washing until free of unreacted oxine, are decomposed to oxine by dilute hydrochloric acid. The liberated oxine is treated with standard aqueous bromine, and it reacts, giving the 5,7-dibromo derivative, and the excess bromine is estimated volumetrically.

Quinoline-2- and quinoline-4-carbaldehydes can be obtained by the selenium dioxide oxidation of the corresponding methylquinolines. They possess many properties of aromatic aldehydes and undergo the Cannizzaro reaction. Quinoline-2-carbaldehyde closely resembles pyridine-2-aldehyde (p. 267) in chemical properties.

A number of quinoline ketones have been obtained by general methods involving Grignard reagents or quinolyl lithiums (cf. p. 256) and do not have unexpected properties.

Quinolinecarboxylic acids can be prepared by many general methods and are similar to the pyridinecarboxylic acids. The 2-carboxylic acids give yellow-red colours with ferrous salts (cf. p. 270), the 8-carboxylic acids give red colours or precipitates, while the other acids give no visible reaction. The colours are due to the formation of chelated complexes.

Quinolinecarboxylic acids can be decarboxylated quite readily, and in the case of polycarboxylic acids this occurs preferentially at positions 2 and 4 and proceeds as in the pyridine series (p. 269).

E. Synthetic methods

(1) The most widely used synthesis of quinoline is that of Skraup,[28] and it can be applied to the synthesis of many derivatives, providing that the substituents present are unchanged by the reaction conditions. Quinoline itself (**51**) is prepared by heating aniline, glycerol, concentrated sulphuric acid, and an oxidizing agent, which may be nitrobenzene, stannic chloride, ferric salts, oxygen, or arsenic pentoxide. The initial stages of the reaction may be very vigorous, and can be moderated by the presence of some ferrous sulphate; yields are often further improved by the addition of boric acid. The glycerol is first converted into acrolein, which with aniline gives the aldehyde (**49**); cyclization to a tetrahydroquinoline (**50**) is then followed by dehydration and oxidation.[29] Many attempts to prepare quinoline directly from acrolein have been made, but high yields, comparable to those of Skraup's procedure, were obtained only when acrolein vapour was passed into a mixture of aniline, sulphuric acid, and an oxidizing agent. Under ordinary conditions it appears that the acrolein is polymerized before it can combine with the aniline.

NH_2 + OHC–CH=CH_2 → **49** → **50** →

→ **51**

The general mechanism of the synthesis follows from the formation of 2- (**52**) and 4-methylquinolines (**53**), when the glycerol or acrolein is replaced by crotonaldehyde or methyl vinyl ketone, respectively. If the carbonyl group combined first with the aniline to give a Schiff's base, and this then cyclized onto the benzene ring, the positions of the methyl groups would be reversed. Intermediates of type **49** have been isolated, as have the two geometric forms of the intermediate corresponding to **50** from 4-toluidine and crotonaldehyde.[29]

NH2 + OHC–CH=CHMe → **52** (2-methylquinoline, Me)

NH2 + MeOC–CH=CH2 → **53** (4-methylquinoline, Me)

In the case of 3-substituted anilines cyclization can occur in two ways. Where the 3-substituent is a strongly activating one, such as a methyl, hydroxyl, or methoxyl group, only a 7-substituted quinoline is obtained from the reaction. With 3-bromo, 3-chloro, or 3-dimethylamino groups both isomers are isolable,

Cl, NH2 —Skraup→ Cl, N + Cl, N

with the 7-substituted quinoline as the major product, but in the case of 3-nitro- and 3-carboxyanilines the ratio of 5- to 7-substituted quinoline produced is about 4 : 1.

When acid-sensitive groups are present the synthesis may fail under ordinary conditions but be successful when very short times (e.g., 90 seconds)[30] are employed. Good yields of nitroquinolines, for example, 6-methoxy-8-nitroquinoline, can be obtained from 4-methoxy-2-nitroaniline if arsenic pentoxide is employed as an oxidizing agent.

(2) The Döbner–von Miller synthesis is closely related to that of Skraup and consists of heating a primary aromatic amine with an aldehyde in the presence of hydrochloric acid. No oxidizing agent is used. Treating acetaldehyde with aniline in EtOD gave *cis*- and *trans*-**55**, no deuterium being incorporated.[31] Under these conditions therefore the acetaldehyde does not self-condense to give crotonaldehyde before reaction with the amine, for if it did an intermediate of

type **49** would have been formed with incorporation of deuterium at position 3 of the resulting quinoline. The Schiff base **54** must be the key intermediate and self-condense as shown. When deuterium oxide–deuterium chloride was employed for the reaction some deuterium exchange had occurred in the *cis*-**55** isolated, but not nearly enough for crotonaldehyde to be an obligatory intermediate. In the ordinary synthesis *N*-ethyl- **(56)** and *n*-butylaniline are

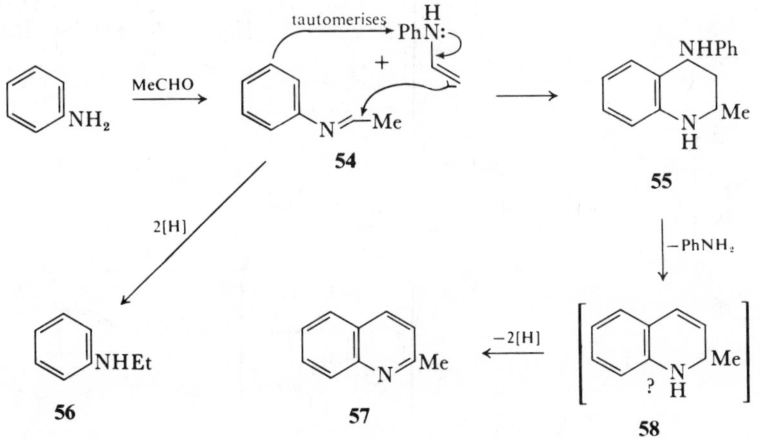

by-products and can be produced in the dehydrogenation of the intermediate dihydroquinoline **(58)** by **54** or the corresponding compound derived from crotonaldehyde. Replacing the acetaldehyde by acetone gives a dihydroquinoline **(59)**, the hydrochloride of which on heating with a trace of *t*-butyl hydroperoxide aromatizes to 2,4-dimethylquinoline in 47% yield.[32] Heating aniline, an aldehyde, and pyruvic acid also leads to quinoline-4-carboxylic acids **(60)**, which have been used for the characterisation of aliphatic aldehydes.

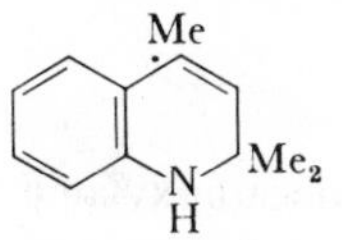

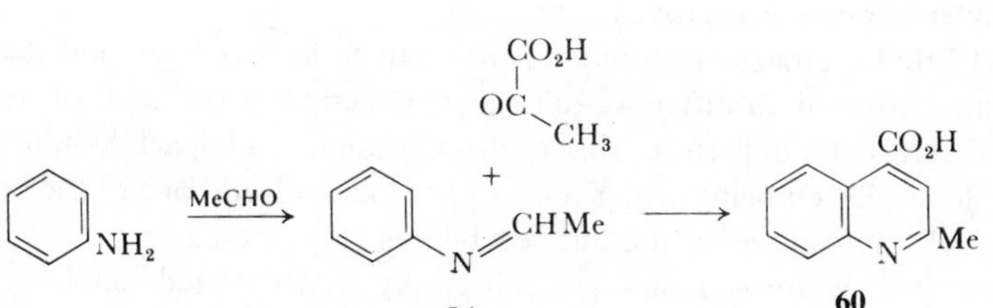

(3) Friedlander's synthesis of quinoline from 2-aminobenzylaldehyde (p. 299) has been extended to ketones, and it is possible to control the direction of cyclisation. In acid conditions the preferred enol form of the aliphatic ketone attacks the carbonyl group of the aromatic component, while under alkaline conditions the preferred anion reacts.[33]

COPh NH$_2$ + MeCOEt

61

H$^+$ Ph OH Me NH$_2$ Me O → Ph Me Me N

OH$^-$ Ph OH NH$_2$ O Et → Ph N Et

Pfitzinger's use of isatin, instead of the rather unstable compounds of type **61**, has given a synthesis of great value and diversity. Many substituted isatins and carbonyl compounds undergo the reaction, and the resulting quinoline-4-

O N H O —NaOH→ CO$_2$Na CO NH$_2$ + H$_2$CR OCR′ → CO$_2$H R R′ N

isatin

carboxylic acids are easily decarboxylated to the corresponding quinolines. Ambiguities in the synthesis can, however, arise if R and R′ contain active methylene or carbonyl groups.

(4) Ethyl acetoacetate combines with aniline in two ways, and the products can be converted to different quinolines. Reaction in the cold, or up to about 100°C, gives the anil (**62**). This cyclizes (Conradt–Limpach synthesis) to the quinolone (**63**) on being dropped into a hot inert solvent, one of the best being a diphenyl–diphenyl ether mixture. On boiling ethyl acetoacetate with aniline the anilide (**64**) is formed and is cyclized by concentrated sulphuric (Knorr's synthesis) or polyphosphoric acid to the quinolone (**65**). A useful variation of this synthesis[34] is to use an acetylenic ester (e.g., **66**), which often gives a

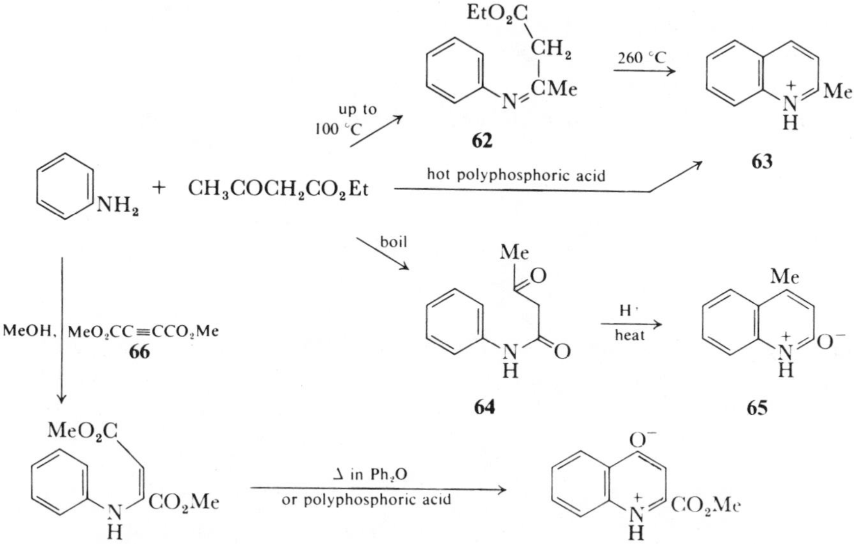

much cleaner product than the corresponding β-keto ester (e.g., $MeO_2CCOCH_2CO_2Me$).

In a similar way the anils **67** and **68**, obtained from aniline with either nitromalondialdehyde[35] or acetylacetone, can be cyclised to the corresponding quinolines by acids.

A most valuable development[36] of this last synthesis consists in converting ketones to corresponding β-chlorovinyl aldehydes (**69**) and condensing these with aromatic amines. Quinolines are obtained in high yield.

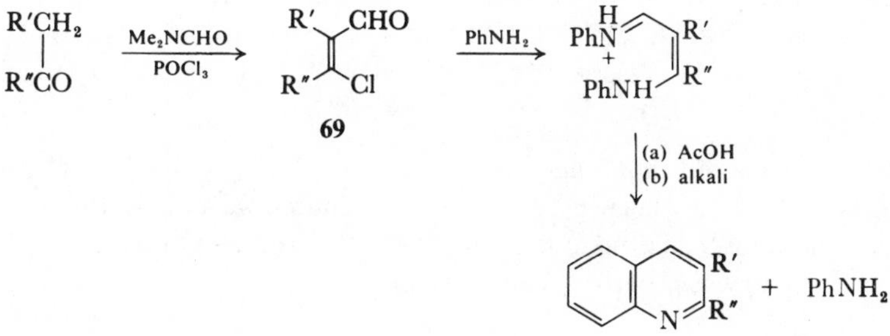

(5) A large number of quinolines have also been obtained by the cyclization of the appropriate aniline, often obtained from the corresponding nitro compound. 2-Quinolone was synthesized in 1852 by the method outlined, and **70**, obtained from 2-nitrobenzaldehyde and acetone, cyclizes similarly in acidic reducing conditions. In the presence of hot alkali 2-nitrobenzaldehyde and acetone give indigo (**71**, see also p. 210). This reaction can be used as a test for 2-nitrobenzaldehydes.

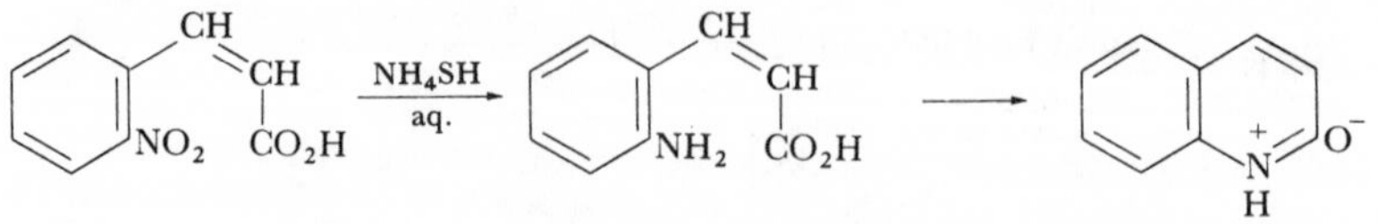

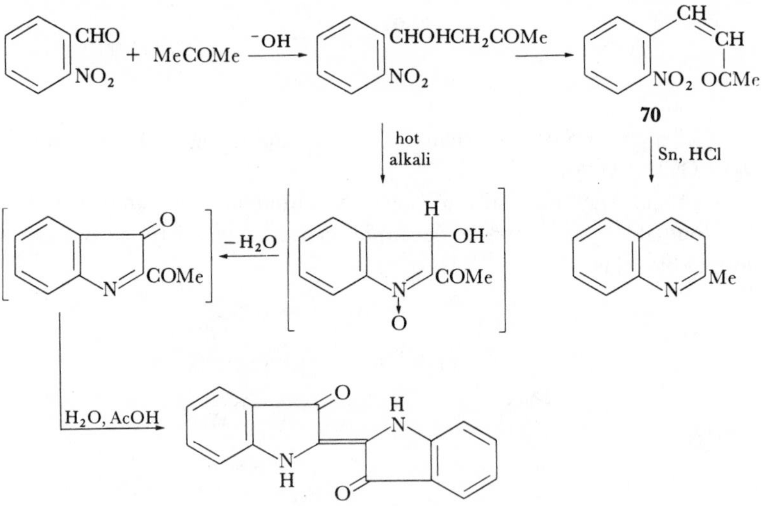

70

71

F. Natural occurrence and compounds of special interest

Quinoline itself does not occur in nature, although many alkaloids derived from the ring system are known and a concise account is available.[37] 8-Hydroxyquinoline, or oxine, is widely used as a chelating agent and has been discussed earlier (p. 310).

1. *Chemotherapeutic quinolines*

Cinchona bark has been used for several hundred years as a treatment for malaria. The active constituent is quinine (**72**), the structure of which (excluding stereochemistry) was proved by degradation in 1906.

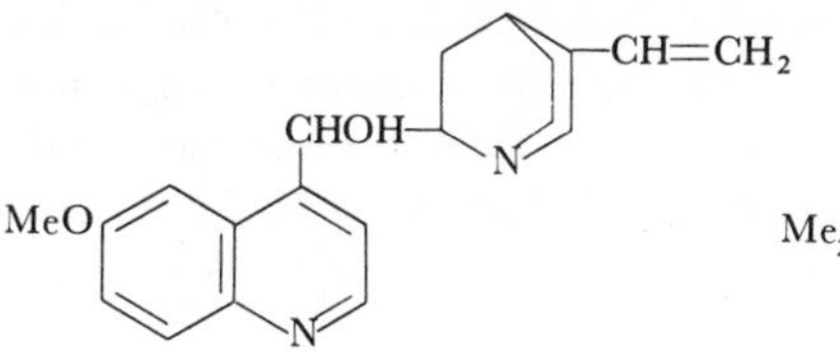
CH=CH2
CHOH
N
MeO
N

72

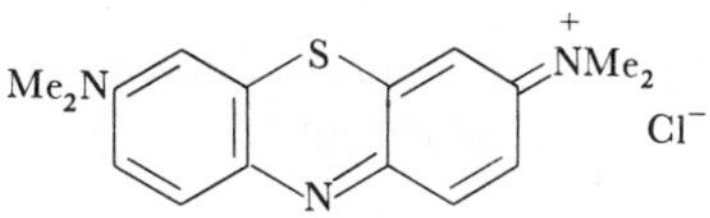
Me2N
S
NMe2
Cl⁻
N

73

On the hypothesis that a structure–activity relationship might exist, many quinolines were synthesized in the hope of finding a better anti-malarial drug than quinine, but in the early stages little success was obtained. Following the observation of Ehrlich that methylene blue (**73**) had some antimalarial activity, Schulemann, Schönhofer, and Wingler in the late 1920s found that replacement of one of its dimethylamino groups by a diethylaminoethylamino side chain increased the activity. The intense colour of the compound militated against its possible use in man, and so replacement of the ring system by that of the less chromophoric quinoline was investigated. This led to the discovery in 1928 of high activity in plasmoquin (**74**). Subsequently, many quinolines were synthesised as potential antimalarial agents, the best proving to be plasmoquin, pentaquine (**75**), and chloroquine (**76**), a synthesis of which is outlined; other powerful antimalarial agents are atebrin (p. 334) and paludrine (p. 434).

MeO
N
NH
MeCH(CH2)3NEt2

74

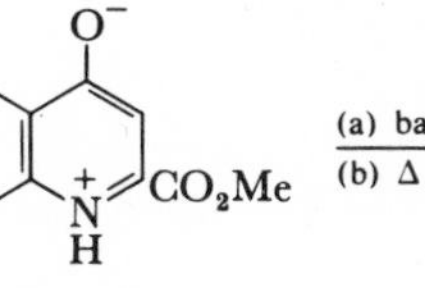
MeO
N
NH
(CH2)5NHi-Pr

75

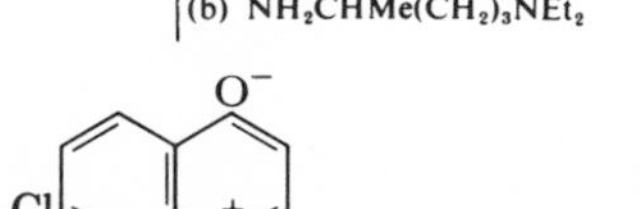
NHCHMe(CH2)3NEt2
Cl
N

76

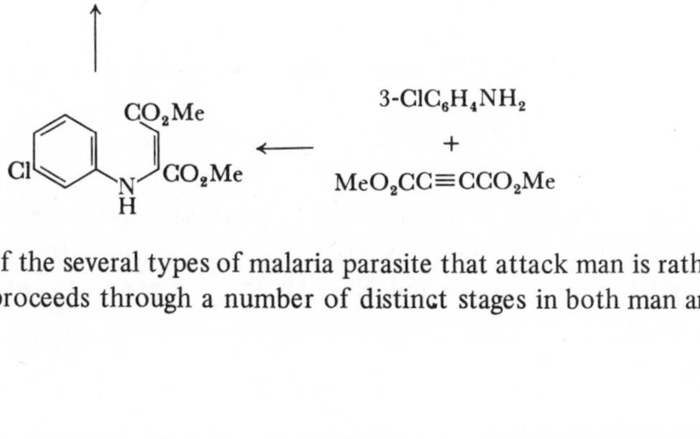
(a) POCl3
(b) NH2CHMe(CH2)3NEt2
O⁻
Cl
N
H
CO2Me
(a) base
(b) Δ
CO2Me
Cl
N
H
CO2Me
3-ClC6H4NH2
+
MeO2CC≡CCO2Me

The life cycle of the several types of malaria parasite that attack man is rather complicated and proceeds through a number of distinct stages in both man and the second host, the anopheles mosquito. It has been clearly established that different antimalarial agents can attack the parasite at different stages in its life cycle. A combination of chemotherapeuticals is therefore the best method of dealing with an established infection, while Paludrine is particularly useful as a prophylactic. Excellent discussions of the chemotherapy of malaria are available.[38]

Nupercaine (**77**) is a powerful, but rather toxic, local anaesthetic, and vioform (**78**) is a useful agent for the treatment of gastrointestinal infection. Camptothecin (**79**) an alkaloid[39] from *Camptotheca acuminata*, is attracting much interest as an anticancer agent.[40]

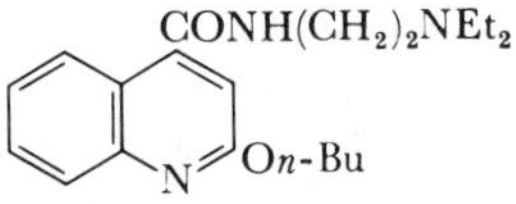

$CONH(CH_2)_2NEt_2$ On-Bu Cl I OH O N O O Et OH

77 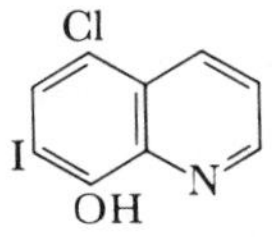 **78** 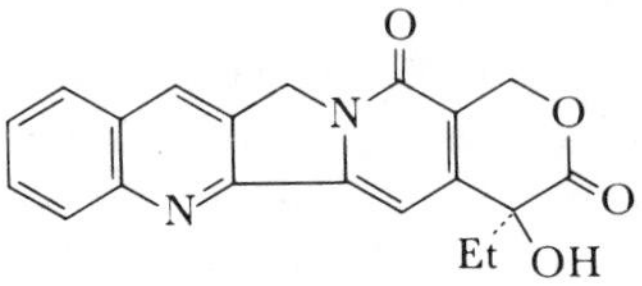 **79**

2. *Quinoline dyes*

Many commercially valuable dyestuffs are compounds with fused quinoline rings. A comprehensive account is available,[41] but of the simple quinolines only quinophthalone (**80**, quinoline yellow) has perhaps still a limited use in varnishes.

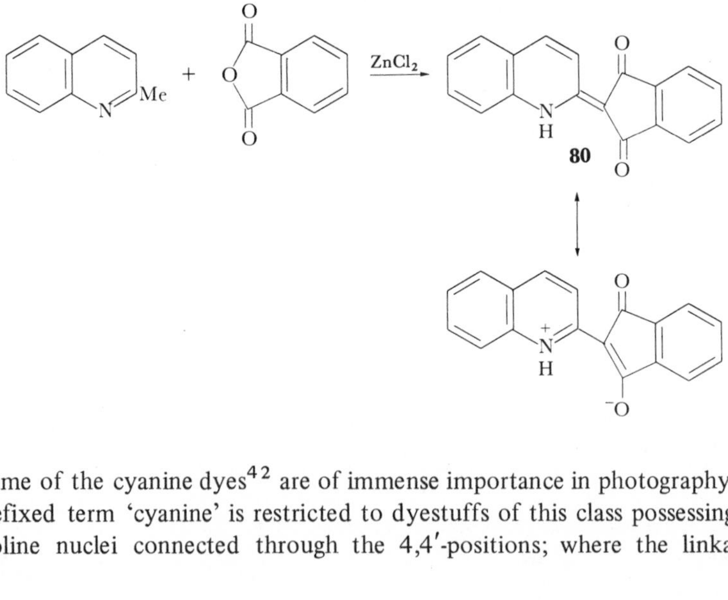

Me + O $ZnCl_2$ N H O O **80**

Some of the cyanine dyes[42] are of immense importance in photography. The unprefixed term 'cyanine' is restricted to dyestuffs of this class possessing two quinoline nuclei connected through the 4,4′-positions; where the linkage is through other positions it should be specified; the prefixes iso, pseudo, xanthoapo, and erythroapo are sometimes used to indicate 2,4′-, 2,2′-, 2,3′-, and 4,3′-linkages, respectively. Cyanines such as **81** and **82** are often named 'methincyanines,' to indicate the presence of a methine group between the rings. On this system **86** and **87** are tri- and pentamethincyanines, respectively. One, or

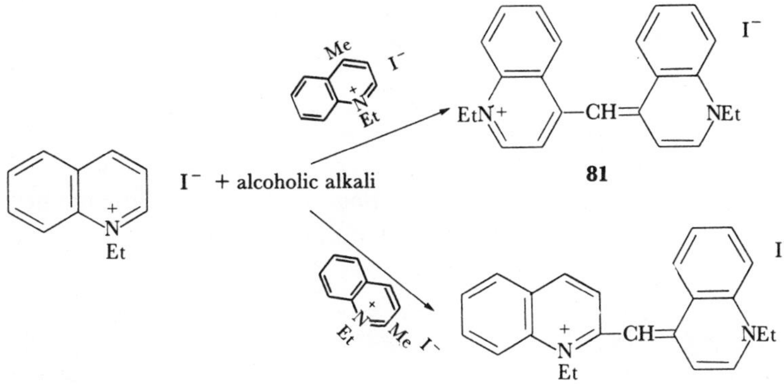

I⁻ + alcoholic alkali

81

82 82% yield

both, of the quinoline rings can be replaced by other heterocyclic systems, the conjugated chain between them can be branched or bear substituents, and very many of these compounds have been made.

The blue dye, cyanine (**81**), was first obtained in 1856 by the action of caustic alkali on a mixture of 1-ethyl-4-methylquinolinium iodide and 1-ethylquinolinium iodide. It is useless as a fabric dyestuff owing to its instability to light, but in 1875 Vogel found that it would sensitize photographic emulsions to green light. However, cyanine also fogged the emulsions. In 1903 a great advance

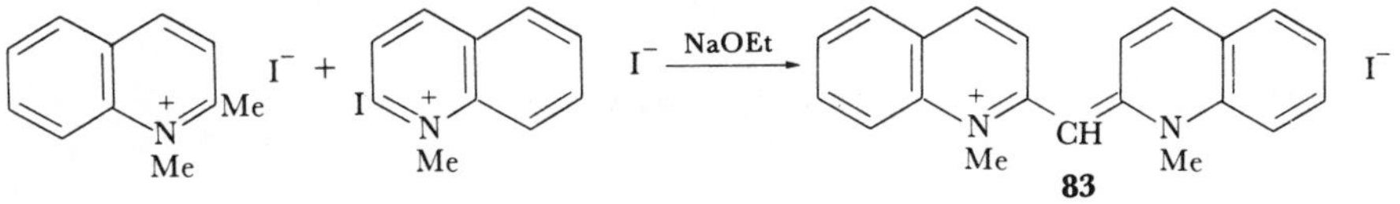

NaOEt

83

was made when the 2,4′-cyanine (**82**) was found to sensitize emulsions without fogging. 2,2′-Cyanines, the first (**83**) of which was prepared by a method that has been used for other cyanines, are also sensitizers, but the apocyanines e.g. **84** and **85**), which are yellow compounds possessing directly linked quinoline rings, have no useful photographic properties.

Although in the formulae of the above cyanines the positive charge is placed on one nitrogen atom, the other being tertiary, the compounds are clearly resonance hybrids, of which the structures given are but one canonical form.

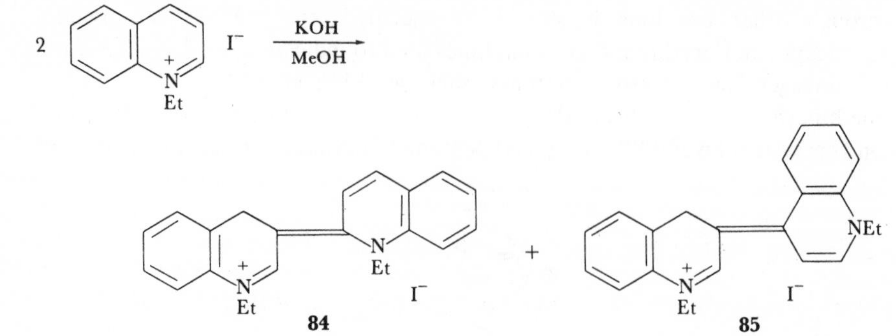

Increasing the conjugation between the rings, such as in the trimethincyanine 'pinacyanol' (**86**), moves the position of maximum absorption of visible light towards the red and gave the first sensitizer to red light. Replacing the ethyl orthoformate of the above synthesis by β-anilinoacraldehyde gives the pentamethincyanine (**87**), while with glutaconic aldehyde (p. 234) dianilide in triethylamine at low temperatures the heptamethincyanine (**88**) is obtained. This last has been used to sensitize film in the 9000–10,000 Å region of the infrared.

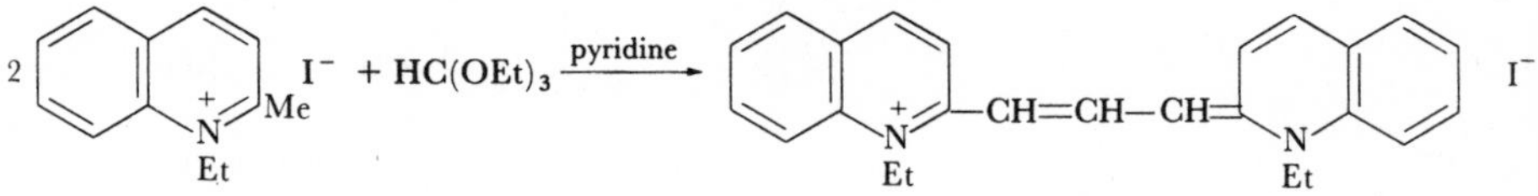

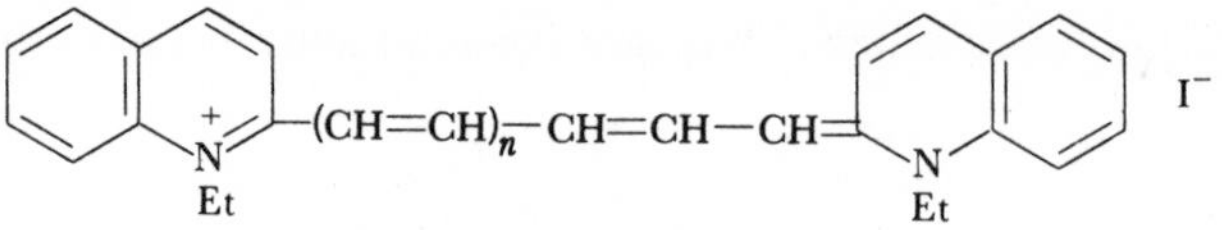

87, $n = 1$ **88**, $n = 2$

2. ISOQUINOLINE

A. Introduction

Isoquinoline (**1**) was first isolated from the quinoline fraction of coal tar by fractional crystallization of its bisulphate in 1885; it has also been obtained from crude petroleum. It does not occur free in nature, but many alkaloids[37] (e.g., papaverine) contain either the aromatic or the reduced isoquinoline system. In physical and chemical properties it closely resembles quinoline. The positions of

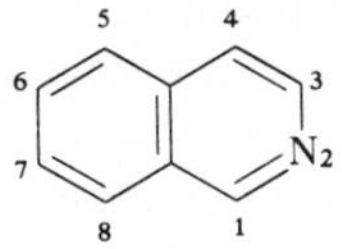

Isoquinoline

the ring are usually numbered as indicated. Isoquinoline has been called 3,4-benzopyridine or 2-azanaphthalene.

B. Physical properties and structure

Isoquinoline is a colourless solid (f.p. 26·5°C, b.p. 243°C) with a smell like that of benzaldehyde. It is volatile in steam, sparingly soluble in water, and soluble in many organic solvents. It turns yellow on normal storage (c.f. pyridine, p. 235).

The bicyclic structure (**1**) was first proposed on the basis of the oxidation to a mixture of phthalic and pyridine-3,4-dicarboxylic acids, and was confirmed by synthesis. The bond lengths for isoquinoline hydrochloride have been determined by X-ray methods,[43] and as for naphthalene (p. 300) there is an alternation in bond length indicative of some 'bond fixation.' The resonance energy of isoquinoline has not been determined experimentally, but theoretical studies suggest that it is close to that of quinoline, and the ultraviolet spectra of these compounds are very similar. The dipole moment of isoquinoline (2·60 D) is slightly higher than those of pyridine and quinoline. These data suggest that isoquinoline is best considered as a resonance hybrid to which uncharged structures (**1**, **2**, and **3**) and charged structures, including **4** and **5**, contribute; other charged structures with the charges in different rings can also be written.

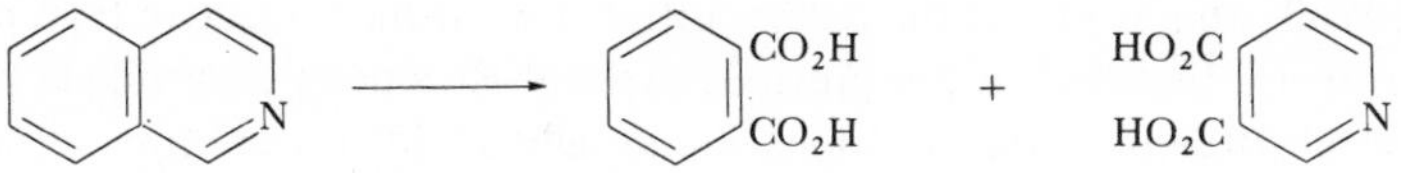

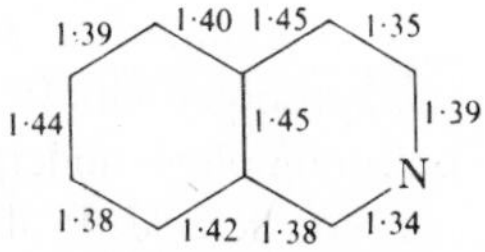

Isoquinoline hydrochloride, bond lengths in Angstroms

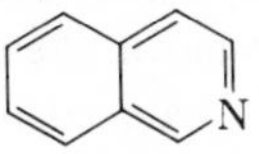

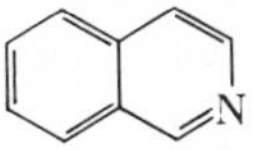

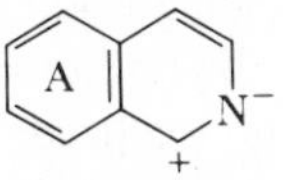

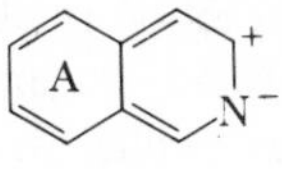

The chemical reactions of isoquinoline, as in the case of quinoline, are in agreement with the postulate of some 'bond fixation' and suggest that the symmetrical double-bond structures **1**, and to some extent, **4** are the major contributors to the hybrid. Of the two charged structures shown, **4** is expected to be the more important, because only in this structure does ring A possess a Kekulé form. This is confirmed by chemical evidence (pp. 325 and 326) and agrees with the results of calculations of the π-electron densities for the ring system by various molecular-orbital methods. All these methods show that position 1 is the most electron deficient in the molecule.

In order to account for certain other reactions of quinolines (pp. 325 and 326), which are perhaps surprising, it is necessary to include smaller contributions to the resonance hybrid from structures such as **5**, in which the carbocyclic ring A must be written in the *ortho* quinonoid form.

C. Chemical properties

Isoquinoline resembles quinoline in many ways. The effect of the nitrogen atom is largely confined to its own ring. By analogy with pyridine both positions 1 and 3 would be expected to be 'positive' and susceptible to nucleophilic attack. While this is in fact the case, position 1 is far more reactive than position 3, for reasons discussed in the previous section. Reactions occurring at position 1, or to substituents at this positions, are usually analogous to those occurring at position 2 in the quinoline series.

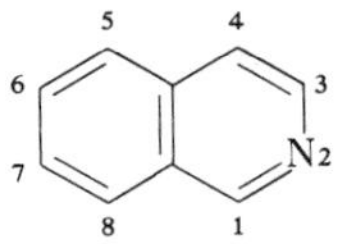

5 4 6 3 N2 7 8 1

Isoquinoline

5 4 6 3 2 7 N 8 1

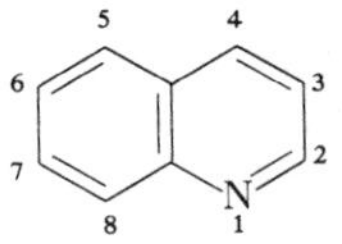

Quinoline

1. *Addition and ring-opening reactions*

Isoquinoline (pK_a 5·14) is a stronger base than quinoline. It forms an *N*-oxide (p. 326) and reacts with alkylating agents at the nitrogen atom to give quaternary salts such as **6**. Some of these salts undergo Diels–Alder reactions across the 1,4-positions and are reduced to 1,2-dihydroisoquinolines by sodium borohydride. The corresponding hydroxide (**7**) is in equilibrium with the pseudohydroxide (**8**, 1,2-dihydro-1-hydroxy-2-methylisoquinoline) and on oxidation gives the isoquinolone (**9**, cf. p. 262). Isoquinoline undergoes the Reissert reaction (p. 302), yielding isoquinoline-1-carboxylic acid. It also reacts with Grignard reagents and alkyllithiums to give 1-substituted-1,2-dihydroisoquinolines, which are readily oxidized to the isoquinolines. Ozonolysis of isoquinoline gives phthalic acid and pyridine-3,4-dicarboxylic acid.

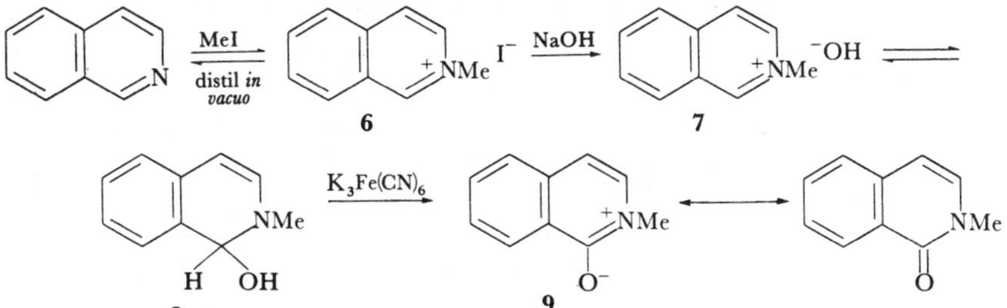

The reduction of isoquinoline by sodium in liquid ammonia[44] gives the reactive 1,2-dihydro derivative (**10**),[45] which polymerizes to a trimer. Reduction with tin and hydrochloric acid, or further reaction with sodium in liquid ammonia, gives 1,2,3,4-tetrahydroisoquinoline (**11**), which has the properties of a secondary amine, while hydrogenation over platinum in the presence of acid fully reduces both rings to a mixture of *cis*- and *trans*-decahydroisoquinolines (**12**). All these compounds can be dehydrogenated to isoquinoline. Treatment of

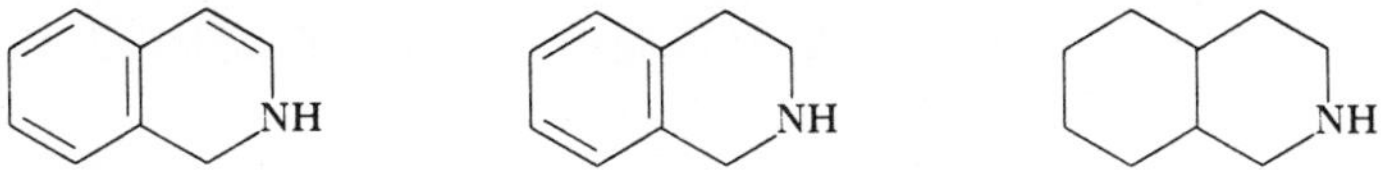

11 with excess methyl iodide followed by alkali gives 1,2,3,4-tetrahydro-2,2-dimethylisoquinolinium hydroxide (**13**), which is interesting, as on degradation by Hofmann's method (p. 244) the ring breaks between positions 2 and 3, giving **14**, while under Emde's conditions (p. 304) an alternative scission leads to **15**. The first reaction probably proceeds by an E_2 elimination of a proton from position 4, facilitated by the positive charge at position 2, while the second may involve the formation of a benzyl anion.

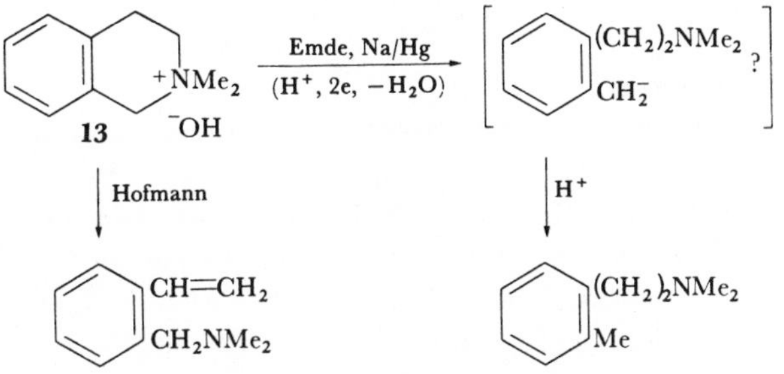

A number of isoquinolines can be converted into naphthalenes through successive ring opening and closure. For example, 1-benzyl-2-methylisoquinolinium iodide with alkali gives a pseudohydroxide (**17**). This can lose water to yield the 'methylene base' (**16**), which is relatively stable in contrast to

NMe CHPh **16** ← NMe HO CH_2Ph **17** —NaOH aq., hot→ [H C=CHNHMe $COCH_2Ph$] → [$CH_2CH{=}NMe$ $COCH_2Ph$] —H_2O, $-MeNH_2$→ [CH_2CHO $COCH_2Ph$] → Ph OH **18**

compounds of this type in the pyridine series, but on heating with aqueous sodium hydroxide methylamine is split out, giving 1-hydroxy-2-phenylnaphthalene (**18**). The addition product (**19**) of isoquinoline and sulphur trioxide with cold alkali gives **20**, while hot alkali gives a dialdehyde (**21**), which self-condenses to **22**.[46]

$\overset{+}{N}SO_3^-$ **19** —NaOH aq. cold→ NSO_3^- H O^- 2 Na^+ **20** —heat, NaOH aq.→

CHO CH_2 CHO **21** OHC CH_2 CHO → CHO CHO **22**

2. *Substitution reactions*

Localization energy calculations[47] for the isoquinolinium cation suggest that electrophilic attack occurs preferentially at position 5, then at position 8. The species actually nitrated in over 71·3% sulphuric acid has been shown experimentally[48] to be the isoquinolinium cation and 5-nitroisoquinoline is formed mainly.[49] Sulphonation, and bromination in sulphuric acid,[50] also take place through the cation and at position 5, while bromination in the presence of aluminium chloride, which must coordinate with the nitrogen atom, gives

NO_2

HNO_3, H_2SO_4, 0°

N + N

NO_2

90·4% 9·6%

$Hg(OAc)_2$

HgOAc

N

successively 5-bromo-, 5,8-dibromo-, and 5,7,8-tribromoisoquinoline. Bromination and mercuration under less acidic conditions, however, take place at position 4. The localization energy approach accounts for this, if the neutral molecule first yields an intermediate 1,2-dihydro derivative which then substitutes at position 4, and also predicts that free-radical and nucleophilic attack will occur at position 1. Bromine at 450°C actually causes some 1-bromination, and examples of nucleophilic attack are shown. The structures of 1-aminoisoquinoline and 1-isoquinolone are similar to those of 2-aminopyridine

$NaNH_2$ NaOH

N N $\overset{+}{N}H$

NH_2 O^-

Isocarbostyril or isoquinolone

(p. 258) and 2-pyridone (p. 262). Isoquinoline is attacked by phenyl radicals at many positions and by the benzyl radical mainly at positions 1 and 3,[19] but the isoquinolinium cation is attacked more readily and by both benzyl[19] and acetyl[20] radicals almost exclusively at position 1.

D. Derivatives of isoquinoline

As already mentioned, substituents at position 1 of the isoquinoline ring have reactivities similar to those of corresponding 2-substituted quinolines or

PhCHO

N N

Me CH=CHPh

23

Me PhCHO CH=CHPh

N N

24 **25**

pyridines, while those at position 3 are much less reactive, and in the past have even been wrongly thought to possess none of the reactivity of their 1-substituted isomers. An example of this is 3-methylisoquinoline (**24**), which, like 1-methylisoquinoline (**23**) will condense with benzaldehyde in the presence of zinc chloride, but requires more vigorous conditions,[51] to give **25**. *N*-Methylation naturally increases the ease of these reactions. 1,3-Dichloroisoquinoline (**26**) on treatment with sodium methoxide gives **27**, thus showing the

Cl N Cl **26** —RNH_2→ Cl N NHR + NHR N NHR

R = 4-ClC_6H_4—

26 —MeONa→ Cl N OMe **27**

26 —HI, HOAc→ Cl N **28** —NaOMe, Me_2SO→ OMe N **29**

greater reactivity of the chlorine at position 1. This is again shown in the reaction with hydrogen iodide, and it is interesting that the chlorine atom of **28** is displaced by sodium methoxide in dimethyl sulphoxide but not in methanol. As the chlorine atoms at positions 1 and 3 of **26** are replaced successively when the compound is boiled with 4-chloroaniline,[52] it is clear that the 3-chlorine atom must be more reactive (cf. 3-bromopyridine, p. 256) than that of chlorobenzene.

$^{+}$N—O^{-} **30** —Ac_2O, boil→ $^{+}$NH O^{-} **31** 50–60% + OH N **32** 8–9%

Isoquinoline 2-oxide (**30**), obtained from isoquinoline with hydrogen peroxide and acetic acid, on treatment with acetic anhydride gives isoquinolone (**31**) and some 4-hydroxyisoquinoline (**32**). No 3-hydroxyisoquinoline is formed, but this compound can be prepared by the hydrolysis of **29**. It is interesting that, in contrast to 1-isoquinolone, the stability of the two possible tautomers for 3-hydroxyisoquinoline is about the same.

E. Synthetic methods

Most syntheses of isoquinolines start from 2-phenylethylamines and involve cyclization through an additional carbon atom provided by the carbonyl group of another compound. The main syntheses described here have all been widely used and have been reviewed.[53]

(1) The Bischler–Napieralski synthesis involves heating an acyl derivative (**34**) of a 2-phenylethylamine (**33**) with phosphorus oxychloride or pentachloride in an inert solvent, or with phosphorus pentoxide in pyridine.[54] A dihydroisoquinoline (**35**) is formed, which may be converted into the isoquinoline (**36**) by dehydrogenation or oxidation. It is also possible to convert the amide (**34**) into the corresponding iminohalide (**39**) with phosphorus halides, and then to cyclise with aluminium chloride, in which case **38** is probably an intermediate. A new variation is to alkylate a nitrile with a suitable halide (**37**) when the cation formed (**38**) cyclises.[55] Poor results are obtained in all these syntheses if R is hydrogen.

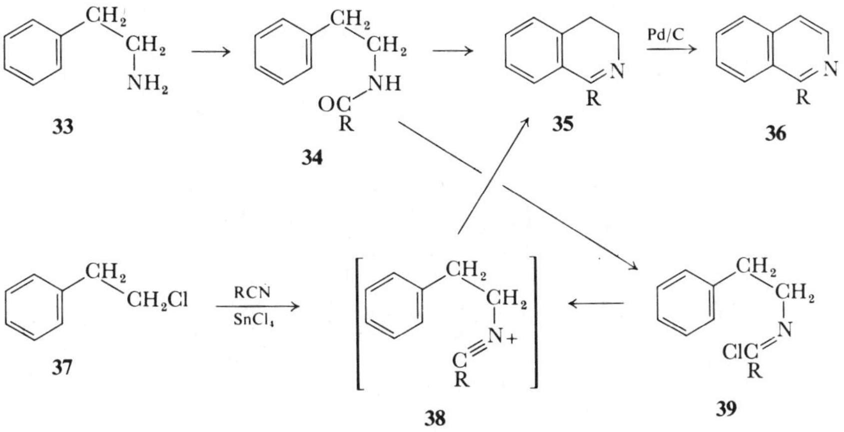

Pictet and Gams applied the cyclisation to hydroxy compounds such as **40** and obtained isoquinolines (**43**) without the disadvantage of an oxidation or dehydrogenation stage. The synthesis proceeds through the formation of the oxazolines[56] (**41**), which can be isolated if short reaction times are employed, and unsaturated amides (**42**).

(2) A useful synthesis of tetrahydroisoquinolines, discovered by Pictet and Spengler, is to cyclize the Schiff's bases (**44**) from 2-phenylethylamines with dilute hydrochloric acid. The tetrahydroisoquinolines can, of course, be dehydrogenated to isoquinolines.

Ph OH Me NHCOR **40** $\xrightarrow[\text{boiling PhMe or decalin}]{P_2O_5}$ Ph H Me O N: → H^+ **41** ⟶ Ph Me NH OCR **42** $\xrightarrow{-H_2O}$ Ph Me N R **43**

(3) Another type of synthesis, that of Pomeranz and Fritsch, starts from aromatic aldehydes. Conversion into Schiff's bases (e.g., **45**) with aminoacetal

CH_2 CH_2 NH_2 $\xrightarrow{RCHO}$ CH_2 CH_2 N HC R **44** $\xrightarrow{H^+}$ NH R

followed by cyclization by acid gives isoquinolines. A variation,[57] which gives high yields if the aryl group possesses activating substituents, involves the conversion of **45** to **46**; the intermediate **47** (5,6-dimethoxy) has been isolated.

CHO $\xrightarrow{H_2NCH_2CH(OEt)_2}$ $(EtO)_2CH$ CH_2 N CH **45** $\xrightarrow{H_2SO_4}$ N

(a) PtO_2/H_2
(b) Tosyl Cl, pyridine

$(EtO)_2CH$ CH_2 NTosyl CH_2 **46** $\xrightarrow{6\text{M HCl}}$ [NTosyl] **47**

Poor results are usually obtained using aromatic ketones in the Pomerantz–Fritsch synthesis, and then the alternative route shown can be employed.[58]

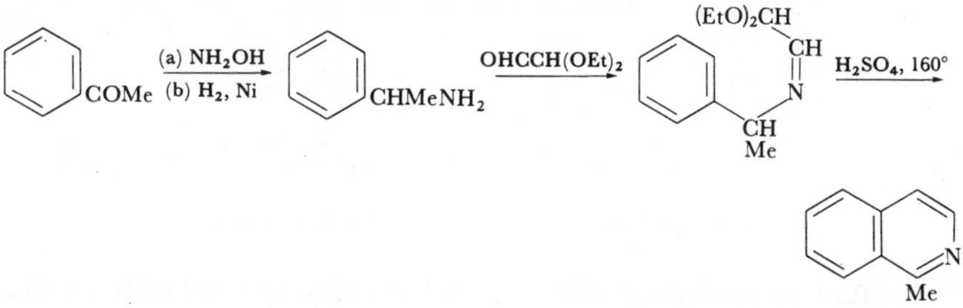

3. QUINOLIZINE

Although a number of alkaloids are complex derivatives of quinolizine, sometimes called pyridocoline, the parent compound has attracted attention only in recent years. The quinolizinium cation (2), which is isoelectronic with naphthalene, has an ultraviolet absorption spectrum similar to those of quinoline and isoquinoline in confirmation of its aromatic nature. Salts have been

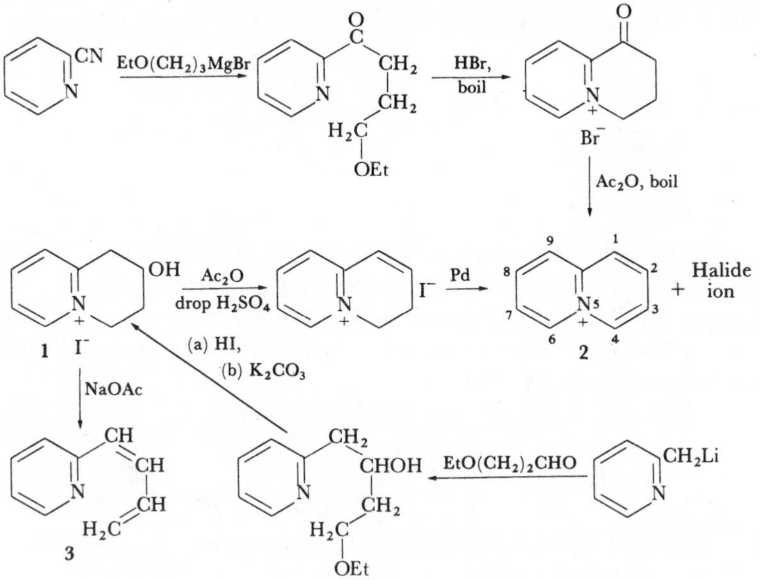

made[59,60] by the two methods outlined, the first synthesis giving a 48% overall yield. The localization energies[61] for electrophilic and nucleophilic attack on the cation are in agreement with the observation that bromine in acetic acid does not cause substitution.[62]

Localization energies ($-\beta$)

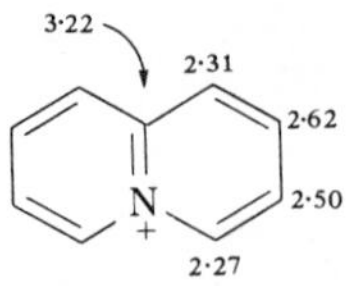

3·22 2·31 2·62 2·50 2·27

Electrophilic attack

2·59 2·31 2·20 2·50 1·97

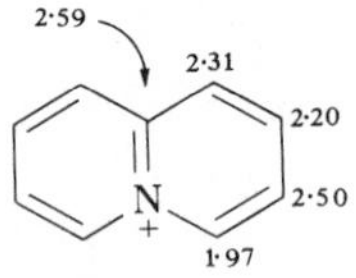

Nucleophilic attack

All efforts to prepare *H*-quinolizine, which can be written in three tautomeric forms (**4**, **5**, and **6**) have been unsuccessful, although complex derivatives of both 4*H*- and 9a*H*-quinolizines are well known.[63] The attempt[59] to aromatize **1** opened the ring and gave 2-butadienylpyridine (**3**), and although nucleophilic attack of lithium hydride[64] and Grignard reagents[65] apparently takes place at position 4 of the quinolizinium ring (**2**), in agreement with the theoretical predictions, the products isolated were again **3** and its δ-substituted derivatives. However as a series of hexahydro- and octahydroquinolizines were isolated[64] from reductions of **2** involving sodium borohydride in ethanol, it appears likely that 4*H*-quinolizine can have a short independent existence. Aliphatic and aromatic amines also attack the quinolizinium cation causing ring opening and the formation of compounds such as **7**.[66]

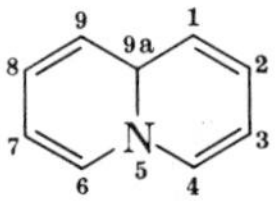

4 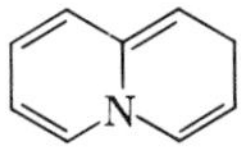 **5** 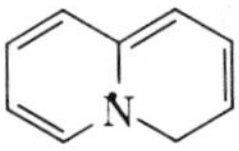 **6** 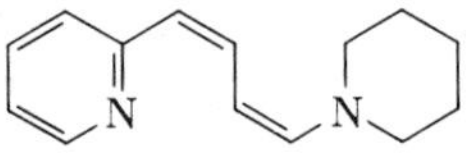 **7**

The tetraester **8** has been obtained from pyridine and dimethyl acetylenedicarboxylate[63] as a very labile red compound which rapidly undergoes a [1,5] sigmatropic shift[67] to the 4*H*-isomer **9**. The stability of the bicyclic system in both **8** and **9** is due[67a] to resonance involving the nitrogen atom and the nearest conjugated ester group (e.g., **10**).

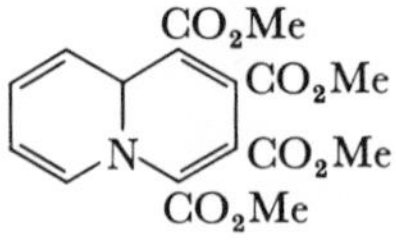

8 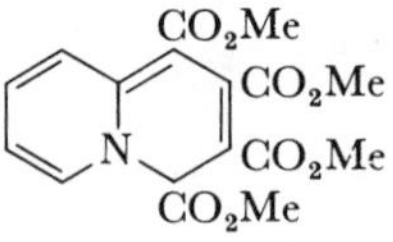 **9** 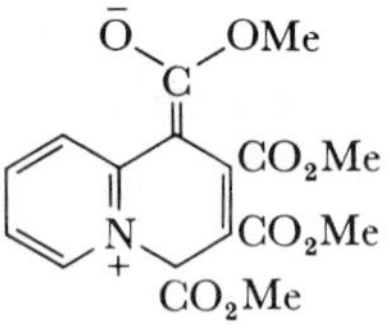 **10**

1- and 3-Hydroxyquinolizinium salts have been synthesized; they brominate readily at positions 2 and 4, respectively and have reactions of typical phenols.[68] In contrast, the 2-hydroxy isomer (**11**)[68] behaves as quite a strong

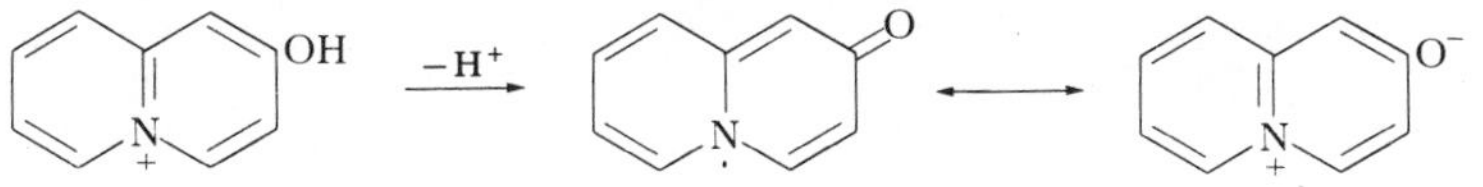

11 $\xrightarrow{-H^+}$ **12** ⟷ **13**

acid and easily loses its proton to give 2-quinolizone, which is in fact a vinylogous amide and can be represented as a resonance hybrid (**12**–**13**). 4-Quinolizone[69] (**14**) appears to be a weaker base than 2-quinolizone and is not reduced by lithium aluminium hydride. It is perhaps best represented by the charged formulation (**16**). Nitration and bromination occur first at position 3 and afterwards at position 1.[70] Successive treatment with methyl sulphate and perchloric acid gives 4-methoxyquinolizinium perchlorate (**15**).[71]

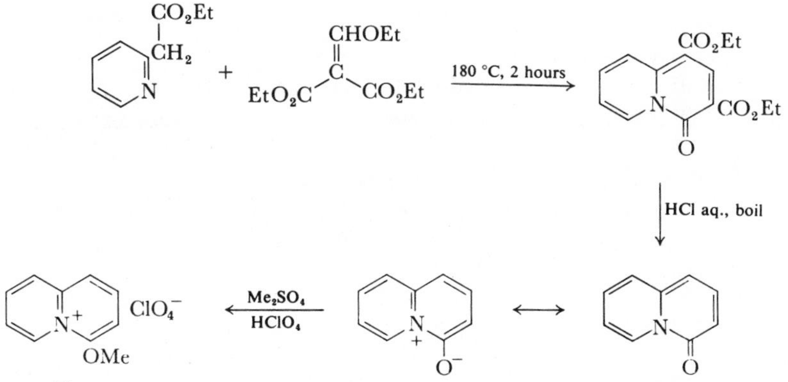

CO_2Et, CH_2, N + CHOEt, C, EtO_2C, CO_2Et —180 °C, 2 hours→ CO_2Et, N, CO_2Et, O

HCl aq., boil

15 (ClO_4^-, OMe) ←Me_2SO_4, $HClO_4$— **16** ⟷ **14**

Quinolizidine (**18**) is an aliphatic tertiary base which undergoes the Hofmann degradation (p. 244). It has been obtained[69] from 1 by complete hydrogenation

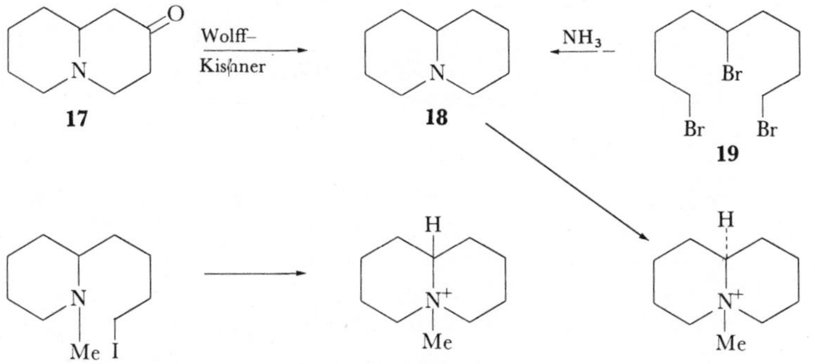

17 —Wolff–Kishner→ **18** ←NH_3— **19** (Br, Br, Br)

20 (Me I) → *cis*, **21**

18 → *trans*, **22**

to the saturated amino alcohol, oxidation to the ketone (**17**), and reduction as indicated, and also from the bromo compound (**19**). Direct methylation gives a quaternary salt with a *trans* ring junction (**22**), while the *cis* isomer (**21**) is obtained from the cyclization of **20**.[72]

4. ACRIDINE AND PHENANTHRIDINE

A. Introduction

Both acridine and phenanthridine are present in high-boiling fractions of coal tar but are most readily obtained by synthesis. They can be regarded, respectively, as aza derivatives of anthracene and phenanthrene, or as 2,3- and 3,4-benzoquinoline. The parent compounds have many properties which can be anticipated from those of pyridine, quinoline, or isoquinoline; these are not considered in detail here. Examples are the reactivity of halogen atoms and methyl groups at positions 9 in acridine and 6 in phenanthridine, and the formation of highly crystalline salts with acids and of quaternary salts with alkyl halides. The quaternary salts are oxidized to the corresponding 'pyridones' with potassium ferricyanide.

Interest in the acridine and phenathridine series is due in part to the useful chemotherapeutic properties of certain derivatives, and also to their very interesting interactions with DNA, which have led to the understanding of the genetic code.

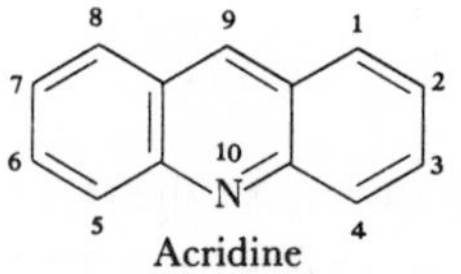

Acridine

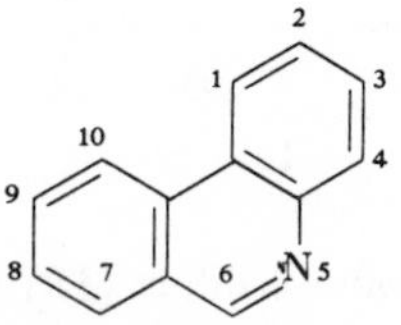

Phenanthridine

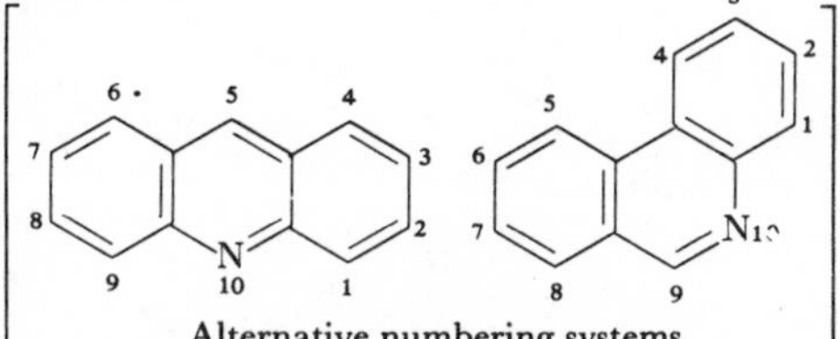

Alternative numbering systems used occasionally

B. Acridine

Acridine (**1**) is a pale yellow comparatively volatile highly crystalline solid of m.p. 110°C. It boils at 345°C, and the vapour is a powerful irritant to the nose and throat. Acridine and many of its derivatives give highly fluorescent

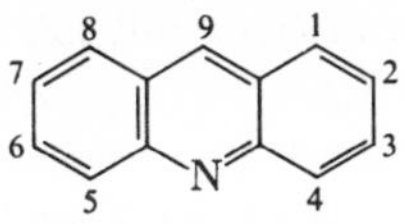

1

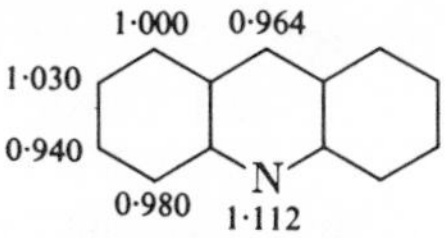

π-Electron density

solutions. The fluorescence is usually blue or green, and acridone (**6**), properly called 9-acridanone, is one of the most powerful fluorescent materials known. Nitration and bromination of acridine give mainly 2,7-di- and 2,4,5,7-tetrasubstitution products; it is impossible to halt the reactions at the mono- or trisubstitution stages. Nucleophilic attack takes place at position 9; acridine and sodamide give 9-aminoacridine (**9**). These results agree with the molecular-orbital picture[72a] of the π-electron densities round the acridine molecule. Acridine 10-oxide, obtained from acridine with perbenzoic acid, both nitrates and brominates at position 9 (cf. pyridine 1-oxide, p. 249).

The most general synthesis of acridines starts with a diphenylamine-2-carboxylic acid and usually gives good results. Diphenylamine-2-carboxylic acids (e.g., **2**) are obtainable from anilines and 2-chlorobenzoic acids in the presence of a base and metallic copper (Ullmann's reaction); the exact function of the copper is unknown. The nucleophilic displacement of the chlorine atom is facilitated by the *ortho* (electron-attracting) carboxyl group. The complementary reaction, that of anthranilic acid and bromobenzene, does give some diphenylamine-2-carboxylic acid, but in poorer yield, as the bromine atom is not activated and the amino group is deactivated. Another valuable synthesis of diphenylamine-2-carboxylic acids involves the Chapman rearrangement, which often succeeds when that of Ullmann fails. In the case of diphenylamine-2-carboxylic acid, the sodium derivative of methyl salicylate is treated with

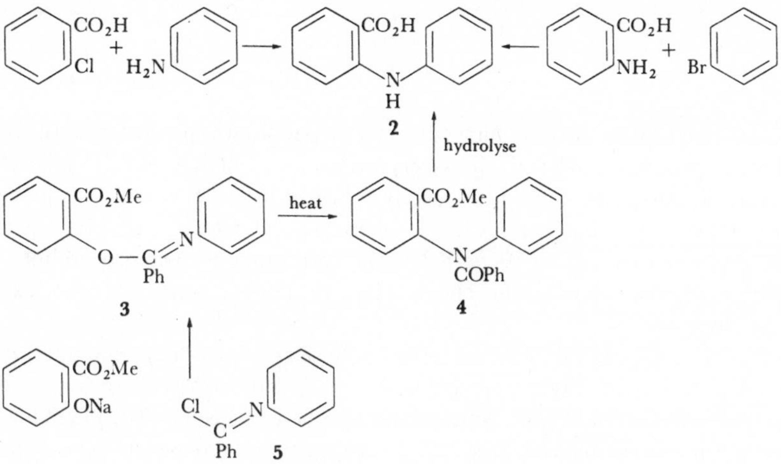

2

hydrolyse

heat

3 **4**

5

N-phenylbenzimido chloride (**5**), obtained from benzanilide and phosphorus pentachloride. The product **3**, a complex imido ether, on heating rearranges to **4**, which on hydrolysis gives the desired acid (**2**).

Cyclization of diphenylamine-2-carboxylic acid (**2**) to 9-acridanone (**6**), the structure of which is similar to that of 4-pyridone (p. 264), occurs with sulphuric acid under carefully controlled conditions, while phosphorus oxychloride gives 9-chloroacridine (**7**) quantitatively. Reduction of acridone or 9-chloroacridine as indicated yields acridan (**8**, 9,10-dihydroacridine), which is oxidized by air, ferric chloride, and many other oxidizing agents to acridine (**1**). The central ring of acridine, like that of anthracene, is readily reduced. The product, acridan (**8**), greatly resembles diphenylamine in chemical properties.

The halogen atom of 9-chloroacridine (**7**) is reactive and is displaced by ammonia in the presence of acid catalysts, giving 9-aminoacridine (**9**). This is a strong base (cf. 4-aminoquinoline, p. 309), and the hydrochloride is a valuable antibacterial agent used mainly for minor injuries and burns.

$POCl_3$
CO_2H N H **2** —H_2SO_4→ O N H **6** —$POCl_3$→ / ←HCl aq., boil— Cl N **7**
6 —Na, AmOH→ **8**; **7** —H_2, Raney Ni, base→ **8**; **7** —$(NH_4)_2CO_3$ in PhOH→ **9**
N **1** ←oxidise— / —H_2, Ni→ H_2 N H **8**; NH_2 N **9**

Atebrin (**12**), also called Quinacrine and Mepacrin, was used as an antimalarial agent during World War II. It is very efficacious but has certain undesirable properties, and it has now been largely replaced by other drugs. It is prepared from 4-methoxyaniline and 2,4-dichlorobenzoic acid, which, through the general synthesis given above, are converted to the corresponding 9-chloroacridine (**11**). This, with the appropriate amine (**10**) in phenol, gives Atebrin as the hydrochloride.

The biacridinium salt **15** gives out an intense green light (chemiluminescence) on oxidation with hydrogen peroxide. It is probably the most powerful chemiluminescent substance known and is called Lucigenin. The chemiluminescence can last for many hours, and its brightness is increased at the expense of

$$CH_3COCH_2CO_2Et \xrightarrow[Et_2NCH_2CH_2Cl]{NaOEt} CH_3COCH(CO_2Et)CH_2CH_2NEt_2 \xrightarrow{\text{'ketonic hydrolysis'}}$$

$$CH_3CO(CH_2)_3NEt_2 \xrightarrow{H_2,\ Ni,\ in\ NH_3\ (liquid)} MeCH(NH_2)(CH_2)_3NEt_2 \quad \mathbf{10}$$

$$\mathbf{11} \xrightarrow[\text{PhOH as solvent}]{\mathbf{10}} \mathbf{12}$$

duration by osmium tetroxide. The light is emitted by an excited form of the end product, 10-methylacridanone (**13**), which is produced along with an equivalent amount of unexcited 10-methylacridanone by an electrocyclic opening of the four-membered ring in **14** (see also p. 385).

13 $\xrightarrow{POCl_3}$ ($\bar{O}_2PCl_2$) $\xrightarrow[\text{acetone}]{Zn}$ $\xrightarrow{HNO_3\text{ aq., boil}}$ **15** ($2NO_3^-$) $\xrightarrow{H_2O_2}$ **14** $\xrightarrow{\text{concerted ring opening}}$ 10-methylacridanone* $\xrightarrow{\text{chemiluminescence}}$ **13**

* = excited

C. Phenanthridine

Phenanthridine (**18**) is a colourless crystalline compound, m.p. 108°C, with a weak blue fluorescence in dilute ethanol.

Nitration in sulphuric acid using 1 mole of nitric acid gives a mixture of 1- and 10-nitrophenanthridine with smaller proportions of the 2-, 3-, 4-, and 8-isomers, and bromination in sulphuric acid, with silver sulphate, gives the 2-, 4-, and 10-bromo compounds in decreasing order of abundance.[73] The electrophilic substitution of this compound is not well understood. Nucleophilic attack with potassium amide in liquid ammonia, however, gives 6-aminophenanthridine, as expected from the isoquinoline analogy. Phenanthridine undergoes the Reissert reaction (p. 302) and is easily reduced (tin and hydrochloric acid) or hydrogenated over Raney nickel to give 5,6-dihydrophenanthridine. This last has the properties of an *N*-alkylaniline and, unlike 9,10-dihydroacridine (acridan, p. 334), is basic and forms *N*-acetyl and *N*-nitroso derivatives.

The best method of preparing phenanthridine (**18**) consists of cyclizing 2-formamidobiphenyl (**16**) with phosphorus oxychloride and stannic chloride in nitrobenzene; the iminochloride (**17**) is an intermediate. Many other 2-acylaminobiphenyls have been cyclised similarly. The photolysis of benzaldehyde anil (**19**) also gives phenanthridine.[74]

NH HC O **16** boil + $POCl_3$ + $SnCl_4$ in $PhNO_2$ N HC Cl **17** 1 10 4 5 N 7 6 **18** $h\nu$ PhCH=NPh **19**

Phenanthridone (**20**) is readily prepared and its structure and properties are similar to those of acridanone (p. 334) and the pyridones (p. 261). It is attacked by electrophilic reagents at positions 2 and 4 and with phosphorus oxychloride it gives 6-chlorophenanthridine (**21**), which possesses an active chlorine atom.

N C O polyphosphoric acid 2 4 NH O **20** $POCl_3$ N Cl **21**

Certain phenanthridinium salts are remarkably active against trypanosome infections, and following the initial discovery of activity by Browning in 1938, several hundred have been prepared and tested. Of these dimidium bromide (**22**)

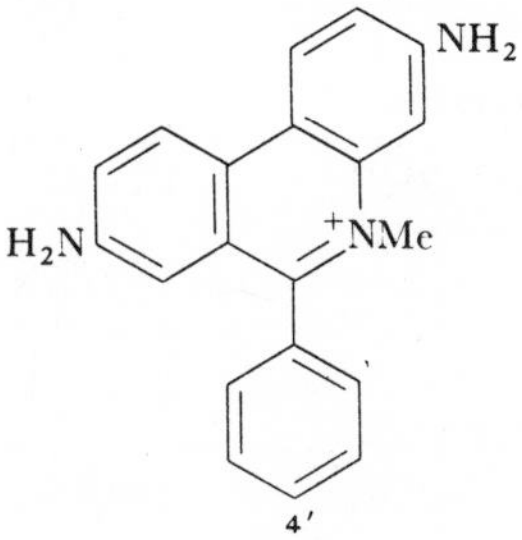

22

is currently used for treating *Trypanosoma congolense* in African cattle; its 4′-amino derivative is more active and is known as trimidium bromide.

5. BENZOPYRYLIUM SALTS AND BENZOPYRANS

A. Introduction

Many compounds belonging to these classes are found in the plant kingdom. Most red and blue flower petals contain anthocyanins, derivatives of the 1-benzopyrylium cation (**1**), as the coloured material, and the yellow benzo-2- (**2**) and benzo-4-pyrones (**3**) are also widely distributed. Sulphur analogues of **1**

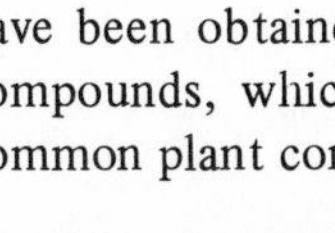

1 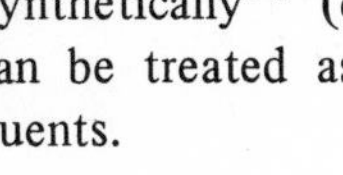 **2** 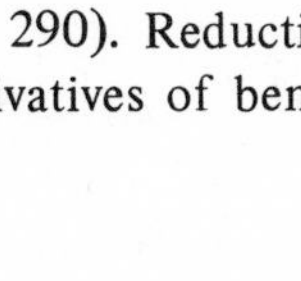 **3** 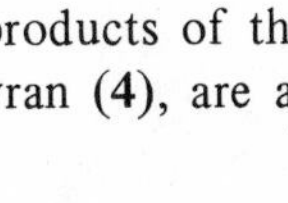 **4**

have been obtained synthetically[75] (cf. p. 290). Reduction products of these compounds, which can be treated as derivatives of benzopyran (**4**), are also common plant constituents.

B. The anthocyanins

The anthocyanidins are all hydroxylated derivatives of the 2-phenylbenzopyrylium (flavylium) cation and can also possess methoxyl groups. The common ones are pelargonidin (**5**), cyanidin (**6**), delphinidin (**7**), peonidin (**8**), and malvidin (**10**); petunidin (**9**) occurs rarely.

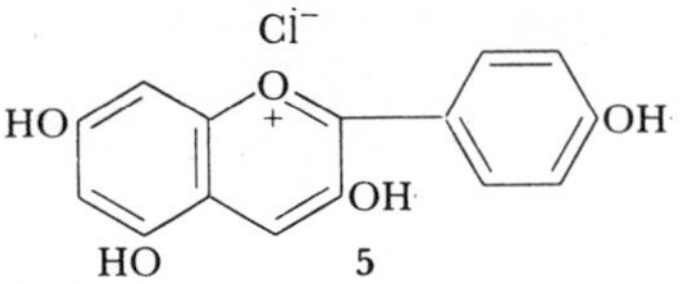

Anthocyanidins do not occur as such in plants but are present in the form of glycosides called anthocyanins. Those derived from cyanidin are most frequently encountered. Glycosidation occurs preferentially at the 3-hydroxyl, and the usual sugars involved are glucose, galactose, and rhamnose. A sugar group can also be attached to the 5-hydroxyl group, as in the case of cyanin (**11**), and other hydroxyl groups can also be glycosidated. Another complication is that the anthocyanin may be partially esterified with malonic, 4-hydroxybenzoic, or 2- or 4-hydroxycinnamic acids; the exact constitutions of most naturally occurring anthocyanins are unknown. Plants often contain mixtures of anthocyanins that may not be derived from the one anthocyanidin. Separation can be difficult, even with chromatographic methods.

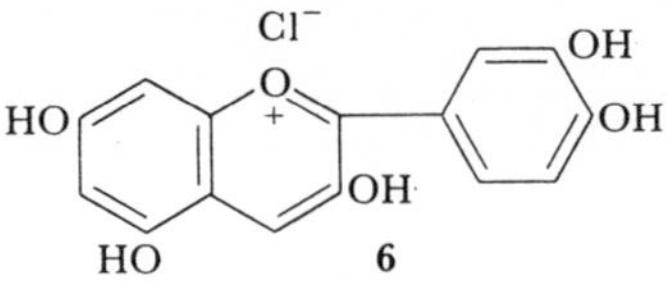

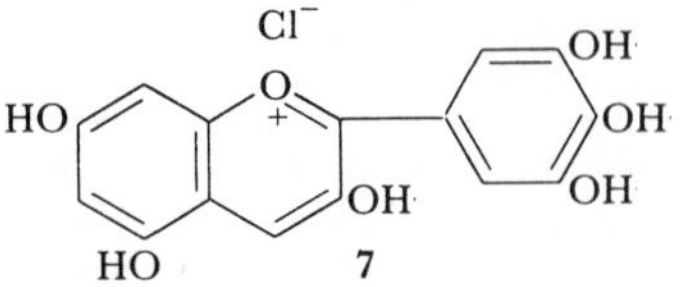

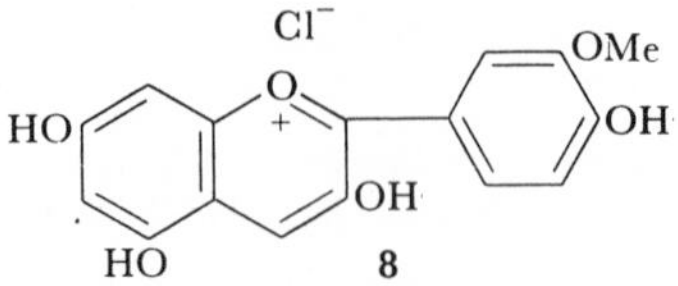

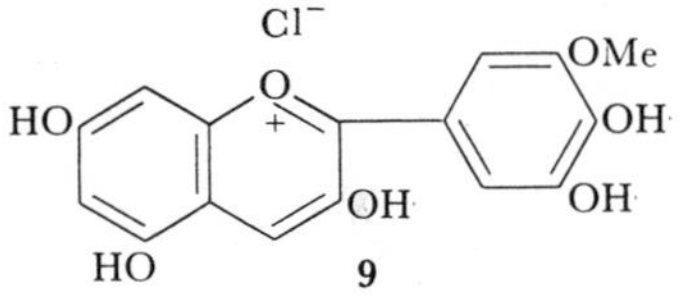

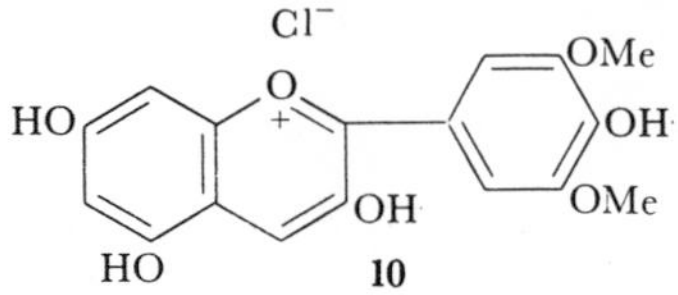

The chemistry of cyanin, cyanidin-3,5-β-diglucopyranoside (**11**), which can be isolated from *Centaurea cyanus* (cornflower) or *Rosa gallica* (rose), very closely resembles that of the other anthocyanins and is treated here in some detail as an example. Cyanin can be extracted from the petals by cold 1% aqueous

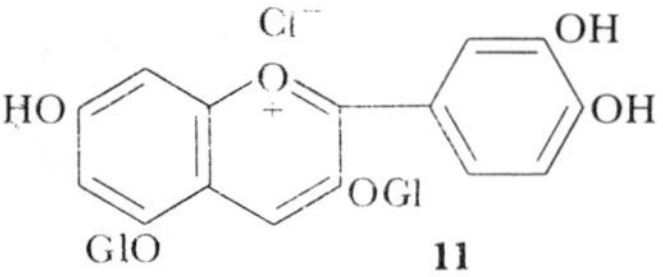

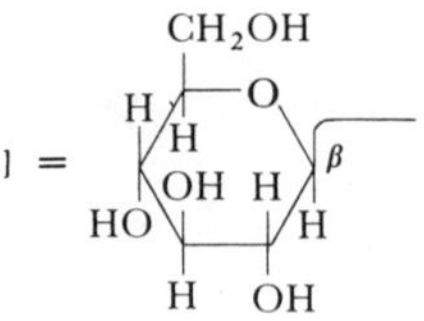

hydrochloric acid, or by acidified ethanol, and after much purification is obtainable as a very dark red, almost black, crystalline solid. Hydrolysis with boiling hydrochloric acid splits off the glucose, giving cyanidin (**6**), which can be extracted from the solution with amyl alcohol. The sugar remains in the aqueous layer. Chromatographic and colour reaction comparisons of both the sugar and anthocyanidin with authentic samples are used for their identification.

Cyanidin (chloride) is a pyrylium salt (cf. p. 281) and gives a deep red solution in dilute aqueous acid (pH 3). The cation (**12**) is best considered as a resonance hybrid to which other charged structures, such as **13**, contribute. On treating the solution with sodium acetate the colour changes to violet (pH 8), which slowly fades. This can be attributed to the formation of a violet quinonoid colour base (**15**), which adds water yielding the less conjugated (colourless) carbinol (**16**). Cyanidin with sodium hydroxide (pH 11) gives a deep blue solution, which is due to the formation of a resonant anion (possibly **14**–**14a**); acidification regenerates cyanidin.

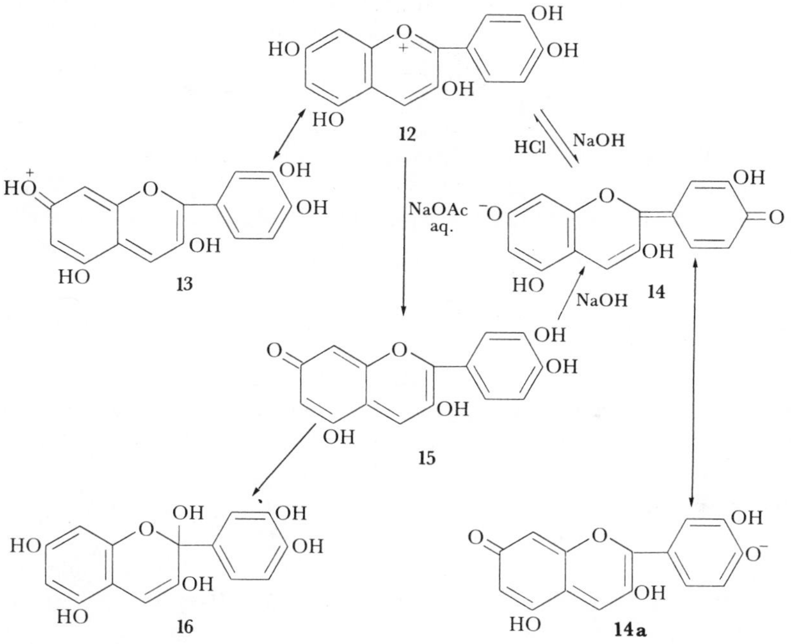

Anthocyanins and anthocyanidins with *ortho* hydroxyl groups give deep blue colours with ferric chloride at a slightly acid pH. This test is a useful diagnostic criterion, and the iron can be replaced by other metals, such as aluminium. The formation of such complexes is probably more important than the pH of the petal sap in determining flower-petal colours; other factors, such as the presence of copigments in the flower, are also important.

The structure of cyanidin (**12**) chloride is established by degradation, by its relationship to quercetin (**17**), and by synthesis. Fusion with alkali gives phloroglucinol and protocatechuic acid (**18**). Methoxy groups are usually split

HO, HO, OH, OH, OH, O **17**

Mg, HCl aq. or (a) $LiAlH_4$, (b) HCl aq.

Cl^-, HO, HO, OH, OH, OH **12**

NaOH fuse

HO, HO, OH + HO_2C, OH, OH **18**

from anthocyanidins under these conditions, and so milder methods, such as the use of aqueous baryta or sodium hydroxide, are used in their presence; malvidin (**10**) then gives 4-hydroxy-3,5-dimethoxybenzoic (syringic) acid. Cyanidin can also be obtained by the reduction of quercetin (**17**) and by Robinson's synthesis, which is the best available for this type of compound. Phloroglucinaldehyde, obtainable from phloroglucinol, hydrogen chloride, and hydrogen cyanide, on careful benzoylation gives the monobenzoyl derivative (**19**). This (**19**), with the triacetoxyacetophenone (**20**) is converted into 5-benzoylcyanidin chloride (**21**) with the loss of the phenolic acetoxy groups. Cold alkali, in the absence of oxygen, hydrolyses the benzoyl group and opens the ring. The product (**22**) is not isolated, but boiling with hydrochloric acid gives cyanidin chloride (**23**). The biosynthesis of cyanidin is discussed on p. 347.

The diglycoside cyanin (**11**) was synthesized similarly. β-Pentaacetyl-D-glucopyranose with hydrogen bromide gives α-bromo-*O*-tetraacetylglucose, which with phloroglucinaldehyde gives the β-tetraacetylglusose derivative (**24**). It is noteworthy that two (Walden) inversions of configuration occur at the 'aldehyde' carbon atom of the glucose molecule in these last two reactions. In a similar way **25** was obtained from the hydroxyacetophenone (**26**), α-bromo-*O*-tetraacetylglucose, and silver carbonate. **24** and **25** combined under the conditions successful in the cyanidin synthesis to give a flavylium salt (**27**) from which the remaining acetyl groups were removed as indicated, giving cyanin (**11**).

The hydrogenation of cyanidin over platinum yields (±)-epicatechin (**28**, substituents at 2,3-position *cis*), which possesses two asymmetric centres. (±)-Epicatechin has been converted into its geometrical isomeride (±)-catechin, and both have been isolated from plant sources in an optically active condition.

Many plants contain colourless materials called leucoanthocyanins or leucoanthocyanidins, which on boiling with dilute hydrochloric acid lose water and are converted into anthocyanidins. It appears that they are flavane-3,4-diols, and leucocyanidin can therefore be written as **29**. Flavane is a trivial name for 2-phenyl-2,3-dihydrobenzo-4-pyran. Proanthocyanidins, exemplified by

HO OH CHO OCOPh **19**
OAc OC OAc $AcOCH_2$ **20**
KOAc, Ac_2O
$ClCH_2CO$ OH OH
$POCl_3$, boil
Catechol + chloroacetyl chloride → OH $OCOCH_2Cl$
EtOAc, HCl gas
Cl^- HO O + OH OH OH OCOPh **21**
NaOH aq., under N_2
HO OH O C COH C H OH OH OH **22**
HCl aq. boil
Cl^- HO O + OH OH OH HO **23**

HO OH CHO OA—Gl **24**
+ OC OAc OAc CH_2OA—Gl **25**
$HOCH_2CO$ OAc OAc **26**
EtOAc, HCl
Cl^- O O + OH OH OA—Gl OA—Gl **27**
A—Gl (tetra-acetyl-glucosyl) = CH_2OAc O H H β OAc H AcO H H OAc
(a) NaOH under H_2
(b) HCl at 60°
Cl^- HO O + OH OH OGl OGl **11**
Gl (glucosyl) = CH_2OH O H H β OH H HO H H OH

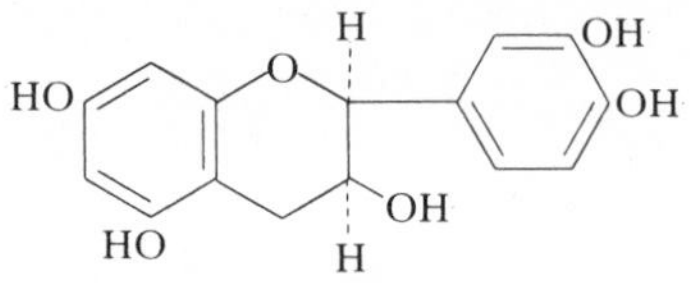

28

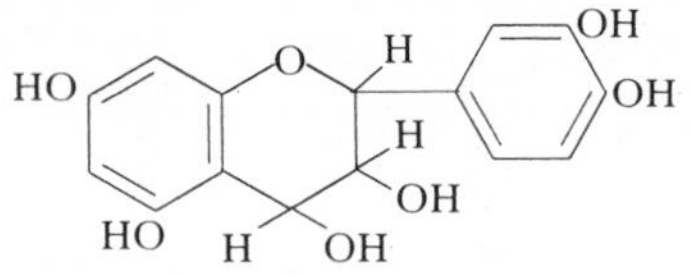

29

proanthocyanidin-A2,[76] form one of the major groups of plant polyphenols. They are complex compounds, and this particular one is hydrolysed by concentrated hydrochloric acid to cyanidin (**11**) and (–) epicatechin (**28**). D-γ-Tocopherol, the most active compound of the vitamin E (antisterility) group, is a complex benzopyran (**30**).[77]

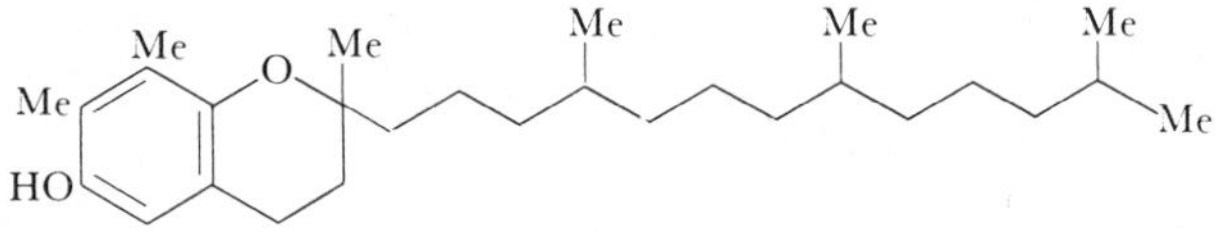

30

C. Benzo-2-pyrones

Benzo-2-pyrone, commonly called coumarin (**32**), is the sweet-smelling constituent of white clover. A considerable number of hydroxy- and methoxy-coumarins, and their glycosides, have been isolated from plant sources. Coumarin is the internal lactone of 2-hydroxy-*cis*-cinnamic acid, and the ring is

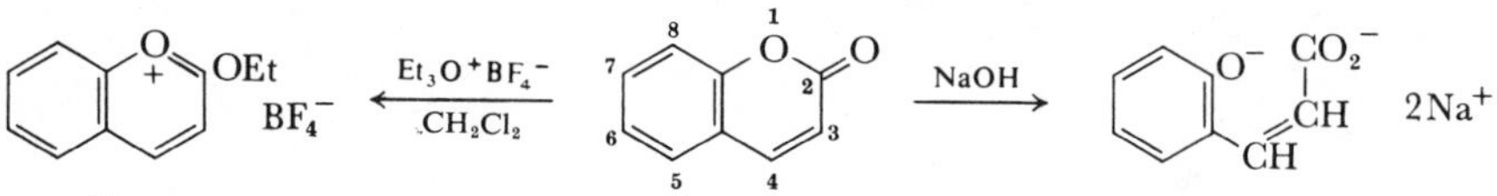

$Et_3O^+BF_4^-$, CH_2Cl_2; NaOH

31 32 33

opened with alkalis, giving salts of coumarinic acid (**33**). Bromine adds easily to the 3,4-double bond of coumarin, giving a dibromide which readily loses hydrogen bromide and can be readily converted into benzofuran (p. 218). The 3,4-double bond has considerable olefinic character and undergoes Diels–Alder additions, for example, with 2,3-dimethylbutadiene. Coumarin is not protonated by aqueous acids, but with triethyloxonium tetrafluoroborate it gives the 2-ethoxy-benzo[*b*]pyrylium salt **31**.[78] Chloromethylation takes place at position 3, the neutral molecule probably being attacked; however nitration and sulphonation cause 6-substitution and it is likely that a cation (cf. **31**) is being substituted. Grignard reagents can attack both the carbonyl group and the 4-position of coumarins, giving mixtures,[79] and Michael addition reactions occur at position 4. Coumarin therefore shows little aromatic character in the heterocyclic ring.

Coumarin can be synthesized from salicylaldehyde by a Perkin reaction. The other product is *O*-acetylcoumaric acid (**34**), which has the *trans* configuration.

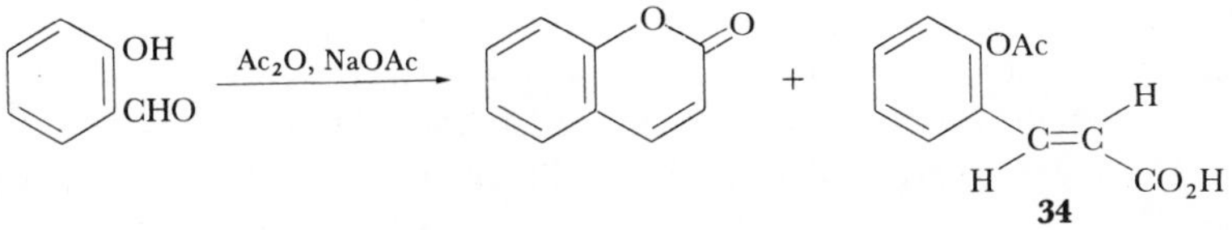

34

Another synthesis, due to von Pechmann,[80] is perhaps the most widely used for coumarins. Chromones can, however, also be obtained (p. 345). It consists of heating a phenol with a β-keto ester in concentrated sulphuric acid. A coumarinic ester (e.g., **35**) can be an intermediate in the synthesis. Best results

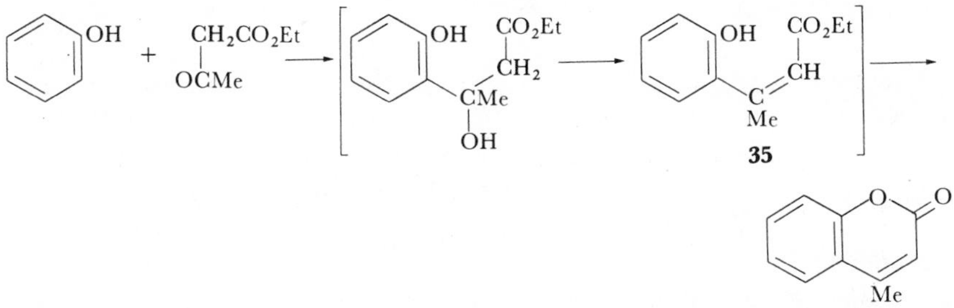

35

are obtained when the phenol possesses activating groups. Coumarin itself can be made in poor yield from phenol, and malic acid in sulphuric acid, a combination that gives β-formylacetic acid (p. 284).

Dicoumarol (**36**) is present in sweet clover. It is a blood anticoagulant and

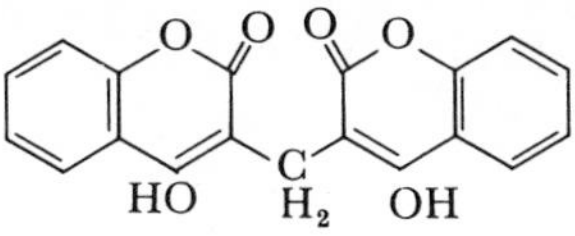

36

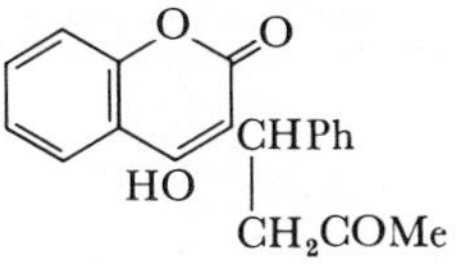

37

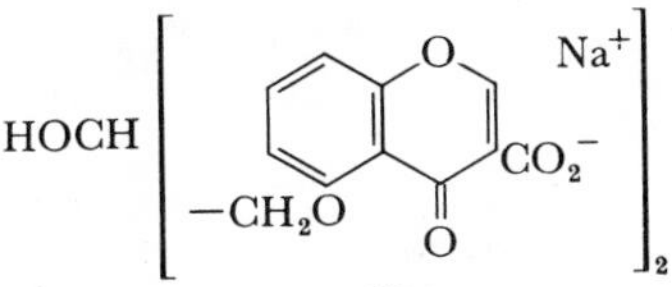

38

leads to the haemorrhagic sweet-clover disease that kills cattle. Warfarin (**37**) is a useful and powerful rodenticide that kills through its haemorrhagic properties. Intal (**38**) is a most valuable drug in the treatment of bronchial asthma.[81]

D. Benzo-4-pyrones

Benzo-4-pyrone, or chromone (**39**), has not yet been isolated from natural sources, although a few hydroxy derivatives occur in certain plants. In contrast, many 2-phenylbenzo-4-pyrones, usually called flavones (old name anthoxanthidins), and their glycosides (anthoxanthins) and a number of isoflavones (3-phenylbenzo-4-pyrones, cf. **59**) have been identified in plant extracts.

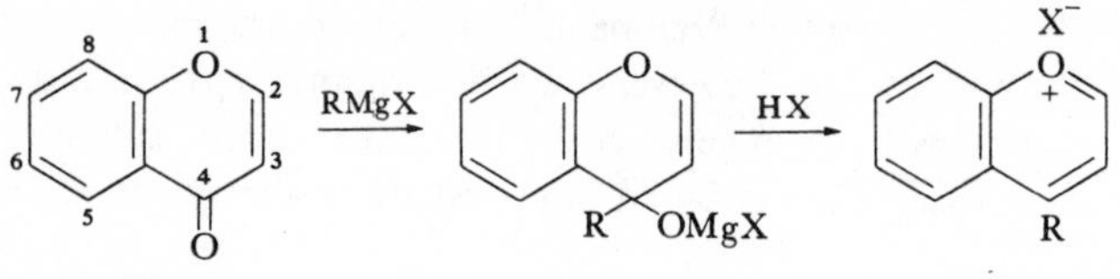

RMgX HX R OMgX X^- R

39 **40**

Chromone is best considered as a resonance hybrid analagous to that which represents 2,6-dimethyl-4-pyrone (p. 287). It behaves as an unreactive α,β-unsaturated ketone and gives an oxime, but not a phenylhydrazone, under ordinary conditions. The methyl group of 2-methylchromone is reactive, as is that of crotonaldehyde, and chromones with Grignard reagents (above) give benzopyrylium salts (e.g. **40**).

Flavone (**41**) has general properties similar to those of chromone. Degradation with alkali has proved a valuable method of structure determination for naturally occurring flavones. Flavone itself gives initially a β-diketone (**42**),

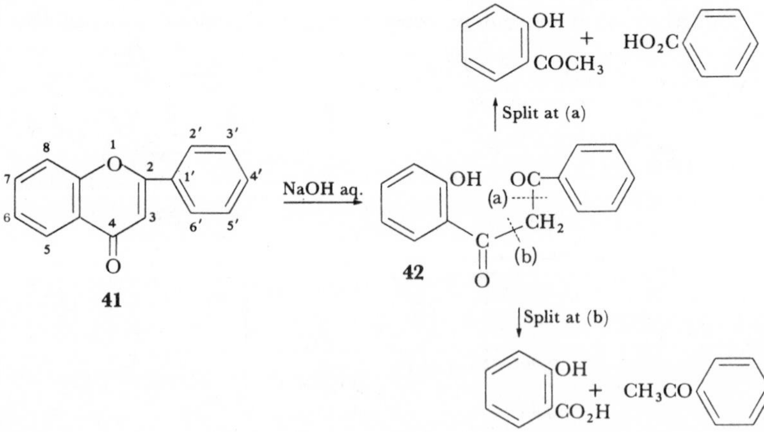

OH + HO_2C $COCH_3$

Split at (a)

NaOH aq. OH OC (a) CH_2 C (b) O

41 **42**

Split at (b)

OH + CH_3CO CO_2H

which then undergoes scission in both possible ways to yield four products. Identification of these products usually leads to a unique structure for the parent flavone.

Chromones and flavones can be synthesized by a number of methods, three of which are shown.

(1) 2-Methoxyacetophenone (**43**) undergoes Claisen reactions with many esters, and the products (**44**) can be cyclized. Chromone itself (**45**, R = H) is made in this way from ethyl formate, and the method has been used for the preparation of many flavones, including luteolin (**50**).

OMe COMe **43** $\xrightarrow[\text{NaOEt}]{RCO_2Et}$ OMe, O=CR, CH_2, C=O **44** $\xrightarrow{\text{HI}}$ O, R, O **45**

(2) 2-Hydroxyacetophenone with benzaldehyde and a basic catalyst gives the unsaturated ketone **46**. Cyclization with acid gives flavanone or 2-phenyl-chromanone (**47**). This type of cyclisation is also effected enzymically.[82] Bromination, presumably at position 3, followed by dehydrobromination, or oxidation by selenium dioxide or phosphorus pentachloride, gives flavone (**48**).

OH CHPh =CH C=O **46** $\xrightarrow[\text{EtOH}]{\text{HCl}}$ 1 O, Ph, 4, 3, O **47** $\xrightarrow{\text{oxidation}}$ O, Ph, O **48**

(3) Some chromones can be obtained from β-keto esters, phenols, and phosphorus pentoxide;[80] when sulphuric acid is used as a condensing agent coumarins are usually formed (p. 343).

OH + O=CMe, CHMe, EtO_2C $\xrightarrow{P_2O_5}$ O, Me, Me, O **49**

Most naturally occurring flavones are hydroxylated at positions 5 and 7. An example is luteolin (**50**), a yellow compound that was at one time extracted from *Reseda luteola* (wild wood) and used as a dyestuff. Its structure was deduced from the products of alkali fusion. These were 3,4-dihydroxyacetophenone, 3,4-dihydroxybenzoic (protocatechuic) acid, and phloroglucinol, the last presumably being derived from phloroglucinolcarboxylic acid and 2,4,6-trihydroxyacetophenone, which would be formed first (cf. alkali degradation of

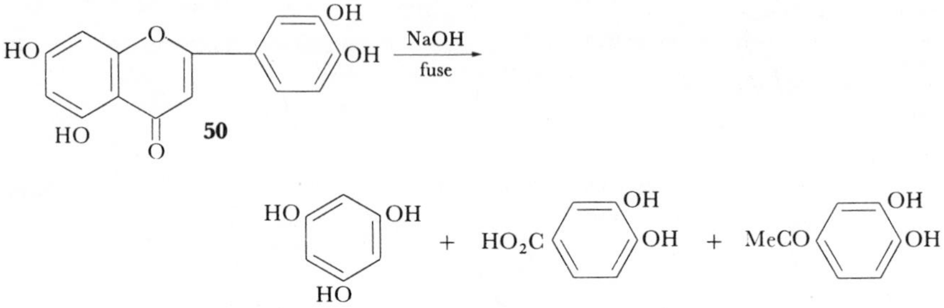

HO, HO, O, O, OH, OH, **50** $\xrightarrow[\text{fuse}]{\text{NaOH}}$ HO, OH, HO + HO_2C, OH, OH + MeCO, OH, OH

flavone, above). Luteolin has been synthesized by method (1) above from 2,4,6-trimethoxyacetophenone and ethyl veratrate (3,4-dimethoxybenzoate). All the methoxy groups are demethylated at the hydrogen iodide stage.

A related pentahydroxyflavone is quercetin (**53**), which is one of the most widely distributed natural yellow pigments. It often occurs in the form of

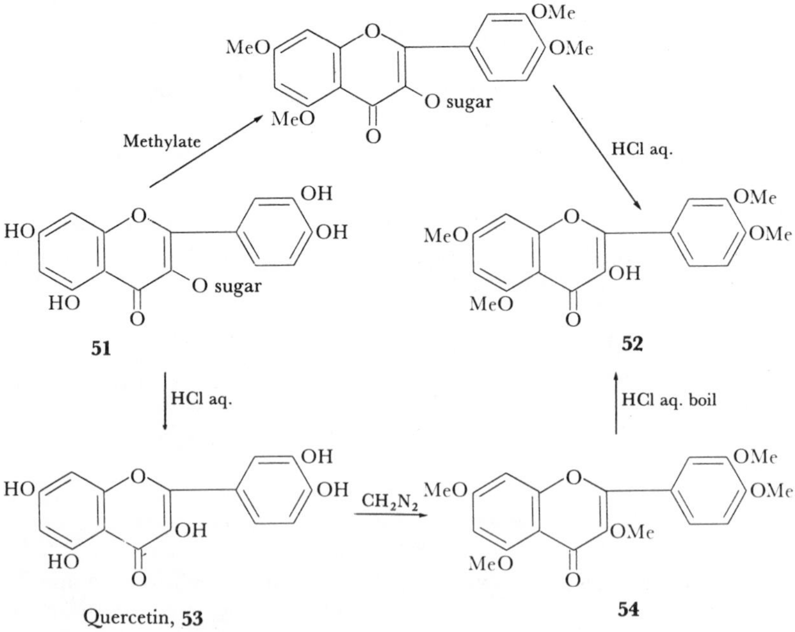

MeO, MeO, O, O, O sugar, OMe, OMe

Methylate

HCl aq.

HO, HO, O, O, O sugar, OH, OH

51

MeO, MeO, O, O, OH, OMe, OMe

52

HCl aq.

HCl aq. boil

HO, HO, O, O, OH, OH, OH

Quercetin, **53**

CH_2N_2

MeO, MeO, O, O, OMe, OMe, OMe

54

glycosides, for instance, the 3-glucoside is found in maize and a 3-rhamnoglucoside, rutin, occurs in many plants. Rutin is of medical interest, as it prevents capillary fragility. The positions of the sugar groups in a flavone glycoside (e.g. **51**) are found by complete methylation of all the free hydroxyl groups, often with diazomethane, followed by acid hydrolysis which removes only the sugar residues. The positions of the free hydroxy groups in the resulting aglycone (**52**) are then determined by degradation or comparison with synthetic materials and correspond to the position(s) of glycosidation. The 3-*O*-methyl group of quercetin pentamethyl ether (**54**) is quite readily hydrolysed. The only recognizable fragments from the alkali fusion of quercetin are phloroglucinol and protocatechuic acid. Quercetin has been synthesized from 3,4-dimethoxybenzaldehyde (veratraldehyde) and 2,4,6-trimethoxyacetophenone, which were

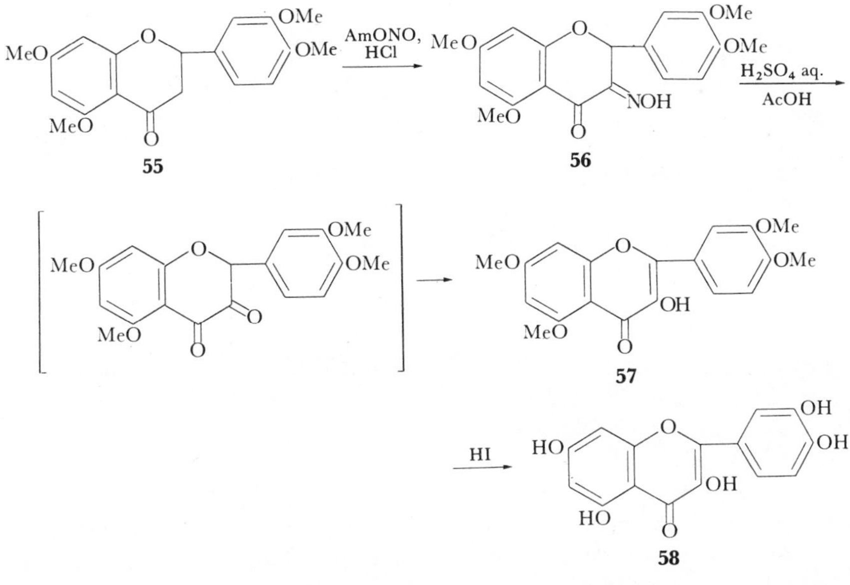

first converted into the flavanone (**55**) by method (2) on p. 345. Amyl nitrite and hydrochloric acid attack the activated methylene group of this flavanone (**55**) in the usual way, forming an oxime (**56**). Hydrolysis gives a diketone which tautomerizes to **57** and on demethylation yields quercetin (**58**).

The positions of the hydroxyl groups of quercetin and cyanidin are identical. Cyanidin has not been converted into quercetin, although the reverse transformation can be effected by powerful reagents (p. 340). Because of these, and other reasons, Robinson suggested many years ago that these compounds were formed by parallel, but not identical, routes in plants, that ring A was derived from phloroglucinol and joined directly onto a C_6–C_3 fragment which completed the molecule. In support of this, Birch showed that added radioactive phloroglucinol and phenylalanine (converted en route to 3,4-dihydroxycinnamic acid) were incorporated into quercetin synthesized by *Chlamydomonas eagamentos*. He also suggested that the phloroglucinol part (ring A) could be built up from three acetic acid molecules, and this has been confirmed[83] for quercetin synthesized by *Fagopyrun tataricum* (buckwheat). By growing wheat cuttings in the presence of methyl-labelled and carboxyl-labelled acetic acid, and

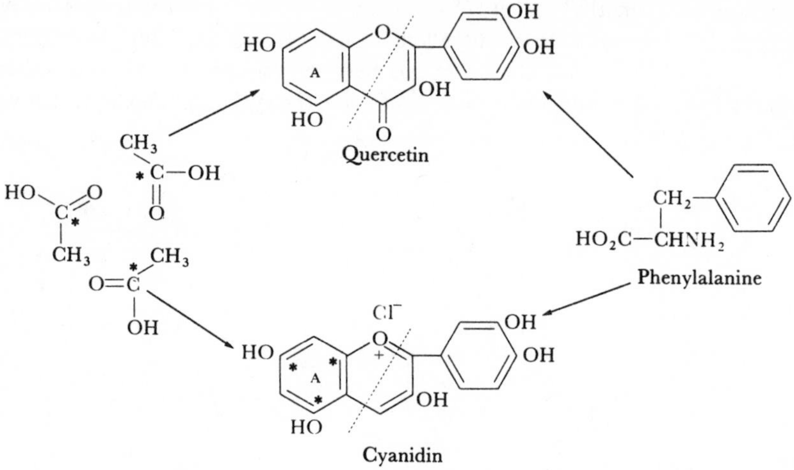

Quercetin

Phenylalanine

Cyanidin

phenylalanine, respectively, it has been shown that all the atoms of ring A can be derived from acetic acid and all the carbon atoms of the rest of the molecule can be derived from phenylalanine. Exactly similar results were obtained in studies of the biosynthesis of cyanidin by red cabbage. In support of Birch's theory it was also shown that the carbon atoms bearing the hydroxyl groups of the phloroglucinol obtained on degradation of the cyanidin were those of the carboxyl groups of labelled acetic acid used. Cyanidin and quercetin are therefore built up from the acetic acid and phenylalanine carbon atoms as shown, but the mechanism by which this is done is not yet fully understood.[84]

Relatively few derivatives of isoflavone (**59**), compared with those from flavone, have been isolated from plant sources. They are formed *in vivo* from phenylalanine, like flavones, but a 1,2-shift of the phenyl group occurs at some stage. They often occur as glycosides, which are readily hydrolysed to the aglycones by dilute acid. The structures of these aglycones have usually been elucidated from those of the products of alkaline degradations; that of isoflavone itself is shown. Formic acid is obtained quantitatively in the first stage.

Isoflavones can be made by a 'reversal' of the alkaline degradation. Poor

59 —(KOH, EtOH, boil)→ [2-hydroxyphenyl–C(=O)–C(Ph)=CHOH] →

HCO_2H + 2-hydroxyphenyl–$COCH_2Ph$ —(NaOH, fuse)→ phenol + HO_2CCH_2Ph

yields are generally obtained, especially if hydroxyl groups are present, as they generally are with naturally occurring isoflavones. An intermediate hydroxy compound (e.g. **60**) has been isolated in a few instances.

2-hydroxyphenyl–C(=O)–CH_2Ph —(HCO_2Et, Na)→ [**60**] → **59**

The synthesis of genistein (**62**, 5,7,4′-trihydroxyisoflavone) by a very much better method, using ethyl oxalyl chloride, is outlined. The first product (**61**) is readily hydrolysed by sodium carbonate and the resulting acid decarboxylated

2,4,6-trihydroxyphenyl–C(=O)–CH_2–(4-hydroxyphenyl) —($ClCOCO_2Et$, pyridine)→ **61** (5,7,4′-trihydroxyisoflavone-2-CO_2Et) —(hydrolyse, heat to near m.p.)→ **62**

to give a 50% overall yield. Genistein is of particular interest in that it is an oestrogen about 10^{-5} times as active as oestrone. This was discovered during an investigation of an outbreak of sheep infertility in Australia. The sheep were eating an unusual amount of clover, which caused the trouble, and the active agent was genistein.

GENERAL BIBLIOGRAPHY

QUINOLINE AND ISOQUINOLINE

H. D. Springall and I. T. Millar, *Sidgwick's Organic Chemistry of Nitrogen*, Oxford University Press, London, 1966.

R. C. Elderfield and W. J. Gensler, in *Heterocyclic Compounds*, ed. R. C. Elderfield, Vol. 4, Wiley, New York, 1952.

N. Campbell, in *Chemistry of Carbon Compounds*, ed. E. H. Rodd, Vol. IVA, Elsevier, New York, 1957.

J. M. Bobbitt, *4-Keto-1,2,3,4-tetrahydroisoquinolines, Adv. Heterocycl. Chem.*, **15**, 99 (1973).

QUINOLIZINE

R. M. Acheson, *Adv. Heterocycl. Chem.* **1**, 125 (1963).

B. S. Thyagarajan, *Adv. Heterocycl. Chem.*, **5**, 291 (1965).

ACRIDINE

R. M. Acheson, *Acridines,* 2nd ed., Interscience, Wiley, New York, 1973.

PHENANTHRIDINE

C. F. H. Allen, *Six-Membered Heterocyclic Nitrogen Compounds with Three Condensed Rings*, Interscience, New York, 1958.

BENZOPYRANS, BENZOPYRONES, AND BENZOPYRYLIUM SALTS

S. Wawzonek, in *Heterocyclic Compounds*, ed. R. C. Elderfield, Vol. 2, Wiley, New York, 1950.

K. P. Link, in H. Gilman, *Advanced Organic Chemistry*, Vol. 2, 2nd ed., Wiley, New York, 1942.

T. A. Geissman (ed.), *The Chemistry of the Flavonoid Compounds*, Pergamon, Oxford, 1962.

R. D. Barry, 'Isocoumarins,' *Chem. Rev.*, **64**, 229 (1964).

L. Jurd, in *Advances in the Chemistry of Plant Pigments*, ed. C. O. Chichester, Supplement 3 to *Advances in Food Research*, Academic Press, New York, 1972.

H. Grisebach, *Biosynthetic Patterns in Microorganisms and Higher Plants*, Wiley, New York, 1967.

J. B. Harborne, T. J. Mabry, and H. Mabry (eds.) *The Flavonoids*, Chapman and Hall, London, 1975.

W. K. Warburton, 'Isoflavones,' *Quart. Rev.*, **8**, 67 (1954).

REFERENCES

1. F. Klages, *Chem. Ber.*, **82**, 358 (1949).
2. D. W. J. Cruickshank, *Acta Crystallogr.*, **10**, 504 (1957).
3. R. D. Brown and R. D. Harcourt, *J. Chem. Soc.*, 3451 (1959).
4. F. D. Popp, *Adv. Heterocycl. Chem.*, **9**, 1 (1968).

4a. R. E. Lyle and P. S. Anderson, *Adv. Heterocycl. Chem.*, **6**, 45 (1966).

5. W. Hückel and L. Hagedorn, *Chem. Ber.*, **90**, 752 (1957).
6. W. A. Remers, G. J. Gibs, C. Pidacks, and M. J. Weiss, *J. Org. Chem.*, **36**, 279 (1971).
7. A. J. Birch and P. G. Lehman, *J. C. S. Perkin I,* 2754 (1973).
8. C. E. Crawforth, O. Meth-Cohen, C. A. Russel, *J. C. S. Perkin I*, 2807 (1972); *Chem. Comm.*, 259 (1972).
9. D. A. Archer, H. Booth, and P. C. Crisp, *J. Chem. Soc.*, 269 (1964) and earlier papers.
10. P. B. D. La Mare, M. Kiamud-din, and J. H. Ridd, *Chem. Ind. (Lond.*), 361 (1958).
11. M. Kiamud-din and M. E. Haque, *Chem. Ind.* (*Lond.*), 1753 (1964); M. Kiamud-din and A. K. Choudhury, *Chem. Ind.* (*Lond.*), 1840 (1963).

12. M. J. S. Dewar and P. M. Maitlis, *J. Chem. Soc.,* 2521 (1957) and earlier papers.
13. M. W. Austin and J. H. Ridd, *J. Chem. Soc.,* 4204 (1963).
14. M. J. S. Dewar and P. M. Maitlis, *J. Chem. Soc.,* 944 (1957).
15. R. D. Brown, B. A. Coller and R. D. Harcourt, *Australian J. Chem.*, **14**, 643 (1961).
16. J. J. Eisch and B. Jaselskis, *J. Org. Chem.,* **28**, 2865 (1963).
17. J. J. Eisch, *J. Org. Chem.,* **27**, 1318 (1962).
18. J. J. Eisch, *J. Org. Chem.,* **27**, 4682 (1962).
19. K. C. Bass and P. Nababsingh, *J. Chem. Soc.* (*C*), 338 (1969).
20. T. Caronna, G. P. Gardini, and F. Minisci, *Chem. Comm.,* 201, (1969).
21. E. Ochiai, *J. Org. Chem.,* 18, 534 (1953); A. R. Katritzky, and J. M. Lagowski, *Chemistry of the Heterocyclic* N-*Oxides*, Academic Press, New York, 1971.
22. O. Burchardt, P. L. Kamler, and C. Lohse, *Acta Chem. Scand.*, **23**, 159 (1969) and earlier papers.
23. K. R. Brower, W. P. Samuels, J. W. Way, and E. D. Amstutz, *J. Org. Chem.,* 18, 1648 (1953).
24. R. E. Lyle and M. J. Kane, *J. Org. Chem.,* **38**, 3740 (1973).
25. R. H. Reitsema, *Chem. Rev.,* **43,** 43 (1948).
26. B. E. Halcrow and W. O. Kermack, *J. Chem. Soc.*, 415 (1945); K. Schofield and T. Swain, *J. Chem. Soc.*, 1367 (1949).
27. J. P. Phillips, *Chem. Rev.,* **56**, 271 (1956).
28. R. H. F. Manske and M. Kulka, *Org. React.,* **7**, 59 (1953).
29. G. M. Badger, H. P. Crocker, B. C. Ennis, J. A. Gayler, W. E. Matthews, W. C. G. Paper, E. C. Samuel, and J. M. Spotswood, *Australian J. Chem.,* **16**, 814 (1963).
30. R. C. Elderfield, W. J. Gensler, T. A. Williamson, J. M. Griffing, S. M. Kupchan, J. T. Maynard, F. J. Kreysa, and J. B. Wright, *J. Amer. Chem. Soc.,* **68**, 1584 (1946).
31. T. P. Forrest, G. A. Dauphinee, and W. F. Miles, *Can. J. Chem.,* **47**, 2121 (1969).
32. E. J. Zobain, W. S. Kelley, and H. C. Dunathan, *J. Org. Chem.,* **29**, 584 (1964).
33. E. A. Fehnel, *J. Org. Chem.,* **31**, 2899 (1966).
34. N. D. Heindel, I. S. Bechara, T. F. Lemke, and V. B. Fish, *J. Org. Chem.,* **32**, 4155 (1967).
35. F. D. Popp and P. Schuyler, *J. Chem. Soc.,* 522 (1964).
36. J. M. F. Gagan and D. Lloyd, *Chem. Comm.*, 1043 (1967).
37. K. W. Bentley, *The Alkaloids*, Interscience, New York 1957.
38. H. R. Ing, in *Organic Chemistry*, ed. H. Gilman, Vol. III, Wiley, New York, 1953, p. 449; D. W. Henry, in *The Acridines*, ed. R. M. Acheson, Interscience – Wiley, New York, 1973.
39. M. E. Wall. M. C. Wani, C. E. Cook, K. H. Palmer, A. T. McPhail, and G. A. Sim. *J. Amer. Chem. Soc.,* **88**, 3888 (1966).
40. L. N. Ferguson, *Chem. Soc. Rev.,* **14**, 289 (1975).
41. K. Venkataraman, Synthetic Dyes, 2nd ed., Academic Press, New York, 1971.
42. F. M. Hamer, *The Cyanine Dyes and Related Compounds*, Interscience, New York, 1963.
43. F. Genet, *Bull. Soc. Fr. Mineral. Crystallogr.,* **88**, 463 (1965).
44. W. Hückel and G. Graner, *Chem. Ber.,* **90**, 2017 (1957).
45. S. F. Dyke, *Adv. Heterocycl. Chem.,* **14**, 279 (1972).
46. K. T. Potts, *J. Chem. Soc.*, 1269 (1956).
47. R. D. Brown and R. D. Harcourt, *Tetrahedron,* 8, 23 (1960).
48. R. B. Moddie, K. Schofield, and M. J. Williamson, *Chem. Ind.* (*Lond.*), 1283 (1963).
49. M. J. S. Dewar and P. M. Maitlis, *J. Chem. Soc.*, 2521 (1957).
50. M. Kiamud-din, in *Physical Methods in Heterocyclic Chemistry*, ed. A. R. Katritzky, Vol. I, Academic Press, New York 1963, p. 124.
51. H. Erlenmeyer, H. Baumann, and E. Sarkin, *Helv. Chim. Acta,* **31**, 1978 (1948).

52. R. D. Haworth and S. Robinson, *J. Chem. Soc.*, 777 (1948).
53. W. M. Whaley and T. R. Govindachari, *Org. React.*, **6**, 74, 151 (1951); W. J. Gensler, *Org. React.*, **6**, 191 (1951).
54. N. Itoh and S. Sugasawa, *Tetrahedron,* **1**, 45 (1957).
55. F. Johnson and R. Madronero, *Adv. Heterocycl. Chem.*, **6**, 95 (1966); A. Hassner, R. A. Arnold, R. Gault, and A. Terada, *Tetrahedron Lett.*, 1241 (1968).
56. A. O. Fitton, J. R. Frost, M. M. Zakaria, and G. Andrew, *Chem. Comm.*, 889 (1973).
57. A. J. Birch, A. H. Jackson and P. U. R. Shannon, *J. C. S. Perkin I*, 2185 (1974).
58. E. Schlittler and J. Müller, *Helv. Chim. Acta,* **31**, 1119 (1948).
59. V. Boekelheide and W. G. Gall, *J. Amer. Chem. Soc.,* **76**, 1832 (1954).
60. E. E. Glover and G. Jones, *J. Chem. Soc.*, 3021 (1958).
61. R. M. Acheson and D. M. Goodall, *J. Chem. Soc.,* 3225 (1964).
62. A. Richards and T. S. Stevens, *J. Chem. Soc.,* 3067 (1958).
63. R. M. Acheson, *Adv. Heterocycl. Chem.,* **1**, 125 (1963).
64. T. Miyadera and Y. Kishida, *Tetrahedron Lett.,* 905 (1965).
65. T. Miyadera, *Chem. Pharm. Bull (Japan)*, **13**, 503 (1965).
66. T. Miyadera and R. Tachikawa, *Tetrahedron,* **25**, 837 (1969); F. Kröhnke and D. Mörler, *Tetrahedron Lett.,* 3441 (1969).
67. R. M. Acheson and B. J. Jones, *J. Chem. Soc. (C)*, 1301 (1970).
67a. P. J. Abbott, R. M. Acheson U. Eisner, D. J. Watkin and J. R. Carruthers, *J. C. S. Perkin I,* (1976).
68. P. A. Duke, A. Fozard and G. Jones, *J. Org. Chem.*, **30**, 526 (1965) and earlier papers.
69. V. Boekelheide and J. P. Lodge, *J. Amer. Chem. Soc.,* **73**, 3681 (1951).
70. B. S. Thyagarajan and P. V. Gopalakrishnan, *Tetrahedron,* **21**, 945 (1965).
71. G. A. Reynolds and J. A. Van Allan, *J. Org. Chem.,* **28**, 527 (1963).
72. T. M. Moynehan, K. Schofield, R. A. Y. Jones, and A. R. Katritzky, *J. Chem. Soc.*, 2637 (1962).
72a. A. T. Amos and G. G. Hall, *Mol. Phys.*, **4**, 25 (1961).
73. G. S. Chandler, *Australian J. Chem.,* **22**, 1105 (1969).
74. C. E. Loader and C. J. Timmons, *J. Chem. Soc. (C)*, 1078 (1966).
75. N. Engelhard and A. Kolb, *Annalen.,* **673**, 136 (1964): I. Degani, R. Fochi, and C. Vincenzi, *Gaz. Chim. Ital.*, **94**, 451 (1964).
76. D. Jacques, *J. C. S. Perkin I,* 2663 (1974); E. Haslam, G. R. Bedford and D. Greatbanks, *Chem. Comm.*, 518 (1973).
77. H. Mayer, P. Schudel, R. Rüegg, and O. Isler, *Helv. Chim. Acta,* **46**, 963 (1963); O. Isler, P. Schudel, H. Mayer, J. Würsch, and R. Rüegg, *Vitamins Hormones,* **20**, 389 (1962).
78. H. Meerwein, G. Hinz, P. Hofmann, E. Kroning, and E. Pfeil, *J. Prakt. Chem.,* **147**, 257 (1937).
79. R. W. Tickle, T. Melton, and J. A. Elvidge, *J. C. S. Perkin I*, 569 (1974).
80. S. Sethna and R. Phadke, *Org. React.*, **7**, 1 (1953).
81. J. Pepys and A. W. Frankland (eds.), *Disodium Chromoglycate in Allergic Airways Disease*, Butterworths, London 1970.
82. E. Wong and E. Moustapha, *Tetrahedron Lett.*, 3021 (1966).
83. J. E. Watkin, E. W. Underhill, and A. C. Neish, *Can. J. Biochem. Physiol.,* **35**, 219, 229 (1957).
84. G. Billek (ed.), *Biosynthesis of Aromatic Compounds*, Pergamon, London, 1966.

CHAPTER VII

COMPOUNDS WITH TWO HETEROATOMS IN A FIVE-MEMBERED RING

The great majority of the known compounds with two heteroatoms in a five-membered ring are aromatic and can be derived formally from pyrrole, furan, or thiophene by the replacement of one of the methine (=CH) groups by a nitrogen atom. They can be derived similarly from pyridine by replacing two methine groups (–CH=CH–) by an imino group (NH), or oxygen, or sulphur atoms, and the properties of the ring systems produced are, very roughly, an amalgamation of those of the parents. Partially reduced and fully reduced derivatives of the above ring systems are known. The dihydro (e.g., dihydroimidazole) and tetrahydro (e.g., tetrahydroimidazole) derivatives are named after

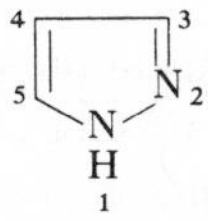

Pyrazole 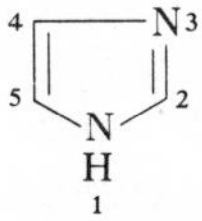 | Imidazole 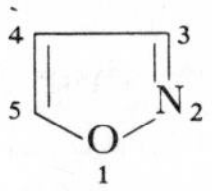 | Isoxazole 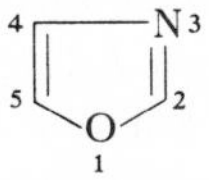 | Oxazole

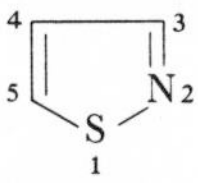

Isothiazole 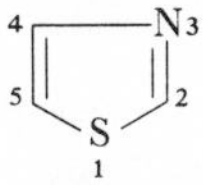 | Thiazole

the corresponding aromatic compound (e.g., imidazole) by replacing the -ole by -oline (e.g., imidazoline) and -olidine (e.g., imidazolidine), respectively. Three isomeric imidazolines can be formulated, differing in the position of the double bond (cf. pyrrolines, p. 90).

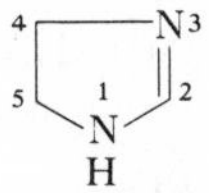

2-Imidazoline 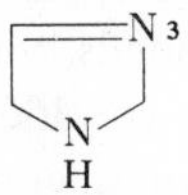 | 3-Imidazoline 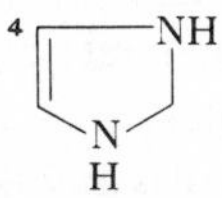 | 4-Imidazoline 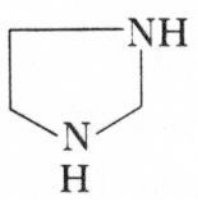 | Imidazolidine

Where both heteroatoms are oxygen or sulphur it is not possible to have an uncharged aromatic ring, and compounds of these types have been studied less intensively than those where the ring contains nitrogen. The basic ring systems are named systematically, the positions of the heteroatoms are always given, and the endings -ole and -olane are used, respectively, for the unsaturated and saturated rings.

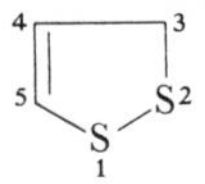

1,2-Dithiole 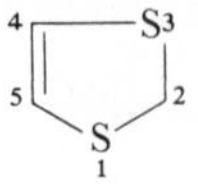 1,3-Dithiole 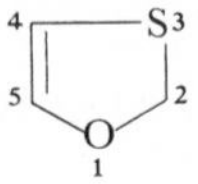 1,3-Oxthiole 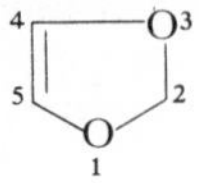 1,3-Dioxole 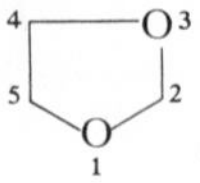 1,3-Dioxolane

The numbering of rings containing two heteroatoms follows the rules that the lowest numbers of the possible alternatives are used and that oxygen, sulphur, nitrogen (NH), and nitrogen (=N–) take this order of precedence (see p. 2). Certain difficulties can, however, arise in naming pyrazoles and imidazoles, and are discussed on the next page, in connexion with these compounds.

1. PYRAZOLE AND IMIDAZOLE

A. Introduction

As early as 1884 Knorr discovered the antipyretic (temperature-reducing) action of a pyrazole derivative in man. He named the compound antipyrine (p. 363). This stimulated interest in pyrazole chemistry, and although antipyrine is now little used, a number of pharmaceuticals and dyestuffs contain the ring system.

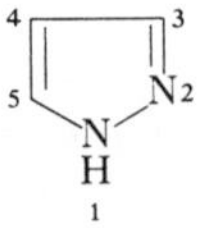

Pyrazole 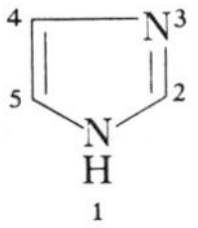 Imidazole

Imidazole was first prepared in 1858 from glyoxal and ammonia, and called glyoxaline. This name is obsolete, and imidazole or iminazole is greatly preferred. In contrast to pyrazoles, which occur infrequently in nature, certain imidazoles are very important in living systems. Examples are the essential amino acid histidine (p. 365) and related compounds, vitamin B_{12} (p. 121), biotin (p. 171), 5-aminoimidazole-4-carboxyamide (p. 366), and the pilocarpine alkaloids. A number of imidazoles are useful chemotherapeutic agents.

B. Physical properties and structure

Pyrazole, m.p. 70°C, and imidazole, m.p. 90°C, are both highly crystalline solids which are soluble in water but almost insoluble in petroleum ether. They are both tautomeric, as either nitrogen atom of each ring can bear the hydrogen atom. An example is methylimidazole, for which structures **1** (4-methylimidazole) and **2** (5-methylimidazole) can be written. As only one substance,

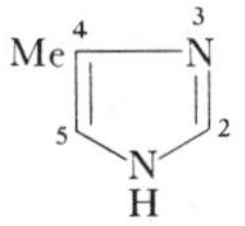

1 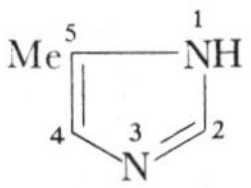 2 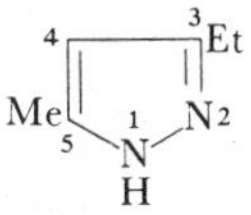 3 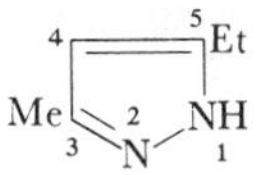 4

which may be a single tautomer or an equilibrium mixture, is known, it would be incorrect to call it either 4- or 5-methylimidazole, as these names imply the definite structures given above. It is therefore called 4(5)-methylimidazole to avoid the difficulty. In a similar way the pyrazole **3**, and/or **4**, is named 3(5)-ethyl-5(3)-methylpyrazole. There is no evidence suggesting that 'pyrrolenine' tautomers (cf. pyrrole, pp. 90 and 201) of pyrazole or imidazole exist in measurable concentrations.

TABLE 1. Boiling point and pK_a values of nitrogen heterocyclic compounds

Compound	Boiling point °C at 760 mm	Basic pK_a	Acidic pK_a
Pyridine	115	5·6	–
Pyrrole	130–131	–3·8	17·5
1-Methylpyrazole	127	2·1	–
Pyrazole	187	2·5	14
1-Methylimidazole	199	7·4	–
Imidazole	256	7·2	14·5

The boiling points in the Table 1 are easily accounted for on the basis of hydrogen bonding, which is absent in pyridine and present to a very small extent in pyrrole. 1-Methylpyrazole and 1-methylimidazole both boil much below the parent compounds, which shows the importance of the hydrogen atom at position 1. The large difference in boiling point between pyrazole and imidazole is remarkable. Both compounds associate in solution, but imidazole and some derivatives can give as little as 5% of the expected freezing-point depression in

5

6

benzene. This shows that aggregates of the order of 20 molecules are formed. The results are best accounted for on the assumption that imidazole associates polymolecularly as in **5**, while pyrazole forms largely a bimolecular complex **6**. Imidazole does not associate in water.

The pyrazole ring is present in two different environments in the crystal, and averaged molecular dimensions from the X-ray data[1] are shown. The dimensions for imidazole itself have not been determined, but those for the imidazole ring in its complex with 5,5-diethylbarbituric acid are noted below.[2] The only C–C

Pyrazole, mean bond lengths (Å)

Imidazole bond lengths (Å) in 1 : 1 complex with 5,5-diethylbarbituric acid

bond in the ring is shorter than the normal benzenoid distance (1·395 Å). The resonance energies of pyrazole and imidazole have been calculated[3] from heats of combustion data as 123 kJ (29·3 kcal)/mole and 59·5 kJ (14·2 kcal)/mole, respectively. The chemical reactions of the compounds are also consistent with the presence of a great deal of aromatic character. They suggest that the compounds are best considered as resonance hybrids to which charged structures such as **7**, and **8** and **8a**, contribute in the two cases. This is in agreement with electron-density calculations, which suggest that electrophilic attack should occur at position 4 in pyrazole and in imidazole.

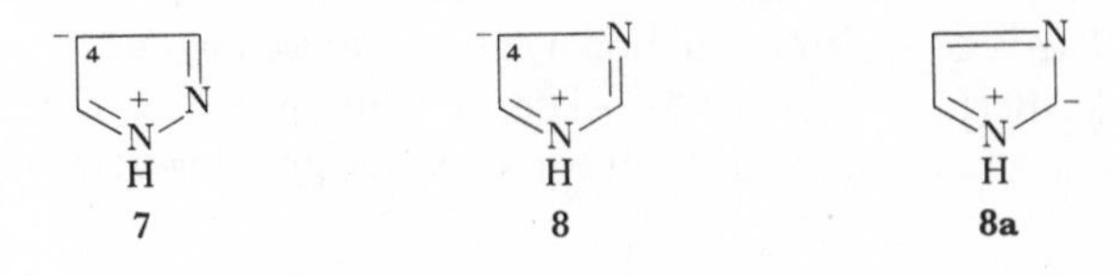

C. Chemical properties and synthetic methods

To the first approximation the chemical properties of pyrazole and imidazole are a combination of those of pyridine and pyrrole. Imidazole forms many stable crystalline salts and is a stronger base than pyridine (see Table 1). This agrees with the assumption that there is a little extra resonance energy associated with the cation because of the contribution of structures **9** and **9a** to the resonance

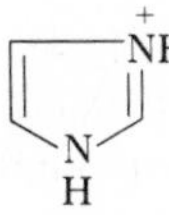

9 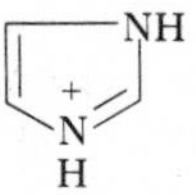 **9a**

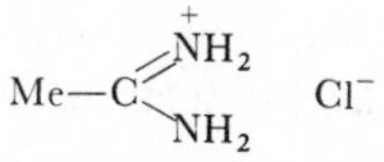

$Me{-}C(=\overset{+}{N}H_2)NH_2\ Cl^-$

Acetamidine hydrochloride

hybrid, which represents the molecule (cf. resonance in the amidine cation, p. 259). Similar structures can be written for the cation derived from pyrazole, but in contrast this compound is a weaker base than pyridine. The acidic properties of pyrazole and imidazole are slightly more pronounced (Table 1) than those of pyrrole owing to the inductive effect of the 'pyridine' nitrogen atom. They react with potassium, and a few other metals, in the same way as pyrrole (p. 96). Imidazole forms a sparingly soluble silver salt with ammoniacal

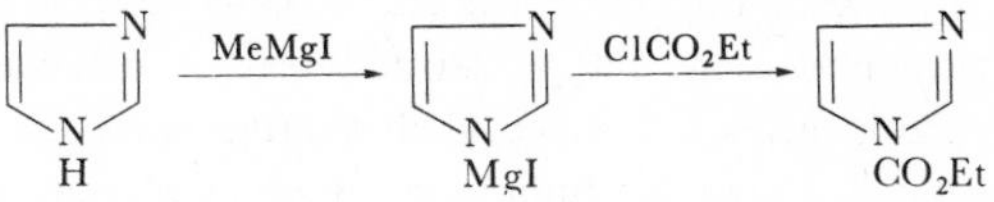

MeMgI → (N-MgI) ; $ClCO_2Et$ → (N-CO_2Et)

10

silver nitrate. These metal derivatives with alkyl or acyl halides give 1- (or *N*-) substituted pyrazoles or imidazoles (e.g., **10**). Trimethyl phosphate with triethylamine *N*-methylates both imidazole and pyrazole.[4] 1-Methylpyrazole behaves in an interesting way with butyllithium, for it must deprotonate from both the methyl group and position 5, as with benzaldehyde[5] the product is a mixture of **11** and **12**.

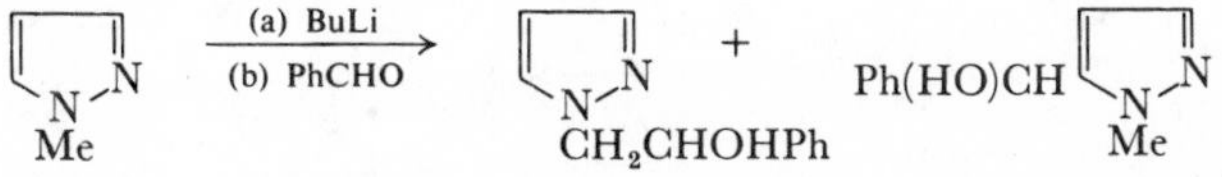

(a) BuLi, (b) PhCHO → $CH_2CHOHPh$ (**11**) + Ph(HO)CH (Me) (**12**)

When the imidazole or pyrazole is not substituted symmetrically, as in the case of 4(5)-methylimidazole, two different tautomeric structures are possible (1 and 2, p. 355), and in fact the compound behaves as a mixture in chemical reactions. It is worth noting that both tautomers give the same aromatic cation (**13**) and anion (**14**) by the respective addition or loss of a proton.

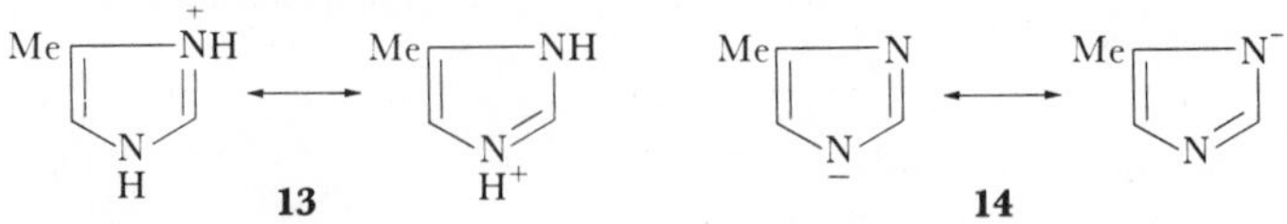

Pyrazole and imidazole are remarkably stable to strong acids and to oxidation. The products of nitration of pyrazole depend very much on the conditions. With acetic anhydride and nitric acid the 1-nitro compound (**15**) is formed, and this isomerises to **16** on heating,[6] possibly by a [1,5] shift.

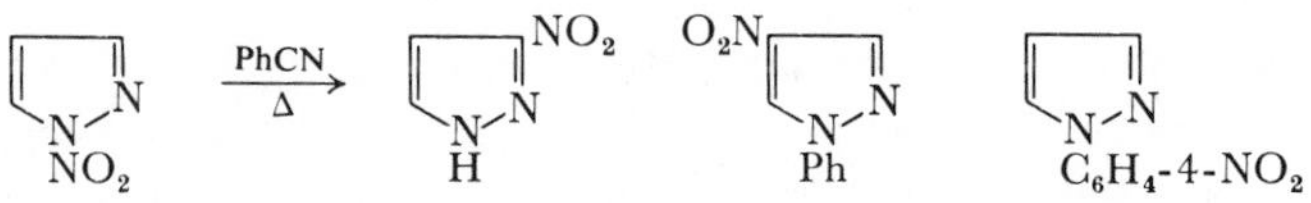

In concentrated sulphuric acid, when the pyrazolium cation is being attacked,[7] 1-methylpyrazole exchanges[8] the 4-hydrogen atom for deuterium (in D_2SO_4) while pyrazole nitrates, brominates, and sulphonates at position 4. 1-Phenylpyrazole in acetic anhydride nitrates as the free base at position 4 to give **17**, but in concentrated sulphuric acid the pyrazolium salt is attacked to give **18**.[9] 1,3,5-Trimethylpyrazole gives the 4-acetyl derivative with acetyl chloride and aluminium chloride.[10] Pyrazole undergoes base-catalysed iodination[11] at position 4, and although it does not couple with aromatic diazonium salts, 3,5-dimethylpyrazole reacts with nitrous acid, followed by sodium hydroxide, to give 3,5-dimethyl-4-diazopyrazole (**19**; cf. diazopyrrole, p. 107), which yields a red dye (**20**) with 2-naphthol.[12] 1-Phenylpyrazole, with phenyl radicals from benzoyl peroxide, gives 30% of 1,3-diphenylpyrazole along with much substitution of the original benzene ring.[13]

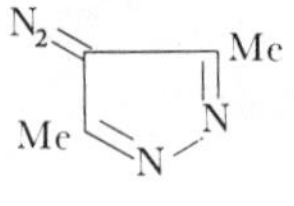

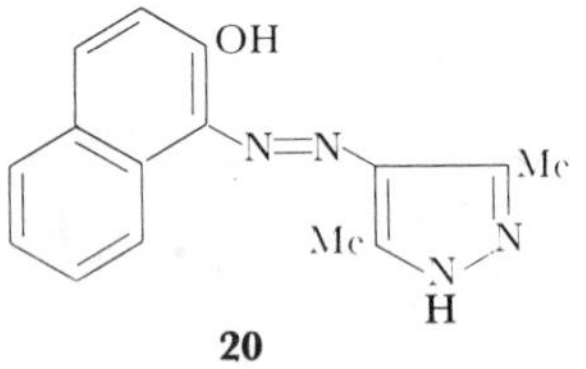

Imidazole (cf. structures **8** and **8a**, p. 356) is more susceptible to electrophilic reagents than pyrazole. Nitration and deuterium exchange[14] take place via the cation in sulphuric acid solution, at position 4(5); iodination takes place initially[15] at the same position, but further reaction, as with bromination, leads to 2,4,5-trisubstitution. Diazo coupling takes place at position 2 in alkaline solution to give bright red azo dyes. The results of kinetic experiments[16] suggest that the imidazole anion (cf. **14**) is the reacting species in iodination, diazo-coupling, and base-catalysed deuterium exchange. 2-Aminoimidazole can be diazotised (p. 367). In view of the great stability of the imidazole ring, its opening under normal benzoylation conditions is remarkable; benzimidazole behaves similarly.

PhCOCl, NaOH / ice → HC(NHCOPh)=HC(NHCOPh) + HCO_2^- Na^+

It is remarkable that irradiation of pyrazole converts it partially to imidazole, and the pyrazole **21** gives the imidazole **22**.[17] Indazole (**45**) is similarly isomerised to benzimidazole (**46**), but the detailed mechanisms of these

21 —$h\nu$→ **22**

transformations are not clear.[18] Imidazole is very resistant to reduction, but some pyrazoles can be reduced to pyrazolines and pyrazolidines.

Pyrazole itself (**25**, R = R′ = R″ = H) is readily synthesized from the commercially available 1,1,3,3-tetramethoxypropane (malondialdehyde acetal, **23**) and hydrazine in the presence of acids. Hydrolysis to malondialdehyde (**24**) occurs and is followed by the formation of pyrazole. This type of synthesis is successful with almost any 1,3-dicarbonyl compound and monosubstituted hydrazine. Many pyrazoles have also been made by the addition of acetylenes with electron-attracting substituents (**26**) to diazoalkanes (**27**).[19]

$CH_2CH(OMe)_2$–$CH(OMe)_2$ (**23**) —H^+→ CH_2CHO–CHO (**24**) + H_2N–NH_2 ⟶ **25** ⟵ R′C≡CR (**26**) + $\bar{C}HR''$–N^+≡N (**27**)

There is no synthetic procedure in the imidazole series so widely applicable as those for simple pyrazoles. Imidazole itself is best obtained[20] from paraldehyde. Bromination in the presence of ethylene glycol gives a cyclic acetal of bromoacetaldehyde (**28**, 2-bromomethyl-1,3-dioxolane). This, on heating with formamide in the presence of ammonia, is converted to imidazole; bromoacetaldehyde is probably an intermediate formed by hydrolysis of the dioxolane (**28**).

BrCH$_2$ (dioxolane) **28** → CHO–CH$_2$Br + H$_2$N–OCH + NH$_3$ → imidazole (NH)

α-Hydroxyaldehydes or α-hydroxyketones (acyloins) give imidazoles similarly. A new route which clearly has possibilities for development has been reported recently.[21] There are many syntheses of limited scope available for imidazoles.

NH_2CSCO_2Et $\xrightarrow[\text{or } FSO_2OMe]{Et_3O^+BF_4^-}$ $H_2N^+{=}C(SR)CO_2Et$ (R = Me or Et) $\xrightarrow[NaOAc]{^+NH_3CH_2COMe\ Cl^-}$ Me-imidazole-CO_2Et (NH)

The most general involves the condensation of an α-amino-aldehyde or an α-amino-ketone hydrochloride (**29**) with hot aqueous potassium thiocyanate. 4-Imidazoline-2-thione (**30**) is formed through the intermediary of an α-thioureidoaldehyde (**29**). Desulphurization of the thione is easily effected with Raney nickel or by oxidation with concentrated nitric acid. In this case a sulphinic acid (**31**) is formed and loses sulphur dioxide.

CHO–CH$_2$–$\overset{+}{N}H_3$ Cl$^-$ $\xrightarrow{KCNS}$ [CHO–CH$_2$–NH–CS–NH$_2$] **29** →

30 (imidazoline-2-thione) $\xrightarrow{HNO_3}$ [imidazole-2-SO_2H] **31** $\xrightarrow{-SO_2}$ imidazole; **30** $\xrightarrow{\text{Raney Ni}}$ imidazole

D. Derivatives of pyrazole and imidazole

Pyrazoline (**32**) has three possible tautomeric structures, but that shown is the most stable. It can be prepared as indicated and is a colourless liquid, b.p. 140°C. In contrast to pyrazole, which is so stable to acids and bases, pyrazoline

$CH_2{=}CH_2$ + $CH_2{=}\overset{+}{N}{=}\overset{-}{N}$ → pyrazoline (NH) **32** ← $HC{=}CH_2$–CHO + H_2N–NH_2

is decomposed by hot water. Olefines combine with diazomethanes to give initially 1-pyrazolines, and the original stereochemistry of the olefine is retained. For instance, methyl tiglate (**33**) with diazomethane yields the *cis*-dimethyl-1-pyrazoline (**34**), in a concerted addition that is at least 98% stereospecific. The photolysis of **34** to the cyclopropane **35** is also a concerted process, the stereochemistry being retained, but the thermolysis of **34** probably proceeds through radical intermediates and gives a mixture of **35** with its geometric isomer.[22] 1-Pyrazolines can isomerise to the more stable 2-pyrazolines with a trace of acid.[23]

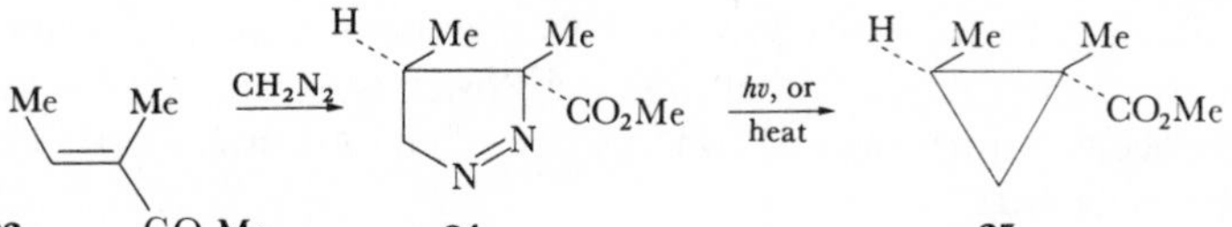

33 $\xrightarrow{CH_2N_2}$ **34** $\xrightarrow{h\nu,\ \text{or heat}}$ **35**

Pyrazolidine (**36**), b.p. 138°C, is obtained, along with other products, from hydrazine and 1,3-dibromopropane. It behaves as a *N,N'*-dialkylhydrazine in forming a *N,N'*-dibenzoyl derivative and in being readily oxidised to the corresponding pyrazoline (**32**). It is usually difficult to oxidise pyrazolines to pyrazoles, but bromine, and lead tetraacetate[24] can be effective.

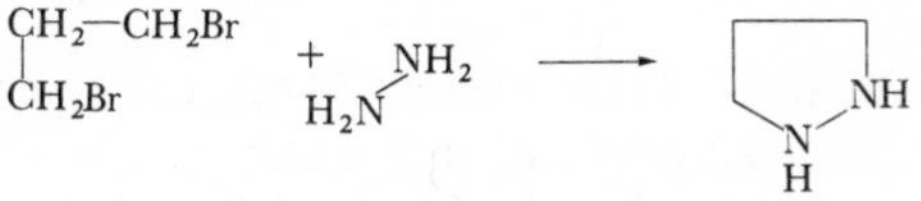

$CH_2BrCH_2CH_2Br + H_2NNH_2 \longrightarrow$ **36**

Pyrazolones have recently been used in a general synthesis of allenes (**37**), and the whole process can be completed in one vessel.[25]

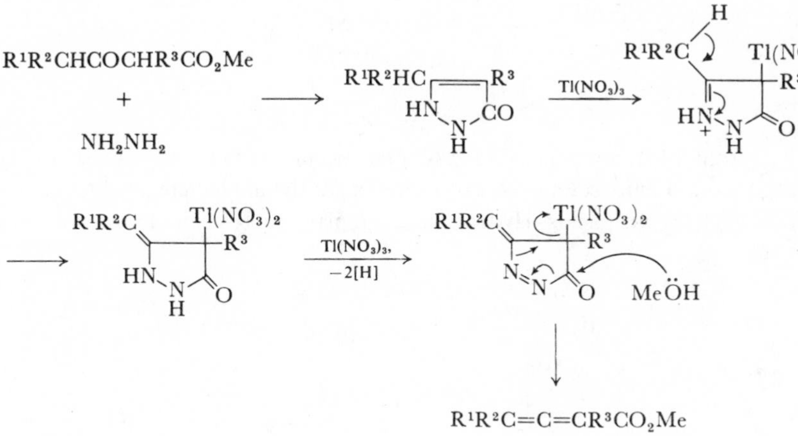

$R^1R^2CHCOCHR^3CO_2Me + NH_2NH_2 \longrightarrow \xrightarrow{Tl(NO_3)_3} \longrightarrow \xrightarrow{Tl(NO_3)_3,\ -2[H]} \xrightarrow{MeOH} R^1R^2C{=}C{=}CR^3CO_2Me$

37

Imidazolines (**38**) and their derivatives are usually synthesized from ethylenediamines and carboxylic acids, or their derivatives. They are not obtainable

$$\begin{matrix} CH_2NH_2 \\ | \\ CH_2NH_2 \end{matrix} + HO(O{=})CR \xrightarrow[\text{heat}]{HCl} \mathbf{38} \xrightarrow[\text{heat}]{H_2O} \begin{matrix} CH_2NH_2 \\ | \\ CH_2NHCOR \end{matrix}$$

38

by reduction or hydrogenation of imidazoles, nor can they be hydrogenated to imidazolidines. Partial hydrolysis takes place with hot water and is completed by alkali. Two imidazolines, Priscol (**39**) and Privine (**40**), are valuable vasodilating and vasoconstricting agents, respectively, and Antistine (**41**) is a useful antihistaminic drug.

39 (2-CH_2Ph) **40** (2-CH_2-1-naphthyl) **41** (2-CH_2NPhCH_2Ph)

Very few imidazolidines (**42**) are known. They can be considered as cyclic aldehyde ammonias and are hydrolysed very readily by dilute acids.

Imidazolid-2-one (**43**) is readily obtained as indicated and is best considered as a substituted urea. Biotin (p. 171) is an important imidazolidone.

$$\begin{matrix} CH_2NHR \\ | \\ CH_2NHR \end{matrix} + OHCR' \underset{H^+ \text{ aq.}}{\rightleftharpoons} \mathbf{42} \qquad \begin{matrix} CH_2NH_2 \\ | \\ CH_2NH_2 \end{matrix} + CO(NH_2)_2 \xrightarrow{\text{heat}} \mathbf{43}$$

42 **43**

Hydantoin (**44**), or imidazolid-2,4-dione, occurs in beet sap and is readily prepared from aminoacetonitrile and cyanic acid; the immediate product of this reaction cyclizes with acids. Hydantoin can replace acylglycines in the synthesis

$$CH_2O \xrightarrow{NH_4CN} NCCH_2NH_2 \xrightarrow{HCNO} NCCH_2NHCONH_2 \xrightarrow{H^+} \mathbf{44}$$

44

of amino and keto acids (p. 374), and allantoin (p. 420), a derivative, is related to uric acid.

Benzopyrazole is often called indazole (**45**) and is prepared from cyclohexanone.[26] It yields benzimidazole (**46**) on photolysis[27] but has little current interest.

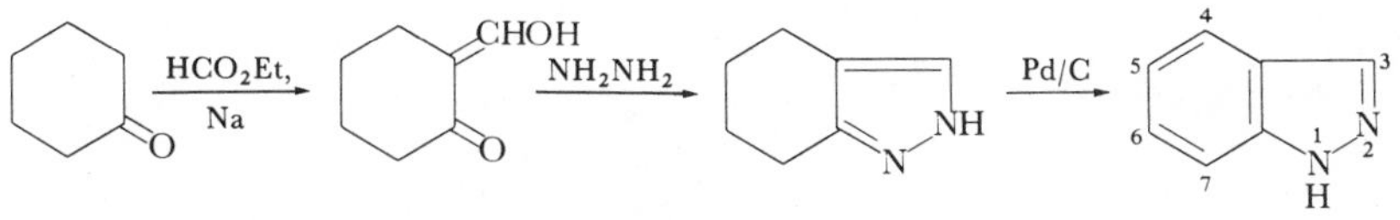

The benzimidazole (**46**) ring is of much greater interest than that of indazole, as it occurs in Vitamin B_{12} (p. 121) and in many biologically active compounds. Benzimidazoles are usually prepared from 1,2-diaminobenzenes by reaction with carboxylic acids or derivatives, such as nitriles or imino ethers, under acid conditions. The benzimidazole system is highly aromatic. Benzimidazole is hard to oxidise, but permanganate does give imidazole-4,5-dicarboxylic acid. 1-Substituted benzimidazoles are reduced by lithium aluminium hydride to 2,3-dihydro derivatives, but air at room temperature causes reoxidation.[28] Like imidazole it forms metal derivatives, and salts with acids. In sulphuric acid the 2-hydrogen atom is exchanged for protons in solution, the ylid **47** being the intermediate.[29] In sulphuric acid it nitrates at position 5(6), while in alkaline solution, where the anion formed by loss of the 1-hydrogen atom from **46** is probably attacked, iodination occurs at position 2.

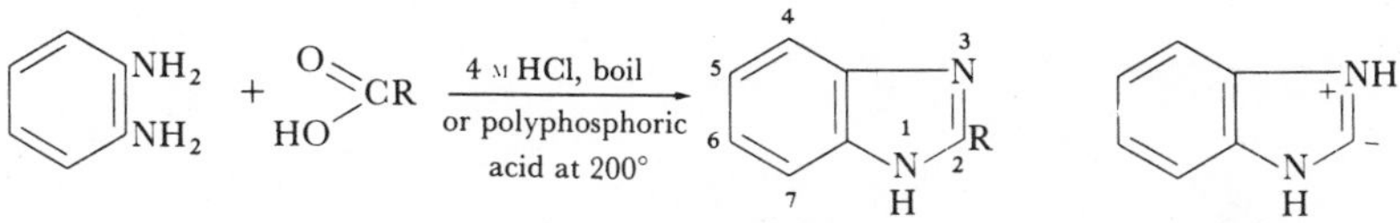

E. Compounds of special interest and natural occurrence

1. *Pyrazoles*

The first natural pyrazole, β-pyrazol-1-ylalanine, occurs in watermelon seeds.[30] The most important derivatives of pyrazole are in fact all pyrazolones. Antipyrine (**49**) is one of the earliest synthetic drugs and is named after its antipyretic properties. It is still used to some extent. Butazolidine (**50**), another tautomeric pyrazolone, is a powerful antiinflammatory drug used in rheumatic

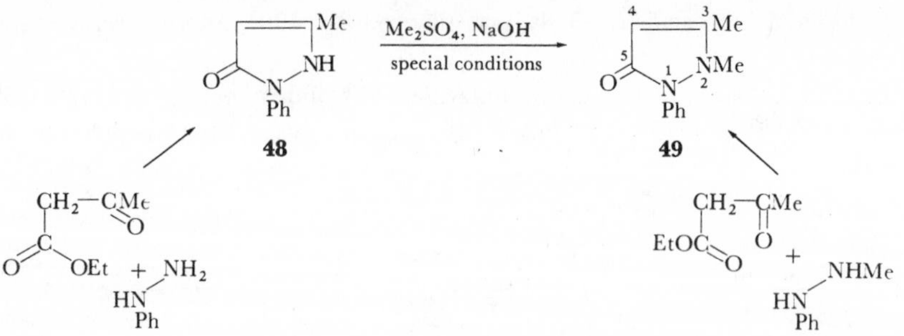

Me_2SO_4, NaOH
special conditions

48

49

conditions, but it has dangerous side reactions. Picrolonic acid (**51**), made by

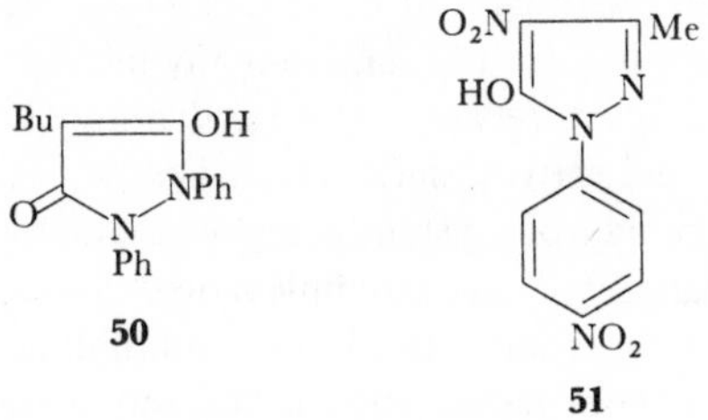

50

51

nitrating **48**, usually forms highly crystalline salts suitable for the characterization of organic bases.

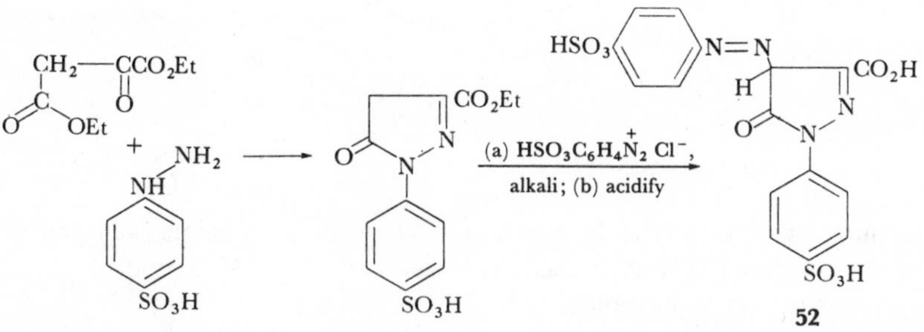

(a) $HSO_3C_6H_4\overset{+}{N}_2$ Cl^-, alkali; (b) acidify

52

Tartrazine (**52**) is a yellow dye for wool, and other pyrazolone dyes have been gaining commercial importance in recent years. 3-Methyl-1-phenylpyrazol-5-one, an intermediate in the synthesis of antipyrine (**49**), can have several tautomeric formulations (e.g., **48** and **53**) and gives a green-black dye (**54**) with 4-nitrosodimethylaniline or with 4-aminodimethylaniline and silver chloride in the presence of light. This type of process is of great importance in colour photography.[31]

Me_2N–C₆H₄–NO or Me_2N–C₆H₄–NH_2 + AgCl + light + **53** ⟶ **54**

Me, N, N, O, Ph, **53**; Me_2N, N=, Me, N, N, O, Ph, **54**

2. *Imidazoles*

Many imidazole derivatives have powerful pharmacological properties (p. 362). The imidazole ring occurs in a number of naturally occurring compounds, discussed in Chapter IX, and in the very widely distributed essential amino acid L-histidine (**57**) and its derivatives. The best synthesis of histidine

Imidazole $\xrightarrow{PhCH_2Cl}$ 1-benzylimidazole (CH_2Ph) $\xrightarrow{CH_2O}$ (CH_2OH, CH_2Ph) $\xrightarrow[-PhMe]{Na,\ liq.\ NH_3}$ (CH_2OH, N–H) $\xrightarrow{HNO_3}$ (CHO, N–H) **55**

55 + **56** (HN, S, O, S) $\xrightarrow{AcOH}$ (CH, N, H, HN, S, O, S) $\xrightarrow{HI,\ red\ P}$ ($CH_2CHNH_2CO_2H$, N, H) **57**

starts from imidazole, which is first converted into imidazole-4(5)-carbaldehyde (**55**). This with 2-mercapto-5-thiazolone (**56**) in acetic acid followed by a reductive ring opening (cf. oxazolones, p. 375) gives histidine (**57**). Histidine is a basic amino acid, as both the primary amino group and the imidazole ring are basic and there is only one carboxyl group. Histidine therefore forms salts with organic acids, and the DL-histidine obtained by synthesis can be resolved through its salt with D-tartaric acid.

The biosynthesis of histidine appears to involve adenosine-5′-phosphate (p. 425), ribose-1-phosphate, and glutamine.[32] 5-Amino-4-imidazolecarboxyamide (**58**) and histidinol phosphate are formed, and the latter gives histidine (**57**) after several transformations. The amide (**58**) can be converted again *in vivo* into adenosine-5′-phosphate.

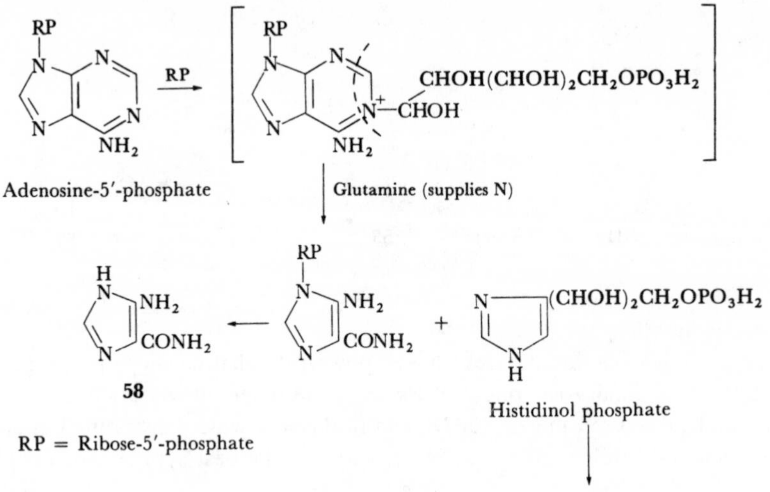
RP
RP
CHOH(CHOH)$_2$CH$_2$OPO$_3$H$_2$
CHOH
NH$_2$
Adenosine-5′-phosphate
Glutamine (supplies N)
NH$_2$
CONH$_2$
(CHOH)$_2$CH$_2$OPO$_3$H$_2$
Histidinol phosphate

58

RP = Ribose-5′-phosphate

57

Several mammalian degradation routes of histidine are known. Perhaps the most important is decarboxylation to histamine (**59**), an amine with very

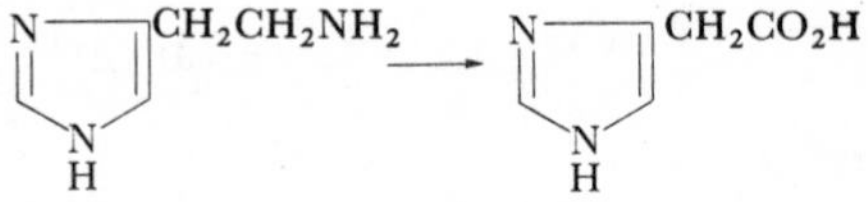
CH$_2$CH$_2$NH$_2$
CH$_2$CO$_2$H

59 **60**

important and powerful pharmacological properties. These include the stimulation of glands and smooth muscle, and dilation of capillaries. It is connected with many pathological conditions, including allergies, and a large number of

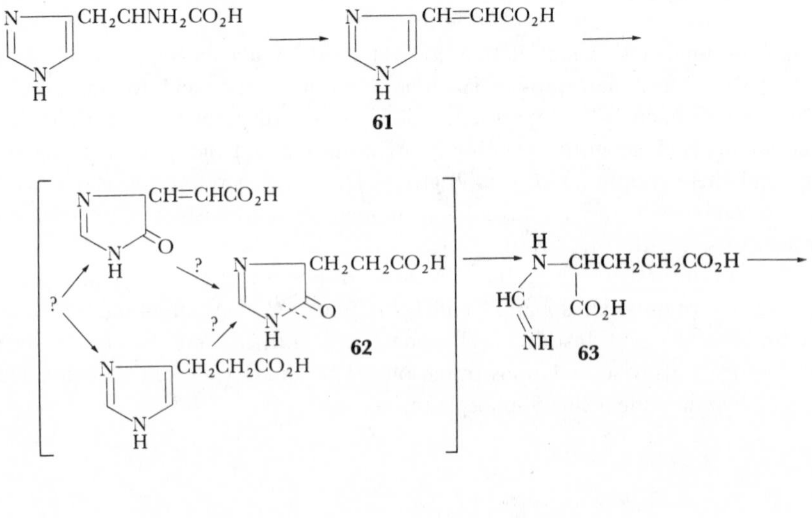
CH$_2$CHNH$_2$CO$_2$H
CH=CHCO$_2$H
CH=CHCO$_2$H
CH$_2$CH$_2$CO$_2$H
CH$_2$CH$_2$CO$_2$H
N—CHCH$_2$CH$_2$CO$_2$H
CO$_2$H
HC
NH

61 **62** **63**

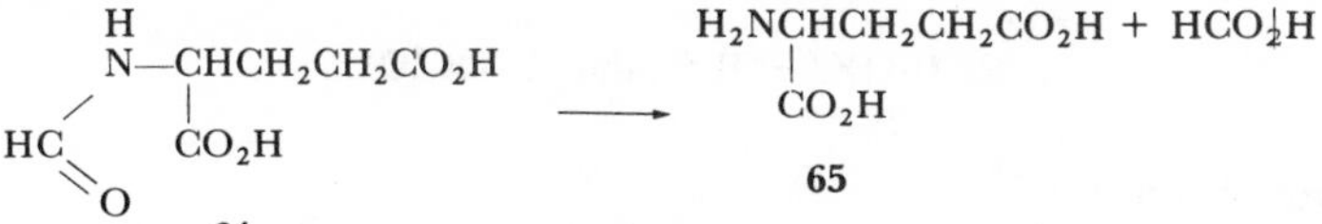

$$\underset{\mathbf{64}}{\mathrm{HC(=O)-NH-CH(CO_2H)CH_2CH_2CO_2H}} \longrightarrow \underset{\mathbf{65}}{\mathrm{H_2NCH(CO_2H)CH_2CH_2CO_2H}} + \mathrm{HCO_2H}$$

compounds ('antihistamines') which antagonize the action of histamine in various biological systems have been found. Some of these have medicinal value. Histamine is further degraded *in vivo* to imidazole-4-acetic acid (**60**), and it is likely that the imidazole ring of this compound is opened in a similar way to that of histidine (opposite page).

Histidine is also broken down *in vivo* through urocanic acid (**61**) with ring scission leading to glutamic acid (**65**). Imidazolones, such as **62**, which has not been isolated, are known[33] to split up very easily. Formamidinoglutaric acid (**63**) is formed from histidine in the rat[34] and breaks up through formamidoglutaric acid (**64**) to glutamic acid (**65**).

Ergothioneine (**66**) occurs in blood and animal tissues, but its function is not yet known. The antibiotic azamycin is 2-nitroimidazole (**67**), which has been synthesized[35] from 2-aminoimidazole, buffered nitrous acid, and copper sulphate.

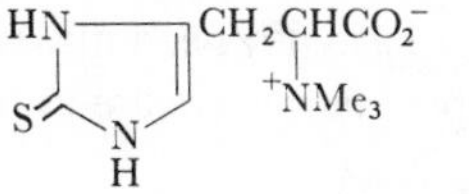

HN—CH$_2$CHCO$_2^-$ / $^+$NMe$_3$ / S / N / H

66

N / NO$_2$ / N / H

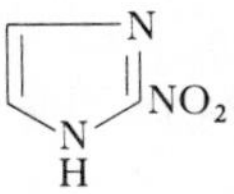

67

Cypridina luciferin is a complex imidazole derivative (**68**) which with oxygen and the appropriate enzyme gives out light via the production of an excited (asterisked) carbonyl group (see p. 385).[36]

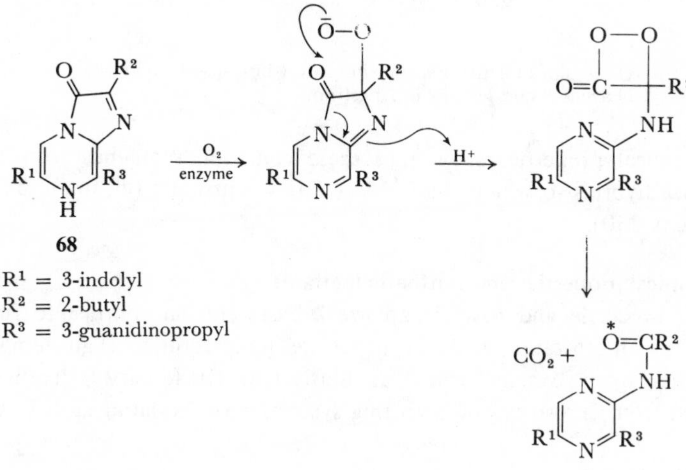

O R^2 N N R^1 R^3 N H $\xrightarrow[\text{enzyme}]{O_2}$ O—O O R^2 N N R^1 R^3 N $\xrightarrow{H^+}$ O—O O=C—R^2 N NH R^1 R^3 N

68

R^1 = 3-indolyl
R^2 = 2-butyl
R^3 = 3-guanidinopropyl

↓

CO$_2$ + $\overset{*}{O}$=CR^2 / NH / N / R^1 R^3 N

2. ISOXAZOLE AND OXAZOLE

A. Introduction

Although isoxazole was prepared in 1903, oxazole was not synthesised until 1947. Only in the last 10 years or so have these ring systems attracted much interest.

B. Physical properties and structure

Isoxazole and oxazole are intermediate between furan and pyridine in boiling point; they are very weak bases[37] (pK_a −2·03 and +0·8, respectively) and much less basic than pyridine owing to the electronegative effect of the oxygen atom.

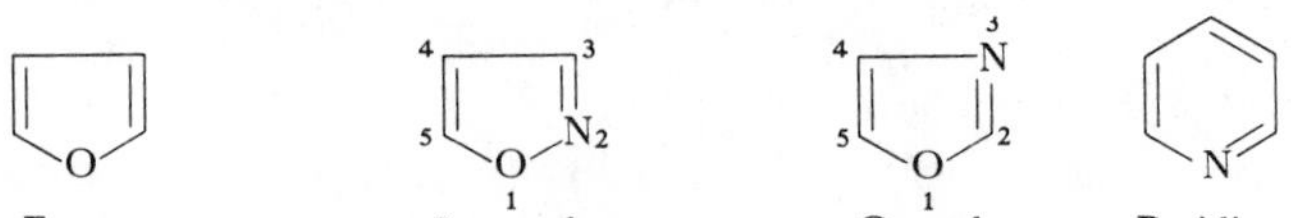

Furan	Isoxazole	Oxazole	Pyridine
b.p. 31°	b.p. 95°	b.p. 69°	b.p. 115°

Simple oxazoles and isoxazoles usually resemble pyridine in odour. The resonances energies for these heterocycles have not been measured, and only the dimensions for the oxazole ring in a complex derivative are available.[38]

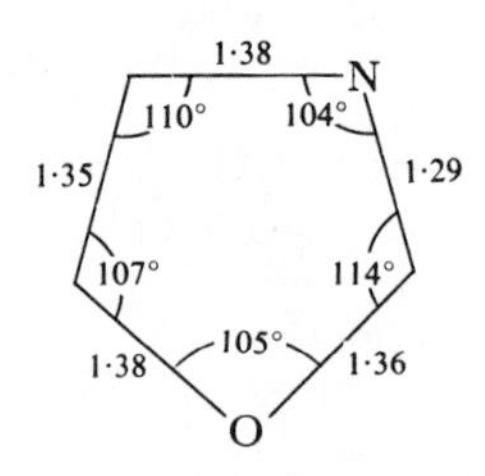

Dimensions for the oxazole ring in 1,4-bis-2-(5-phenyloxazolyl) benzene, bond lengths in Angstroms

The chemical properties show that these compounds are best considered as resonance hybrids to which charged structures contribute (cf. furan, p. 125 and pyridine p. 230).

C. Chemical properties and synthetic methods

Both isoxazole and oxazole are weak bases and, in contrast to furan, are stable to concentrated acids at moderate temperatures. High temperatures sometimes open oxazole rings, as illustrated. Quaternary salts have been obtained from derivatives of both ring systems with alkylating agents. Although oxazoles are stable to alkali, the isoxazole ring is opened easily under these conditions; the direction and ease of opening depends on the substituents present.

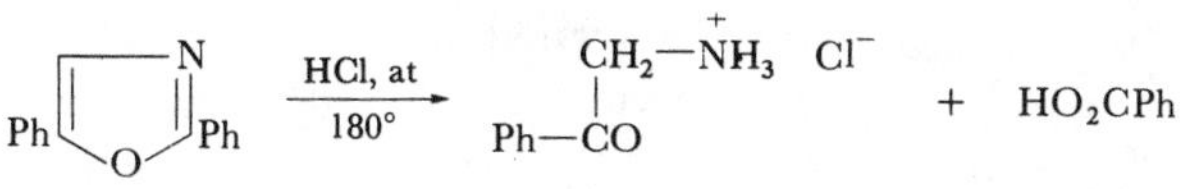

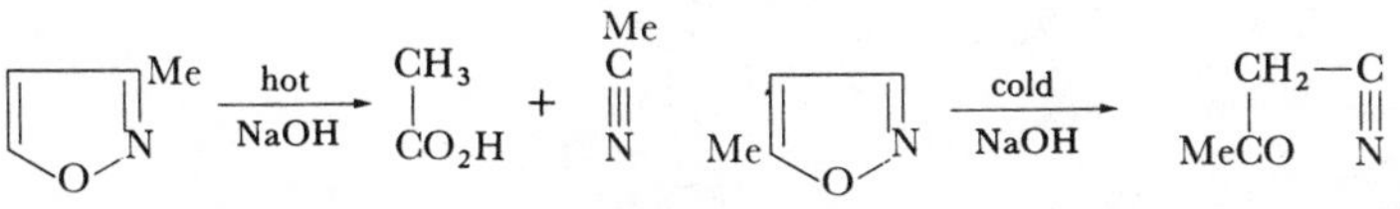

The 3,4,5-triphenylisoxazole ring (**1**, synthesis page 370) was opened by Meisenheimer by ozonolysis. On the assumption that the remaining double bond of the isoxazole ring between carbon and nitrogen atoms does not isomerize, and this is very unlikely to occur under the mild conditions used, the product has the structure shown. It is the benzoyl derivative (**2**) of an oxime of known stereochemical configuration. Meisenheimer showed that this derivative corresponded to an oxime (**3**) which on treatment with phosphorus pentachloride in ether underwent the Beckmann rearrangement to give **4** and thereby proved for the first time that the migrating group was *trans* to the hydroxyl group of the oxime.

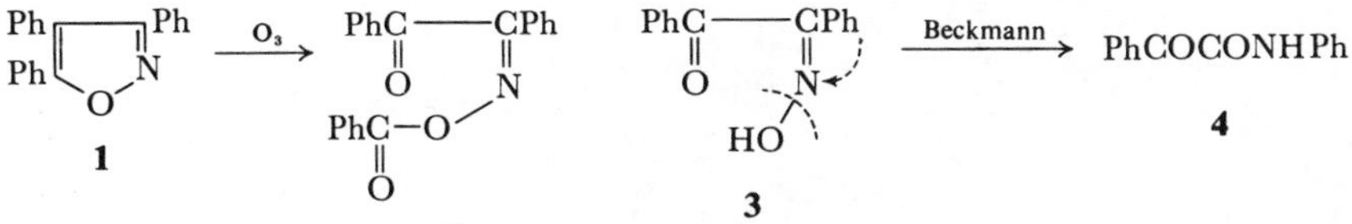

Phenylisoxazoles and phenyloxazoles are nitrated preferentially in the phenyl ring, thus showing that the deactivating effect of the 'pyridine' nitrogen atom overcomes the activating properties of the 'furan' oxygen atom. Electrophilic substitution, for example, the chloromethylation of 3,5-dimethylisoxazole (**5**), takes place[39] at position 4 to give a valuable synthetic intermediate (p. 274), and deuterium exchange in sulphuric acid also takes place at this position.[40]

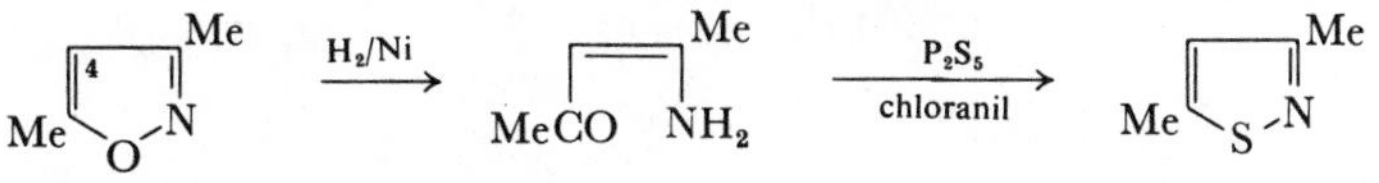

The N–O bond of isoxazoles (e.g., **5**) is easily split by hydrogenation to give enaminoketones, which have been converted into dihydropyridines (p. 274) and isothiazoles (e.g., **6**).[41] Isoxazolium salts (e.g., **7**) with bases yield very reactive ketenimines (**8**). These combine with NH_2-protected amino acids to give products (**9**) which *N*-acylate amino acids, forming new peptides (**10**).[42,43]

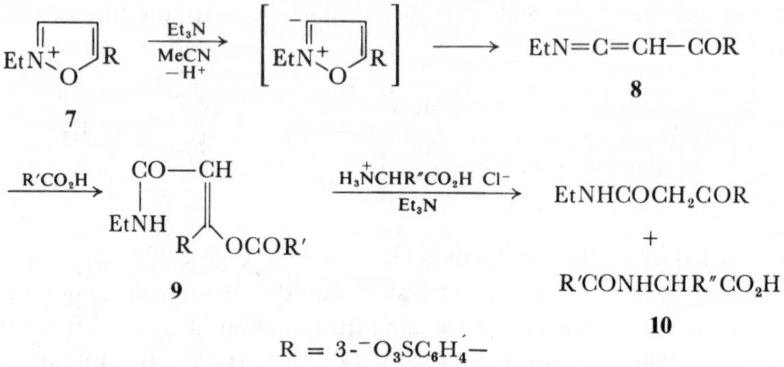

$EtN{=}C{=}CH{-}COR$ (**8**)

$R'CONHCHR''CO_2H$ (**10**)

$R = 3\text{-}^{-}O_3SC_6H_4\text{—}$

The photolysis of 2-phenyloxazole (**11**) causes isomerisation[44] to **12**, **13**, and **14**, but the mechanisms whereby these conversions occur have not been firmly established.[18]

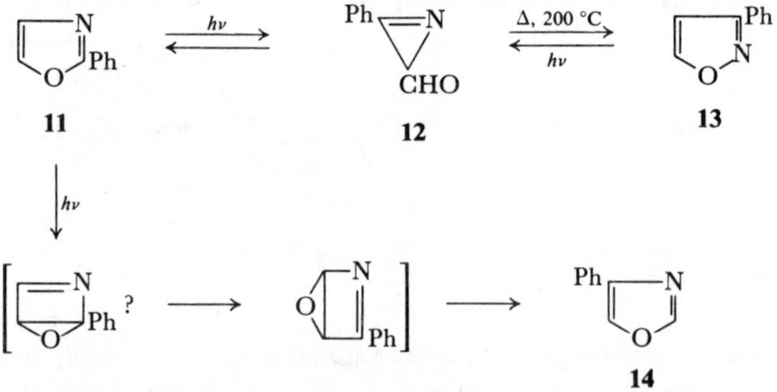

Oxazoles can undergo the Diels–Alder reaction (p. 274), which is consistent with the idea that the short 2,3- and 3,4-bonds (cf. pyridine, p. 230) suggest that the molecule will possess diene rather than aromatic properties.

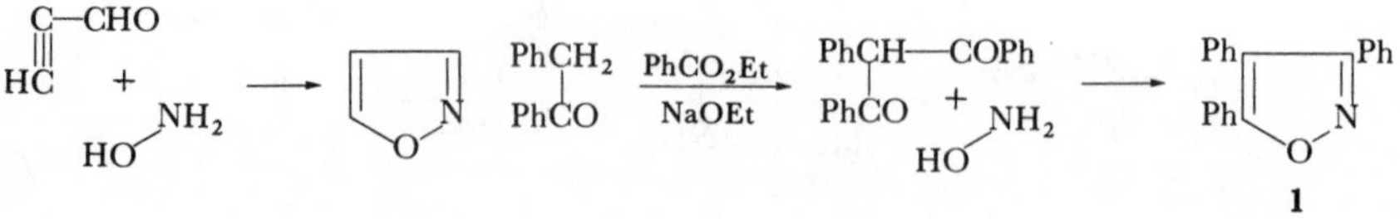

Isoxazoles can often be used instead of the less controllable 1,3-dicarbonyl compounds in syntheses,[43] and most have in fact been prepared from such compounds, or the equivalent acetylenic derivatives, with hydroxylamine. Other synthetic methods include the interaction of acetylenic ketones with hydrazoic acid,[45] and nitrile oxides with activated acetylenes.[46]

$$PhCOC{\equiv}CPh + HN_3 \longrightarrow \text{3-Ph, 5-Ph isoxazole}$$

$$MeO_2CC{\equiv}CCO_2Me + PhC{\equiv}N^+{-}O^- \longrightarrow \text{4,5-}(MeO_2C)_2\text{-3-Ph-isoxazole}$$

Oxazole was prepared by Cornforth by a method which can be used for substituted oxazoles.

$$HCN + i\text{-}PrOH + HCl \longrightarrow H_2\overset{+}{N}{=}CHOPr\text{-}i\ Cl^-$$

$$+ EtO_2CCH_2NH_2 \longrightarrow EtO_2CCH_2{-}N{=}CHOPr\text{-}i \xrightarrow[KOEt]{HC(OEt)_3} EtO_2CC(=CHO^-\,K^+){-}N{=}CHOPr\text{-}i$$

$$\xrightarrow{\text{boil + HOAc}} \text{4-EtO}_2\text{C-oxazole} \xrightarrow{\text{hydrolysis}} \text{4-HO}_2\text{C-oxazole} \xrightarrow[\text{quinoline + CuO}]{\text{boil in}} \text{oxazole}$$

Another and much older synthesis is the cyclisation of acylamido ketones, which are often obtained from oximes by reductive acylation. Studies using ^{18}O have shown that the amide oxygen atom is retained completely in the final oxazole,[47] and that the other oxygen (asterisked) is completely lost.

$$MeCO{-}CH_2CO_2Et \xrightarrow[(c)\ H_2SO_4]{(a)\ NaOH,\ (b)\ NaNO_2} MeCO{-}CH{=}NOH \xrightarrow[Ac_2O]{Zn,\ HOAc} MeCO^*{-}CH_2{-}NH{-}OCMe \xrightarrow[SOCl_2]{H_2SO_4,\ or}$$

$$\left[\text{5-Me-5-HO}^*\text{-2-Me-oxazoline}\right] \xrightarrow{-H_2O^*} \text{2,5-Me}_2\text{-oxazole}$$

D. Derivatives and compounds of special interest

A few isoxazole derivatives (e.g., **15**) have been isolated from plants and all possess *N*-substituents.[48]

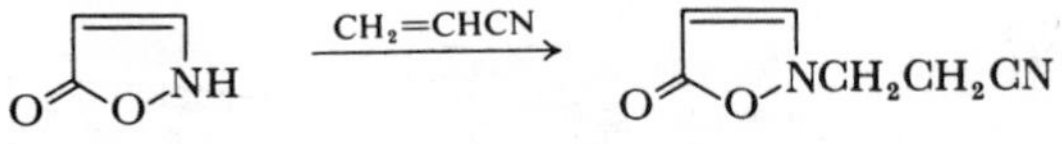

$CH_2{=}CHCN$ → NCH_2CH_2CN

15

2-Isoxazolines, such as **16**, are insoluble in alkali, in contrast to the isomeric unsaturated oximes (**17**), which are usually incapable of cyclization. This is

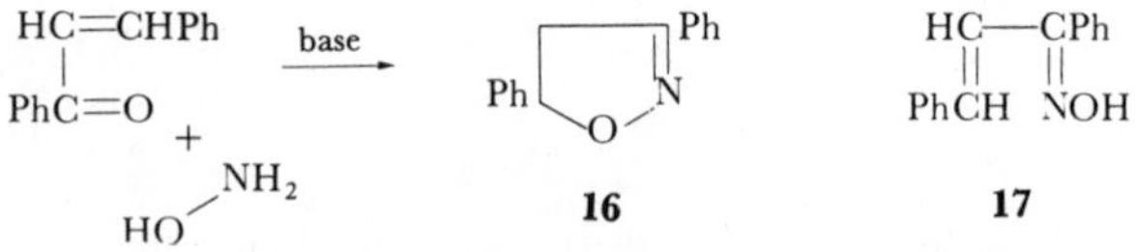

HC=CHPh, PhC=O + NH_2 HO — base → Ph, Ph, N, O **16**; HC—CPh, PhCH NOH **17**

perhaps because they have the wrong geometric configuration or that in isoxazoline formation the hydroxylamine first adds to the carbon–carbon double bond (Michael addition) and cyclization follows. The 2-methyl derivative of isoxazolidine (**18**), which can be considered as a *N,O*-dialkylhydroxylamine, undergoes slow inversion[49] of the nitrogen atom (cf. oxaziridines, p. 47). The antibiotic oxamycin, or cycloserine, is D-4-amino-3-isoxazolidone (**19**).

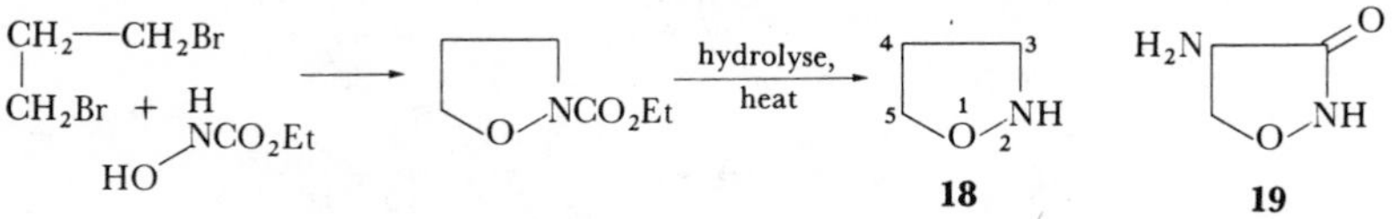

CH_2—CH_2Br, CH_2Br + H NCO_2Et HO → NCO_2Et — hydrolyse, heat → NH **18**; H_2N, O, NH **19**

The isoxazole ring can be fused to a benzene ring in two ways, yielding benzo[*d*]isoxazole (**20**), which can have a number of uncharged canonical

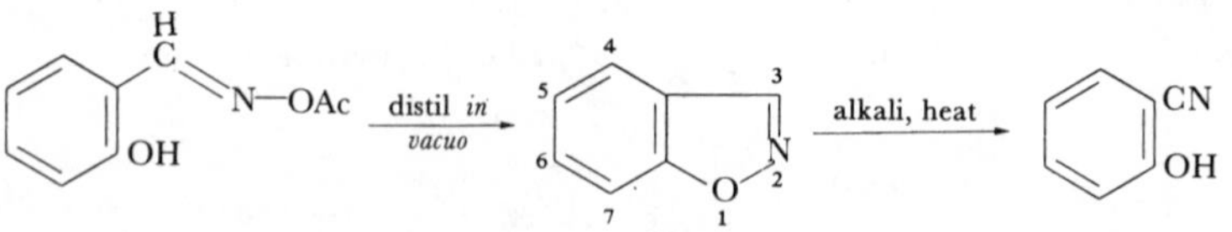

H C N—OAc OH — distil *in vacuo* → **20** — alkali, heat → CN OH **21**

structures, and anthranil (**22**), which can be represented only by one uncharged formulation. Charged formulations can be written for both molecules and contribute, particularly in the case of anthranil (**23** and **24**), to the resonance hybrids which best represent these compounds. Benzoisoxazole is a very feeble base and nitrates at position 5. Treatment with alkali removes a proton from position 3, and the resulting ylid cleaves to **21**.[50] Anthranil (**22**) is readily obtained from 2-nitrobenzaldehyde. Nitration[51] gives mainly the 5-nitro derivative with a little of the 7-isomer, while with warm sodium hydroxide it yields anthranilic acid.

CHO, NO_2 —(Sn + HCl)→ **22**   **23**   **24**

Three positions are possible for the double bond in the dihydrooxazole, or oxazoline ring, but only 2-oxazoline (**26**) and derivatives are known. 2-Oxazoline, b.p. 98°C, has most of the reactions of an imino ether (e.g., **28**). Examples are hydrolysis with boiling water to **27**, and the formation of an unstable hydrochloride which slowly rearranges to **25**.

CH_2—NH / CH_2Cl CHO (**25**) —(KOH 50%)→ **26** —(H_2O, boil)→ CH_2—NH / CH_2OH CHO (**27**)   RC(=NH)OMe (**28**)

Oxazolines have been greatly exploited in recent years as synthetic intermediates and can be easily obtained from aziridines. In the example given the carboxyl group of a bromobenzoic acid is protected by oxazoline formation, the halide is converted into the Grignard derivative, employed, and the carboxyl function is regenerated by acid hydrolysis. The oxazoline **30** is easily converted to an ylid **29**, and quenching with deuterium oxide gives the 2-deuterated

Ar = BrC_6H_4—

Me_2-aziridine NH + ArCOCl —(Et_3N)→ Me_2-aziridine NCOAr —(H^+, CH_2Cl_2)→ Me_2-oxazoline-Ar

Me_2CNH_2 / CH_2OH + HO_2CAr → Me_2C(NHCOAr)–H_2C–OH —($SOCl_2$)→ Me_2-oxazoline-Ar —(H^+ aq.)→ Me_2CNH_2 / CH_2OH

oxazoline. Methylation to **31** followed by reaction with a Grignard reagent and hydrolysis gives a valuable synthesis of deuteriated aldehydes (**32**). The 2-methyl

Me_2 N O (**29**) ⇌ (BuLi / H^+) Me_2 N O (**30**) → (MeI) Me_2 $\overset{+}{N}$Me O H/D (**31**)

31 → ((a) RMgX uncomplexed (b) H^+) $Me_2CNMeCHR_2$ / CH_2OH

31 → ((a) RMgX, 2HMPA (b) H^+ aq.) $Me_2CNHMeCH_2OH$ + RC=O (H/D) (**32**)

group of oxazolines such as **33** is activated and the lithium derivative can be alkylated by primary halides at –78°C and by secondary ones at –50°C. This leads to a useful synthesis of carboxylic acids (e.g., **34**) with considerable

Me_2 N O Me (**33**) → ((a) BuLi, –78 °C (b) $MeCHI(CH_2)_3I$) Me_2 N O $(CH_2)_4CHIMe$ → ($LiN(i\text{-}Pr)_2$, –78 °C, warm to 20 °C) Me_2 N O Me, *cis* (90%) → (H^+) Me_2CNH_2 / CH_2OH + HO_2C Me, *cis* (**34**)

stereochemical control attributed to the covalent nature of the nitrogen–lithium bond and the more rapid cyclisation of the least hindered rotamer intermediate.[52]

Certain derivatives of oxazoline, the oxazolones (azlactones[53]), are valuable intermediates in the Erlenmeyer synthesis of aromatic amino and keto acids. An example is the condensation of benzaldehyde with hippuric acid (benzoylglycine; acetylglycine can also be used) in the presence of acetic anhydride and sodium acetate. The reaction proceeds through the formation of 2-phenyl-5-oxazolone (**35**), which can be considered as the anhydride of hippuric acid. This compound is very reactive and condenses at position 4 *in situ* with the aromatic aldehyde to give the benzylidene oxazolone (**36**). Aliphatic aldehydes self-condense preferentially under the usual reaction conditions, but they usually react like aromatic aldehydes with 2-phenyl-5-oxazolone in the absence of acetic

$PhCONHCH_2CO_2H$ —(Ac_2O, NaOAc)→ 35

PhCHO + 35 (2-phenyloxazol-5-one, Ph at 2) → 36 (PhCH= oxazolone, Ph) → $PhCH{=}C(NHCOPh)CO_2H$ 37

36 —(HI, red P)→ $PhCH_2CHNH_2CO_2H$ 39

37 —(H_2, Ni)→ $PhCH_2CH(NHCOPh)CO_2H$ —(hydrolysis)→ 39

38 (hydantoin)

anhydride or other catalysts to give compounds analogous to **36**. The benzylidene oxazolone (**36**) can be converted into an α-amino acid (**39**) by reduction and hydrolysis. This can be done in several ways, including treatment with hot red phosphorus and iodine. Less drastic procedures are the opening of the ring with cold alkali or boiling water to **37**, followed by successive hydrogenation and hydrolysis. Hydrolysis of the acylamidocinnamic acid (**37**) by alkali, or better by acid, is a useful route to phenylpyruvic acid (**40**). The hippuric acid can be replaced by hydantoin (**38**), which reacts at position 5.

$PhCH{=}C(NHCOPh)CO_2H$ (**37**) —(H^+)→ [$PhCH{=}C(NH_2)CO_2H$ → $PhCH_2C({=}NH)CO_2H$] → $PhCH_2COCO_2H$ **40** ($+NH_3$)

Oxazolidines (e.g., **41**) are usually obtained from 2-amino alcohols and carbonyl compounds (see also p. 15). When the nitrogen atom is unsubstituted (e.g., **41**) the compounds are tautomeric with the corresponding Schiff's bases (e.g., **42**), and in this case infrared absorption spectrum studies have shown that the equilibrium mixture contains ca. 20% of **42**. Oxazolidines are readily hydrolysed by acid.

$HOCH_2CH_2NH_2$ + $O{=}C(Me)Pr$ —($-H_2O$, azeotrope with benzene)→ **41** (2-MePr-oxazolidine, NH) ⇌ $CH_2OH{-}CH_2{-}N{=}CMePr$ **42**

Benzoxazoles are usually prepared by heating 2-aminophenols with acid anhydrides, although carboxylic acids and other acid derivatives can be used.

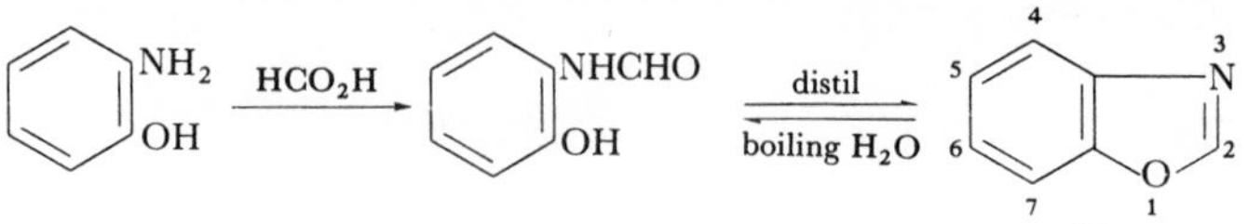

NH_2 HCO_2H NHCHO distil / boiling H_2O

OH OH

43

Benzoxazole (**43**), b.p. 182°C, is a weak base; it forms a methiodide and is readily hydrolysed by boiling water or dilute acid. An interesting synthesis of 2-phenylbenzoxazole, which may involve a benzyne intermediate (**44**), is also

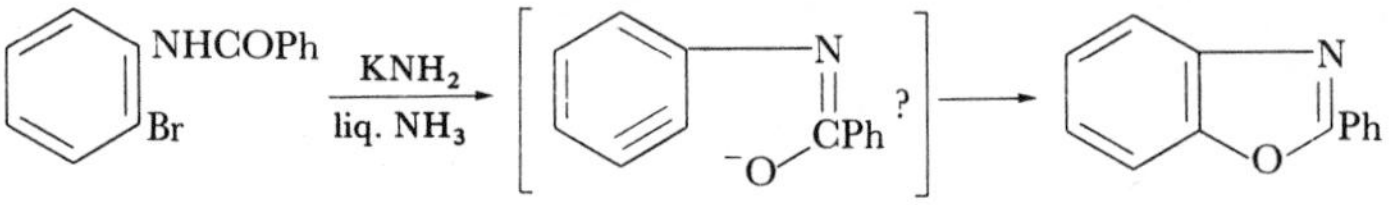

NHCOPh KNH_2 / liq. NH_3 N CPh ? N Ph

Br $^{-}$O O

44

outlined. 6-Methoxybenzoxazolone (**45**) is of interest, as it occurs in corn and is

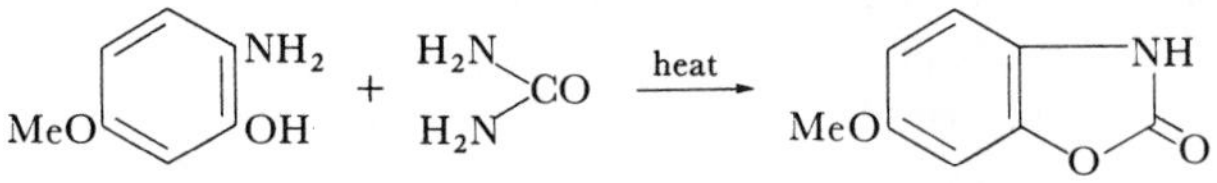

NH_2 + H_2N CO H_2N heat NH

MeO OH MeO O O

45

the compound responsible for resistance to the corn borer.[54] Infrared studies have shown that benzoxazol-2-ones (e.g., **45**) exist predominantly as the keto tautomers. An alkaloid (**46**)[55] and a mould metabolite (**47**)[56] are the first discovered naturally occurring oxazoles.

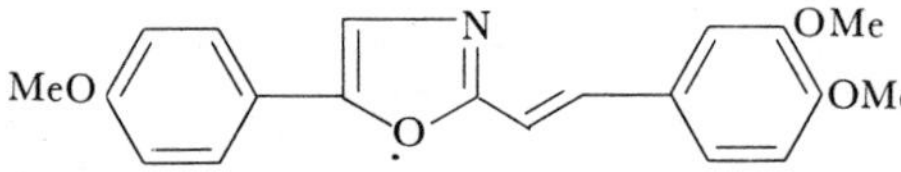

MeO N OMe OMe N Me O

O N H

46 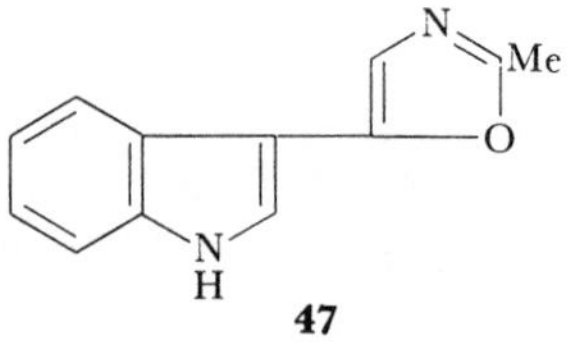 **47**

3. ISOTHIAZOLE AND THIAZOLE

A. Introduction

The penicillins (p. 64), vitamin B_1 (p. 381) and the antibacterial sulphathiazole (p. 385) are all derivatives of thiazole, and in consequence the chemistry of this ring system is well developed. Interest in isothiazole was singularly lacking before 1956.

B. Physical properties and structure

Both isothiazole and thiazole are weak bases and resemble pyridine in odour and boiling point. Their structures have not been investigated by

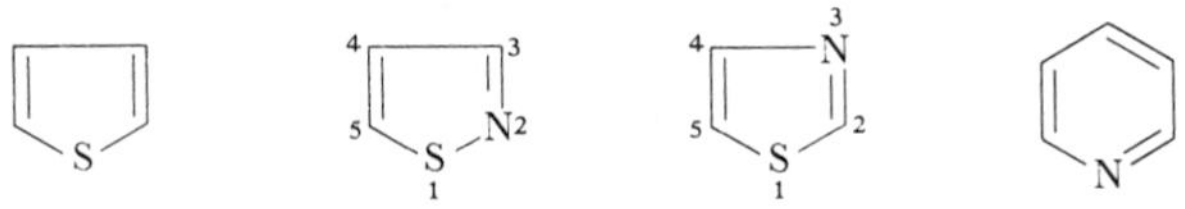

Thiophene b.p. 84°

Isothiazole b.p. 112°

Thiazole b.p. 117°

Pyridine b.p. 115°

physical methods, but the highly aromatic nature of both ring systems is shown in their chemistry. This, to a rough approximation only, is an amalgamation of that of pyridine and thiophene.

C. Isothiazoles

Isothiazole (**2**) can be obtained from propylene, sulphur dioxide, and ammonia at 200°C over an aluminium catalyst, from propynal (**1**), and from the benzene derivative (**3**) which constituted the first synthesis. It is a colourless liquid which forms a crystalline chloroaurate and chloroplatinate. Isothiazoles can also be made from isoxazoles (p. 369).

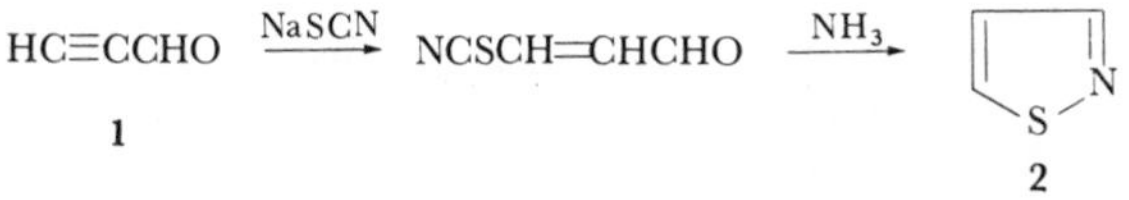

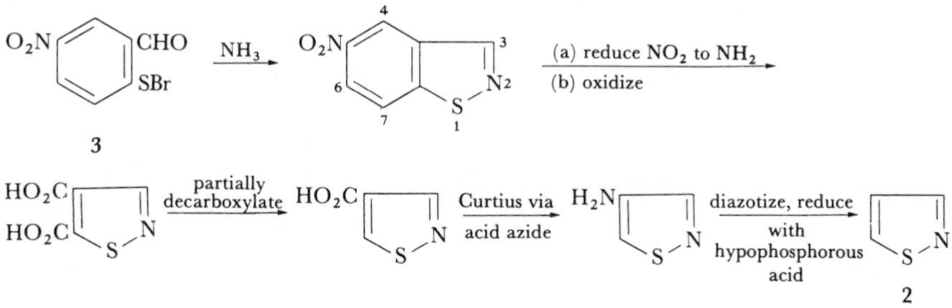

$$HC{\equiv}CCHO \xrightarrow{NaSCN} NCSCH{=}CHCHO \xrightarrow{NH_3} \mathbf{2}$$

1 **2**

O_2N, CHO, SBr (**3**) $\xrightarrow{NH_3}$ O_2N-benzisothiazole $\xrightarrow[\text{(b) oxidize}]{\text{(a) reduce } NO_2 \text{ to } NH_2}$

HO_2C, HO_2C $\xrightarrow{\text{partially decarboxylate}}$ HO_2C $\xrightarrow[\text{acid azide}]{\text{Curtius via}}$ H_2N $\xrightarrow[\text{with hypophosphorous acid}]{\text{diazotize, reduce}}$ **2**

Isothiazole with butyllithium at −70°C give the 5-lithium derivative, which has normal properties. Nitration and halogenation occur at position 4, and phenyl radicals, from benzoyl peroxide, attack all three positions. 3-Methyliso-thiazole quaternizes slowly with methyl iodide, but it does not condense with benzaldehyde nor react with selenium dioxide. Its 3-methyl group is inactive and unlike that of 2-methylpyridine. Chromic acid oxidation does, however, give the

3-carboxylic acid, which like the 5-isomer, but not the 4-isomer, is readily decarboxylated to isothiazole.[57] Isothiazole-3-carbaldehyde has the normal properties expected of an aromatic aldehyde.

Benzo [*c*] isothiazole (**4**) is easily prepared.[58] Nitration and bromination occur[59] mainly at position 5, the nitrogen atom can be alkylated, and the resulting salts (**5**) are readily hydrolysed[60] to 2-aminobenzaldehydes (**6**).

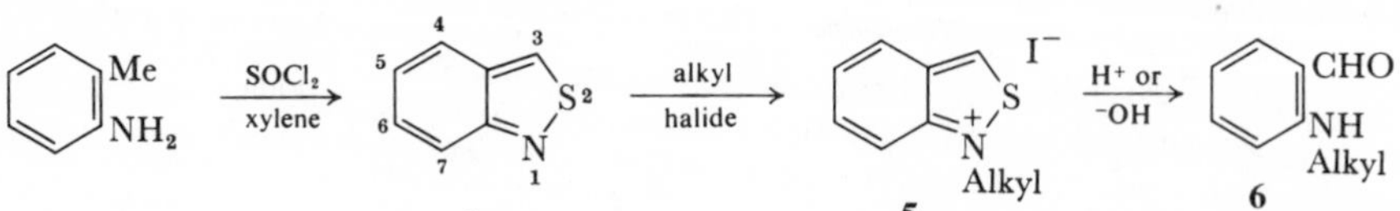

Me, NH_2 —($SOCl_2$, xylene)→ **4** —(alkyl halide)→ **5** (I^-, N-Alkyl) —(H^+ or ^-OH)→ **6** (CHO, NH Alkyl)

Isoselenazole (**7**), a colourless liquid, b.p. 68°C at 46 mm, has a pyridine-like smell and has been obtained from propynal (**1**) by successive reaction with selenocyanic acid and ammonia.[61]

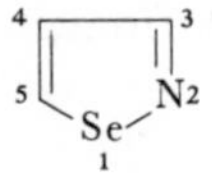

7 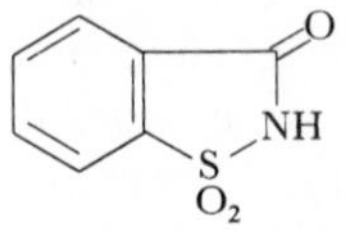 **8**

The only isothiazole of present importance is saccharin (**8**), which is prepared by the permanganate oxidation of 2-methylbenzenesulphonamide. On account of its sweet taste it is used as a sucrose substitute.

D. Thiazole, chemical properties and synthetic methods

Thiazole (**9**) is a weaker base (pK_a 2·5) than pyridine (pK_a 5·2), but it forms a crystalline hydrochloride and an aurichloride. It also forms thiazolium salts (e.g., **12**) with alkyl halides, and these (see thiamine p. 381), like 1-alkyl-pyridinium salts, are decomposed by alkali.

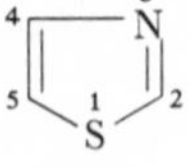

9 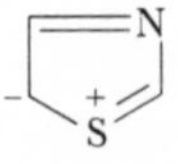 **10** 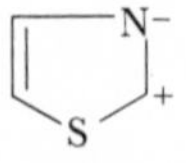 **11** 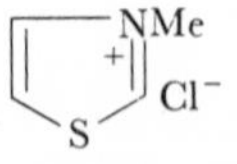 **12** 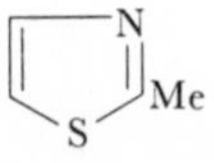 **13**

The major contributor to the resonance hybrid which represents thiazole is the uncharged canonical structure **9**, and chemical reactions show that charged structures such as **10** (cf. thiophene, p. 153) and **11** (cf. pyridine, p. 230) must have significance (Table 2). The fact that electrophilic substitution (sulpho-

TABLE 2. Calculated localization energies for thiazole in β units[61]

Position	Type of attack		
	Electrophilic	Nucleophilic	Radical
2	2·134	1·879	2·006
4	2·679	2·670	2·674
5	2·222	2·104	2·163

nation) occurs preferentially at position 5, otherwise at position 4, is not in agreement with the above localization energies and suggests that another species, the conjugate acid of thiazole, is probably being substituted. As the nitration of 2-, 3-, and 4-phenylthiazole effects substitution only at position 4 of the benzene rings, it is clear that the thiazole ring is comparatively unreactive. The calculations are in excellent agreement with experiment for nucleophilic and radical attack. Thiazole with sodamide gives 2-aminothiazole, with butyllithium it gives thiazole 2-lithium, and when the 2-position is alkylated it gives the 5-lithium derivative.[62] Hydrogen atoms at position 2 of thiazolium salts (e.g. **14**, see also thiamine, p. 384) are particularly easily exchanged via the formation of ylids. There has been much discussion as to why this is so. Molecular orbital calculations[63] suggest that the sulphur atom bears a substantial amount of positive charge, so structure **15** is important. It is also probable that the ylid **16** is stabilised by resonance involving **17** and other structures involving the sulphur *d* orbitals. Phenyl radicals, from benzoyl peroxide, give the 2-, 4-, and

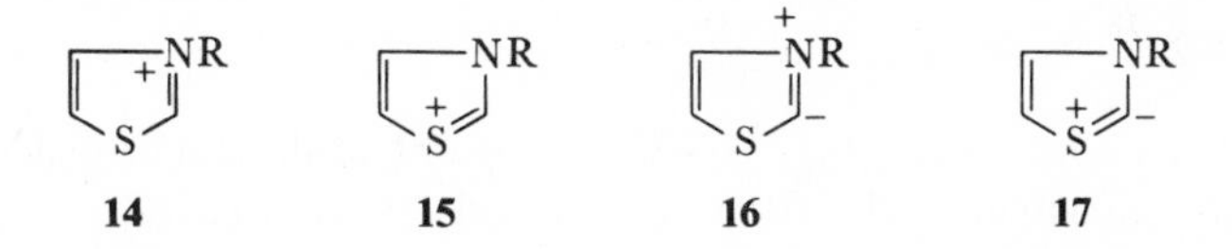

5-phenylthiazoles in 55 : 15 : 30 ratio,[61] while bromine at high temperatures (radical?) forms 2-bromothiazole.

2-Methylthiazole (**13**), but not its 4- and 5-isomers, reacts with benzaldehyde and zinc chloride to give 2-styrylthiazole, and the halogen atom of 2-bromothiazole is reactive. Only substituents at the 2-position of thiazole, therefore, are comparable in reactivity to those at position 2 or 4 of pyridine (pp. 230 and 246).

In contrast to 2-aminopyridine (p. 260) a diazonium salt is obtained from 2- (and incidentally 5-) aminothiazole under conditions successful for aniline. This shows the pronounced effect of the sulphur atom in partially neutralizing the electron-demanding character of the carbon atom (position 2), which is most strongly under the influence of the ('pyridine') nitrogen atom.

Thiazoles are generally resistant to reduction, and to oxidation by nitric acid; potassium permanganate often opens the ring.

The most valuable thiazole synthesis[64] is that of Hantzsch and consists of heating an α-halogenated aldehyde or ketone with a thioamide. Unpurified

$$CHO\text{—}CH_2Cl + H_2N\text{—}CH\text{=}S \xrightarrow{\text{heat}} \text{thiazole}$$

thioamides, including thioformamide, which are often obtained from the amides and phosphorus pentasulphide, can be used successfully in the synthesis. The thioamide can even be thiourea, when a 2-aminothiazole is formed.

E. Derivatives of thiazole

Although three tautomeric structures can be written for dihydrothiazole, only 2-thiazoline (**19**), b.p. 139°C, is known (cf. 2-oxazoline, p. 373). It has been obtained by the cyclization of the formyl derivative of 2-mercapto-ethylamine (**18**) with phosphorus pentoxide. Thiazolines are readily hydrolysed, like iminothioethers, and have been used[43] in the same way as oxazolines (p. 374) for preparing aldehydes. Thiazolines can be reduced to thiazolidines by aluminium amalgam.

$$\underset{\mathbf{18}}{HS\text{—}CH_2\text{—}CH_2\text{—}NH\text{—}CH\text{=}O} \xrightarrow{P_2O_5} \underset{\mathbf{19}}{\text{2-thiazoline}} \xrightarrow{H^+ \text{ aq.}} HS\text{—}CH_2\text{—}CH_2\text{—}NH_2 + HCO_2H$$

Thiazolidine itself (**20**), b.p. 165°C, can be obtained from formaldehyde and 2-mercaptoethylamine, and it forms an *N*-acetyl derivative (**21**).

$$HS\text{—}CH_2\text{—}CH_2\text{—}NH_2 + O\text{=}CH_2 \longrightarrow \underset{\mathbf{20}}{\text{thiazolidine (NH)}} \qquad \underset{\mathbf{21}}{\text{thiazolidine (NCOMe)}}$$

The stability of thiazolidines depends greatly on the substituents present, but the synthetic reaction shown is easily reversible. This accounts for the decomposition of thiazolidines by mercuric chloride or by mild oxidizing agents when the mercaptide or disulphide corresponding to the original mercaptan is obtained, respectively. Acylation increases the stability of the thiazolidine ring.

Interest in thiazolidines stems largely from the presence of an acylated thiazolidine ring in the penicillins (p. 65).

Rhodanine, 4-oxothiazolidine-2-thione (**22**) is obtained from chloroacetic acid and ammonium dithiocarbamate. It reacts with aromatic aldehydes at

CO_2H / CH_2Cl + $NH_2CS\text{-}S^-$ $^+NH_4$ → **22** —PhCHO→ PhCH= … → $PhCH_2CSCO_2H$ **23**

position 5, and on alkaline hydrolysis the products yield thiopyruvic acids (**23**). This can be compared with certain reactions of oxazolones (p. 375).

Benzothiazole (**24**), b.p. 234°C, is prepared in a similar way to benzoxazole (p. 376). It is a weak base which forms salts with strong acids and quaternary salts

NH_2 / SH + HCO_2H → **24**   **25** ($\overset{+}{N}Me$, Me, I^-)

with alkyl halides. The methyl group at position 2 of 2,3-dimethylbenzothiazolinium iodide (**25**) is highly reactive, and this compound has been converted to cyanine-type dyes (p. 318), which are valuable photographic sensitizers. The tautomeric 2-mercaptobenzothiazole (**26**) is an extremely good accelerator for rubber vulcanization.

NO_2 / Cl —Na_2S_2, CS_2, heat→ [NO_2 / SH → NH_2 / SH + S=C=S] → **26** (N, SH, S)

F. Thiazoles of natural occurrence and special interest

Vitamin B_1 (**27**), often called thiamine and occasionally aneurin, is a thiazolium salt. Its pyrophosphate (**28**) is a coenzyme which has been isolated in crystalline form from yeast and rice polishings and is involved in the decarboxylation of pyruvic acid to acetaldehyde. Thiamine can be converted into the pyrophosphate by heating with orthophosphoric acid and sodium pyrophosphate. Thiamine deficiency in man produces beriberi and polyneuritis.

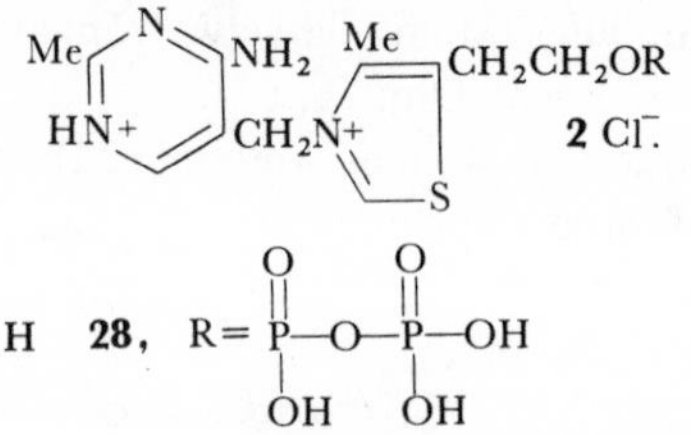

27, R = H   28, R = P(O)(OH)—O—P(O)(OH)—OH

The structure of the vitamin was deduced from the products of degradation and confirmed synthetically by R. R. Williams in 1935–1936. Thiamine is quite stable at pH 3–6, but it decomposes quite rapidly in alkaline solution, with the opening of the thiazolium ring. The final product (**29**) can be reduced back to thiamine by stannous chloride in hydrochloric acid.

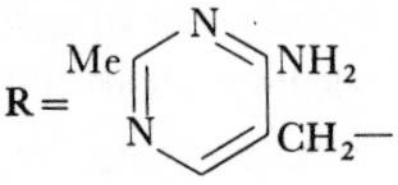

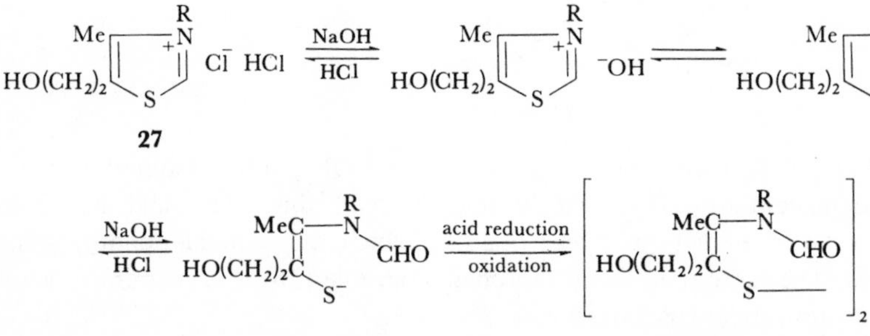

27

29

The key degradation which quickly led to the structure of the vitamin was effected by sodium sulphite. The products were a pyrimidine (**30**) and a thiazole (**33**), which on nitric acid oxidation gave the known 4-methylthiazole-5-carboxylic acid (**34**); this acid was also obtained by the direct nitric acid oxidation of thiamine. The pyrimidine (**30**) was converted into a known compound (**31**) on reduction, but the position of linkage between the two ring systems in the vitamin was not established. This point was clarified when thiamine and liquid ammonia were found to give the usual thiazole (**33**) and a new pyrimidine (**32**), which was prepared from **35** and ammonia.

Thiamine has been synthesized by a number of routes, of which the most important is outlined below. It consists in preparing the pyrimidine **35** and the thiazole **36** by standard methods and condensing them together. The product is thiamine bromide hydrobromide (**37**), which is converted into the chloride hydrochloride (**27**) by silver chloride.

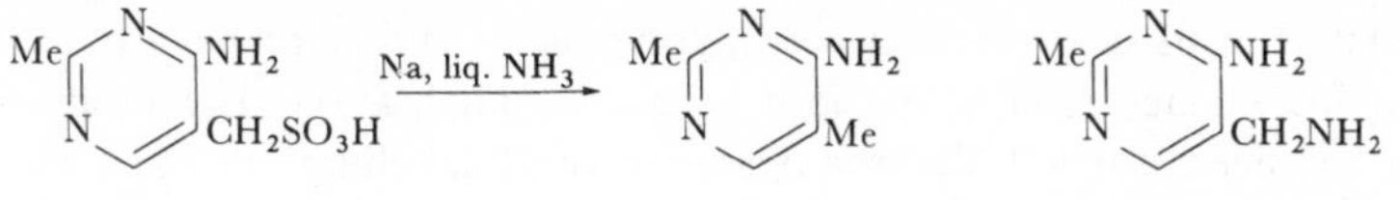

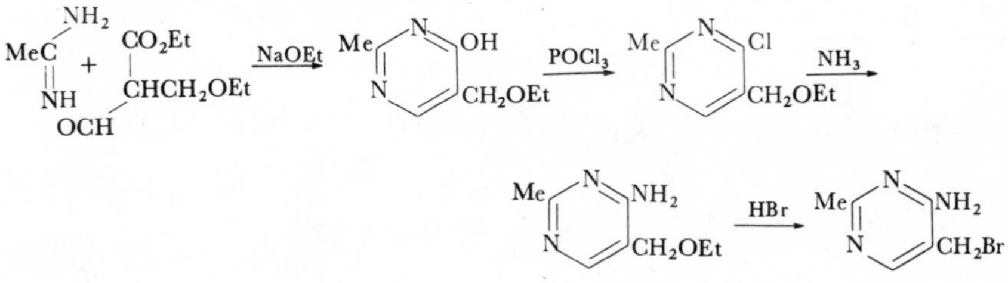

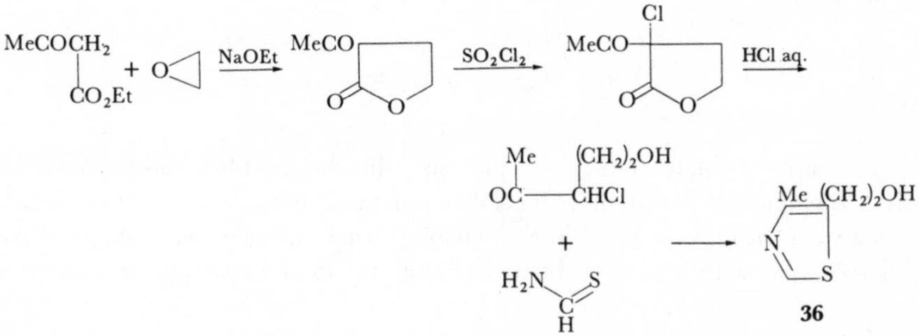

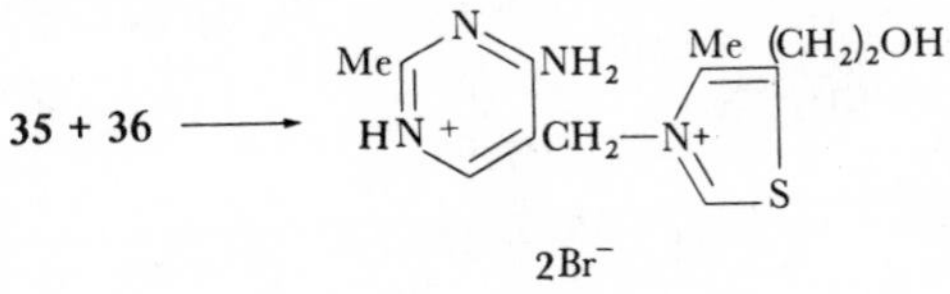

Thiamine pyrophosphate (**38**) is a coenzyme which is much involved in carbohydrate metabolism. It loses a proton from position 2 to give a resonance-stabilised (see p. 379) ylid, or zwitterion (**39**), which behaves as an anion (R^-) in its next reaction with pyruvic acid. This anion behaves as though it were a cyanide anion acting in a benzoin condensation. The product (**40**), which has been isolated, decarboxylates to the thiazolium salt (**42**), which has also been isolated. This last compound can split up into acetaldehyde and thiamine pyrophosphate via **41** and **39**, but, reacting again via the ion **41**, can also combine with acetaldehyde yielding acetoin (**43**) and reforming thiamine pyrophosphate.[65]

RH= Me N NH$_2$ Me (CH$_2$)$_2$ OPOPOH (O O; HO O$^-$) N CH$_2$N+ S
38
$\xrightleftharpoons[+H^+]{-H^+}$
Me N NH$_2$ Me (CH$_2$)$_2$ OPOPOH (O; HO O$^-$) N CH$_2$N+ S$^-$ = R^-
39

R^- + MeCOCO$_2$H ⟶ MeC(O$^-$)(R)CO$_2$H ⟶ MeC(OH)(R)CO$_2^-$ $\xrightarrow{-CO_2}$ MeC$^-$(OH)(R) $\xrightarrow{-MeCHO}$ R^-
39 **40** **41**

41 $\xrightleftharpoons[H^+]{-H^+}$ MeCHROH **42**

41 $\xrightarrow{MeCHO}$ MeC(OH)(R)—C(H)(Me)—O$^-$ ⟶ MeC(O$^-$)(R)—C(H)(Me)—OH ⟶ R^- + MeC(=O)CH(OH)Me
39 **43**

A yellow strongly fluorescent pigment, thiochrome (**46**), can be obtained from thiamine by oxidation with alkaline potassium ferricyanide and also occurs along with thiamine in yeast. It is probably formed through the opening of the thiazole ring with alkali to **44**, cyclization to **45**, followed by an oxidative

Me N NH$_2$ N CH$_2$—N(CHO) C=C(CH$_2$)$_2$OH (S$^-$ Na$^+$) Me
44
⟶
Me N N N N—C(Me) C(CH$_2$)$_2$OH ($^-$S Na$^+$)
45
⟶
Me N N N S (CH$_2$)$_2$OH N N Me
46

cyclization, which probably involves a sulphur radical eventually giving thiochrome (**46**). Sulphathiazole (**47**) is a useful antibacterial agent.

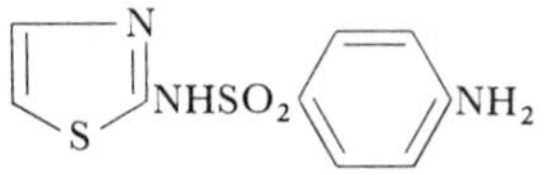

47

Luciferin, the chemiluminescent compound of the American firefly (**48**), has been synthesised[66] and may be formed *in vivo* from benzoquinone and cysteine.[67] The bioluminescence reaction takes place[68] through the formation

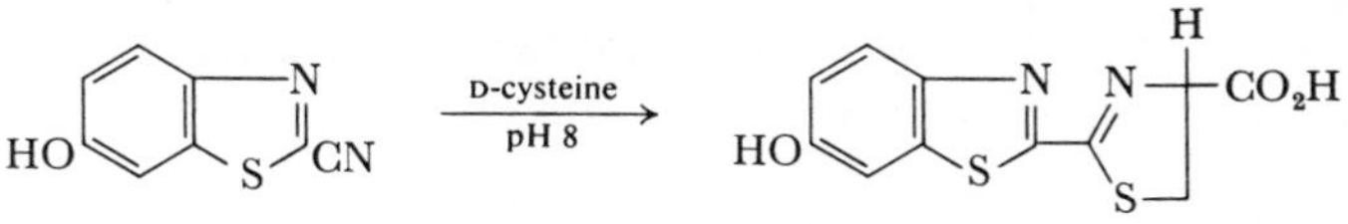

D-cysteine, pH 8

48

of the mixed anhydride (**49**) by reaction with adenosine triphosphate (p. 425). This with oxygen yields the dioxetanone **50**, which undergoes an electrocyclic ring opening[69] to give carbon dioxide and the actual light emitter, the excited thiazolinone anion **51**.[70]

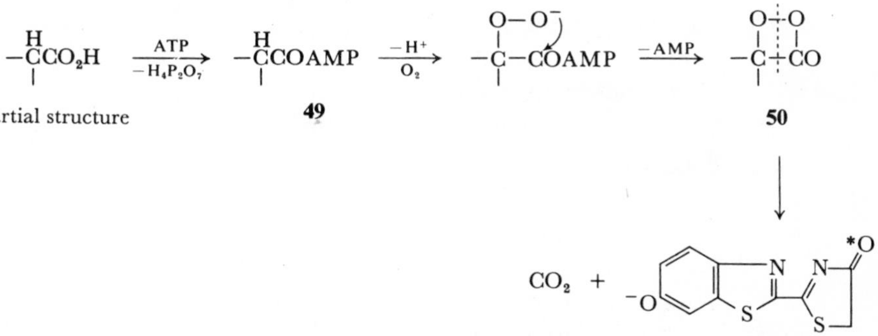

48, partial structure   **49**   **50**

CO_2 + **51**

4. DIOXOLE, DIOXOLANE, AND THEIR SULPHUR DERIVATIVES

1,3-Dioxole (**4**) is a colourless liquid, b.p. 51–52·5°C, which is very sensitive to oxygen and must be prepared and distilled under nitrogen. It was obtained[71] from the Diels–Alder adduct of anthracene (**1**) and vinylidene carbonate. Alkaline hydrolysis of this cyclic ester (**2**) gives the corresponding diol, converted

(a) HO⁻
(b) CH_2O/H^+

Δ → 1 + 4

1 2 3

to the acetal by paraformaldehyde and 4-toluenesulphonic acid. Pyrolysis of this acetal (**3**) causes a reverse Diels–Alder reaction yielding 1,3-dioxole. The benzo derivative, methylenedioxybenzene (**5**), can be obtained from catechol and methylenedichloride in 91% yield if the best conditions are employed.[72] Another route involves a desulphurisation.

CH_2Cl_2, Me_2SO, NaOH

(a) $CSCl_2$ + base
(b) Raney Ni

5

1,3-Dioxolane (**6**) is obtained from ethylene glycol, or oxirane, and formaldehyde. In general, dioxolanes of this type are prepared from the aldehyde or ketone and ethylene glycol in the presence of an acid catalyst under such conditions that the water formed is removed. This is commonly done by azeotropic distillation with benzene or toluene. 1,3-Dioxolanes are readily

$$\begin{array}{c} CH_2OH \\ | \\ CH_2OH \end{array} + O{=}CH_2 \rightleftharpoons H_2O + \mathbf{6}$$

hydrolysed by dilute aqueous acid but are stable to alkali. Dioxolane formation is therefore a very valuable and easy method of protecting carbonyl groups against alkali. Conversly, dioxolane formation, usually with acetone as the carbonyl component, is a very useful procedure for protecting hydroxyl groups which are suitably oriented for ring formation. This application is widely employed in sugar chemistry.

1,3-Dioxolan-2-one, or ethylene carbonate (**7**), is readily obtainable. It is very much more stable to acids than 1,3-dioxolane, but is decomposed to ethylene glycol and carbon dioxide by alkali. The formation of such carbonates is therefore a complementary protection procedure for suitably oriented (1,2-*cis*)

$$\begin{array}{l} CH_2OH \\ | \\ CH_2OH \end{array} + \begin{array}{l} Cl \\ Cl \end{array}\!\!>CO \xrightarrow{\text{pyridine}} \textbf{7}$$

7

hydroxyl groups where stability to acidic reagents is required and, again, has much use in sugar chemistry.

1,3-Oxathiolanes (e.g., **8**) and 1,3-dithiolanes (e.g., **10**) can be obtained from 2-mercaptoethanol and ethane-1,2-dithiol in the same way as 1,3-dioxolanes, and they have similar properties. 1,3-Dithiolanes on boiling with Raney nickel in alcohol are desulphurized, a very useful way of converting a ketone (**9**) into the corresponding hydrocarbon (**11**). 1,3-Oxathiolanes on similar treatment give

$$\textbf{8} \qquad \begin{array}{l} CH_2SH \\ | \\ CH_2SH \end{array} + O{=}CR_2 \longrightarrow \textbf{10} \xrightarrow[\text{Ni}]{\text{Raney}} R_2CH_2$$

8 **9** **10** **11**

back the original ketone. Although the cyclic peroxide 1,2-dioxolane is not known, its tetramethyl derivative (**12**) is obtainable from 2,5-dimethylpentan-2,5-diol with hydrogen peroxide and is stable enough for distillation on a small scale at atmospheric pressure. Few compounds of this type have been prepared. In contrast, 1,2-dithiolanes are well known. The parent compound **13** is an

$$\textbf{12} \qquad \begin{array}{l} CH_2{-}CH_2Br \\ | \\ CH_2Br \end{array} \xrightarrow{Na_2S_2} \textbf{13} \underset{FeCl_3}{\overset{\text{Zn, HOAc}}{\rightleftharpoons}} \begin{array}{l} CH_2{-}CH_2SH \\ | \\ CH_2SH \end{array}$$

12 **13**

unstable yellow oil with the properties of an aliphatic disulphide. A derivative, (+)-α-lipoic or 6,8-thioctic acid (**14**), is most interesting biochemically. It is a

$$ClCO(CH_2)_4CO_2Et \xrightarrow[AlCl_3]{CH_2{=}CH_2,} CH_2{=}CHCO(CH_2)_4CO_2Et \xrightarrow[\text{HOAc, } H_2]{\text{cobalt polysulphide,}}$$

$$\begin{array}{l} CH_2CH_2CH(CH_2)_4CO_2Et \\ |\qquad\quad | \\ SH \qquad SH \end{array} \xrightarrow{FeCl_3} \text{(1,2-dithiolan-3-yl)}(CH_2)_4CO_2Et \xrightarrow{\text{hydrolysis}}$$

$$\text{(1,2-dithiolan-3-yl)}(CH_2)_4CO_2H \xrightarrow{\text{Raney Ni}} Me(CH_2)_6CO_2H$$

14

growth factor for certain microorganisms, it is involved in the oxidative decarboxylation of pyruvic acid, and it may be involved in photosynthesis. Desulphurization with Raney nickel gives octanoic acid. The best synthesis of racemic 6,8-thioctic acid is outlined.

The aromatic 1,2- (**15**) and 1,3-dithiolium cations (**16**) have been prepared and much investigated.[73] Reduction of 3,5-diphenyl-1,2-dithiolium salts gives, reversibly, a stable radical (**17**) which slowly dimerises.[74]

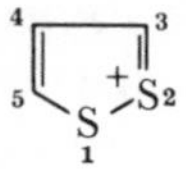

15 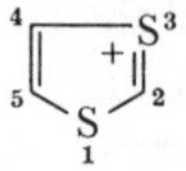 **16** 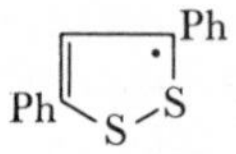 **17**

GENERAL BIBLIOGRAPHY

R. H. Wiley, *Five- and Six-Membered Compounds with Nitrogen and Oxygen (Excluding Oxazoles)*, Interscience, New York, 1962.

J. D. Loudon, in *Chemistry of Carbon Compounds*, Vol. IVA, ed. E. H. Rodd, Elsevier, New York, 1957.

PYRAZOLE

A. N. Kost and I. I. Grandberg, 'Progress in Pyrazole Chemistry,' *Adv. Heterocycl. Chem.*, **6**, 347 (1966).

R. H. Wiley (ed.), *Pyrazoles, Pyrazolines and Condensed Rings*, Wiley-Interscience, New York, 1967.

R. H. Wiley and P. Wiley, *Pyrazolones, Pyrazolidones and Derivatives*, Interscience, New York, 1957.

T. L. Jacobs, in *Heterocyclic Compounds*, Vol. 5, ed. R. C. Elderfield, Wiley, New York, 1956.

IMIDAZOLE

K. Hofmann, *Imidazole and its Derivatives*, Interscience, New York, 1953.

E. S. Schipper and A. R. Day, in *Heterocyclic Compounds*, Vol. 5, ed. R. C. Elderfield, Wiley, New York, 1956.

M. R. Grimmet, 'Advances in Imidazole Chemistry,' *Adv. Heterocycl. Chem.*, **12**, 104 (1970).

P. N. Preston, 'Benzimidazoles,' *Chem. Rev.*, **74**, 279 (1974).

L. B. Townsend, 'Imidazole Nucleosides and Nucleotides,' *Chem. Rev.*, **67**, 533 (1967).

R. J. Ferm and J. L. Riebsomer, 'Imidazoles,' *Chem. Rev.*, **54**, 593 (1954).

ISOXAZOLE

N. K. Kochefkov and S. D. Sokolov, *Advan. Heterocycl. Chem.*, **2**, 365 (1963).

R. A. Barnes, in *Heterocyclic Compounds*, Vol. 5, ed. R. C. Elderfield, Wiley, New York, 1956.

K.-H. Wünsch and A. J. Boulton, 'Indoxazenes and Anthranils,' *Adv. Heterocycl. Chem.*, **8**, 277 (1967).

OXAZOLES

J. W. Cornforth, in *Heterocyclic Compounds*, Vol. 5, ed. R. C. Elderfield, Wiley, New York, 1956.

R. Filler, 'Recent Advances in Oxazolone Chemistry,' *Adv. Heterocycl. Chem.*, **4**, 75 (1965).

R. Lakhan and B. Ternai, 'Advances in Oxazole Chemistry,' *Adv. Heterocycl. Chem.*, **17**, 99 (1974).

THIAZOLES

J. M. Sprague and A. H. Land, in *Heterocyclic Compounds*, Vol. 5, ed. R. C. Elderfield, Wiley, New York, 1956.

SELENAZOLES

E. Bulka, *Adv. Heterocycl. Chem.*, **2**, 344 (1965).

ISOTHIAZOLES

R. Slack and H. R. H. Wooldridge, *Adv. Heterocycl. Chem.*, **4**, 107 (1965).

M. Davis, 'Benzisothiazoles,' *Adv. Heterocycl. Chem.*, **14**, 43 (1972).

1,3-DIOXOLANES

R. C. Elderfield and F. W. Short, in *Heterocyclic Compounds*, Vol. 5, ed. R. C. Elderfield, Wiley, New York, 1956.

CYCLIC PEROXIDES

M. Schulz and K. Kirschke, 'Cyclic Peroxides,' *Adv. Heterocycl. Chem.*, 8, 165 (1967).

REFERENCES

1. H. W. W. Ehrlich, *Acta Cryst.*, **13**, 946 (1960); see also E. N. Maslen, J. R. Cannon, A. H. White, and A. C. Willis, *J.C.S. Perkin II*, 1298 (1974).
2. I-N. Hsu and B. M. Craven, *Acta Crystallogr.*, **B30**, 988 (1974).
3. A. F. Bedford, P. B. Edmondson, and C. T. Mortimer, *J. Chem. Soc.*, 2927 (1962).
4. K. Yamauchi and M. Kinoshita, *J.C.S. Perkin I*, 2506 (1973).
5. D. E. Butler and S. M. Alexander, *J. Org. Chem.*, **37**, 215 (1972).
6. J. W. A. M. Janssen, H. J. Koeners, C. G. Kruse, and C. L. Habraken, *J. Org. Chem.*, **38**, 1777 (1973).
7. M. W. Austin, J. R. Blackborow, J. H. Ridd, and B. V. Smith, *J. Chem. Soc.*, 1051 (1965).
8. S. Clementi, P. P. Forsythe, C. D. Johnson, and A. R. Katritzky, *J.C.S. Perkin II*, 1675 (1973).
9. A. G. Burton, A. R. Katritzky, M. Konya, and H. O. Tarban, *J.C.S. Perkin II*, 389 (1974).
10. I. I. Grandberg, L. G. Vasina, A. S. Volkova, and A. N. Kost, *Zh. Obshch. Khim.*, **31**, 1887 (1961).
11. J. D. Vaughan, D. G. Lambert, and V. L. Vaughan, *J. Amer. Chem. Soc.*, **86**, 2857 (1964).
12. H. P. Patel and J. M. Tedder, *J. Chem. Soc.*, 4589 (1963).
13. B. M. Lynch and M. A. Khan, *Can. J. Chem.*, **41**, 2086 (1963).
14. T. D. Breeze and J. H. Ridd, footnote 4 in *Tetrahedron*, **22**, 835 (1966).
15. A. Grimison and J. H. Ridd, *Proc. Chem. Soc.*, 256 (1958); M. S. R. Naidu and H. B. Bensusan, *J. Org. Chem.*, **33**, 1307 (1968).
16. R. D. Brown, H. C. Duffin, J. C. Maynard, and J. H. Ridd, *J. Chem. Soc.*, 3937; J. H. Ridd, *ibid*., 1238 (1955).
17. P. Beak, J. L. Miesel, and W. R. Messer, *Tetrahedron Lett.*, 5315 (1967).
18. J. A. Barltrop and A. C. Day, *Chem. Comm.*, 177 (1973).
19. H. Reimlinger, *Annalen,* **713**, 113 (1968); J. Bastide, J. Lemaitre, and J. Soulier, *Compt. Rend.*, **266**, *C*, 1393 (1968).

20. H. Bredereck, R. Gompper, R. Bangert, and H. Herlinger, *Angew. Chem.*, **70**, 269 (1958).
21. J. E. Oliver and P. E. Sonnet, *J. Org. Chem.*, **38**, 1437 (1973).
22. T. V. Van Auken and K. L. Reinhart, *J. Amer. Chem. Soc.*, **84**, 3736 (1962); T. V. Van Auken, *Diss. Abstr.*, **22**, 3413 (1962).
23. C. G. Overberger and J. P. Anselme, *J. Amer. Chem. Soc.*, **84**, 869 (1962).
24. W. A. F. Gladstone and R. O. C. Norman, *J. Chem. Soc. (C)*, 1536 (1966).
25. E. C. Taylor, R. L. Robey, and A. McKillop, *J. Org. Chem.*, **37**, 2797 (1972).
26. C. Ainsworth, *J. Amer. Chem. Soc.*, **79**, 5242 (1957).
27. H. Tiefenthaler, W. Dörscheln, H. Göth, and H. Schmid, *Tetrahedron Lett.*, 2999 (1964).
28. R. Garrer, G. O. Garrer, and H. Suschitzky, *J. Chem. Soc. (C)*, 825 (1970).
29. J. A. Elvidge, J. R. Jones, C. O'Brien, E. A. Evans, and J. C. Turner, *J.C.S. Perkin II*, 432 (1973).
30. F. F. Noe and L. Fowden, *Biochem. J.*, **77**, 543 (1960).
31. C. E. K. Mees, *The Theory of the Photographic Process*, 2nd ed., Macmillan, New York, 1954, p. 592.
32. H. S. Moyed and B. Magasanik, *J. Biol. Chem.*, **235**, 149 (1960).
33. B. Witkop and H. Kny, *Int. Congr. Biochem., 4th, Vienna*, 1958, Section 12–26.
34. M. Kraml and L. P. Bouthillier, *Can J. Biochem. Physiol.*, **34**, 783 (1956).
35. A. G. Beaman, W. Tanz, T. Gabriel, and R. Duschinsky, *J. Amer. Chem. Soc.*, **87**, 389 (1965).
36. H. Stone, *Biochem. Biophys. Res. Comm.*, **31**, 386 (1968); S. Inoue, S. Sugiura, H. Kakoi, and T. Goto, *Tetrahedron Lett.*, 1609 (1969).
37. D. J. Brown and P. B. Ghosh, *J. Chem. Soc. (B)*, 270 (1969).
38. I. Ambats and R. E. Marsh, *Acta Crystallogr.*, **19**, 942 (1965).
39. G. Stork and J. E. McMurray, *J. Amer. Chem. Soc.*, **89**, 5461 (1967).
40. S. Clementi, P. P. Forsythe, C. D. Johnson, A. R. Katritzky, and B. Terem, *J.C.S. Perkin II*, 399 (1974).
41. D. N. McGregor, U. Corbin, J. E. Swigor, and L. C. Cheney, *Tetrahedron*, **25**, 389 (1969).
42. R. B. Woodward, R. A. Olofson, and H. Mayer, *Tetrahedron*, Suppl. No. 8, Part I, 321 (1966).
43. A. I. Meyers, *Heterocycles in Organic Synthesis*, Wiley, New York, 1974.
44. M. Maeda and M. Kojima, *Chem. Comm.*, 539 (1973).
45. U. Türck and H. Behringer, *Chem. Ber.*, **98**, 3020 (1965).
46. M. Christl and R. Huisgen, *Tetrahedron Lett.*, 5209 (1968).
47. H. W. Wasserman and F. J. Vinick, *J. Org. Chem.*, **38**, 2407 (1973).
48. L. V. Rompuy, N. Schamp, N. O. Kimpe, and R. U. Parijs, *J. Chem. Soc. Perkin I*, 2503 (1973).
49. F. G. Riddell, J. M. Lehn, and J. Wagner, *Chem. Comm.*, 1403 (1968).
50. M. A. Casey, D. S. Kemp, K. G. Paul, and D. D. Cox, *J. Org. Chem.*, **38**, 2294 (1973).
51. Altaf-Ur-Rahman and A. J. Boulton, *Tetrahedron*, Suppl. 7, 49 (1966).
52. A. I. Meyers, E. D. Mihelich, and K. Kamata, *Chem. Comm.*, 768 (1974).
53. E. Baltazzi, *Quart. Rev.*, **9**, 150 (1955); H. E. Carter, *Org. React.*, **3**, 198 (1946).
54. E. E. Smissman, J. B. LaPidus, and S. D. Beck, *J. Amer. Chem. Soc.*, 79, 4697 (1957).
55. R. S. Karimoto, B. Axelrod, J. Wolinsky, and E. D. Schall, *Phytochemistry*, **3**, 349 (1964).
56. B. S. Joshi, W. I. Taylor, D. S. Bhate, and S. S. Karmarkar, *Tetrahedron*, **19**, 1437 (1963).

57. D. L. Pain and E. W. Parnell, *J. Chem. Soc.,*, 7283 (1965).
58. M. Davis and A. W. White, *J. Org. Chem.,* **34**, 2985 (1969).
59. M. Davis and A. W. White, *J. Chem. Soc.* (*C*), 2189 (1969).
60. M. Davis, E. Homfeld, and K. S. L. Srivastava, *J.C.S. Perkin I*, 1863 (1973).
61. J. Vitry-Raymond and J. Metzger, *Bull. Soc. Chim. France*, 1784 (1963).
62. J. Beraud and J. Metzger, *Bull. Soc. Chim. Fr.,* 2072 (1962).
63. P. Haake and W. B. Miller, *J. Amer. Chem. Soc.,* **85**, 4044 (1963); L. P. Bausher, *Diss. Abstr.,* **28**, B, 565 (1967).
64. R. H. Wiley, D. C. England, and L. C. Behr, *Org. Reactions,* **6**, 367 (1951).
65. E. P. Steyn-Parvé and C. H. Monfoort, *Comprehensive Biochem.,* **11**, 3 (1963).
66. E. H. White, F. McCapra, and G. F. Field, *J. Amer. Chem. Soc.,* **85**, 337 (1963).
67. F. McCapra and Z. Razavi, *Chem. Comm.*, 42 (1975).
68. P. J. Plant, E. H. White, and W. D. McElroy, *Biochem. Biophys. Res. Comm.,* **31**, 98 (1968); T. A. Hopkins, H. H. Seliger, E. H. White, and M. W. Cass, *J. Amer. Chem. Soc.,* **89**, 7148 (1967).
69. F. McCapra, *Proc. Chem. Soc.*, 155 (1968).
70. E. H. White, E. Rapaport, T. A. Hopkins and H. H. Seliger, *J. Amer. Chem. Soc.,* **91**, 2178 (1969).
71. D. N. Field, *J. Amer. Chem. Soc.,* **83**, 3504 (1961).
72. W. Bonthrone and J. W. Cornforth, *J. Chem. Soc.* (*C*), 1202 (1969).
73. H. Prinzbach and E. Futterer, *Adv. Heterocycl. Chem.,* **7**, 39 (1966).
74. C. T. Pederson, K. Bechgaard, and U. D. Parker, *Chem. Comm.,* 430 (1972).

CHAPTER VIII

COMPOUNDS WITH TWO HETEROATOMS IN A SIX-MEMBERED RING

These compounds can be formally derived from benzene, and its reduction products, by suitable substitutions of carbon (and hydrogen) atoms by nitrogen, oxygen, and sulphur. The most thoroughly studied ring system of this class is that of pyrimidine, in which the two nitrogen atoms bear the same positional relationship to each other as in the amidines and in the imidazoles (p. 354). This is largely because it occurs in many compounds vital to living systems. A

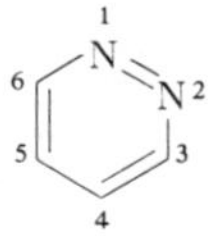

Pyridazine 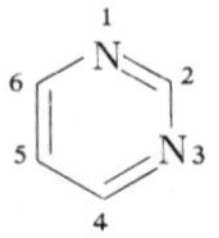 Pyrimidine 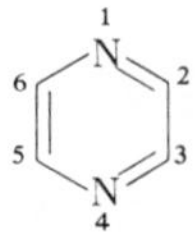 Pyrazine

substantial number of synthetic pyrimidines, and other compounds with the ring systems described in this chapter, are valuable chemotherapeutic agents.

Six-membered rings containing one heterocyclic oxygen or sulphur atom cannot be aromatic unless they bear a positive charge (p. 228), and very few of these are known with a second heteroatom in the ring. More fully reduced rings with these atoms have been examined in greater detail, and a number are considered. When oxygen and nitrogen atoms are present the name oxazine is

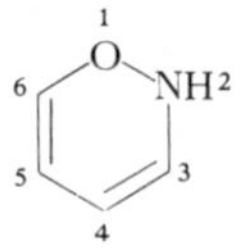

2H-1,2-Oxazine 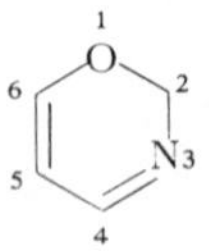 2H-1,3-Oxazine 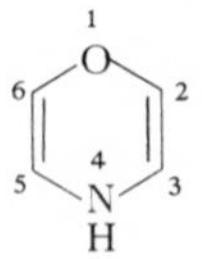 4H-1,4-Oxazine

used and the positions of the atoms are indicated by numbers. A similar nomenclature is used for the thiazines, which contain sulphur and nitrogen

1,2-Dioxane 1,3-Oxathiane 1,4-Dithiane

atoms. The term dioxane is used for the saturated ring with two oxygen atoms; dioxene and dioxadiene refer, respectively, to the comparable rings with one and two double bonds, and a similar nomenclature applies to sulphur analogues.

1. PYRIDAZINE

Pyridazine is a colourless liquid, miscible with both water and benzene but almost insoluble in cyclohexane. Its high boiling point (207·4°C/762 mm) is attributed to large dipole–dipole interactions. The dipole moment is 3·94 D, and the compound shows a highly aromatic type of absorption spectrum. Complete molecular dimensions are not available, but the lengths of the 1,2- and 4,5-bonds (1·330 and 1·375 Å) have been deduced from the microwave spectrum.[1] Pyridazine is therefore clearly aromatic and is best considered as a resonance hybrid to which the nonequivalent structures **1** and **2** are the major contributors. The resonance energy calculated[2] from heat of combustion data is 51·5 kJ (12·3 kcal)/mole. This value is probably too low (cf. pyrazine, p. 408).

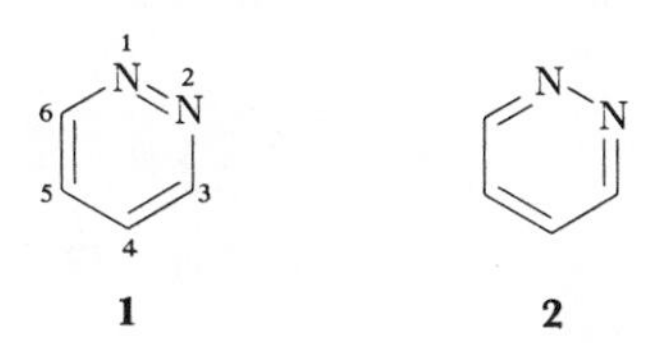

1 **2**

Pyridazine is a weak base (pK_a 2·33; pyridine has pK_a 5·23) which forms crystalline salts. It accepts only one proton, and even triethyloxonium tetrafluoroborate causes only monoalkylation. It is not attacked by electrophilic reagents under conditions where the ring system remains intact, but the mono *N*-oxide, obtained as in the pyridine series, undergoes the usual electrophilic and other reactions normal for such compounds (p. 249). Methyl groups and halogen atoms at positions 3 and 6 are active, like those at position 2 of pyridine (pp. 246 and 255). The pyridazine ring (**3**) can be converted into the pyrazine system (**6**) by a remarkable photorearrangement. In the case of certain perfluoroalkylpyridazines both intermediates **4** and **5** have been isolated.[3] Pyridazines are usually obtainable from hydrazine and suitable dicarbonyl compounds (e.g., **7**, often obtainable from furans, pp. 129 and 130). Pyridazine can be synthesised from maleic dialdehyde as below, but is best prepared from maleic anhydride and hydrazine. The first product is the tautomer **8**, there being

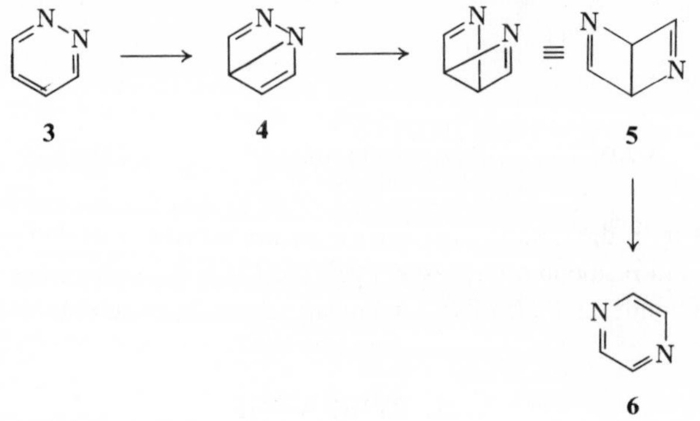

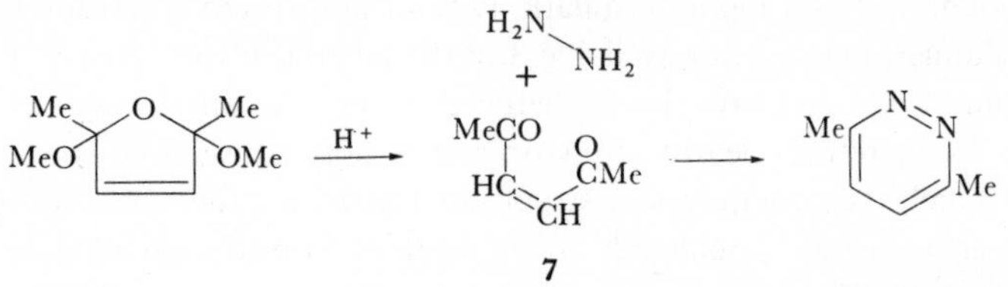

no detectable amounts of **9** and **10** present.[4] The halogen atoms of 3,6-dichloropyridazine (**11**), being reactive, are readily replaced by hydrogen.

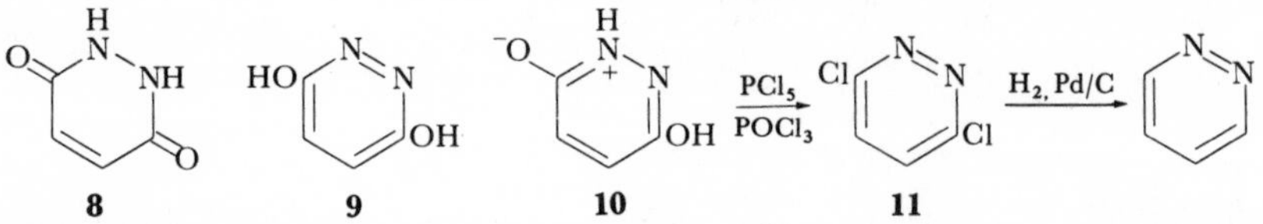

Maleic hydrazide, with *t*-butyl hypochlorite at −50 to −77°C, is oxidized to the extremely reactive 2,3-diazaquinone (**12**). On warming it loses nitrogen yielding **13**, and with butadiene it gives **14**.[5]

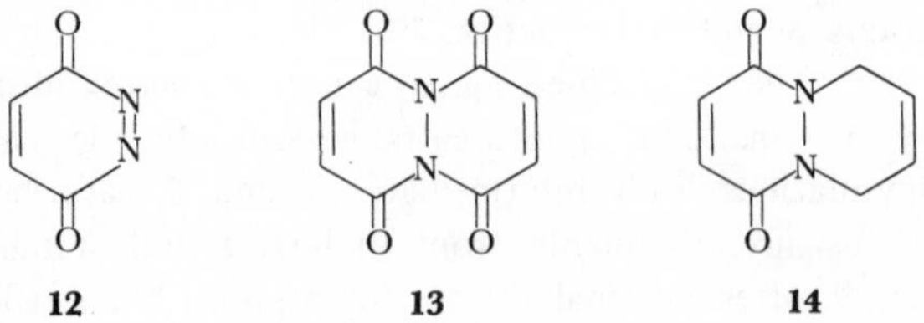

The pyridazine ring can be fused onto a benzene ring in two ways, giving cinnoline (**15**) or phthalazine (**16**). Phthalhydrazide (**17**), the structure of which

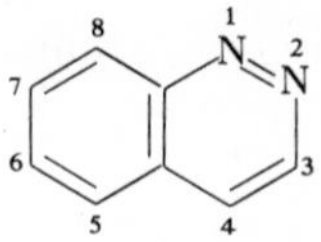

15 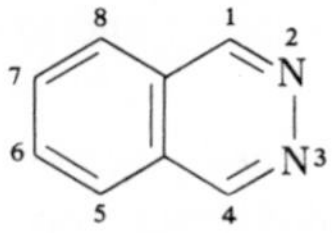 **16**

offers a problem similar to that of maleic hydrazide (**8, 9, 10**), is prepared from phthalimide, or phthalic anhydride, and hydrazine. It is also obtained when hydrazine is used instead of hot concentrated hydrochloric acid to decompose *N*-substituted phthalimides in Gabriel's synthesis of primary amines. Its 5-amino

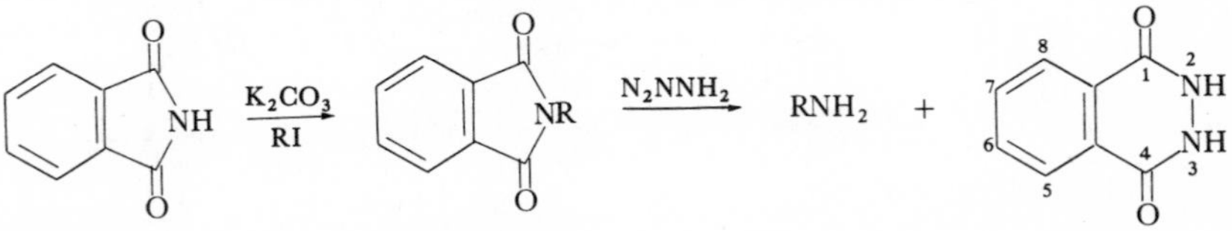

$\xrightarrow[RI]{K_2CO_3}$ $\xrightarrow{N_2NNH_2}$ RNH_2 +

R = alkyl **17**

derivative (luminol) is especially interesting, as with a number of oxidizing agents in the presence of alkali it emits a blue light (chemiluminescence).

2. PYRIMIDINE

A. Introduction

Pyrimidine derivatives, which strictly include the purines (p. 419), occur very widely in living organisms and are among the first compounds to have been studied by organic chemists. Uric acid (p. 420) is a complex pyrimidine, and its oxidation to alloxan (p. 421), a simple pyrimidine, was described in 1845. The barbiturates, valuable soporific and hypnotic drugs, and a number of useful

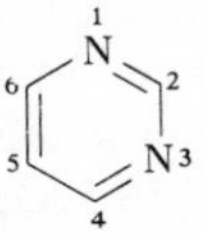

Pyrimidine

antibacterial and antimalarial drugs also contain pyrimidine rings. Vitamins B_1 (p. 381) and B_2 (p. 278) and some of the nucleotides (p. 406) are also pyrimidines. The function and structure of a number of coenzymes possessing the pyrimidine ring have been established. Of these many contain the pyrimidines uracil and cytosine and are known to be concerned principally with the biosynthesis of complex carbohydrates and lipids. At present great interest is being taken in the chemistry and biochemistry of pyrimidines.

B. Physical properties and structure

Pyrimidine is a colourless compound, m.p. 22·5°C, b.p. 124°C. Its dimensions have been determined[6] by an X-ray diffraction study of a crystal at −2°C and closely resemble those of pyridine (p. 230). A number of derivatives, for example, 4-amino-2,6-dichloropyrimidine have been examined similarly and in this case the results were obtained with sufficient accuracy to establish that the

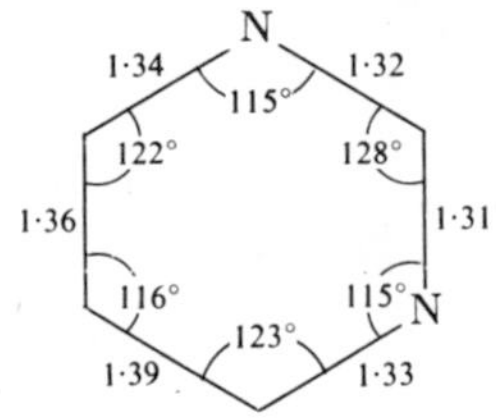

Pyrimidine, bond lengths in Angstroms

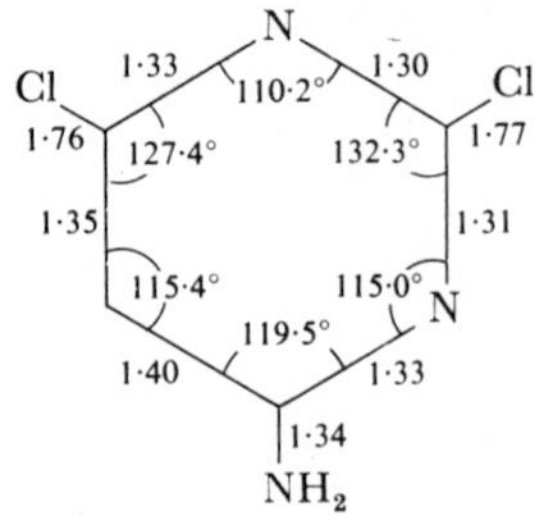

4-Amino-2,6-dichloropyrimidine

nitrogen atom outside the ring is attached to two hydrogen atoms. This excludes alternative tautomeric structures with one of these hydrogen atoms attached to a ring nitrogen atom (cf. 2- and 4-aminopyridines, p. 258). Although the rings of these pyrimidines are not symmetrical in the crystals examined, none of the carbon–carbon bond distances (1·35–1·40 Å) approach that of the aliphatic carbon–carbon single bond distance (1·54 Å). They are similar to that for benzene (1·40 Å) or between this length and that for ethylene (1·33 Å). The general similarity of the size and shape of the pyrimidine ring to that of benzene and pyridine (p. 230) is consistent with its highly aromatic chemical characteristics. The differences between the bond angles and distances of these ring systems suggest that that of pyrimidine is the least aromatic. This is in agreement with the resonance energies calculated[7] by molecular-orbital methods: benzene 150·5 (36), pyridine 129·5 (31), and pyrimidine 109 (26) kJ (kcal)/mole. The resonance energy for pyrimidine of 33·5 kJ (8 kcal)/mole, calculated from heat of combustion data,[8] is probably too low (cf. pyrazine, p. 408).

Pyrimidine is best considered as a resonance hybrid to which the uncharged equivalent Kekulé structures **1** and **2** and the charged structures **3**–**8** contribute. The self-consistent π-electron densities, calculated for the ground state of pyrimidine, are 0·776, 0·825, and 1·103 for positions 2, 4, and 5, respectively.[9]

C. Chemical properties of pyrimidine and its derivatives

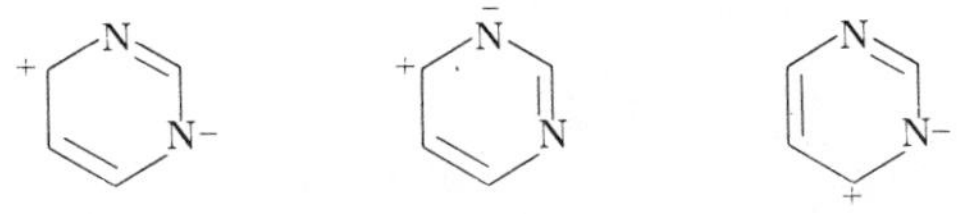

Although an enormous amount of work has been done with pyrimidine derivatives, most of which have been obtained directly from the excellent syntheses available, the chemical reactions of pyrimidine itself (**1**) have been investigated only recently. This is partly because pyrimidine was not readily available, but it can now be obtained quite easily by the decarboxylation of pyrimidine-4,6-dicarboxylic acid[10] or by the catalytic dechlorination of 2,4-dichloropyrimidine (**15**, p. 400).[11]

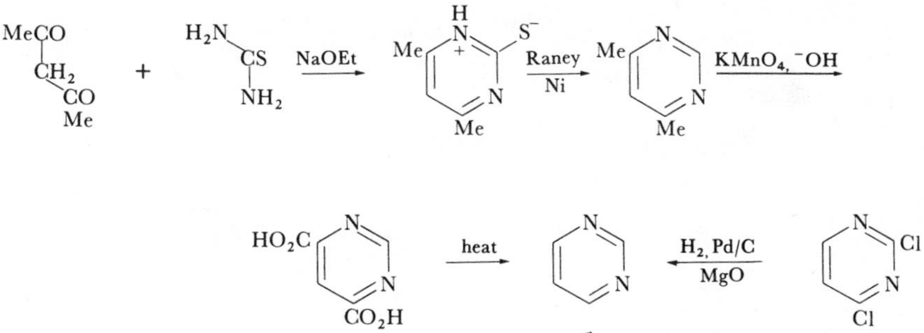

Pyrimidines can be considered best as derivatives of pyridine and, to a lesser extent, as cyclic amidines. Pyrimidine, which accepts two protons under extremely acidic conditions (pK_{a_1} 1·3, pK_{a_2} −6·9), is a much weaker base than pyridine (pK_a 5·23), imidazole (pK_a 7·2), or amidines in general. This is because, unlike imidazole and the amidines, the addition of a proton does not increase the possibilities for resonance and hence the resonance energy. It is a surprisingly weaker base than pyridazine (pK_a 2·33). Only one of the nitrogen atoms of pyrimidine is alkylated by alkylating agents, such as methyl sulphate,[10] but the much more powerful agent triethyloxonium borofluoride alkylates both nitrogen atoms to give a ring bearing two positive charges.[12] Pyrimidine forms only a mono *N*-oxide,[10] which unlike pyridine 1-oxide (p. 250) has not been nitrated successfully.[10]

From a consideration of the charged structures contributing to the resonance hybrids which represent pyrimidine (p. 397) and pyridine, and the π-electron densities (p. 396), it is clear that position 5 of pyrimidine should correspond to position 3 of pyridine and be the most susceptible in the ring to electrophilic

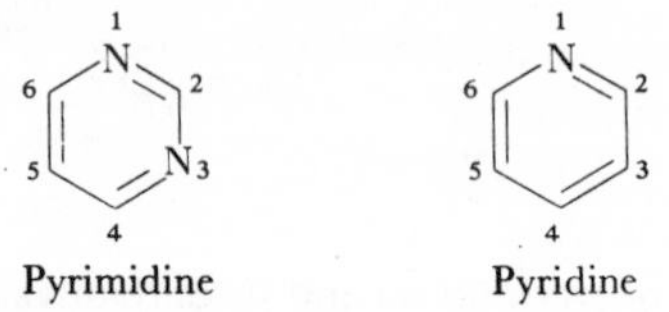

Pyrimidine Pyridine

attack. Pyrimidine (like pyridine, p. 238) hydrochloride is brominated at position 5, but no other electrophilic substitution of pyrimidine itself has been claimed. If activating groups, such as hydroxyl or amino, are present at other positions in the molecule, electrophilic substitution (nitration, nitrosation, diazo coupling) usually occurs, but only at position 5. Groups present at position 5 generally have properties similar to those exhibited by the same substituents at position 3 of pyridine, or present in a benzene ring. An example is 5-hydroxypyrimidine (**9**), which has been obtained as outlined.[13] It has phenolic properties and gives a red ferric chloride colour. 5-Aminopyrimidine,

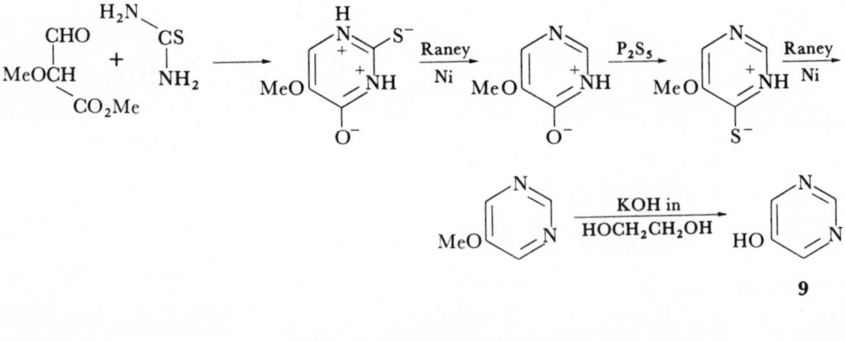

Raney Ni; P_2S_5; Raney Ni; KOH in $HOCH_2CH_2OH$

9

however, unlike 3-aminopyridine (p. 257), reacts with nitrous acid, but neither 5-hydroxypyrimidine nor a diazonium salt could be detected in the product, which remains unidentified. The synthesis of 5-hydroxypyrimidine shows two instances of the Raney nickel desulphurization of mercaptopyrimidines, and as mercaptopyrimidines are easily synthesized, this desulphurization procedure has had many applications. 5-Nitropyrimidine (**11**) has been synthesised[14] by oxidising the bishydrazino derivative **10** with silver oxide, and on hydrogenation yields 5-aminopyrimidine. With water it undergoes covalent hydration to **12**, the first example of this type of addition in the pyrimidine series.[14]

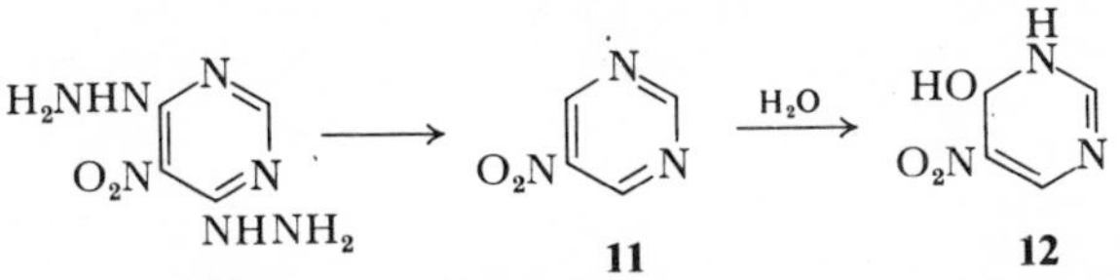

In a similar way positions 2, 4, and 6 of pyrimidine formally correspond to those of 2 or 4 of pyridine and, in the few cases investigated, are attacked by nucleophilic reagents such as sodamide and phenylmagnesium bromide[13];

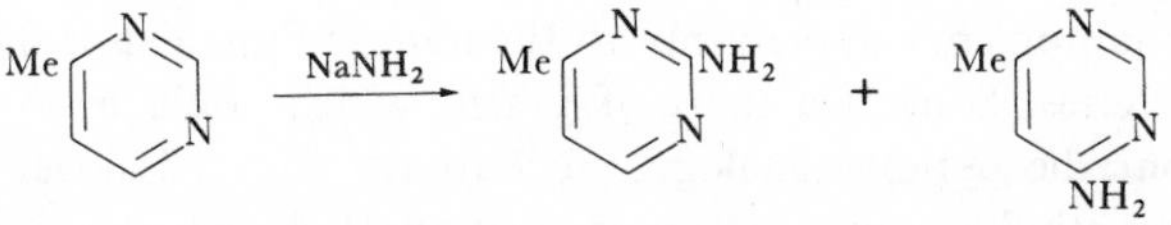

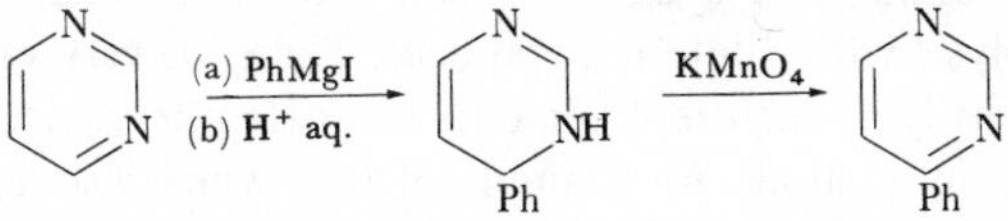

however, pyrimidine itself is decomposed by hot aqueous alkali. Substituents at these positions also have comparable reactivities and properties in the two series. For instance, 2,4,6-trimethylpyrimidine (cf. 2-methylpyridine, p. 247) reacts with benzaldehyde to give 2,4,6-tristyrylpyrimidine (**13**), and 2-aminopyrimidine, like 2-aminopyridine (p. 260) can be converted to the 2-fluoro derivative by sodium nitrite in 43% tetrafluoroboric acid.[15]

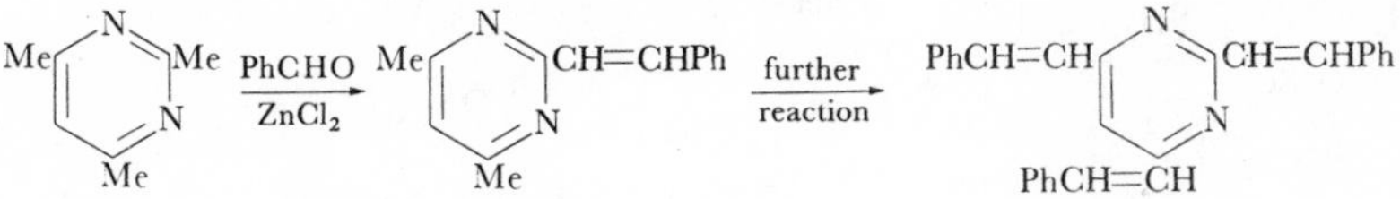

Pyrimidine is attacked at the 2- and 4-positions by the 4-nitrophenyl radical.[16]

X-Ray diffraction studies[17] have shown that the arrangement of the atoms in crystalline uracil (2,4-dihydroxypyrimidine, **14**) is analogous to that in 2-pyridone (p. 262) and that, likewise, there is considerable hydrogen bonding between molecules. In general, often by ultraviolet absorption spectrum comparisons, it has been shown that hydroxy- and mercaptopyrimidines adopt the tautomeric keto structure as long as the aromaticity of the ring can be maintained by a suitable electron distribution (e.g. as in **14**), while aminopyrimidines do not tautomerize appreciably to the imino forms. This is the same as in the pyridine series. Some reactions of uracil, which could easily have been predicted from the pyridine analogy, are outlined above. Halogen atoms and methoxy and methylmercapto groups at positions 2, 4, or 6 of a pyrimidine can be replaced by amino groups on treatment with ammonia, or by hydroxyl groups on hydrolysis with dilute mineral acid. These are very valuable synthetic procedures, especially as such replacements can often be carried out stepwise in a polysubstituted pyrimidine. An example of this, which also illustrates another use of the mercapto group in pyrimidine chemistry, is the synthesis of cytosine [**19**, 4-amino-2-(1*H*)pyrimidone].[18] 4-Hydroxy-2-mercaptopyrimidine (**17**), prepared from thiourea and ethyl formylacetate (p. 150) and which is probably most accurately written in the zwitterionic form shown (cf. **14** above), gives the *S*-ethyl, and not the *N*-ethyl, derivative (**18**) with ethyl iodide and sodium ethoxide. Cytosine is then obtainable as indicated; an alternative synthesis is hydrolysis of **16**.

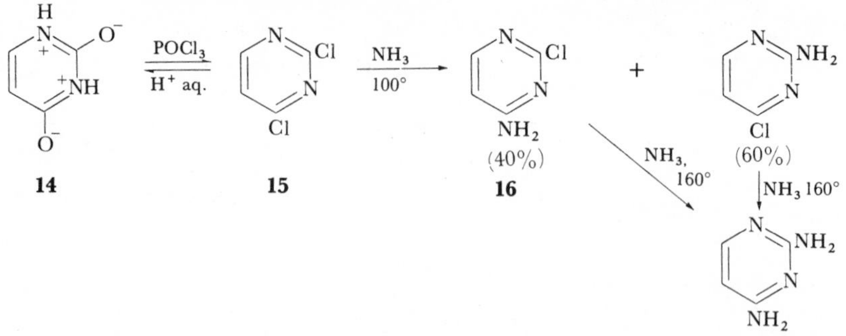

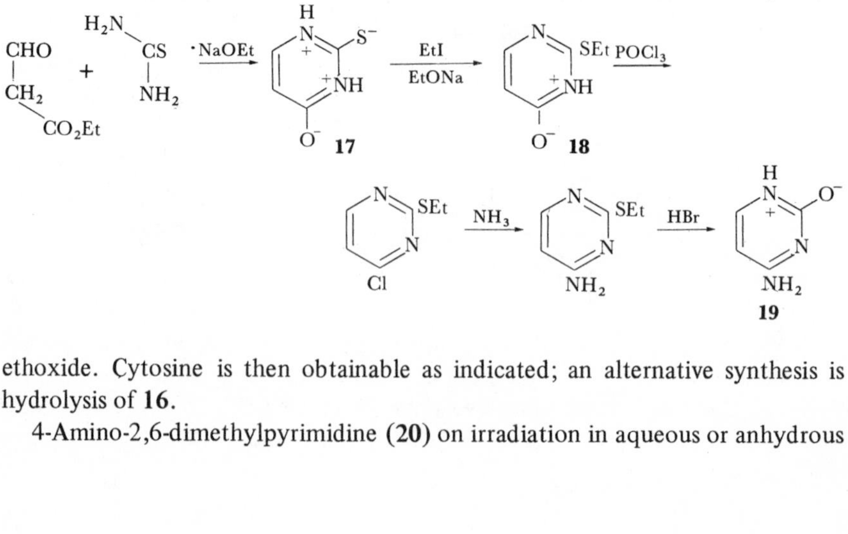

4-Amino-2,6-dimethylpyrimidine (**20**) on irradiation in aqueous or anhydrous media gives 2-amino-3-cyanopent-2-en-4-imine (**22**) in quantitative yield. It is thought that the primary photoproduct is **21**, which then breaks down.[19]

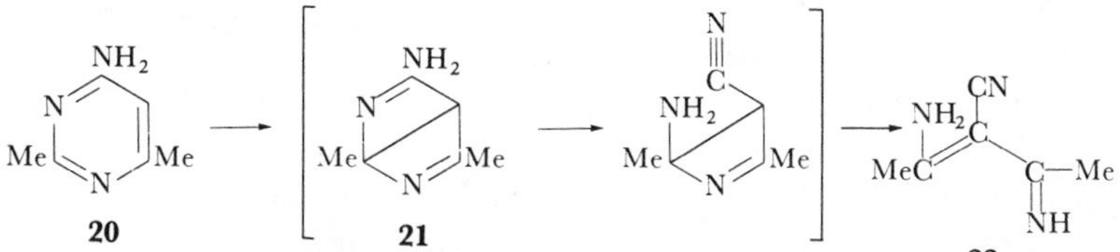

20 → [**21** →] → **22**

In aqueous solution thymidine (**23**) and other pyrimidines undergo[20] a reversible photodimerisation to mainly *cis–syn–cis* derivatives (**24**). This is an extremely important type of reaction, for dimerisation of such pyrimidine residues in DNA strands can lead to faulty transcriptions when replication occurs and consequently causes mutations. Compounds such as **24** can be reduced with sodium borohydride[21] and can be a useful source of cyclobutanes.

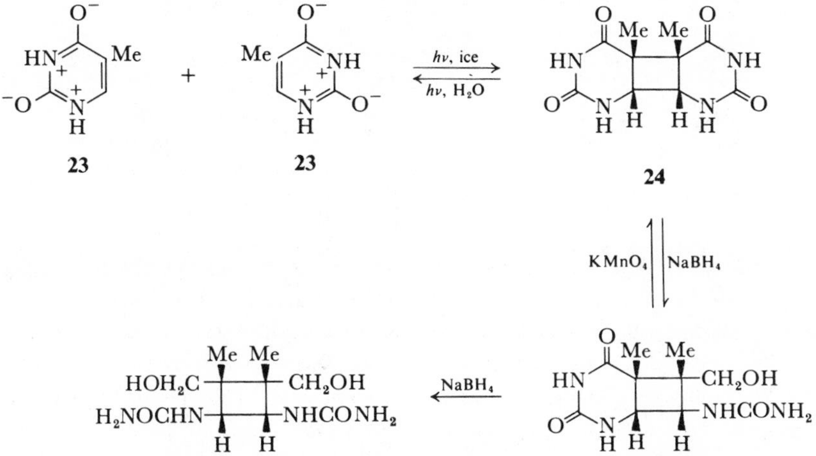

23 + **23** ⇌ **24** ($h\nu$, ice; $h\nu$, H_2O)

24 ⇌ (KMnO$_4$; NaBH$_4$); NaBH$_4$

Barbituric acid (synthesis p. 404) is a tautomeric substance which can be formulated in many ways. X-Ray studies[22] of the solid have shown that it has structure **27** although it is often represented as 2,4,6-trihydroxypyrimidine (**25**)

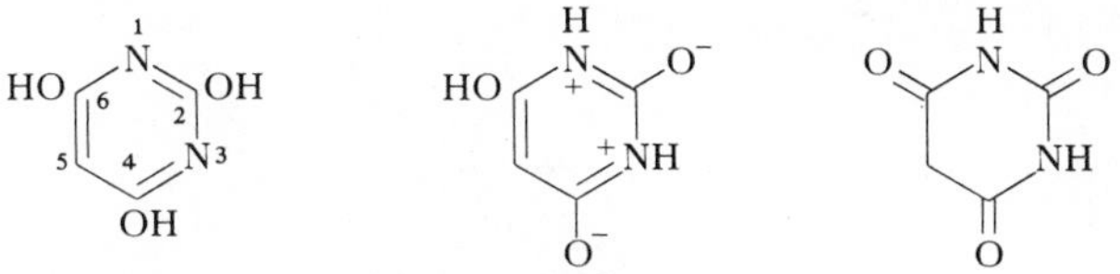

25 **26** **27**

or as **26**. It is a much stronger acid than its 5,5-dialkyl derivatives and has a different ultraviolet absorption spectrum. from them. Alkali removes protons successively from positions 1 and 3 of **27** giving anions, although alkylation under basic conditions gives 5,5-dialkyl derivatives (see also, p. 407) before further reaction takes place at the nitrogen atoms. 5-Arylbarbituric acids cannot be prepared by a similar arylation but are synthesized from phenylmalonic ester by the standard procedure (p. 404).

On nitrosation barbituric acid **28** yields the tautomeric violuric acid (**29**), which forms purple salts. Violuric acid is also formed from alloxan (**30**), a degradation product of uric acid (p. 421), and hydroxylamine. Alloxan has the unusual property of causing a type of diabetes in experimental animals. It is normally obtained as the monohydrate (**31**), which is difficult to dehydrate. The

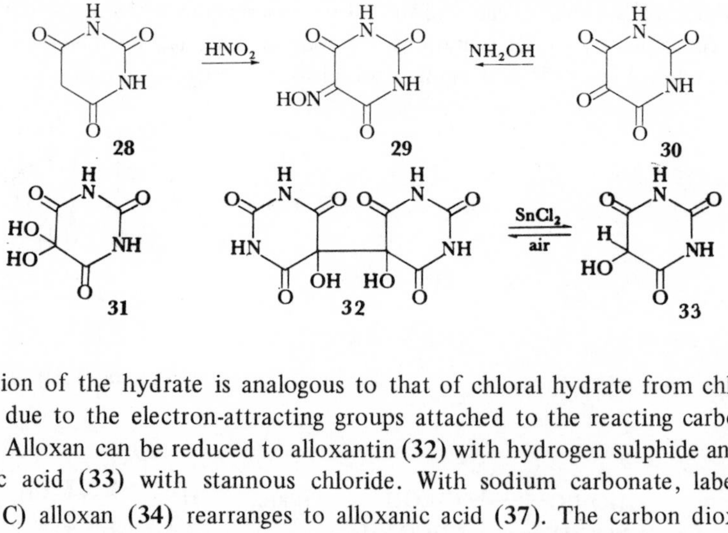

formation of the hydrate is analogous to that of chloral hydrate from chloral and is due to the electron-attracting groups attached to the reacting carbonyl group. Alloxan can be reduced to alloxantin (**32**) with hydrogen sulphide and to dialuric acid (**33**) with stannous chloride. With sodium carbonate, labelled (* = ^{14}C) alloxan (**34**) rearranges to alloxanic acid (**37**). The carbon dioxide formed in the nitric acid oxidation of **37** to a parabanic acid (**38**) contains no label,[23] so ring contraction can proceed via a N → C migration (**35**) or by hydrolysis to **36** and recyclisation.

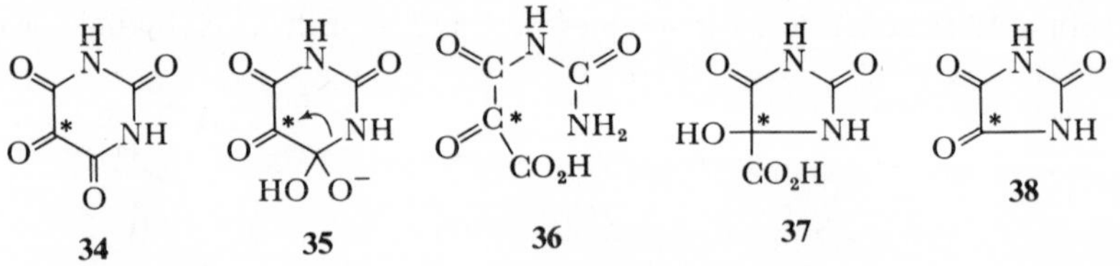

Murexide (**39**), a purple substance formed from alloxantin, ammonium acetate, and glacial acetic acid, is the colouring matter produced in Wöhler and Liebig's test for uric acid (p. 421). The 'murexide test' consists of evaporating the suspected uric acid with concentrated nitric acid and subsequently adding ammonia. Murexide is used as an indicator in complexometric titrations.[24]

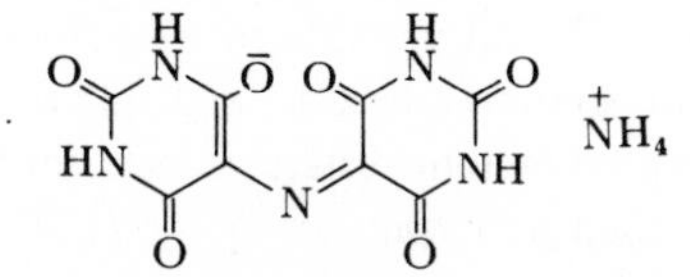

$\overset{+}{N}H_4$

39

Hydrogenation of pyrimidine, and often of derivatives, under acid conditions gives 1,4,5,6-tetrahydropyrimidine (**40**), which has the properties of an aliphatic

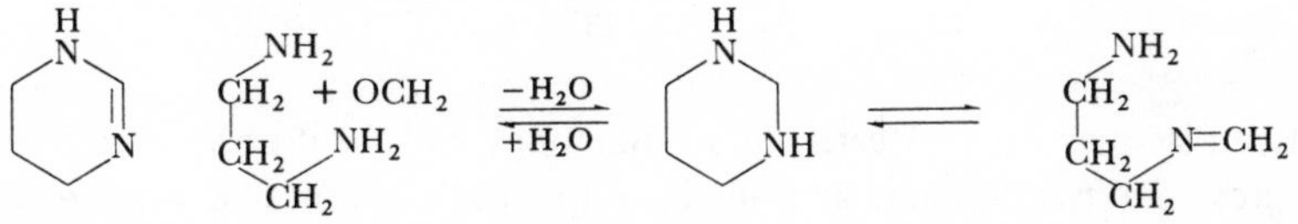

$-H_2O$ / $+H_2O$

40 41 42

amidine. It has also been prepared from propylene-1,3-diamine with formamidine, or ethyl formate followed by pyrolysis.[25] Hexahydropyrimidine (**41**) can be obtained from formaldehyde and 1,3-diaminopropane. It appears to exist in equilibrium with the tautomeric Schiff's base (**42**) and is very readily hydrolysed to the starting materials.

Benzopyrimidine (**43**) is usually called quinazoline, and a few alkaloids contain this ring system. Quinazoline is stable to both cold aqueous acid and alkali, but in hot acid or alkali the ring opens to give 2-aminobenzaldehyde and polymers. It is readily reduced to 3,4-dihydroquinazoline (**44**).

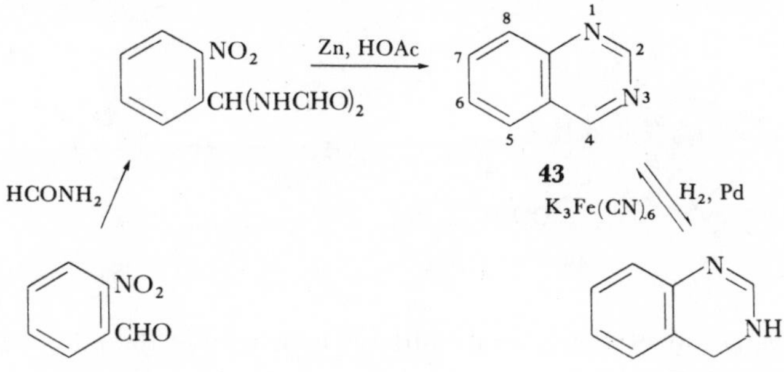

NO_2, $CH(NHCHO)_2$ — Zn, HOAc → 43

$HCONH_2$; NO_2, CHO

43 ⇌ 44 ($K_3Fe(CN)_6$; H_2, Pd)

D. Synthetic methods

Although the pyrimidine ring system has been built up in a number of ways, the most common and versatile method is that in which the ring is formed from two compounds which contribute the N–C–N and C–C–C atoms.

(1) The earliest synthesis of this type, published in 1879, is of barbituric acid (**45**) from urea, malonic acid, and phosphorus oxychloride. An improvement is to use ethyl malonate and sodium ethoxide, and this variation, which also works with mono- and disubstituted malonic esters, is used for the preparation of many of the barbituric acids used as hypnotics (p. 407). The malonic ester can be

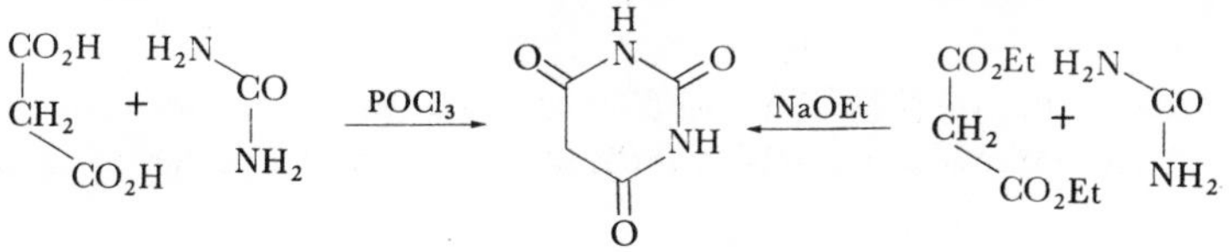

CO_2H H_2N CH_2 + CO CO_2H NH_2 $\xrightarrow{POCl_3}$ H N O O NH O $\xleftarrow{NaOEt}$ CO_2Et H_2N CH_2 + CO CO_2Et NH_2

45

replaced by many *β*-keto esters and *β*-diketones, and the urea by guanidine (**47**), thiourea, *S*-alkylthioureas, and amidines (Traube's synthesis). Amidine hydrochlorides (**46**) are readily available from the corresponding nitriles, as indicated,

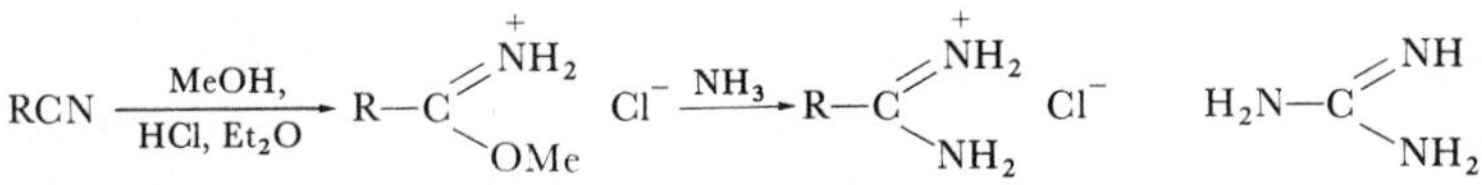

$RCN \xrightarrow[HCl,\ Et_2O]{MeOH,} R{-}C(=\overset{+}{N}H_2)OMe\ Cl^- \xrightarrow{NH_3} R{-}C(=\overset{+}{N}H_2)NH_2\ Cl^-$ $\qquad H_2N{-}C(=NH)NH_2$

46 **47**

but formamidine is more conveniently obtained by the Raney nickel desulphurization of thiourea. Pyrimidines of general structure **48** can therefore be obtained directly, and substituents can often be inserted later at position 5 (p. 398). The synthesis of uracil (**50**) from urea, malic acid, and sulphuric acid is worth special mention; the malic acid decomposes *in situ* to formylacetic acid (**49**, see p. 284).

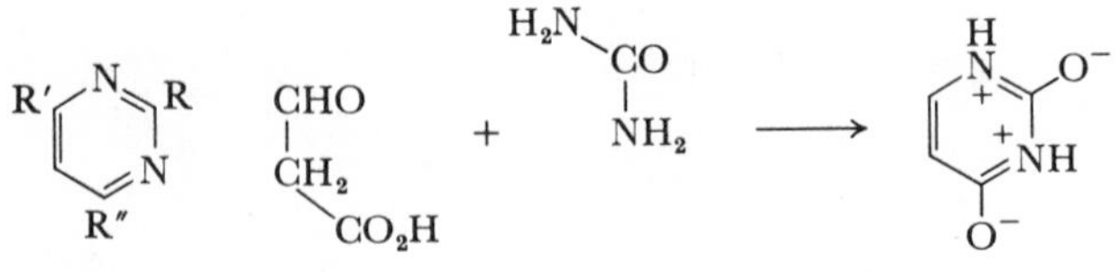

R′ N R N R″ CHO CH_2 CO_2H + H_2N CO NH_2 ⟶ H N O⁻ + + NH O⁻

48 **49** **50**

(2) A variation of the general synthesis is to replace the ethyl malonate by ethyl cyanoacetate, when the reaction takes place in two stages.

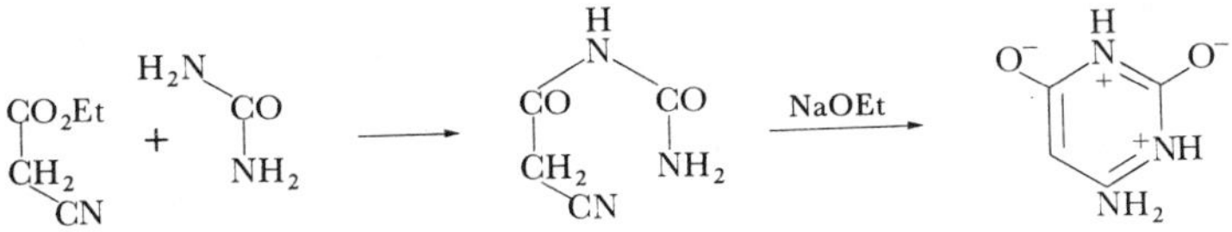

If a hydrogen atom is required at position 4 it is often preferable to use an ethoxymethylene compound, such as **51**, which can be obtained from ethyl cyanacetate and ethyl orthoformate in the presence of sodium ethoxide, instead

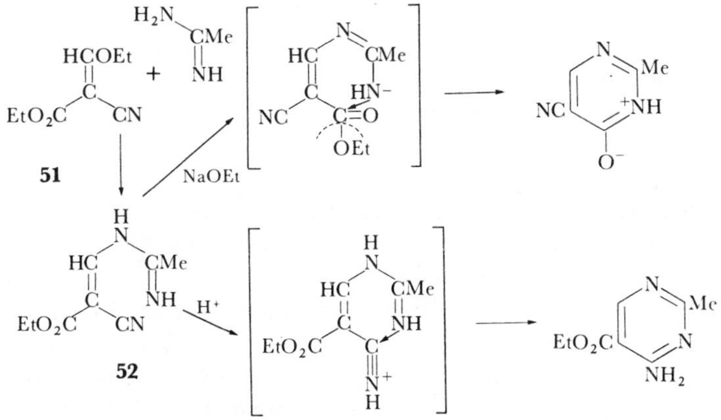

of the corresponding β-formyl ester. The differing modes of cyclization of **52**, according to the conditions, are of interest.

(3) Another synthesis involves the condensation of amidines, or ureas, with unsaturated compounds, such as ethyl crotonate, under basic conditions. It is possible that a Michael addition to the double bond is followed by cyclization. The resulting dihydropyrimidine is readily oxidized.

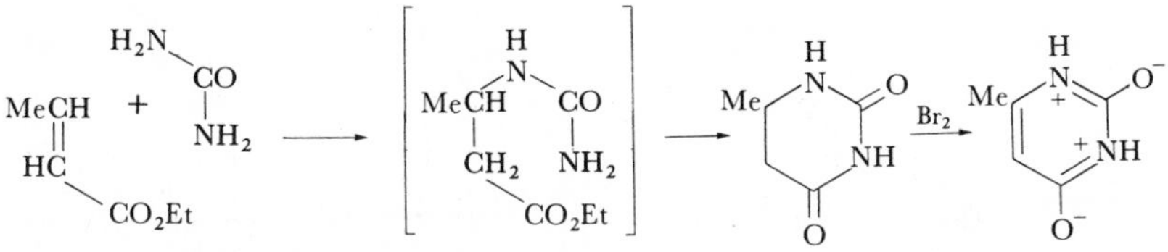

(4) A synthesis of another type, discovered[26] in 1953, appears to be quite general and is very convenient. Ethyl orthoformate reacts with ureas or thioureas to give the corresponding formamidines (**53**), and these, on refluxing in an inert solvent with compounds possessing an active methylene group, yield complex ethylenes (**54**); often both reactions can be carried out simultaneously by

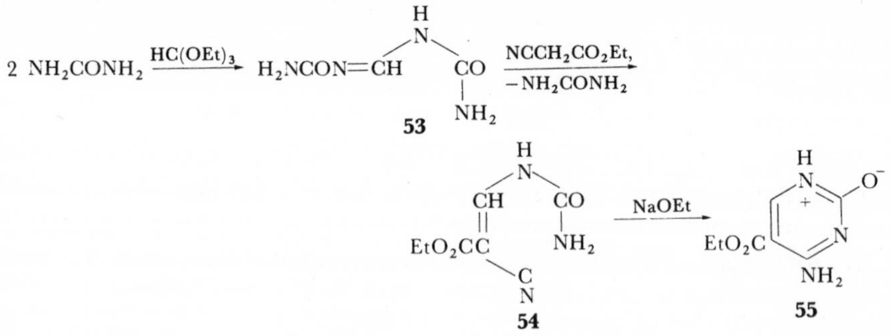

$$2\ NH_2CONH_2 \xrightarrow{HC(OEt)_3} H_2NCON{=}CH{-}NH{-}CO{-}NH_2\ (\mathbf{53}) \xrightarrow[-NH_2CONH_2]{NCCH_2CO_2Et,} (NC)(EtO_2C)C{=}CH{-}NH{-}CO{-}NH_2\ (\mathbf{54}) \xrightarrow{NaOEt} \mathbf{55}$$

heating the three reactants together in an inert solvent. Cyclization of these ethylenes to the pyrimidines (**55**) is effected by sodium ethoxide, and the overall yields are very good.

E. Natural occurrence and compounds of special interest

Although pyrimidine itself has not been found in nature, many substituted pyrimidines, and compounds in which the pyrimidine ring is a part of a more complex ring system, are very widely distributed. Vitamins B_1 (p. 381) and B_2 (p. 431) are pyrimidines. Certain pyrimidine (and purine) ribosides and deoxyribosides, called nucleosides, occur as phosphoric esters, the nucleotides, in most living cells. Some coenzymes are nucleotides and play a key role in metabolic processes. The nucleic acids[27] are macromolecules (cf. starch) which consist of many nucleotide molecules linked together through phosphate groups and can

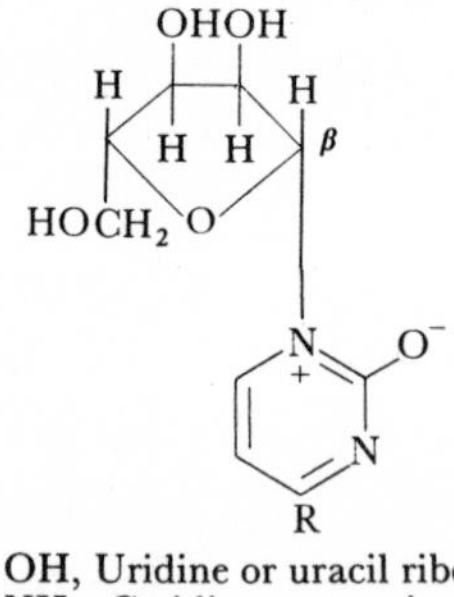

R = OH, Uridine or uracil riboside
R = NH_2, Cytidine or cytosine riboside

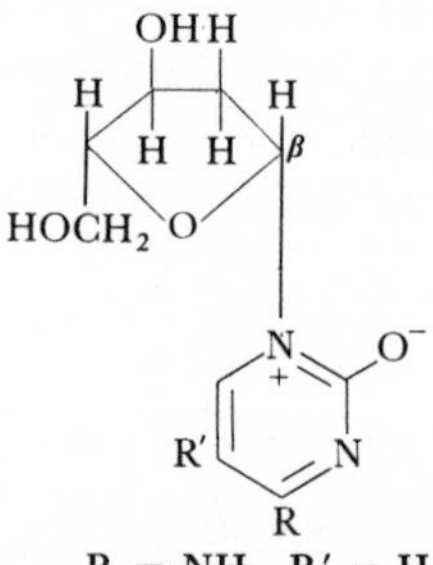

R = NH_2, R′ = H, Cytosine deoxyriboside
R = OH, R′ = Me, Thymidine
R = NH_2, R′ = Me, 5-Methylcytosine deoxyriboside

vary considerably in their purine–pyrimidine ratios. They occur as such, or in association with proteins, in all living cells and are constituents of chromosomes and viruses. The common pyrimidine nucleosides, which are obtained on alkaline hydrolysis of ribonucleic (RNA) and deoxyribonucleic (DNA) acids, are formulated above. Their structures have been determined by degradation, and in some cases confirmed by synthesis. The methods used are quite similar to those employed in the case of the purine glycoside adenosine (p. 424), which is treated in detail as an example.

Derivatives of barbituric acid (p. 401) are perhaps the most widely used pyrimidines in medicine. Veronal (**56**) and Luminal (**57**) are valuable hypnotics, while Pentothal (**58**) is used as a quick-acting anaesthetic. Other pyrimidines of

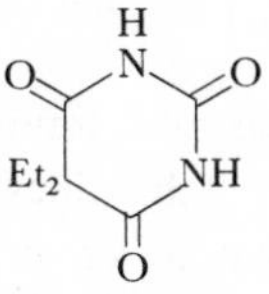

H N O O Et_2 NH O **56**

H N O O Et Ph NH O **57**

H N O S Et PrMeCH NH O **58**

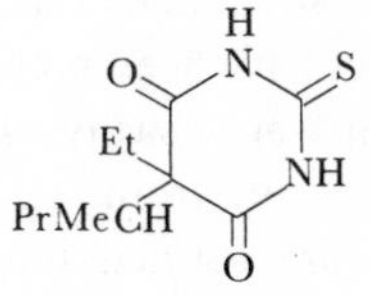

current medicinal interest are the antibacterial agents sulphadiazine (**59**), sulphamerazine (**60**), and sulphamezathine (**61**). The antibiotic 'bacimethrin' is a comparatively simple pyrimidine (**62**) and has been synthesized.[28]

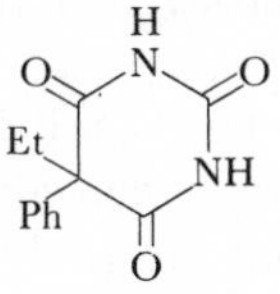

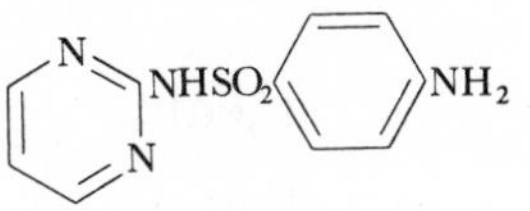

N $NHSO_2$ NH_2 N **59**

N $NHSO_2$ NH_2 N Me **60**

Me N $NHSO_2$ NH_2 N Me **61**

N OMe $HOCH_2$ N NH_2 **62**

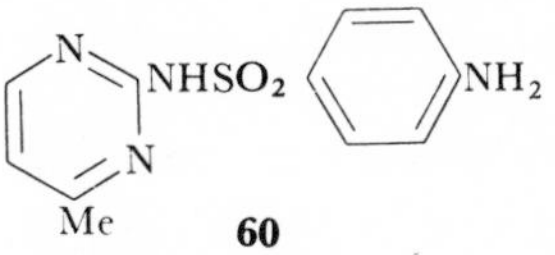

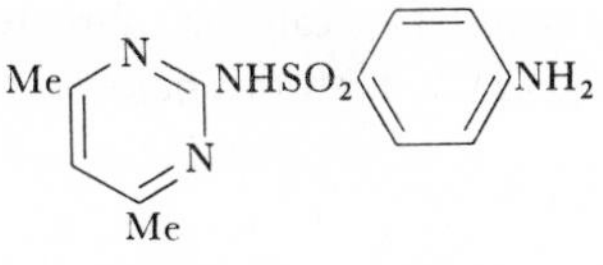

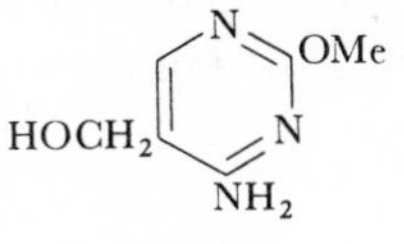

3. PYRAZINES

Pyrazine (**1**) is a colourless solid, m.p. 54°C, b.p. 121°C. Its dipole moment is zero, and the molecular dimensions calculated[29] from the rotational fine structure of its ultraviolet spectrum are shown and resemble those for pyridine

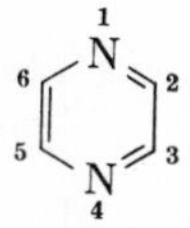

1 N 2 3 N 4 5 6

1, Pyrazine

N 1·341 Å 115° 1·395 Å 122·5° N

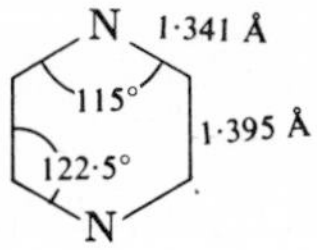

(p. 230). The resonance energy for pyrazine has not been determined accurately, but it appears to be about 102 kJ (24·3 kcal)/mole,[30] the earlier value of 33·5 kJ (8·0 kcal)/mole obtained by a different procedure[8] being much too low.

Pyrazine is a weak base (pK_{a_1} 0.6, pK_{a_2} −6.2) which is monoalkylated by the usual alkylation reagents. Triethyloxonium tetrafluoroborate, however, ethylates[12] both nitrogen atoms (cf. pyrimidine p. 397 and contrast pyridazine p. 393). Pyrazine is either not attacked or it is decomposed by electrophilic reagents. With chlorine at 400°C it (radical mechanism?) gives 2-chloropyrazine and with sodamide 2-aminopyrazine is formed; both pyrazine mono- and di-*N*-oxides have been prepared.

Pyrazines are usually obtained by the self-condensation of α-amino-carbonyl compounds or from 1,2-dicarbonyl derivatives with 1,2-diamines; dihydropyrazines are formed initially but are easily oxidised by air or are dehydrogenated. This type of synthesis is unsatisfactory for pyrazine itself, owing to

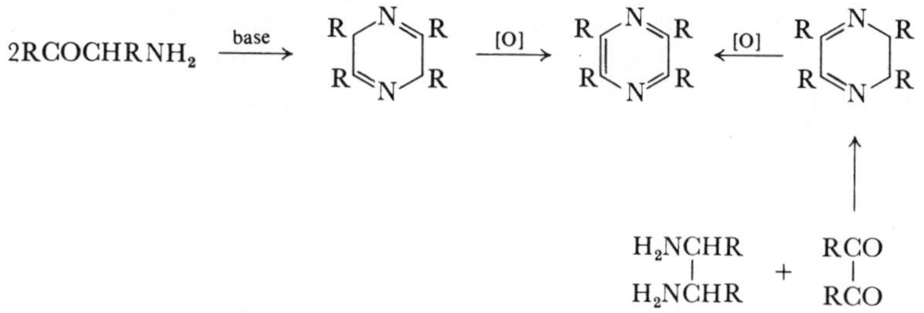

2RCOCHRNH$_2$ —base→ (dihydropyrazine, R,R,R,R) —[O]→ (pyrazine, R,R,R,R) ←[O]— (dihydropyrazine, R,R,R,R) ← H$_2$NCHR–CHRNH$_2$ + RCO–COR

competing alternative reactions, but pyrazine (**1**) can be obtained by the vapour-phase dehydrogenation of piperazine (**2**) or as shown below.

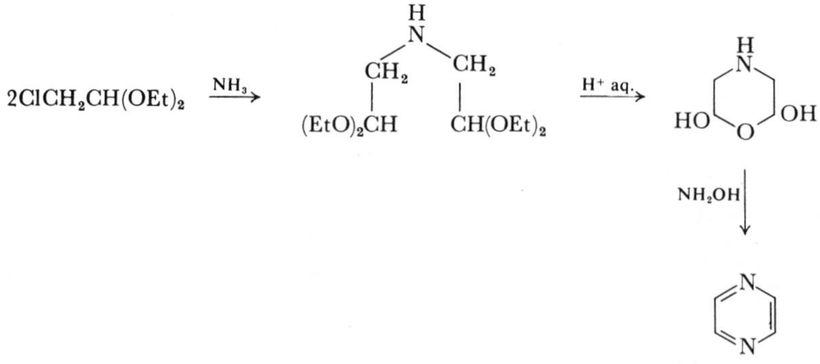

2ClCH$_2$CH(OEt)$_2$ —NH$_3$→ HN(CH$_2$CH(OEt)$_2$)$_2$ —H⁺ aq.→ (2,6-dihydroxymorpholine, HO, OH) —NH$_2$OH→ (pyrazine) **1**

Piperazine (**2**) is a typical aliphatic secondary diamine and can be prepared in the laboratory from aniline by the general secondary aliphatic amine synthesis shown. It is widely used for treating certain parasitic infections of the pig.

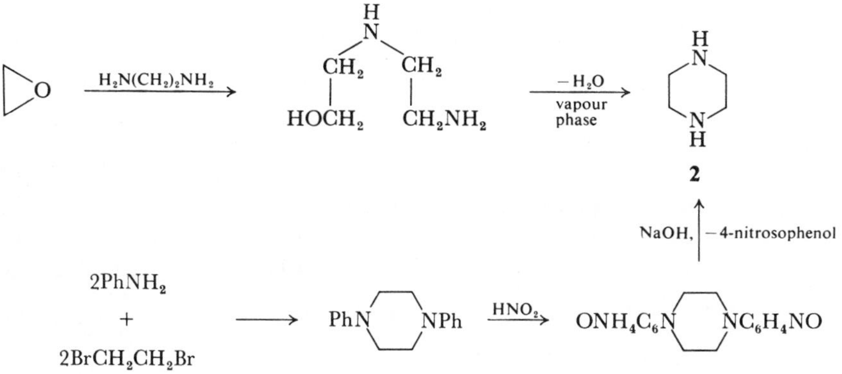

2,5-Dioxopiperazines, such as **3**, which is perhaps more accurately represented as **4**, can be obtained by the thermal dehydration of α-amino acids. 2,5-Dioxopiperazine on careful partial hydrolysis yields glycylglycine (**5**), and it

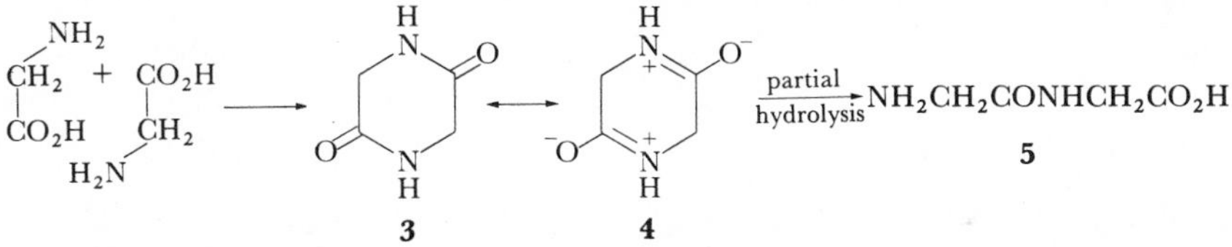

$NH_2CH_2CONHCH_2CO_2H$

5

also gives a silver salt (**6**). This with methyl iodide gives the *N,N'*-dimethyl derivative (**7**), while treatment with benzyl chloride yields **8**, which is an imino ether and is very readily hydrolysed to benzyl alcohol and 2,5-dioxopiperazine by dilute aqueous acid.

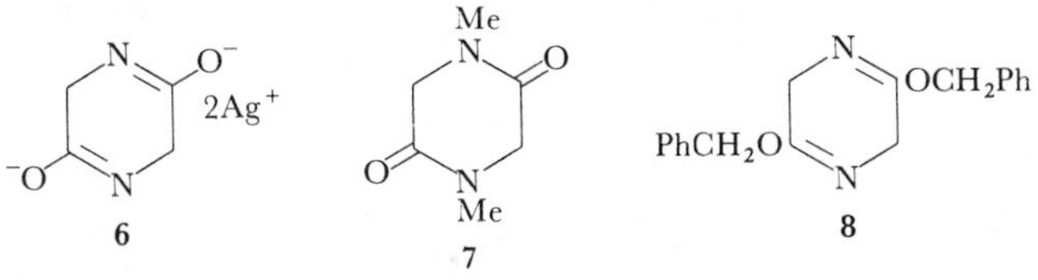

The pyrazine ring is present in a number of naturally occurring compounds. Most of these are pteridines (p. 429), where the ring is fused with that of pyrimidine, but the mould *Aspergillus flavus* produces both **9** and the antibiotic aspergillic acid (**10**), which has been synthesised.[31] *Cypridina* luciferin (p. 367) is a complex pyrazine.

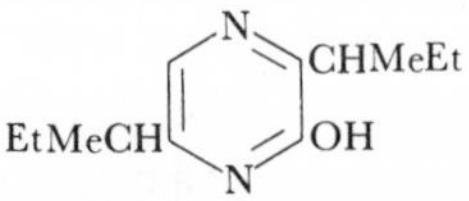

9

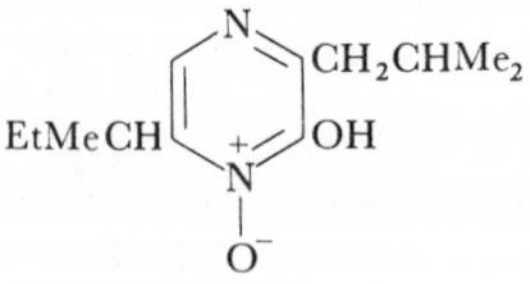

10

Benzopyrazine is usually called quinoxaline (**11**). Quinoxalines are readily formed from 1,2-dicarbonyl compounds and aromatic 1,2-diamines. As quinoxalines are usually highly crystalline, this reaction is valuable diagnostically for both types of reactant.

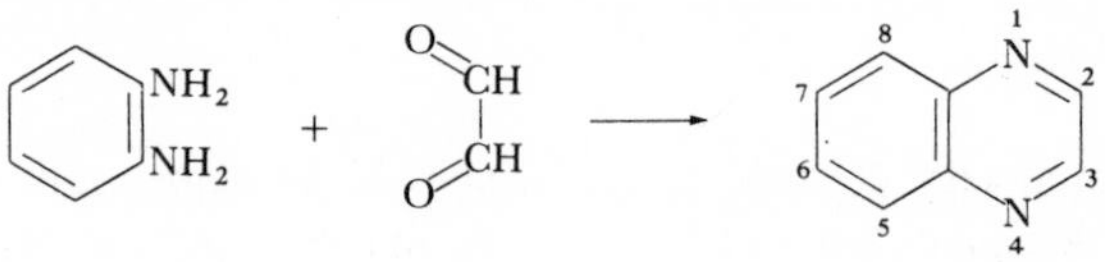

11

Dibenzopyrazine is called phenazine (**12**).

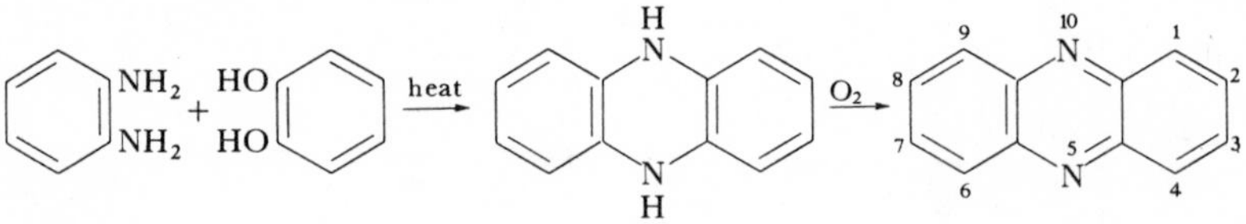

12

4. OXAZINES AND THIAZINES

A. Introduction

Comparatively little work has been done on simple derivatives of these ring systems, and most of this concerns the reduced 1,3- and 1,4-compounds.

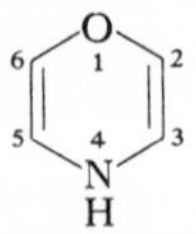

4H-1,4-Oxazine 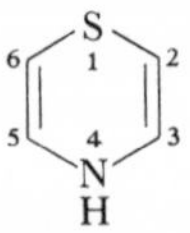 4H-1,4-Thiazine

Cephalosporin (p. 68) is a fused 1,3-thiazine, and in the last few years 1,3-oxazines have proved to be valuable synthetic intermediates.[32]

B. 1,3-Oxazines

4*H*-1,3-Oxazines (e.g., **2**) can be obtained from acylated β-amino ketones (**1**) by treatment with phosphorus oxychloride. They are oxidisable to 1,3-oxazinium salts (**3**), which, like pyrylium salts (p. 282), yield the corresponding nitrogen-containing analogues (e.g., **4**) with ammonia.

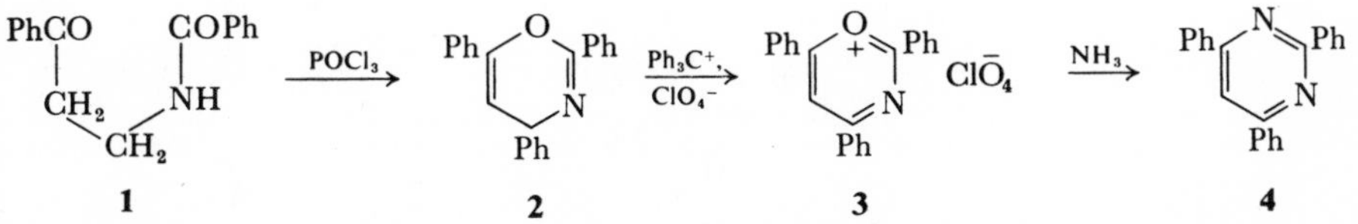

5,6-Dihydrooxazines, exemplified by 5,6-dihydro-2,4,4,6-tetramethyl-1,3-4*H*-oxazine (**6**), are of great interest as synthetic intermediates.[32,33] This compound is easily obtained[34] from the diol **5** and acetonitrile, and the 2-methyl group loses a proton to butyl- or phenyllithium but not to Grignard reagents. Alkyl bromides or iodides attack the resulting carbanion, and subsequent reduction and hydrolysis give aldehydes (**7**) in high yields.[35]

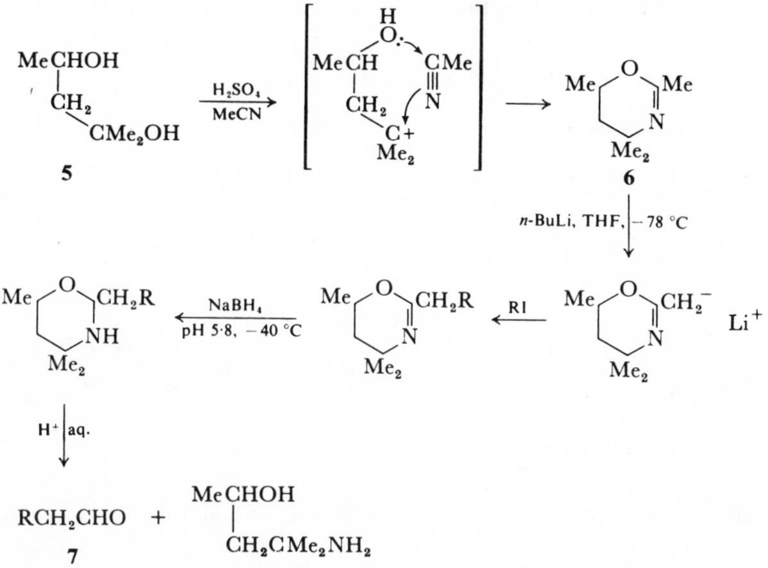

Methylation of the oxazine **6** gives a compound (**8**) with an even more reactive 2-methyl group, and then formation of the enamine **9**, alkylation to **10**, formation of another enamine (**12**), and reduction to **11** can be effected in one reaction vessel. Hydrolysis of **11** then gives an aldehyde.[36] The synthesis can also

6 —MeI→ **8** —NaH, THF or DMF, $-H^+$→ **9** —RI→ **10** —$-H^+$, NaH→ **12** —$NaBH_4$→ **11** —H^+ aq.→

RCH_2CHO + $MeCHOH$ | CH_2CMe_2NHMe

be used to prepare ketones, for the salt **10** is attacked by Grignard reagents at position 2, and hydrolysis of the product (**13**) gives the ketone (**14**).[37]

10 —R′MgX→ **13** —H^+ aq.→

$MeCHOH$ | CH_2CMe_2NHMe + $R'COCH_2R$ **14**

C. 1,4-Oxazines and Thiazines

The most important simple 1,4-oxazine is morpholine (**16**), or tetrahydro-1,4-oxazine, which is prepared from diethanolamine (**15**). Morpholine is a colourless

15 —H_2SO_4→ **16** or Chair form

liquid, b.p. 128°C, which is miscible with water. It is an extremely powerful solvent and has many uses because of this property. It is a moderately strong base (pK_a 8·7), and from calculations of the dipole moments for various conformations and the observed value (1·48 D) it has been concluded that the molecule adopts the chair form with the heteroatoms at the extremities. Morpholine is not attacked at 160°C by either concentrated hydrochloric acid or 10% aqueous sodium hydroxide and has the chemistry of a typical aliphatic secondary amine. Thiamorpholine (**18**), obtainable from mustard gas (**17**) and ammonia, is similar.

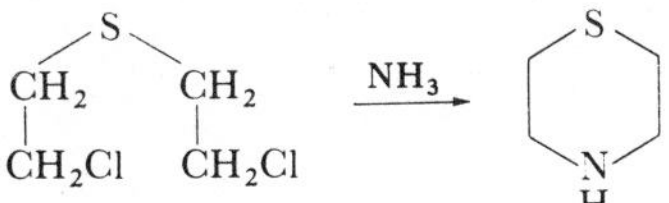

17 **18**

While the monobenzoxazines and monobenzothiazines are of little interest, derivatives of phenoxazine (**19**) and phenothiazine (**22**) have attracted attention.

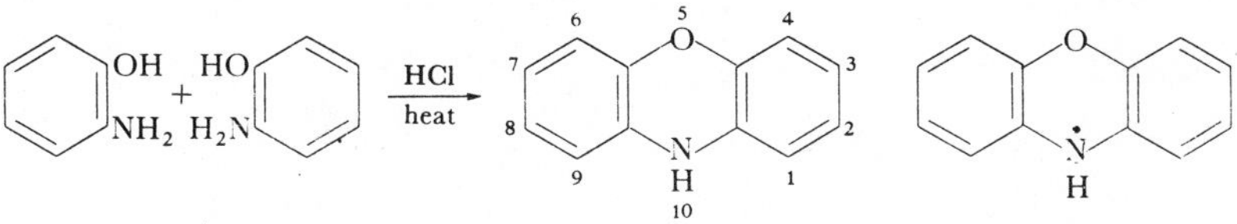

19 **20**

Phenoxazine is a colourless solid, m.p. 156°C, which is readily oxidized to a blue-green free radical (**20**). Many highly coloured materials, such as Meldola's blue (**21**), are phenoxazonium salts and have been used as dyestuffs. The positive charge is not localized on the oxygen atom, as might be inferred from structure **21**. This structure is but one of many which can be written; the compound is

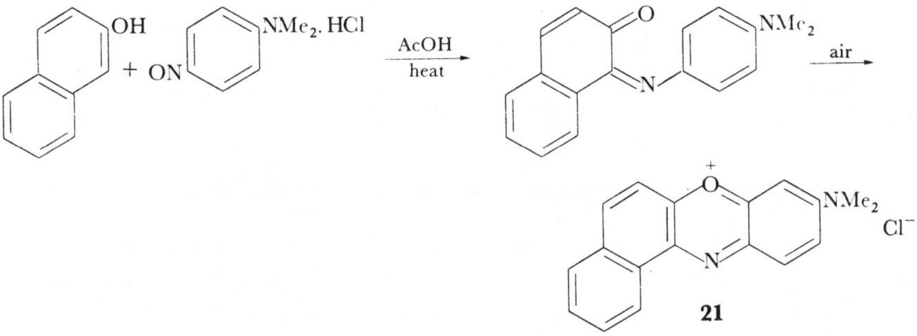

21

best considered as a resonance hybrid. The phenoxazine ring also occurs in the complex peptide containing the antibiotic actinomycin and in the omnochromes, which are insect pigments formed from tryptophan (p. 209).

Phenothiazine (**22**), m.p. 185°C, b.p. 371°C, has a structure in which the carbocyclic rings subtend an angle of 153·5° to each other. The compound is very valuable in the control of mosquitoes, as it kills their larvae. It is used as a fruit spray and as an anthelminthic in farm animals. Methylene blue (**23**), which

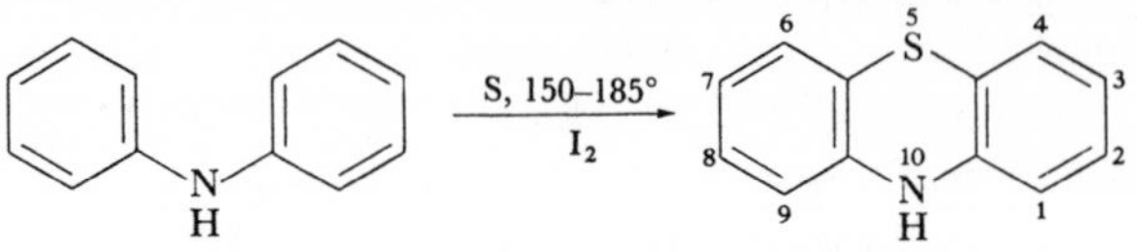

S, 150–185°
I_2

22

can be made as indicated and structurally resembles Meldola's blue (**21**), is a most valuable biological stain. It also has mild antiseptic properties. Reduction gives a colourless leuco compound (**25**), and this provides the basis of a number of quantitative colorimetric methods of estimating reducing agents in biological systems.

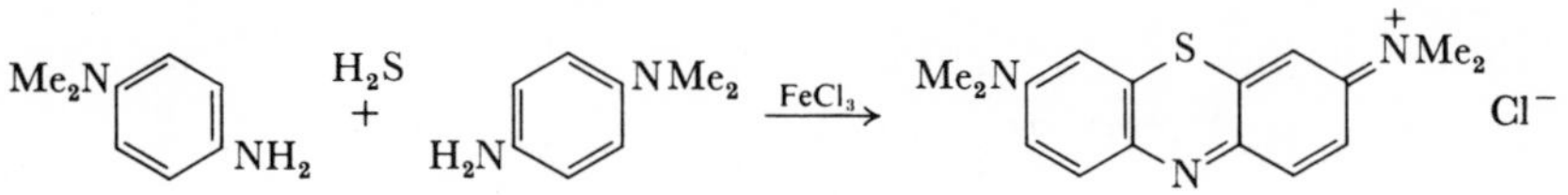

H_2S + ; $FeCl_3$

23

Oxidation | Reduction

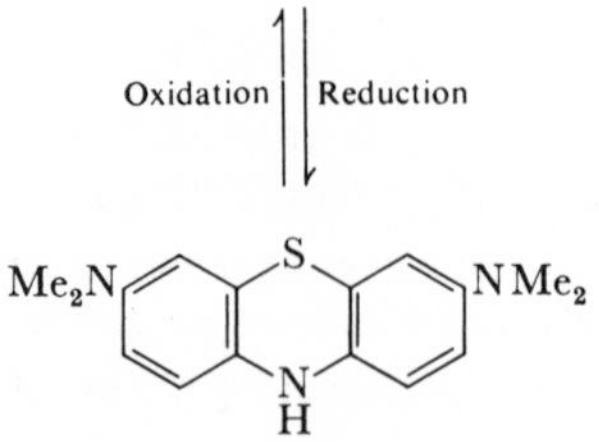

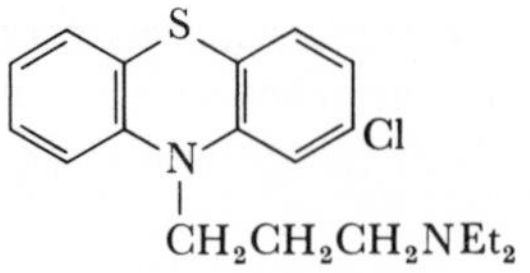

24

25

Chlorpromazine, or Largactil (**24**), is a most useful tranquillizing agent and has a large number of other physiological effects.

5. DIOXANES AND THEIR SULPHUR ANALOGUES

By far the most important compound of this class, 1,4-dioxane, usually called dioxane (1), is a colourless toxic hygroscopic compound of m.p. 12°C and b.p. 102°C. Electron-diffraction studies have shown that, in the vapour, the ring adopts the chair conformation. Dioxane is miscible with water and most organic liquids, and it is a most powerful and widely used industrial solvent. It dissolves many dyes, fats, resins, celluloid, and cellulose derivatives, and can even be used as a substitute for water in the iodoform test when the substrate is insufficiently

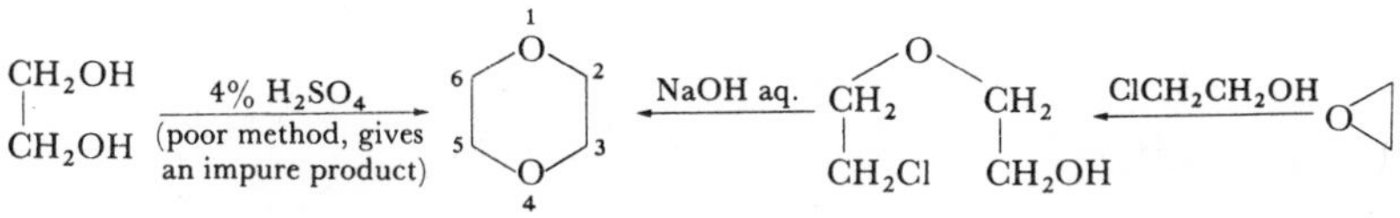

CH_2OH–CH_2OH —(4% H_2SO_4 (poor method, gives an impure product))→ 1 ←(NaOH aq.)— CH_2Cl–CH_2–O–CH_2–CH_2OH ←($ClCH_2CH_2OH$)— oxirane

1

water soluble. Dioxane is prepared industrially from oxirane (ethylene oxide, p. 27) and as usually available contains impurities such as water, acetaldehyde, glycol, and 2-methyldioxolane. It is best purified by boiling with dilute hydrochloric acid, followed by treatment with sodium hydroxide and then with sodium. It has many of the properties of an aliphatic ether. The dioxane ring is opened with cold hydrogen bromide, giving 2,2′-dibromoethyl ether. Dioxane dibromide, m.p. 65°C, formed with bromine, is a useful brominating agent (p. 131) and is much more stable than ether dibromide, m.p. –40°C. Dioxane also forms a complex with sulphur trioxide, which is a mild sulphonating agent. Chlorine reacts readily with dioxane to give a mixture of substitution products which can be separated. 2-Chlorodioxane (**2**) has a reactive chlorine atom in the same sense as 2-chlorotetrahydrofurans (p. 134) and has similar uses. 2,3-Dichlorodioxane (**3**) and 2,3,5,6-tetrachlorodioxane are converted into

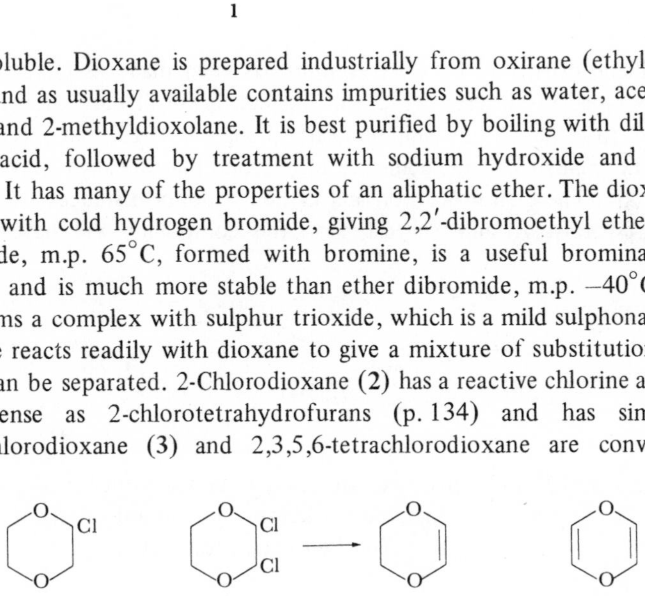

2   3 → 4   5

dioxene (**4**) and dioxadiene (**5**), respectively, by a mixture of magnesium chloride and magnesium. Both **4** and **5** behave as unsaturated ethers.

1,3-Dioxane (**6**), which can be prepared from formaldehyde and propane-1,3-diol in the presence of an acid catalyst, is quite similar to 1,3-dioxolane (p. 386). However, 1,3-dioxanes can be prepared in water solution and, in general, are highly resistant to acid hydrolysis.[38] 1,2-Dioxane (**7**) is a cyclic dialkyl peroxide.[39]

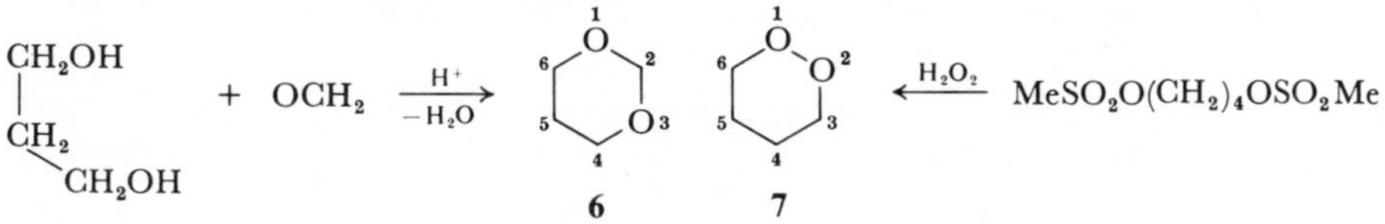

$HOCH_2CH_2CH_2OH$ + OCH_2 —(H^+, $-H_2O$)→ 6   7 ←(H_2O_2)— $MeSO_2O(CH_2)_4OSO_2Me$

6   7

1,4-Oxathiane (**8**) and 1,4-dithiane (**9**) are known. They differ from their oxygen analogues only in the special reactions of the sulphur atom, for instance, in their oxidation to sulphones. X-Ray studies[40] on crystalline 1,4-dithiane have shown that the ring adopts the chair form. The 1,4-disulphoxide is *trans.*[41]

$O(CH_2CH_2I)_2 \xrightarrow{Na_2S}$ **8** $2\ ClCH_2CH_2Cl \xrightarrow{Na_2S}$ **9**

1,4-Dithiadiene (**10**), b.p. 80°C at 20 mm, has been made as indicated.[42] Its structure is especially interesting, as the molecule is not planar. X-Ray studies[43] have shown that the carbon–sulphur–carbon angle is 101°, that the individual halves of the boat-shaped molecule are planar, and that the angle between these planes is 137°. The carbon–carbon and carbon–sulphur distances are 1·29 and 1·78 Å, respectively, and suggest that the compound will have aliphatic properties.

$HSCH_2CH(OEt)_2 + (EtO)_2CHCH_2SH \xrightarrow{H^+}$ 2,5-diethoxy-1,4-dithiane $\xrightarrow[300°]{Al_2O_3}$ **10**

1,3-Dithiane (**12**) and derivatives are easily obtained from 1,3-propanedithiol (**11**) and the appropriate aldehyde or ketone, or suitable derivative, in the presence of an acid catalyst.[44] The hydrogen atoms at position 2 are acidic

$HSCH_2CH_2CH_2SH$ (**11**) $+ CH_2(OMe)_2 \xrightarrow{BF_3-MeOH}$ **12** $\xrightarrow[THF,\ -20\ °C]{BuLi}$ **13** (2-Li) $\xrightarrow{RI}$ **14** (2-R) $\xrightarrow{(a)\ BuLi,\ (b)\ R'I}$ **16** (2-RR')

16 $\xrightarrow[aq.]{H^+}$ **11** + RCOR' (**15**)

16 $\xrightarrow{Raney\ Ni}$ $MeCH_2Me + RCH_2R'$ (**17**)

enough to react with butyllithium, the resulting salt (**13**) can be alkylated giving **14**, and the process can be repeated leading to **16**. This type of reaction cannot be achieved in the 1,3-dithiole series. Compound **16** can then be desulphurised by Raney nickel, or hydrolysed in the presence of a mercaptan scavenger, to give widely applicable syntheses of hydrocarbons (**17**) or ketones (**15**).[45]

GENERAL BIBLIOGRAPHY

PYRIDAZINES

T. L. Jacobs, in *Heterocyclic Compounds*, Vol. 6, ed. R. C. Elderfield, New York, 1956.

M. Tisler and B. Stanovnik. *Adv. Heterocycl. Chem.,* **9**, 211 (1968).

CINNOLINES, PHTHALAZINES, QUINAZOLINES, AND QUINOXALINES

J. C. E. Simpson, *Condensed Pyridazine Rings*, Interscience, New York, 1953.

T. L. Jacobs, R. C. Elderfield, T. A. Williamson, and S. L. Wythe, in *Heterocyclic Compounds*, Vol. 6, ed. R. C. Elderfield, Wiley, New York, 1956.

G. W. H. Cheeseman, 'Quinoxalines,' *Advan. Heterocycl. Chem.,* **2**, 204 (1963).

PYRIMIDINE

D. J. Brown, *The Pyrimidines*, Wiley-Interscience, New York, 1962; D. J. Brown, *The Pyrimidines*, Supplement I, Wiley-Interscience, New York, 1970.

G. W. Kenner and A. R. Todd, in *Heterocyclic Compounds*, Vol. 6, ed. R. C. Elderfield, Wiley, New York, 1956.

PYRAZINES

Y. T. Pratt, in *Heterocyclic Compounds*, Vol. 6, ed. R. C. Elderfield, Wiley, New York, 1956.

G. W. H. Cheeseman and E. S. G. Werstiuk, *Adv. Heterocycl. Chem.,* **14**, 99 (1972).

PHENAZINES

G. A. Swan and D. G. I. Felton, *Phenazines*, Interscience, New York, 1957.

OXAZINES

N. H. Cromwell, in *Heterocyclic Compounds*, Vol. 6, ed. R. C. Elderfield, Wiley, New York, 1956.

Z. Eckstein and T. Urbanski, *Advan. Heterocycl. Chem.,* **2**, 311 (1963).

PHENOXAZINES

M. Ionescu and H. Mantsch, *Adv. Heterocycl. Chem.,* **8**, 83 (1967).

THIAZINES

R. C. Elderfield and E. E. Harris, in *Heterocyclic Compounds*, Vol. 6, ed. R. C. Elderfield, Wiley, New York, 1956.

DIOXANES AND THEIR SULPHUR ANALOGUES

C. B. Kremer, L. K. Rochen, and R. C. Elderfield, in *Heterocyclic Compounds*, Vol. 6, ed. R. C. Elderfield, Wiley, New York, 1956.

M. Schultz and K. Kirsche, 'Cyclic Peroxides,' *Adv. Heterocycl. Chem.,* **8**, 165 (1967).

PHENOTHIAZINES

C. Bodea and I. Silberg, *Adv. Heterocycl. Chem.,* **9**, 322 (1968).

REFERENCES

1. W. Werner, H. Dreizler, and H. D. Rudolph, *Z. Naturforsch.,* **A22**, 531 (1967).
2. J. Tjebbes, *Acta Chem. Scand.,* **16**, 916 (1962).
3. R. D. Chambers, J. R. Maslakiewicz, and K. C. Srivasta, *J.C.S. Perkin I*, 1130 (1975) and earlier papers.

4. A. R. Katritzky and A. J. Waring, *J. Chem. Soc.*, 1523 (1964).
5. T. J. Kealey, *J. Amer. Chem. Soc.*, **84**, 966 (1962).
6. P. J. Wheatley, *Acta Cryst.*, **13**, 80 (1960).
7. M. J. S. Dewar, *Electronic Theory of Organic Chemistry*, p. 36, Oxford University Press, London, 1949.
8. J. Tjebbes, *Acta Chem. Scand.*, **16**, 916 (1962).
9. R. L. Miller, P. G. Lykos, and H. N. Schmersing, *J. Amer. Chem. Soc.*, **84**, 4623 (1962).
10. R. R. Hunt, J. F. W. McOmie, and E. R. Sayer, *J. Chem. Soc.*, 525 (1959).
11. N. Whittaker, *J. Chem. Soc.*, 1646 (1953).
12. T. J. Curphey, *J. Amer. Chem. Soc.*, **87**, 2064 (1965).
13. J. H. Chesterfield, J. F. W. McOmie, and M. S. Tute, *J. Chem. Soc.*, 4590 (1960).
14. M. E. C. Biffin, D. J. Brown, and T. C. Lee, *J. Chem. Soc. (C)*, 573 (1967).
15. D. J. Brown and P. Waring, *J.C.S. Perkin II*, 204 (1974).
16. B. Lythgoe and L. S. Rayner, *J. Chem. Soc.*, 2323 (1951).
17. G. S. Parry, *Acta Cryst.*, **7**, 313 (1954).
18. A. R. Katritzky and A. J. Waring, *J. Chem. Soc.*, 3046 (1963).
19. K. L. Wierzchowski and D. Shugar, *Photochem. Photobiol.*, **2**, 377 (1963); K. L. Wierzchowski, D. Shugar, and A. R. Katritzky, *J. Amer. Chem. Soc.*, **85**, 827 (1963).
20. G. M. Blackburn and R. J. H. Davies, *J. Chem. Soc. (C)*, 2239 (1966); E. Ben-Hur, D. Elad, and R. Ben-Ishai, *Biochim. Biophys. Acta*, **149**, 355 (1967).
21. T. Kunieda and B. Witkop, *J. Amer. Chem. Soc.*, **89**, 4232 (1967).
22. W. Bolton, *Acta Crystallogr.*, **16**, 166 (1963).
23. H. Kwart, R. W. Spayd, and C. J. Collins, *J. Amer. Chem. Soc.*, **83**, 2579 (1961).
24. G. Schwarzenbach and H. M. N. H. Irving, *Complexometric Titrations*, Interscience, New York, 1957.
25. D. J. Brown and R. F. Evans, *J. Chem. Soc.*, 527 (1962).
26. C. W. Whitehead, *J. Amer. Chem. Soc.*, **75**, 671 (1953).
27. J. N. Davidson, *The Biochemistry of the Nucleic Acids*, 7th ed., Chapman and Hall, London 1972.
28. H. C. Koppel, R. H. Springer, R. K. Robins, and C. C. Cheng, *J. Org. Chem.*, **27**, 3614 (1962).
29. J. A. Merritt and K. K. Innes, *Diss. Abstr.*, **20**, 4291 (1960).
30. A. F. Bedford, A. E. Beezer, and C. T. Mortimer, *J. Chem. Soc.*, 2039 (1963).
31. M. Masaki, Y. Chigira, and M. Ohta, *J. Org. Chem.*, **31**, 4143 (1966).
32. A. I. Meyers, *Heterocycles in Organic Synthesis*, Wiley, New York, 1974.
33. R. R. Schmidt, *Synthesis*, **7**, 333 (1972).
34. E. J. Tillermanns and J. J. Ritter, *J. Org. Chem.*, **22**, 839 (1957).
35. A. I. Meyers, A. Nabeya, H. W. Adickes, I. R. Politzer, G. R. Malone, A. C. Kovelesky, R. N. Nolen, and R. C. Portvoy, *J. Org. Chem.*, **38**, 36 (1973).
36. A. I. Meyers and N. Nazarenko, *J. Amer. Chem. Soc.*, **94**, 3243 (1972).
37. A. I. Meyers and E. M. Smith, *J. Org. Chem.*, **37**, 4289 (1972).
38. F. R. Galiano, D. Rankin, and G. J. Mantell, *J. Org. Chem.*, **29**, 3424 (1964).
39. R. Criegee and G. Müller, *Chem. Ber.*, **89**, 238 (1956).
40. R. E. Marsh, *Acta Crystallogr.*, **8**, 91 (1955).
41. H. M. M. Shearer, *J. Chem. Soc.*, 1394 (1959).
42. W. E. Parham, H. Wynberg, and F. L. Ramp, *J. Amer. Chem. Soc.*, **75**, 2065 (1953).
43. P. A. Howell, R. M. Curtiss, and W. N. Lipscomb, *Acta Crystallogr.*, **7**, 498 (1954).
44. D. Seebach, N. R. Jones, and E. J. Corey, *J. Org. Chem.*, **33**, 300 (1968).
45. D. Seebach, *Synthesis*, **1**, 17 (1969).

CHAPTER IX

SOME COMPOUNDS WITH MORE THAN TWO HETEROATOMS

A very large number of compounds possessing more than two heteroatoms in one, or more, rings is known. Most of these contain nitrogen atoms. In this chapter five of the more important systems, shown below, are considered briefly.

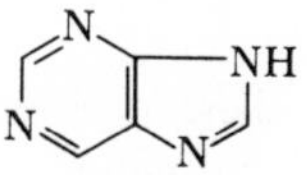

Purine

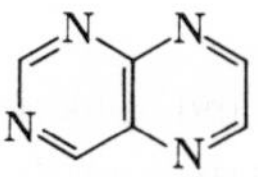

Pteridine

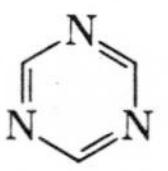

1,3,5-Triazine

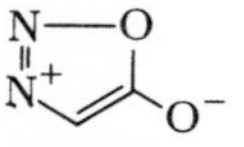

Sydnone

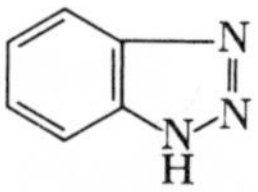

Benzotriazole

1. PURINES

A. Introduction

The purine ring system is one of the most important present in living systems, but as in the case of pyrimidine (p. 406), the parent compound has not been found in nature. The least substituted naturally occurring derivative of purine known is the 9-ribofuranoside, which is present in the fungus *Agaricus nebularis.*[1] The purine ring is present in many natural products, including several nucleotides (hydrolysis products of nucleic acids p. 406), a number of coenzymes, the antibiotic puromycin, uric acid, caffeine and related compounds (the stimulants of tea, coffee, and cocoa), the plant growth regulator zeatin,[2] and several types of compounds which are valuable in the treatment of certain types of cancer.

B. The structure and chemical reactions of purine and its simple derivatives

Purine itself is a colourless compound, m.p. 212°C, which is very soluble in water but sparingly soluble in organic solvents. Its structure[3] and that of 6-aminopurine (adenine) hydrochloride[4] have been deduced from X-ray diffraction data. The interatomic distances show that the purine ring is aromatic. Purine can formally be derived from a fusion of pyrimidine and imidazole rings.

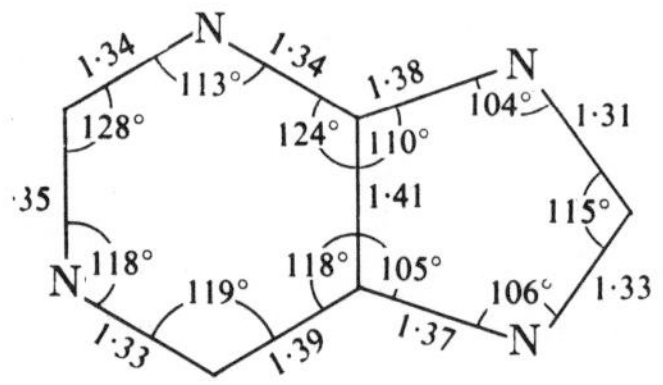

Bond lengths in Angstroms.

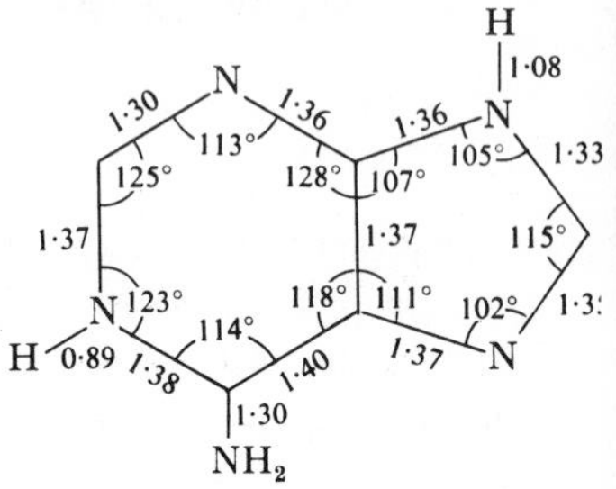

Purine.

Adenine hydrochloride

By analogy positions 2 and 6 (and substituents at these positions) of purine should be similar chemically to positions 2 and 4 of pyrimidine, and position 8 to position 2 of imidazole, and in general this is the case. Purine (pK_a 2·4) is a

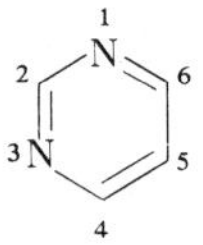

Pyrimidine

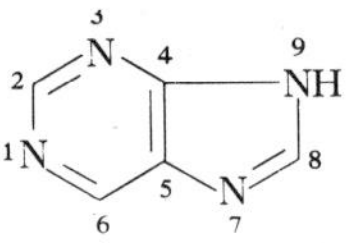

Purine

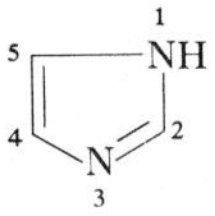

Imidazole

stronger base than pyrimidine (pK_a 1·3). It is also a much weaker base but a much stronger acid (acidic pK_a 8·9) than imidazole (p. 355); these properties are consistent with the 'pyrimidine' ring withdrawing electrons from the 'imidazole' ring of purine. Possibly because of this, and in contrast to imidazole, purine does not couple with benzenediazonium salts.

Uric acid (**3**) is excreted in human urine (ca. 600 mg/day) and is the major nitrogen-containing excretion product of birds and reptiles. It was isolated in 1776, and its general structure was deduced from degradation in 1875; the key degradations were with nitric acid to alloxan (**1**) and urea, and with alkaline permanganate to allantoin (**6**).

The halogen atoms of 2,6,8-trichloropurine (**4**), obtainable from uric acid (**3**), are replaced successively in the order 6, 2, and 8 by nucleophilic reagents. Examples are the synthesis of guanine via **7** and **10**, and of adenine (**8**) via **5**. Adenine is perhaps the most important purine base in living systems, and some

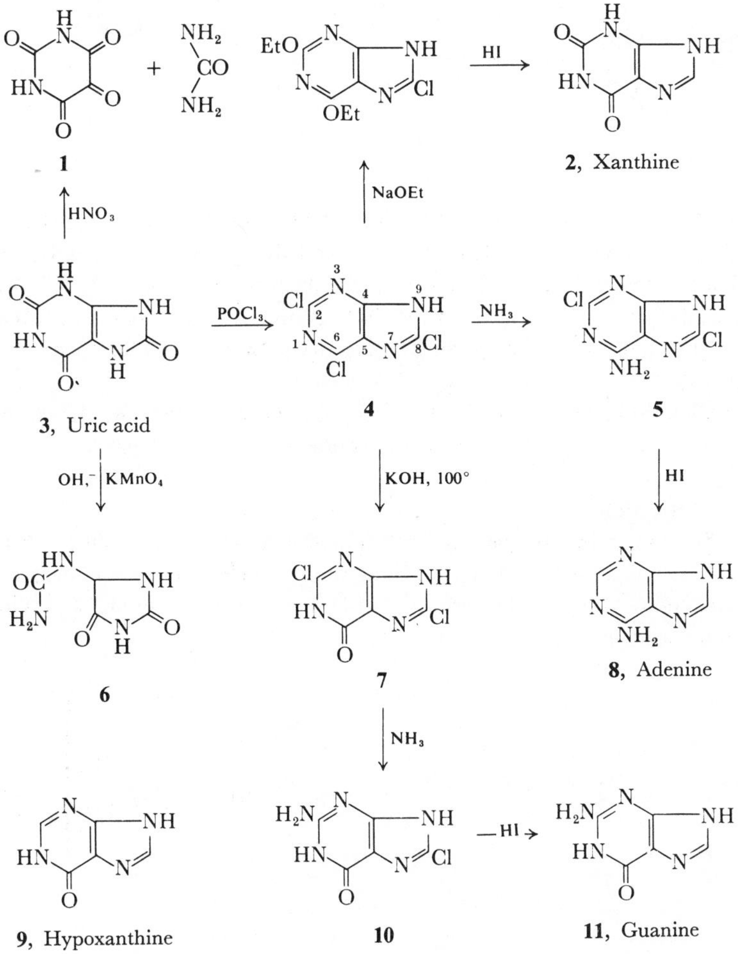

1

2, Xanthine

3, Uric acid

4

5

6

7

8, Adenine

9, Hypoxanthine

10

11, Guanine

of its derivatives are considered on pp. 424–428. The successive replacement of chlorine atoms proved a most useful synthetic procedure in the earlier stages of purine chemistry, but most of the compounds in the flowsheet are more readily available from Traube's synthesis (p. 422). The reactivity of halogens in other purines, for example, in 2,8-dichloropurine, does not always follow the order above, and in this case the 8-atom is the more easily displaced.[5]

The first synthesis of uric acid, in confirmation of its structure, was in 1888, but the best preparation is that of Traube (p. 423). Heating uric acid with formamide gives xanthine (**12**), while on heating with water to 190°C a remarkable reaction occurs, giving some 2,4,7-trihydroxypteridine (**13**). The murexide test for uric acid is described on p. 403.

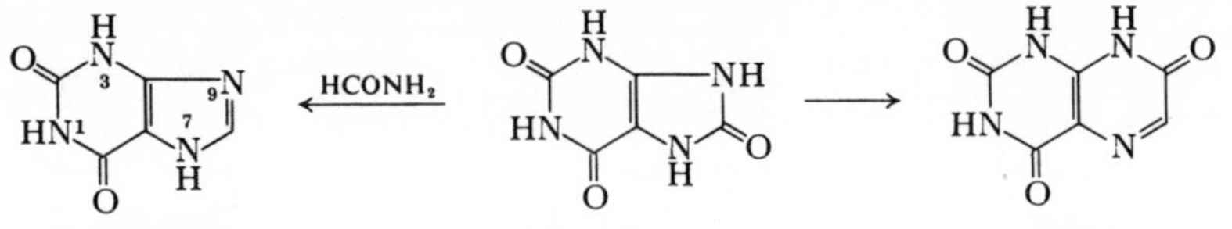

12, Xanthine ← $HCONH_2$ — 3 → 13

Caffeine (1,3,7-trimethylxanthine), obtained commercially from the methylation of xanthine with methyl chloride or sulphate and alkali, is the major stimulant in tea and coffee. In the formula given for xanthine (**12**), a hydrogen atom is shown at position 7. This is the predominant tautomer, in contrast to adenine (**8**) and most other purines where the predominant tautomer is the 9*H* compound.

1,3-Dimethylxanthine, or theophylline, also occurs in tea, and 3,7-dimethylxanthine, or theobromine, is the major alkaloid of the cocoa bean.

C. Synthetic methods

(1) Purine can be obtained in a remarkable manner and in a remarkable 71% yield by heating formamide alone[6] and adenine (**15**) can be obtained by the catalytic reduction of the product (**14**) from cyanacetamide and phenyldiazonium salts in ammonia–formamide.[7]

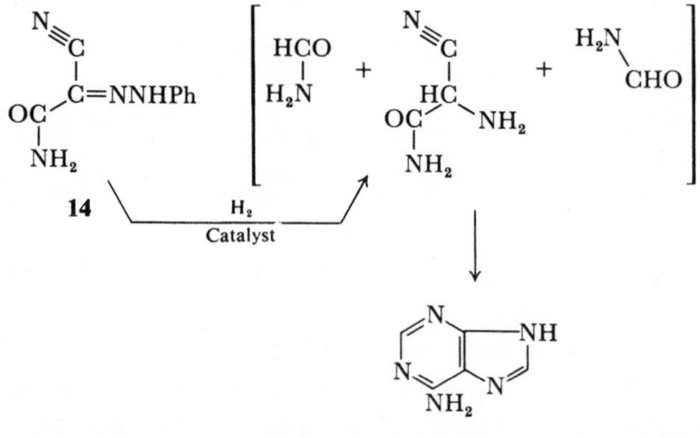

14 —(H_2, Catalyst)→ [HCO–H_2N + N≡C–HC(NH_2)–OC–NH_2 + H_2N–CHO] → **15**

(2) The most important method of purine synthesis is that of Traube, who showed that 4,5-diaminopyrimidines could be cyclized to purines. Purine itself can be obtained from 4,5-diaminopyrimidine and formic acid. 4-Aminopyrimidines, such as **16** are readily available, and on nitrosation, or diazo coupling (p. 398), followed by reduction of the product, give the 4,5-diamino compounds

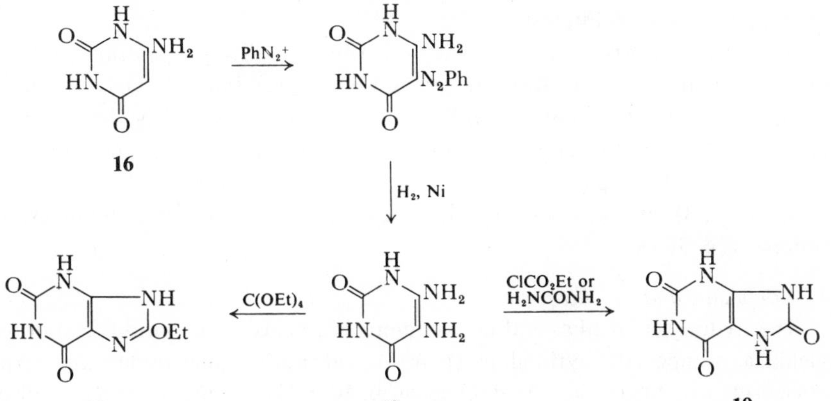

$\xrightarrow{PhN_2^+}$; $\xrightarrow{H_2, Ni}$; $\xleftarrow{C(OEt)_4}$; $\xrightarrow{ClCO_2Et \text{ or } H_2NCONH_2}$

16 **17** **18** **19**

(e.g., **18**). In the case shown, heating with urea, or better with ethyl chloroformate, gives uric acid (**19**). Cyclization of **18** with formic acid, or with the more vigorous mixture of acetic anhydride and ethyl orthoformate,[8] gives xanthine (**12**). With ethyl orthocarbonate or tetrakismethylthiomethane[9] 8-ethoxy- (e.g., **17**) and 8-methylthiopurines are obtained. These substances have synthetic potential, as the 8-substituent can be displaced by nucleophiles. 4,5-Diaminopyrimidines are most easily cyclized to purines unsubstituted at position 8 with sodium dithioformate. In the synthesis of adenine (**21**) outlined, the 5-thioformamidopyrimidine (**20**) is first formed and easily loses hydrogen sulphide. This type of synthesis is particularly valuable, for if a 4,5,6-triamino-pyrimidine is involved and either the 4- or 6-amino group is alkylated, cyclization always occurs onto the alkylated nitrogen atom. A 9-substituted purine is therefore formed, and many naturally occurring purines are of this type. Other syntheses of purines from pyrimidines are available.[10]

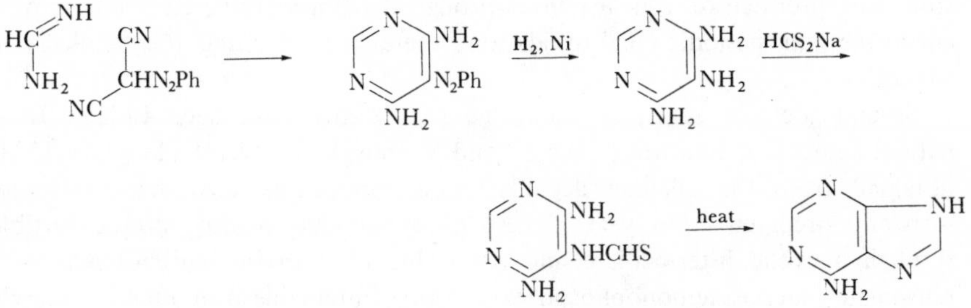

$\xrightarrow{H_2, Ni}$; $\xrightarrow{HCS_2Na}$; $\xrightarrow{heat}$

20 **21**

D. Purines of special interest

Several purine bases are very widely distributed in living organisms and often occur as complex derivatives of greatest biological interest. Of these bases adenine (**21**) is the most common, and the chemistry of its derivatives is considered here as an example. Adenosine derivatives are involved in oxidation–reduction (p. 276), phosphorylation (p. 425), methylation (p. 428), sulphonation (p. 428), and acylation (p. 427) systems *in vivo*, and in bioluminescent processes (p. 385).

1. *Adenosine and its phosphates*

The hydrolysis of ribo- and deoxyribonucleic acids under various conditions yields a number of pyrimidine (p. 406) and purine nucleosides; the terms nucleoside and nucleotide are defined on p. 406. The common purines involved

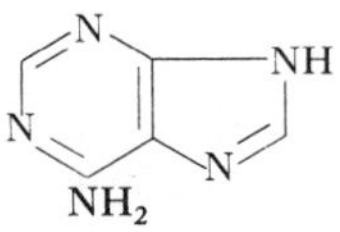

21 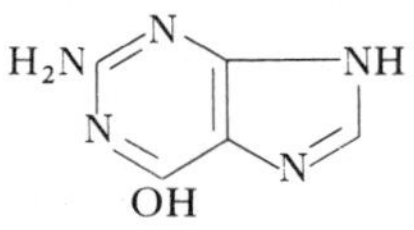 **22**

are adenine (**21**) and guanine (**22**), and from the hydrolysis of ribonucleic acid, the nucleoside adenosine (**25**) can be obtained. The further hydrolysis of adenosine gives adenine and D-ribose, and the ribose was shown to be attached to position 9, and not to position 7, of the ring by a series of ultraviolet absorption spectrum comparisons. Periodate oxidation of adenosine gives the dialdehyde (**26**) and no formic acid, showing that the sugar ring is furanose. The first synthesis of adenosine (**25**) is outlined[11] below and starts with **5**. The ribose is connected to the purine moiety with a β-link; this has been confirmed by X-ray studies. An earlier chemical proof depended on the supposition that α-bromotetraacetylglucose, of known configuration, would undergo a Walden inversion in reaction with the silver salt (**23**). Adenine-9-β-glucoside, prepared from the product of this reaction through dehalogenation, gives **26**, as does adenosine, on periodic acid oxidation, thereby confirming the β-linkage in adenosine.

Several adenylic acids, or adenosine phosphates, have been isolated from natural sources. A mixture of the 2′- and 3′-phosphates, 'yeast adenylic acid,' is obtained from the alkaline degradation of ribonucleic acid, which is commercially prepared from yeast. These phosphates are readily interconvertible through a cyclic intermediate and are stable to periodic acid. Adenosine-5′-phosphate (adenosinemonophosphate or AMP), obtainable from muscle ('muscle adenylic acid') or enzymically from other adenosine phosphates, is attacked by periodic acid and readily forms a boric acid complex in confirmation of the presence of two *cis* hydroxyl groups.

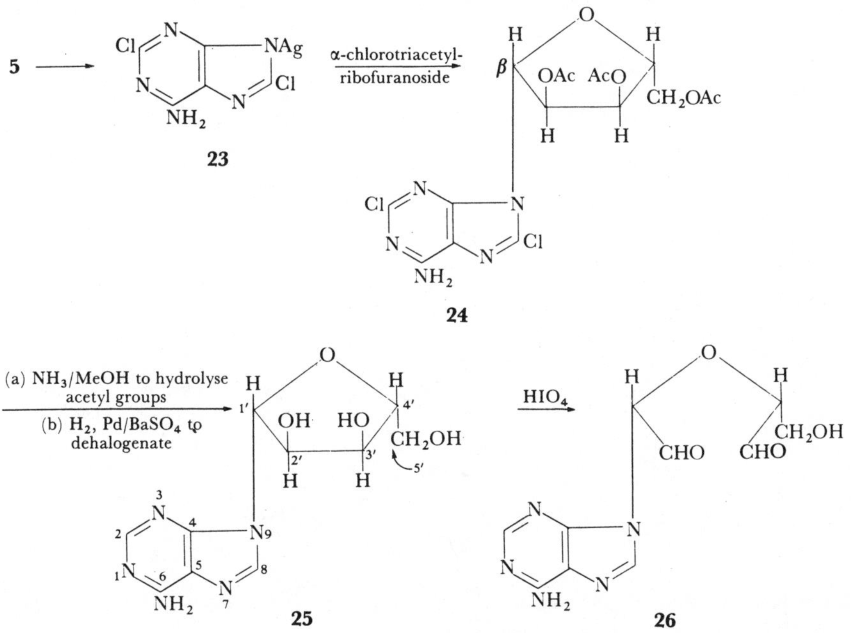

The best synthesis of adenosine-5′-phosphate, of the several described, starts with adenosine (**25**). This, after conversion into the 2′,3′-isopropylidene derivative (**27**) with acetone under acidic conditions, is phosphorylated with dibenzylphosphorochloridate (readily prepared from phosphorus oxychloride and two moles of benzyl alcohol). The product (**28**) is hydrogenated, when the benzyl groups are split off as toluene, and then mild acid hydrolysis yields acetone and adenosine-5′-phosphate (**29**).

Adenosine-5′-diphosphate, ADP (**32**), and adenosine-5′-triphosphate, ATP (**34**), are involved in the reversible phosphorylation of many compounds *in vivo*, and many other polyphosphates derived from guanine, uracil, cytosine, or thymine nucleosides occur in smaller quantities and play specific roles in metabolic processes. ATP is usually isolated from muscle, and its structure was deduced originally from degradative studies. Acid hydrolysis yields adenine and ribose-5-phosphate, while alkaline hydrolysis gives adenosine-5′-phosphate and pyrophosphoric acids. ATP both increases the conductivity of boric acid and consumes 1 mole of periodic acid. These results prove that only the 5-hydroxyl group of the ribose moiety bears the phosphate groups and suggest a linear triphosphate. This last point is in agreement with titration studies, which show

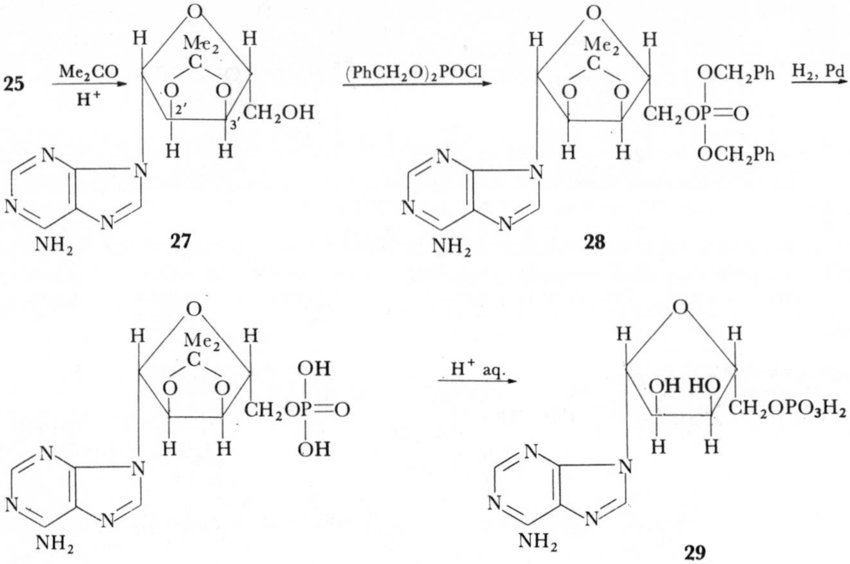

25 $\xrightarrow[H^+]{Me_2CO}$ 27 $\xrightarrow{(PhCH_2O)_2POCl}$ 28 $\xrightarrow{H_2, Pd}$ $\xrightarrow{H^+ \text{ aq.}}$ 29

three primary and one secondary dissociation, and the structures of both ADP and ATP have been confirmed by several syntheses.

Compound **30** is obtained by careful acid hydrolysis of **28** (above), when

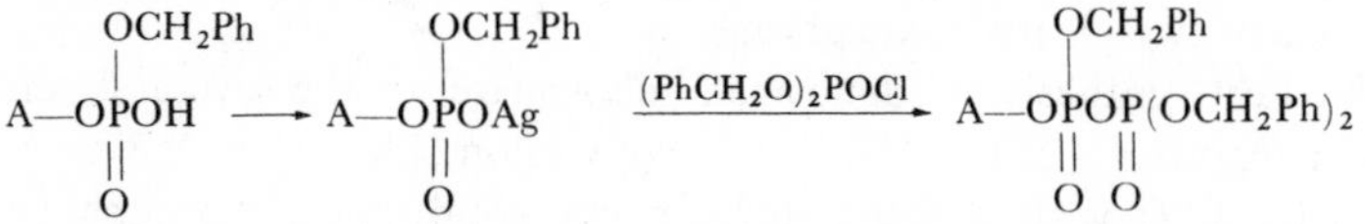

A—OP(O)(OCH$_2$Ph)OH (**30**) ⟶ A—OP(O)(OCH$_2$Ph)OAg $\xrightarrow{(PhCH_2O)_2POCl}$ A—OP(O)(OCH$_2$Ph)OP(O)(OCH$_2$Ph)$_2$ (**31**)

31 $\xrightarrow{H_2, Pd}$ A—OP(O)(OH)OP(O)—(OH)$_2$ (**32**, ADP)

31 ⟶ A—OP(O)(OCH$_2$Ph)OP(O)(OCH$_2$Ph)OH (**33**) $\xrightarrow{(PhCH_2O)_2POCl}$ A—OP(O)(OCH$_2$Ph)OP(O)(OCH$_2$Ph)OP(O)(OCH$_2$Ph)$_2$

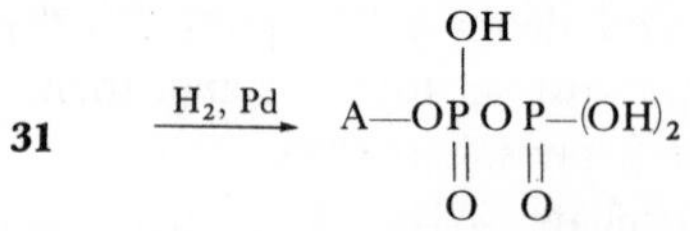

$\xrightarrow{H_2\ Pd}$ A—OP(O)(OH)OP(O)(OH)OP(O)(OH)$_2$ (**34**, ATP)

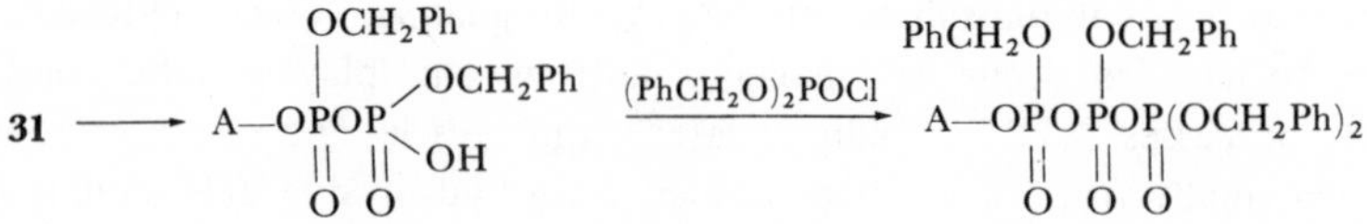

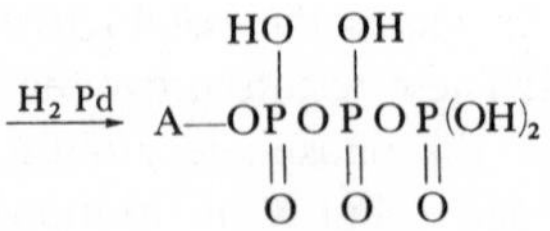

A = Adenosine-5′-

both the isopropylidene and one benzyl group are removed. The silver salt of **30** with dibenzylphosphorochloridate gives **31**, which on catalytic debenzylation yields ADP (**32**). Heating **31** with 4-methylmorpholine (a tertiary base with powerful solvent properties) splits off one benzyl group, giving **33**, which, through the usual procedure with its silver salt followed by hydrogenation, gives ATP (**34**). It is interesting that a much better yield of ATP can be obtained from disilver adenosine-5′-phosphate and excess dibenzylphosphorochloridate followed by hydrogenolysis. A rearrangement occurs, probably through a cyclic intermediate after or during the debenzylation.

2. *Coenzyme A*

Coenzyme A (**35**) is a derivative of adenosine responsible for the transfer of acetyl and other acyl groups in living systems. An *S*-acyl coenzyme A is first

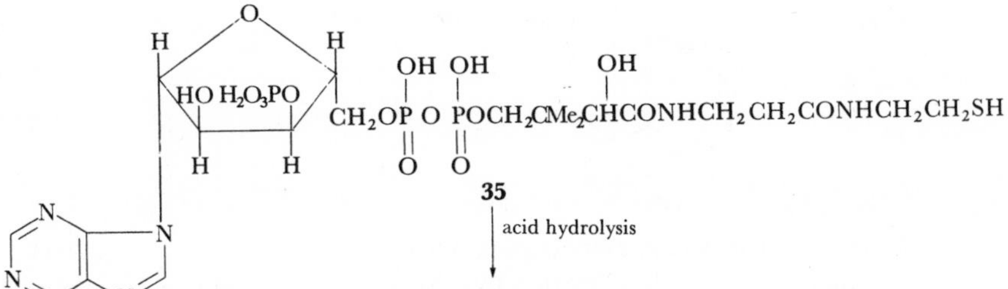

[Structure **35**: adenosine 3′-phosphate (HO, H_2O_3PO on ribose) with $CH_2OP(O)(OH)OP(O)(OH)OCH_2CMe_2CH(OH)CONHCH_2CH_2CONHCH_2CH_2SH$]

35

↓ acid hydrolysis

Adenosine-5′-phosphate + $H_2O_3POCH_2CMe_2CH(OH)CONHCH_2CH_2CO_2H$ + H_3PO_4

29 Pantothenic acid 4′-phosphate

\+

$H_2NCH_2CH_2SH$

formed, and this acylates another molecule with the re-formation of coenzyme A. The structure of the coenzyme has been firmly established by degradation,[12] the major outlines of which are shown, and a chemical synthesis has been achieved.[13]

3. *S-Adenosyl-L-methionine*

S-Adenosyl-L-methionine, earlier called thetin (**37**), is formed from adeno-

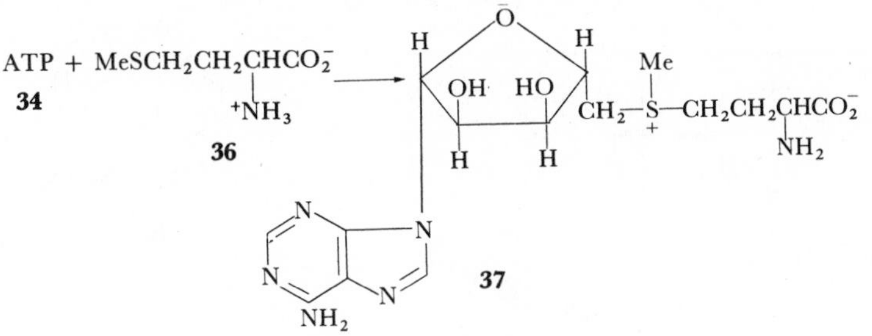

$$\text{ATP} + MeSCH_2CH_2CH(NH_3^+)CO_2^- \longrightarrow \text{Adenosyl-}CH_2\overset{+}{S}(Me)CH_2CH_2CH(NH_2)CO_2^-$$

34 **36** **37**

sine-5′-triphosphate and methionine (**36**) *in vivo* and is concerned in the biological transfer of methyl groups.

4. *Phosphoadenosinephosphosulphate*

The nucleotide coenzyme involved in the conversion of phenols into their *O*-sulphates *in vivo* has been identified as 3′-phosphoadenosine-5′-phosphosulphate (**38**). It has been synthesized[14] from adenosine-3′,5′-diphosphate and 1-proto-1-pyridiniumsulphonate (p. 234) in the presence of sodium bicarbonate.

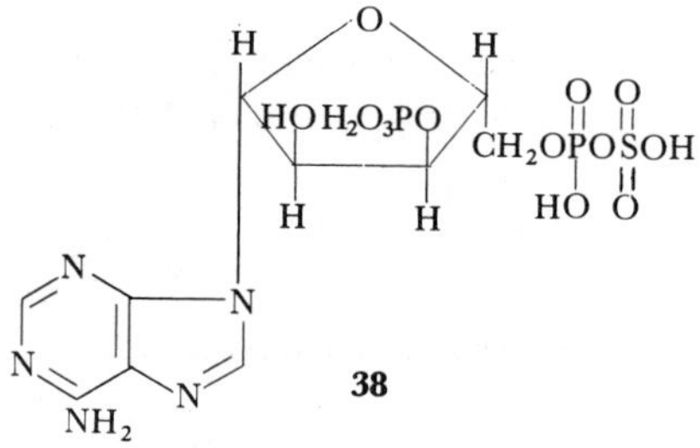

H O H HO H_2O_3PO $CH_2OPOSOH$ O O HO O H H N N N N NH_2

38

5. *Oxidation–reduction coenzymes of adenine*

Nicotinamide–adenine dinucleotide (NAD) and flavine–adenine dinucleotide are discussed on pp. 276 and 432, respectively.

6. *Puromycin*

This is a wide-spectrum fungal antibiotic derived from adenine and has structure **39**. It is a ribose derivative.

7. *Zeatin*

The plant growth regulator zeatin[2] (**40**), an adenine derivative, has been identified and synthesised. Only 0·2–1·0 mg was isolable from 60 kg of corn. The compound induces cell division in plants.

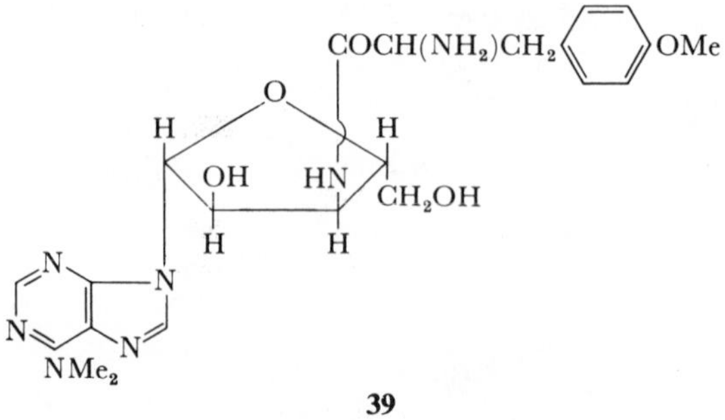

$COCH(NH_2)CH_2$ OMe O H H OH HN CH_2OH H H N N N N NMe_2

39

N NH N N HN $CH_2CH{=}CMeCH_2OH$

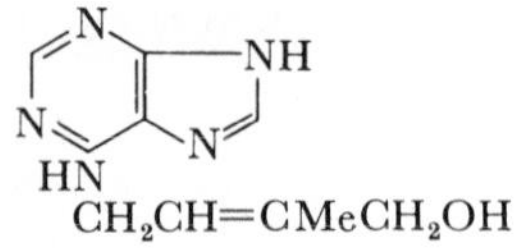

40

8. *Chemotherapeutic purines*

Very many purines have been synthesized in recent years as potential antitumour agents, and of these 6-mercaptopurine has limited value. It may interfere with the utilization of folinic acid (p. 431) by the growing tumour.

2. PTERIDINES

Xanthopterin (2-amino-4,6-dihydroxypteridine) and leucopterin (2-amino-4,6,7-trihydroxypteridine), derivatives of pteridine (**1**), were obtained from the wings of butterflies and other insects in very small quantities in 1891, but their constitution was not established until 1940. This was because only very small quantities of the materials were available, and the early analytical data were erroneous owing to combustion difficulties. Pteridines are usually colourless or yellow and account for the yellow colour of bees and wasps. Other pteridines, including 'folic acid,' which is the name of a group of antipernicious anaemia factors, folinic acid, and vitamin B_2, are of immense importance in living systems, and a recent review is available.[15] Certain pteridines are useful against some types of leukaemia.

Pteridine (**1**) is obtainable from glyoxal and 4,5-diaminopyrimidine. This is an example of the general pteridine synthesis, where a dicarbonyl compound is treated with a 4,5-diaminopyrimidine, and which has been used to prepare leucopterin and xanthopterin. Pteridine is a yellow solid, m.p. 139°C, soluble in

$(CHO)_2$; H_2O ; $-H_2O$

1 **2** **3**

water and organic solvents. It adds water reversibly across the 3,4-positions, to give **2**, an example of a general type of behaviour of many heterocyclic compounds.[16] Ammonia at low concentrations (<1 M) behaves similarly, but at higher concentrations addition to the other ring with the formation of **3** takes preference.[17] With hot acid or alkali the pyrimidine ring is broken with the formation of 2-aminopyrazine-4-carbaldehyde.

Pteridine can be considered as pyrimidinopyrazine, and its properties are consistent with this. For example, hydrogenation under alkaline conditions, which reduces pyrazine but not pyrimidine, reduces the 'pyrazine' ring of pteridine. The molecular dimensions of pteridine have been determined by X-ray methods[18] and are in agreement with its aromatic characteristics. The 7,8-nitrogen–carbon bond is particularly short.

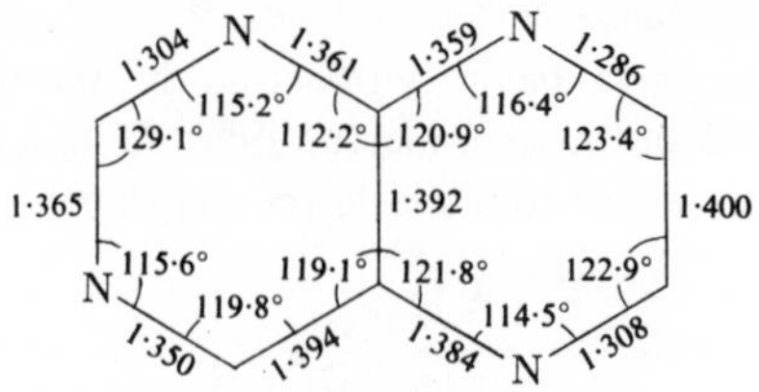

Pteridine, bond lengths in Angstroms

One of the antianaemia factors present in the mixture originally called folic acid, obtained from liver and yeast, is pteroylglutamic acid (**5**). Both it and pteroic acid (**4**) can be synthesized as outlined.

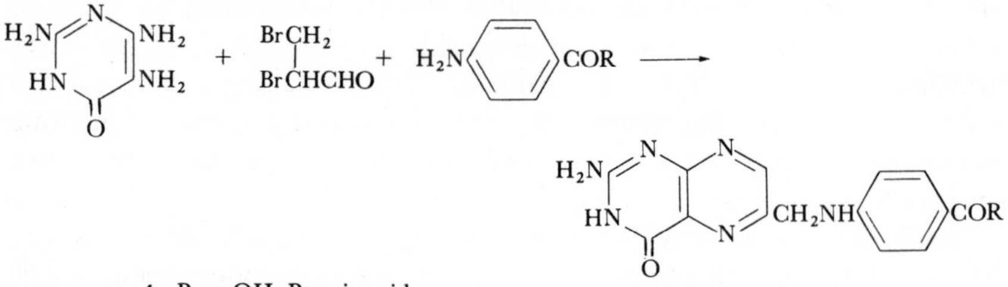

4, R = OH, Pteroic acid
5, R = $NHCH(CO_2H)CH_2CH_2CO_2H$, Pteroylglutamic acid

Oxidation takes place *in situ* and the yields are poor because many competing reactions can occur. A new type of synthesis,[19] outlined below, gives an unambiguous orientation of the substituents in the two rings and has considerable potential.

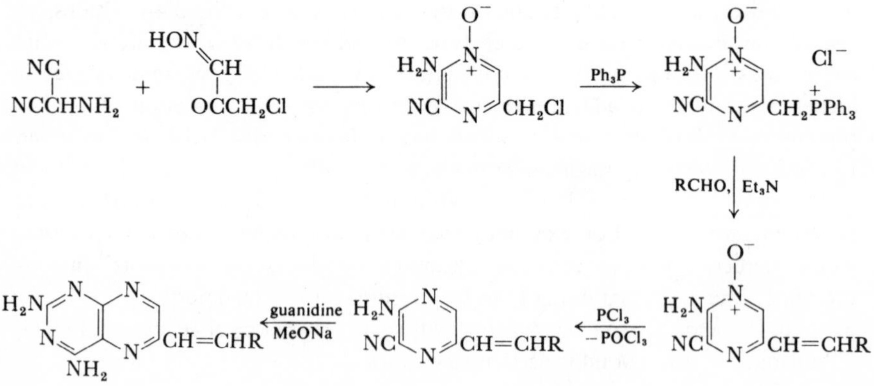

The 5-formyl derivative (**6**) of tetrahydropteroylglutamic acid is known as folinic acid, and this and the related 10-formyl derivative are involved in methylation in purine synthesis. It is essential for mammalian cell division. This function is inhibited by 'aminopterin,' the 4-amino analogue of pteroylglutamic

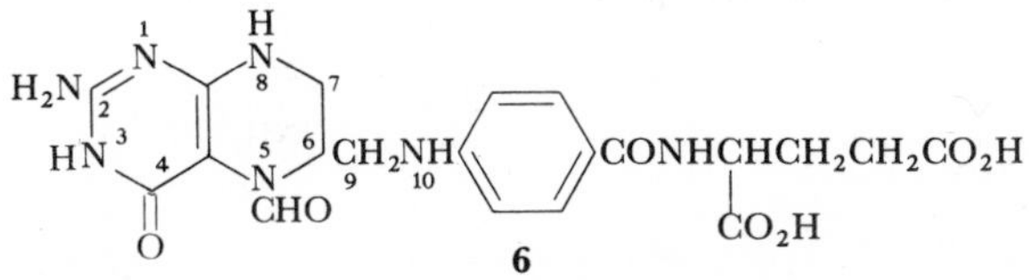

acid, and by 'amethopterin' (4-amino-10-methylpteroylglutamic acid). These two compounds are of value in treating acute leukaemia, as are several other related pteridines.

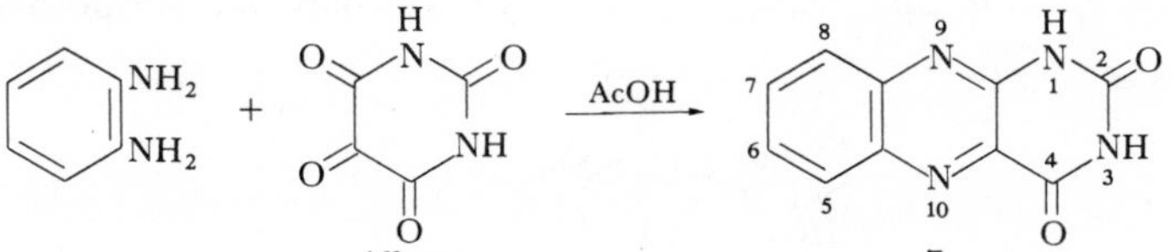

2,4-Dihydroxybenzopteridine is usually called alloxazine (**7**), and vitamin B_2 or riboflavin (**9**) is a derivative. Riboflavin[20] is a yellow compound which gives highly fluorescent solutions and occurs in plant and animal tissue. It is sensitive

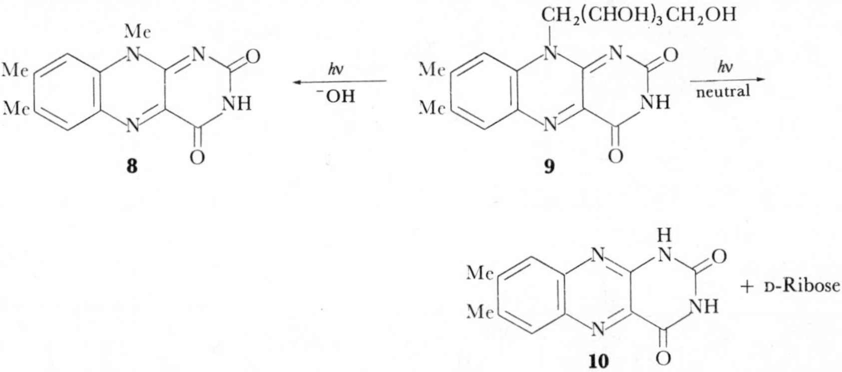

to light, and its structure was deduced from those of its decomposition products, lumiflavin (**8**) and lumichrome (**10**), and was confirmed by synthesis. Riboflavin and lumiflavin are derived from isoalloxazine, the hypothetical tautomer of alloxazine 7 in which the hydrogen atom at position 1 is moved to position 9.

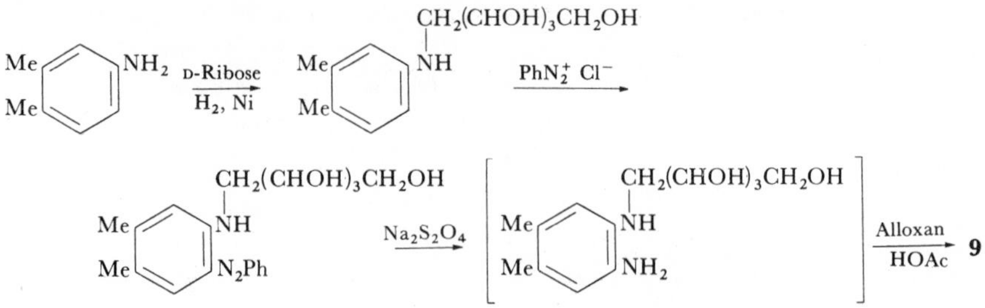

Riboflavin is synthesised *in vivo* from 2 moles of the pteridine **11**, the uracil **12** being the other product. The mechanism of both the *in vitro* and *in vivo* reactions, which proceed through the loss of a proton from the 7-methyl group and attack on a hydrated molecule of **11**, has been elucidated.[21]

The coenzyme, flavin–adenine dinucleotide (**13**), is important in biological oxidation systems, and reduction, which affects only the isoalloxazine ring,

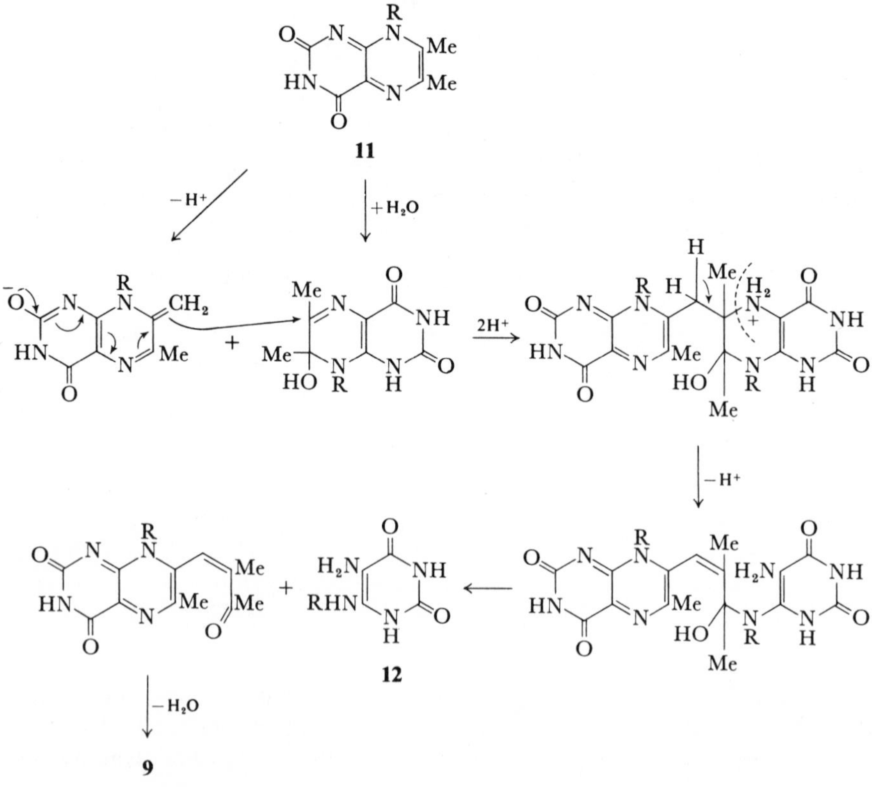

R = D-Ribityl

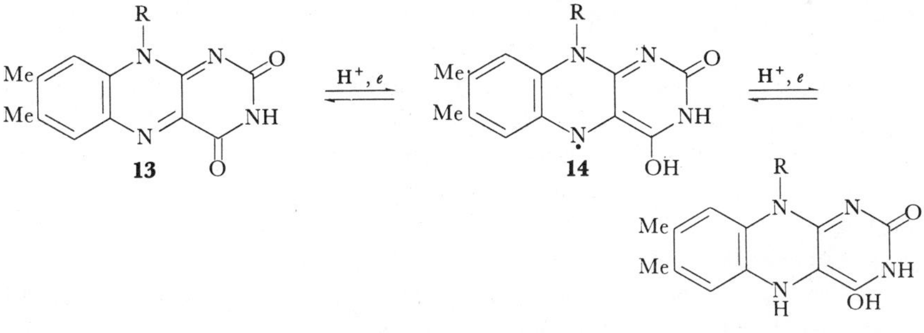

13 14

proceeds in two one-electron steps as outlined. It involves the formation of a free radical (**14**). Riboflavin can act as a photosensitizing agent, and the light-excited molecule can act as an electron donor. The structure of the dinucleotide (**15**) is based on several degradations, including its hydrolysis to adenosine-5′-phosphate (p. 424) and riboflavin-5′-phosphate, and has been confirmed by synthesis.

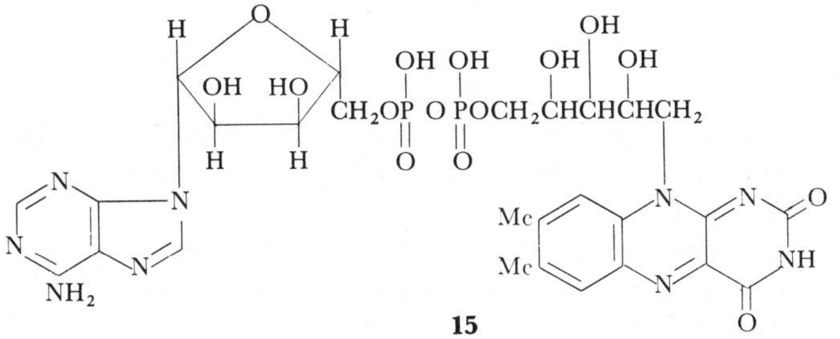

15

3. TRIAZINES

Of the three triazines, only the 1,3,5-triazine is considered here. X-Ray studies[22] on this compound have shown that the carbon–nitrogen distances are all 1·319 Å and that the ring is planar but not a regular hexagon. The angles at whose apexes lie nitrogen and carbon atoms are 113·2 and 126·8°, respectively. 1,3,5-Triazine is best prepared[23] by the thermal or base-catalysed decomposition of formamidine hydrochloride (**2**) and is hydrolysed almost instantly by dilute acids to formic acid and ammonia. This type of ring opening by acids and bases

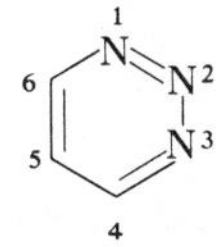

1,2,3-Triazine 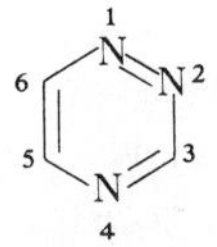 1,2,4-Triazine 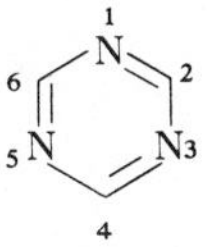 **1**, 1,3,5-Triazine 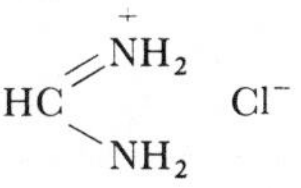 **2**

is very easy, and unlike the case of pyrimidine (p. 399), nucleophilic attack with sodamide opens the ring, giving sodium cyanide and ammonia. Cyanuric chloride, or 2,4,6-trichloro-1,3,5-triazine (**4**), is obtained industrially by the vapour-phase polymerization of cyanogen chloride on charcoal and is a valuable dyestuff intermediate. Its chlorine atoms are very reactive (cf. 2-chloropyridine) and can be displaced readily, and successively if suitable conditions are used, by nucleophilic reagents. For example, hydrolysis with glacial acetic acid, which gives better results than water, which has also been used, yields 2,4,6-trihydroxy-1,3,5-triazine, or cyanuric acid (**3**). This can also be obtained by the trimerization of cyanic acid or, among other products, by heating urea. Cyanuric acid gives cyanuric chloride with phosphorus pentachloride.

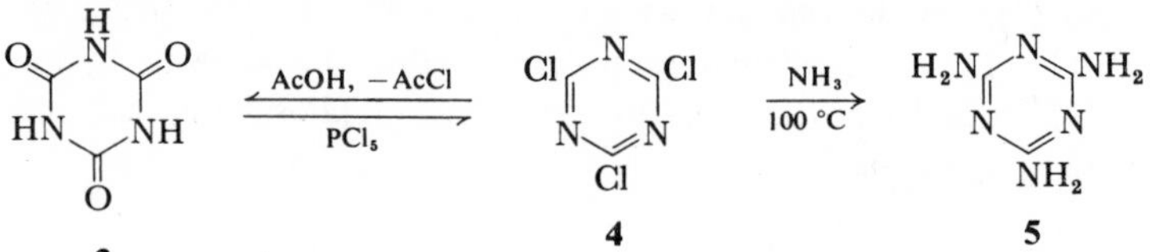

AcOH, −AcCl ; PCl_5 ; NH_3, 100 °C

3 **4** **5**

The halogen atoms of cyanuric chloride are replaced by ammonia (cf. 2-chloropyridine and ammonia, p. 255), giving 2,4,6-triamino-1,3,5-triazine, or melamine (**5**), at 100°C. Melamine can be made in many other ways, one of the most important industrially being the thermal decomposition of urea. Melamine–formaldehyde resins are excellent electrical insulators and are used to make tableware. The valuable high explosive RDX, or Cyclonite, is 1,3,5-trinitrohexahydro-1,3,5-triazine (**6**) and is obtained by nitrating hexamethylenetetramine, and various herbicides contain the triazine ring.[24]

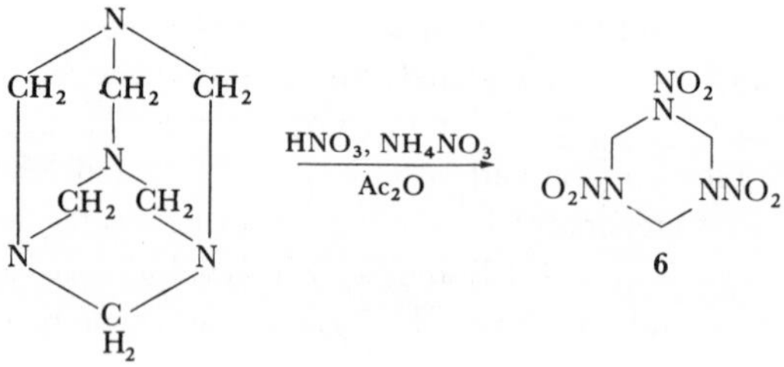

HNO_3, NH_4NO_3 ; Ac_2O

6

The drug Paludrine, also called proguanil (**7**), is widely used as an antimalarial agent (see also p. 317). It has, however, been shown that Paludrine itself has no action on the malaria parasite, but that it is converted into the active agent, a triazine (**8**), by the host.

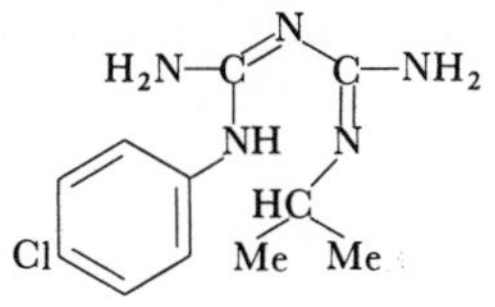

7

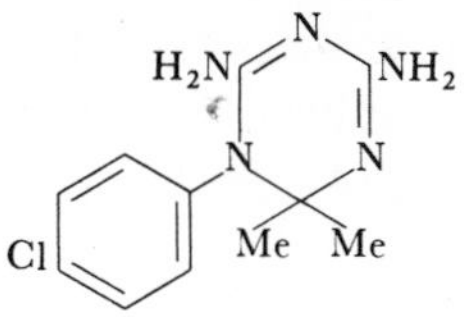

8

4. SYDNONES

N-Nitrosophenylglycine (**1**) on treatment with acetic anhydride yields 3-phenylsydnone (**4**) in a general reaction. The mixed anhydride (**2**) is an intermediate and the rate-determining step is the cyclisation stage forming **3**.[25] This sydnone (**4**). is colourless, has m.p. 135°C, and is easily soluble in benzene but sparingly soluble in water. It is hydrolysed to *N*-nitrosophenylglycine by alkali, but with dilute acid phenylhydrazine, formic acid, and carbon dioxide are obtained. Bromination and nitration of 3-phenylsydnone take place

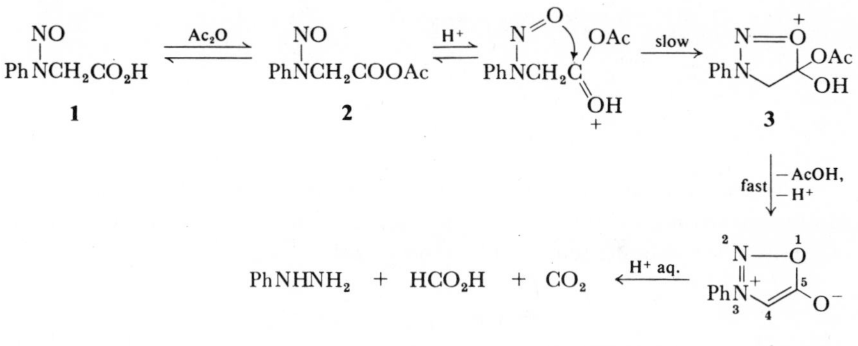

$$\underset{\mathbf{1}}{PhN(NO)CH_2CO_2H} \underset{}{\overset{Ac_2O}{\rightleftharpoons}} \underset{\mathbf{2}}{PhN(NO)CH_2COOAc} \overset{H^+}{\rightleftharpoons} PhN(N{=}O)CH_2C(OAc){=}\overset{+}{O}H \xrightarrow{slow} \underset{\mathbf{3}}{\mathrm{PhN{-}N{=}\overset{+}{O}\ ring,\ C(OAc)(OH)}}$$

$$\mathbf{3} \xrightarrow[fast]{-AcOH,\ -H^+} \mathbf{4}$$

$$PhNHNH_2 + HCO_2H + CO_2 \xleftarrow{H^+\ aq.} \mathbf{4}$$

in the sydnone ring, presumably at position 4. The structure, and particularly the best mode of representation of the sydnone ring, has aroused controversy. 3-Alkylsydnones have a well-defined aromatic type of absorption spectrum, and the ring has a large dipole moment (ca. 7 D) towards the exocyclic oxygen atom. Molecular-orbital calculations have shown that the ring and the exocyclic oxygen atom bear fractional positive and negative charges, respectively. If one electron is in fact transferred from the ring to this oxygen atom, the ring then possesses the six π-electrons required for aromaticity. This would be consistent with its chemistry, but the dipole moment is far too low for such a complete transfer. The molecular parameters for 3-(4-bromophenyl)sydnone have been determined from an X-ray investigation.[26] The N–C and C–C bonds are a little longer than those of pyridine (1·34 Å, p. 230) and benzene (1·40 Å), respectively, and N–N bond is similar to that of pyridazine (1·33 Å, p. 393). These lengths are clearly indicative of an aromatic molecule, even though the C–O lengths are very similar to those found in ethers (1·416 Å) and ketones (1·23 Å). Altogether sydnones

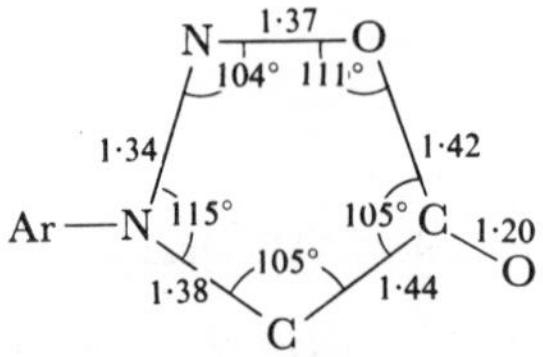

3-(4-Bromophenyl)sydnone, bond lengths in Angstroms

are therefore best considered as resonance hybrids, to which structure **5** and others, including **6**, **7**, and **8** contribute, and are best represented by **5**, or by such other contributing structure as is convenient. Sydnones, like oxazoles

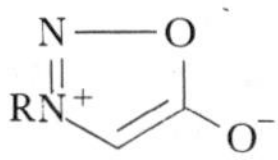

5 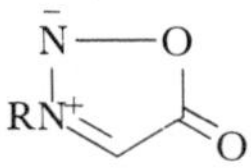 **6** 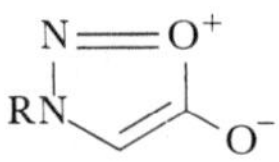 **7** 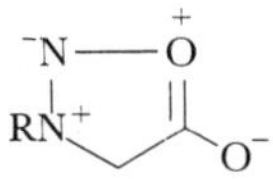 **8** 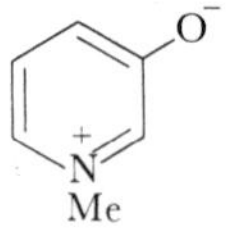 **9**

(p. 274) do, however, show some butadienelike character and undergo cycloaddition reactions of the Diels–Alder type, followed by retro-Diels–Alder reactions in the opposite sense, leading to pyrazoles.[27]

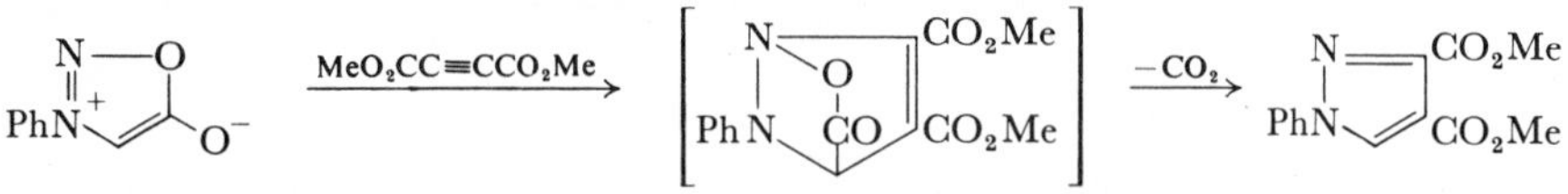

$MeO_2CC{\equiv}CCO_2Me$ $-CO_2$

The term 'mesoionic' has been put forward to describe the sydnones and other aromatic compounds which possess a degree of charge separation. It is difficult to exclude zwitterionic structures such as **9** from the definition, except in a purely arbitrary way, and because of this the new term does not appear to be essential. A new antimalarial agent (**10**) is a sydnone derivative.[28]

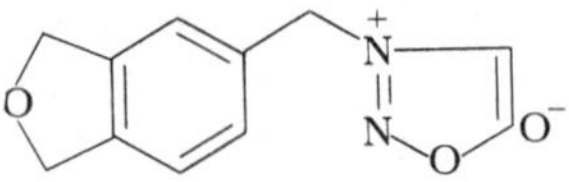

10

5. BENZOTRIAZOLES

Benzo[*d*]-1,2,3-triazole (**1**), usually known as benzotriazole, is a very stable material readily prepared by the diazotisation of 1,2-phenylenediamine. With boiling aqua regia, a mixture of concentrated nitric and hydrochloric acids, it yields the 4,5,6,7-tetrachloro derivative.[29] 1-Alkyl derivatives can be obtained

NH_2 NH_2 H^+, $NaNO_2$ aq. Zn, HCl 7 1 NH N_2 N 4 3 **1**

Me_2SO_4 ^-OH

NMe N N + N NMe N

2 **3**

by the same synthetic route using the appropriate primary–secondary diamine, and on alkylation of the parent compound (**1**) using dimethyl sulphate and sodium hydroxide a mixture of **2** and **3** in 2 : 1 ratio is obtained.[30] 1-Chlorobenzotriazole, readily obtained from **1** with sodium hypochlorite and acetic acid, is a valuable[31] oxidizing agent for the conversion of primary and secondary alcohols to aldehydes and ketones. 1-Aminobenzotriazole (**4**) is obtained, along with about 20% of the 2-isomer, from the parent heterocycle (**1**) with hydroxylamine-*O*-sulphonic acid. Oxidation with lead tetraacetate yields the nitrene (**5**), which fragments too quickly to be trapped, and further decomposition gives benzyne (**6**), which dimerises to diphenylene (**7**) in 60% yield.[32]

NNH_2 N N $Pb(OAc)_4$ [N—N: N N $-N_2$]

4 **5** **6**

7

The most widely used benzotriazole derivative is the 1-hydroxy compound (**8**), which is perhaps the most valuable substance available for activating carboxyl groups for peptide synthesis.[33] It is obtained by heating 2-chloronitrobenzene with hydrazine, an internal oxidation–reduction and cyclisation taking place.

NO_2 Cl $\xrightarrow[-HCl]{NH_2NH_2}$ [$\bar{O}$ O N + NH$_2$ N H] $\xrightarrow{-H_2O}$ NOH N=N

8

This hydroxybenzotriazole with a protected peptide and dicyclohexylcarbodimide gives the corresponding ester. Esters of this type (**9**) show a carbonyl absorption at 1815–1820 cm^{-1} in solution, but in most cases this disappears (reversibly) when the compound solidifies and is replaced by a maximum at ca. 1730 cm^{-1} also shown by *N*-acylbenzotriazoles. This suggests an equilibrium between structures **9** and **10**. Treatment of **9**–**10** with primary or secondary amines, including suitably protected amino acids or polypeptides, gives amides (**11**) very rapidly, at low temperatures, and in exceptionally high yield. The very powerful solvent dimethylformamide can be used for the reaction, racemisation is minimal, and the method is being very widely used for the synthesis of polypeptides.[34]

8 $\xrightarrow[C_6H_{11}N{=}C{=}NC_6H_{11}]{RCO_2H}$ $(C_6H_{11}NH)_2CO$ + NOCOR N=N ⇌ N O$^-$ + N–NCOR

9 **10**

$\downarrow R^1NH_2$

R'NHCOR + **8**

11

GENERAL BIBLIOGRAPHY

PURINES

J. H. Lister, 'Fused Pyrimidines, Pt. II, Purines,' *The Chemistry of Heterocyclic Compounds,* Vol. 24, ed. D. J. Brown, Wiley-Interscience, New York, 1971.

J. N. Davidson, *The Biochemistry of The Nucleic Acids,* 7th ed., Chapman and Hall, London, 1972.

G. E. W. Wolstenholme and C. M. O'Connor (eds.), *Chemistry and Biology of the Purines*, Ciba Foundation Symposium, Churchill, London, 1957.

B. Pullman and A. Pullman, 'Electronic Aspects of Purine Tautomerism,' *Adv. Heterocycl. Chem.*, **13**, 77 (1971).

J. H. Lister, 'Physicochemical Aspects of the Chemistry of Purines,' *Adv. Heterocycl. Chem.*, **6**, 1 (1966).

PTERIDINES

K. Iwai, M. Akino, M. Goto, and T. Iwanami, *Chemistry and Biology of the Pteridines*, Int. Academic Printing Co. Ltd., Tokyo, 1970.

G. R. Penzer and G. K. Radda, 'The Chemistry and Biological Function of Isoalloxazines (Flavines),' *Q. Rev.*, **21**, 43 (1967).

W. Pfleiderer, 'Recent Developments in the Chemistry of Pteridine,' *Angew. Chem. Int. Ed.*, **3**, 114 (1964).

R. C. Elderfield and A. C. Mehta, *The Pteridines*; J. P. Lambooy, 'The Alloxazines,' in *Heterocyclic Compounds*, ed. R. C. Elderfield, Vol. 9, Wiley, New York, 1967.

H. Pembold and W. L. Gyure, *Angew. Chem. Int. Ed.*, **11**, 1061 (1972).

TRIAZINES

J. E. Erickson, P. F. Wiley, and V. P. Wystrach, *The 1,2,3- and 1,2,4-Triazines, Tetrazines and Pentazines*, Interscience, New York, 1956.

R. L. Jones and J. R. Kershaw, 'Chemistry of as-Triazines,' *Rev. Pure Appl. Chem.*, **21**, 23 (1971).

E. M. Smolin and L. Rapoport, *s-Triazines and Derivatives*, Interscience, New York, 1959.

C. H. Grundmann, *Angew. Chem. Int. Ed.*, **2**, 309 (1963).

SYDNONES

W. Baker and W. D. Ollis, *Q. Rev.*, **11**, 15 (1957).

F. H. C. Stewart, *Chem. Rev.*, **64**, 129 (1964).

REFERENCES

1. N. Löfgren and B. Lüning, *Acta Chem. Scand.*, **7**, 225 (1953); G. B. Brown and V. S. Weliky, *J. Biol. Chem.*, **204**, 1019 (1953).
2. D. S. Letham, J. S. Shannon, and I. R. C. McDonald, *Tetrahedron*, **23**, 479 (1967).
3. D. G. Watson, R. M. Sweet, and R. E. Marsh, *Acta Crystallogr.*, **19**, 573 (1965).
4. W. Cochran, *Acta Crystallogr.*, **4**, 81 (1951).
5. A. F. Lewis, A. G. Beaman, and R. K. Robins, *Can. J. Chem.*, **41**, 1807 (1963).
6. H. Yamada and T. Okamoto, *Chem. Pharm. Bull. (Japan)*, **20**, 623 (1972).
7. M. Sekiya and J. Suzuki, *Chem. Pharm. Bull. (Japan)*, **20**, 209 (1972).
8. J. A. Montgomery, *J. Amer. Chem. Soc.*, **78**, 1928 (1956).
9. D. J. Brown and R. K. Lynn, *J.C.S. Perkin I*, 349 (1974).
10. J. H. Lister, *Rev. Pure Appl. Chem. (Australia)*, **11**, 178 (1961); E. C. Taylor and E. E. Garcia, *J. Amer. Chem. Soc.*, **86**, 4720 (1964).
11. J. Davoll, B. Lythgoe, and A. R. Todd, *J. Chem. Soc.*, 967 (1948).
12. J. Baddiley, *Adv. Enzymol.*, **16**, 1 (1956).
13. J. G. Moffatt and H. G. Khorana, *J. Amer. Chem. Soc.*, **83**, 663 (1961).
14. J. Baddiley, J. G. Buchanan, R. Letters, and A. R. Sanderson, *J. Chem. Soc.*, 1731 (1959).
15. H. Rembold and W. C. Gyure, *Angew. Chem.*, **11**, 1061, (1972).
16. A. Albert and W. L. F. Armarego, *Adv. Heterocycl. Chem.*, **4**, 1 (1965).

17. B. E. Evans, *J.C.S. Perkin I,* 357, (1974).
18. C. D. Shirrell and D. E. Williams, *J.C.S. Perkin II,* 41 (1975).
19. E. C. Taylor and T. Kobayashi, *J. Org. Chem.,* **38**, 2817, (1973).
20. G. R. Penzer and G. K. Radda, *Q. Rev.,* **21**, 43, (1967).
21. T. Paterson and H. C. S. Wood, *J.C.S. Perkin I,* 1057 (1972).
22. P. J. Wheatley, *Acta Crystallogr.,* **8**, 224 (1955).
23. A. Kreutzberger, *Fortschr. Chem. Forsch.,* **4**, 273 (1963).
24. H. Gysin, *Chem. Ind. (Lond.),* 1393 (1962).
25. Y. Ogata, A. Kawasaki and H. Kojoh, *J. Org. Chem.,* **39**, 3676 (1974).
26. H. Barnighausen, F. Jellinek, J. Murnik, and A. Vos., *Acta Crystallogr,* **16**, 471 (1963).
27. R. Huisgen, R. Grashey, H. Gotthardt, and R. Schmidt, *Angew. Chem.,* **74**, 29 (1962).
28. W. H. Nyberg and C. C. Cheng, *J. Med. Chem.,* **8**, 531 (1965).
29. D. E. Burton, A. J. Lambie, D. W. J. Lane, G. T. Newbold, and A. Percival, *J. Chem. Soc.,* (*C*), 1268 (1968).
30. P. J. Abbott, R. M. Acheson, M. W. Foxton N. R. Raulins, and G. E. Robinson, *J.C.S. Perkin I,* 1791 (1972).
31. C. W. Rees and R. C. Storr, *Chem. Comm.,* 1305 (1968).
32. C. D. Campbell and C. W. Rees, *J. Chem. Soc.,* (*C*), 752 (1969) and earlier papers.
33. W. König and R. Geiger, *Chem. Ber.,* **103**, 788 (1970).
34. G. Wendelberger, *Methoden der Organischen Chemie*, Synthesen von Peptiden II, 4th ed. (Houben-Weyl), ed. E. Wünsch, Georg Thieme, Stuttgart, 1974.

CHAPTER X

HETEROCYCLIC COMPOUNDS WITH SEVEN-MEMBERED AND LARGER RINGS

Interest in seven-membered ring heterocycles, with three double bonds in the ring, has sprung up in the last few years in connexion with studies of aromaticity, of certain rearrangements, and of the valuable therapeutic properties of some compounds of this group. Many seven-membered ring compounds possessing two or more of the same or different heteroatoms have been synthesised, but this large and very recent area of investigation is very briefly mentioned here. Compounds in which the ring is saturated, or partially saturated, have properties which are predictable from a knowledge of those of similar aliphatic compounds.

Six π-electrons are provided by the six-carbon atoms of the three formal double bonds, so that if no more electrons of this type are available from the heteroatom, as in borepine (**1**), the system might be expected to possess aromatic character. This type of structure would be similar to that of the relatively stable tropylium cation (**2**). Azepine (**3**), oxepine (**4**), and thiepin (**5**),

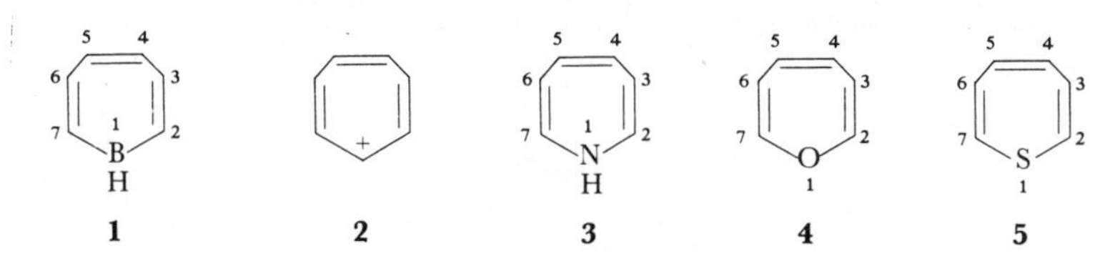

on the other hand, do not fit Hückel's $4n + 2$ rule and so would be expected to have high reactivity and little aromatic character, a situation that appears to pertain. Valency tautomerism has been demonstrated in both azepines and oxepines.

1. BOREPINE

Although borepanes (**1**, R = Cl, Et, or OMe) have been synthesized, the first attempts to prepare borepines have given noncyclic products.[1] However, dibenzo[*bf*]borepine (2) has been obtained as an unstable easily oxidized pyridine derivative. It has been little investigated.[2]

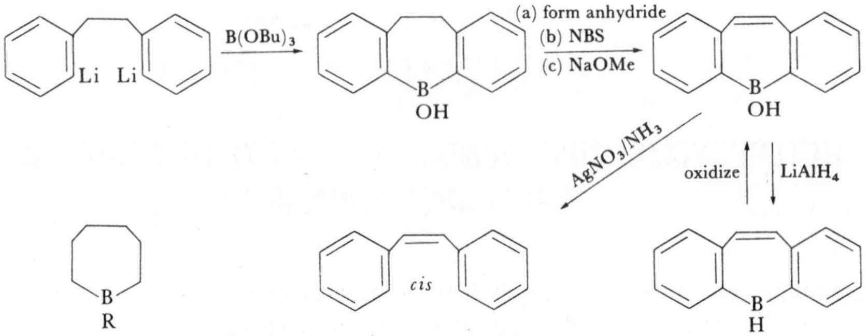

$B(OBu)_3$ (a) form anhydride (b) NBS (c) NaOMe

$AgNO_3/NH_3$ oxidize $LiAlH_4$

Li Li B OH B OH B R B H *cis*

1 NBS = *N*-bromosuccinimide 2

2. AZEPINE

A. Introduction and structure

Since the synthesis of 1-ethoxycarbonylazepine in 1963 by Hafner and König,[3] and by Lwowski et al.,[4] and the discovery of the tranquilising properties of certain diazepines (p. 450), a very great deal of research in this area has been carried out.

Azepines can theoretically exist in various tautomeric forms, a hydrogen atom being present on the nitrogen atom in 1*H*-azepine (**1**) or elsewhere in the ring. Azepine itself appears to be too reactive for isolation, but its 1-(4-bromo-

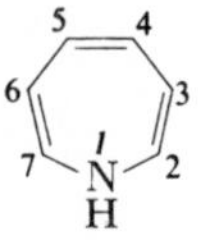

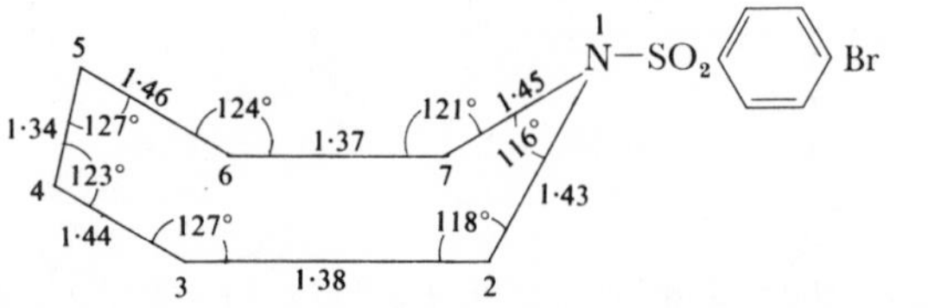

1-(4-Bromobenzenesulphonyl)azepine, bond lengths in Angstroms

benzenesulphonyl) derivative has been subject to an X-ray diffraction analysis.[5] The azepine ring is boat shaped, carbon atoms 2, 3, 6, and 7 are effectively coplanar (maximum deviation 0·02 Å), while the nitrogen atom and the other carbons are between 0·5 and 0·6 Å away from this plane. Although the 3,4 and 5,6 bonds are shorter than the pure C–C single bond (1·54 Å), the other C–C bonds are similar to that of ethylene (1·34 Å) and the C–N bond to that in trimethylamine (1·47 Å). Very little delocalisation of electrons is present and aliphatic-type properties are therefore expected.

B. Chemical Properties and Reactions

The most readily available, and most investigated, simple azepine is the 1-ethoxycarbonyl derivative (**2**), and attempts have been made[3] to prepare 1*H*-azepine from this compound. Careful alkaline hydrolysis followed by acidification is thought to give the 1*H*-compound (**3**), but like other enamines bearing no *N*-substituent it isomerises rapidly to the 3*H*-isomer (**4**); neither

H_2/Pd (a) HO^- (b) H^+

N CO_2Et N CO_2Et **2** N H **3** N **4**

compound has actually been isolated. Reduction of the ester with lithium aluminium hydride at 35°C gave the very unstable 1-methylazepine (**13**). In contrast to alkali, acid hydrolysis of **5** causes aromatisation to the urethane **8**,[3] and in a similar way the amide **6** gives the urea **9**.[6] Heating the 2-methyl

N COR **5**, R = OEt **6**, R = NH_2 N COR **7** NHCOR **8**, R = OEt **9**, R = NH_2

derivative of **5** to 200°C causes a similar rearrangement. The most likely course for these reactions is cyclisation of the azepines to the valency tautomers **7** followed by ring opening. There is as yet no evidence (contrast oxepines) for the existence of such tautomers as stable species in the azepine series, and even when four methylene groups bridge the 2,7-positions of **5** and force these positions together, physical measurements give no evidence for the presence of any of the aziridine tautomer (cf. **7**).

1-Ethoxycarbonylazepine (**5**), acting as a dieneophile across the 4,5-double bond, combines with isobenzofuran (prepared *in situ*) to give the Diels–Alder adduct **10**. It reacts in a similar way with 5,5-dimethoxy-1,2,3,4-tetrachlorocyclopentadiene.[7] Acting as a diene, **5** combines with the reactive dieneophiles tetracyanoethylene[3] and *N*-phenylmaleimide to give normal 1 : 1 molar adducts (e.g., **11**), although it does not react with dimethyl acetylenedicarboxylate at 100°C. At moderate temperatures the azepine **5** dimerises by a concerted

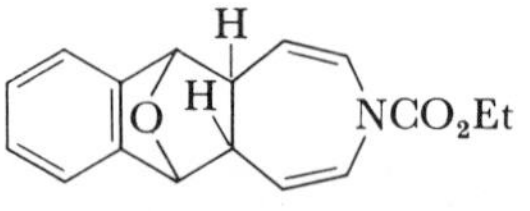

10 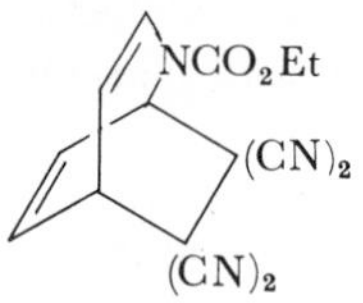 11 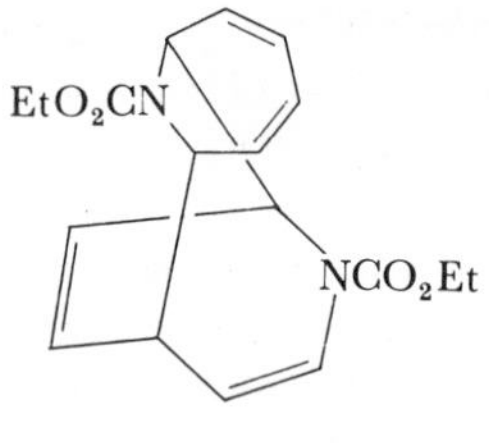 12

$6\pi + 4\pi$ process to give **12**.[8] This last compound on heating isomerises to a symmetrical adduct **15**, which can also be obtained[8] by a multistage process by heating 1-ethoxycarbonylazepine (**5**) to 200°C. 1-Methylazepine (**13**) dimerises, but at 0°C, in a similar way to give **14**, the structure of which was established by an X-ray investigation.[9] The photolysis[10] of **5** causes intramolecular cyclisation to **16**, a process reversed by heating to 100°C.

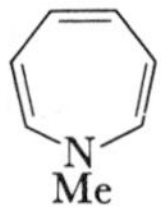

13 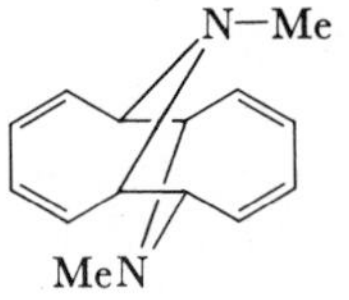 14 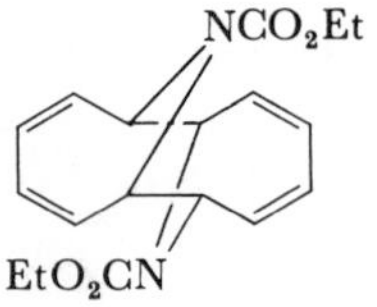 15 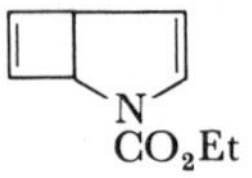 16

An interesting series of azepines has been obtained from the dihydropyridine (**17**), and some of their reactions are outlined below.[11] It is clear that the azepine and dihydropyridine systems must have the same order of stability, and

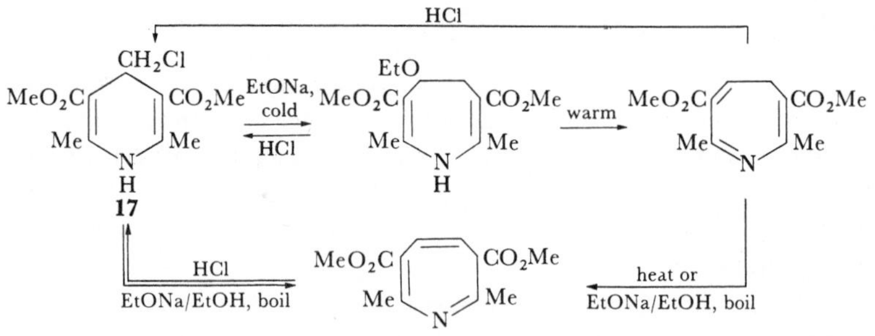

17 18

that of the azepines the 3*H*-tautomer (**18**) is the most stable. The 1-methyl derivative (**19**) of the dihydropyridine (**17**) is also of special interest as with potassium *t*-butoxide, at different concentrations, either the azepine (**21**) or its valency tautomer (**20**) can be obtained; **20** isomerizes in carbon tetrachloride solution (trace of HCl present?) to **21**.[12]

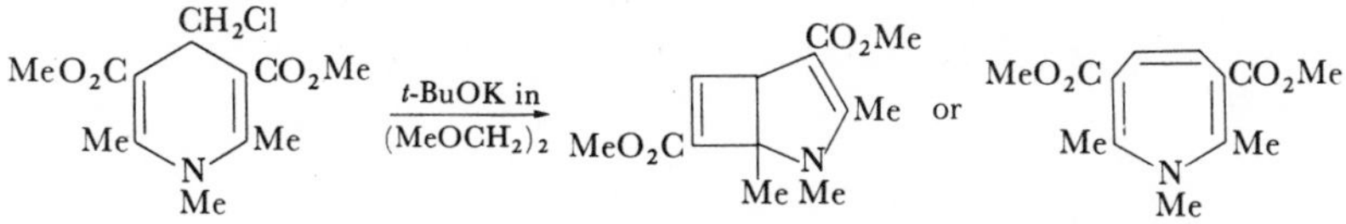

1,3-Dihydro-2*H*-azepin-2-ones (**23**) are very readily available[13] from the sodium salts of 2,6-dialkylphenols (e.g., **22**) with chloramine or methylchloramine at –70°C. Hydrogenation over platinum gives the expected cyclic amide (**26**). The NHCO group of the azepinone (**23**) is amidic in character, as the compound forms sodium salts which alkylate on the nitrogen atom, is hydrolysed by acid eventually leading to the lactone (**27**), is reduced by lithium aluminium hydride to the corresponding amine (**25**), and with triethyloxonium borofluoride gives the corresponding imino ether, the 2-ethoxyazepine (**24**).[14] The double-bond system of the azepinone (**23**) possesses considerable butadienelike

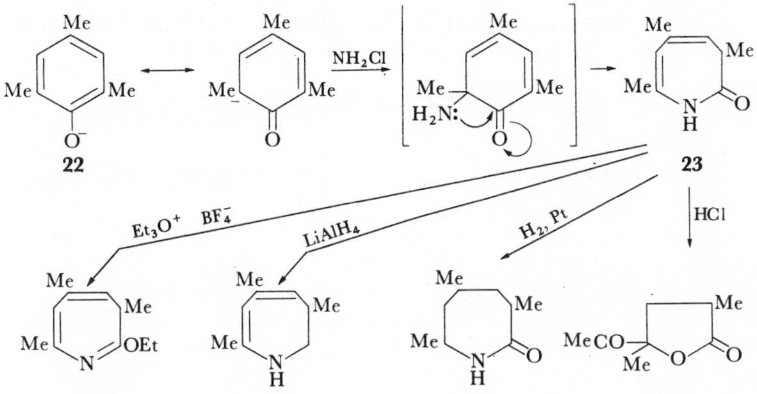

character and adds to dimethyl acetylenedicarboxylate yielding **28**. However, it does not react with the less vigorous *N*-phenylmaleimide, which combines with the amine **25** in the normal way. Photolysis causes isomerization of **23** to **29**.[13] One azepinedione (**30**) has marked activity against Crocker sarcoma.[15]

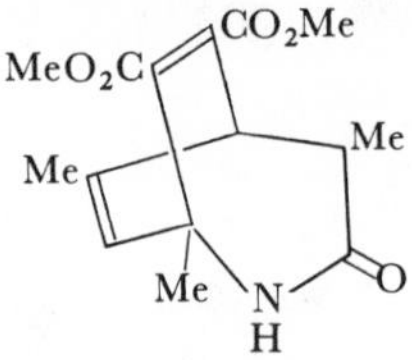

MeO$_2$C CO$_2$Me Me Me Me N H O

28

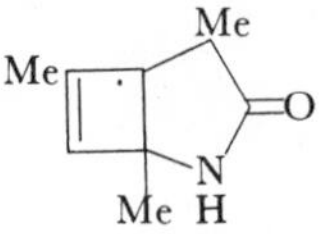

Me Me O N Me H

29

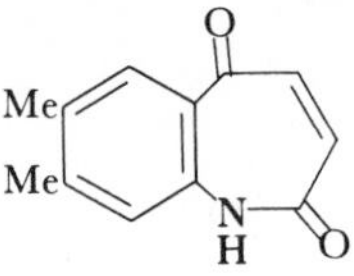

O Me Me N H O

30

C. Synthesis of Azepines

(1) The most direct method is the combination of a nitrene, formed by pyrolysis[16] or photolysis[3,4] of an azido compound, in the presence of a benzene derivative. It has been established[17] that only the singlet nitrene reacts with the benzene. Mixtures result if the benzene is substituted.

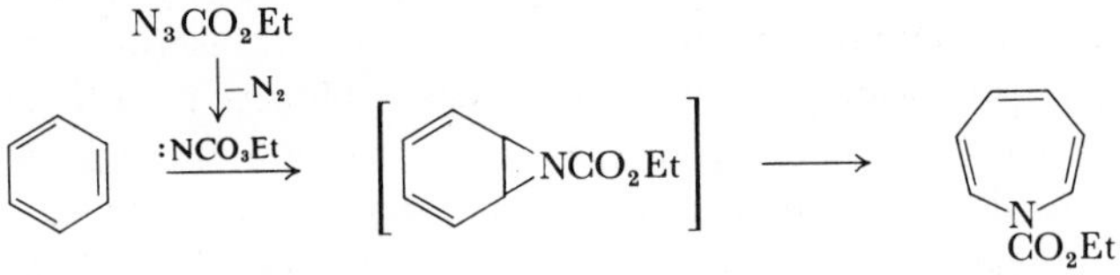

N_3CO_2Et $-N_2$ $:NCO_2Et$ → [NCO_2Et] → N CO_2Et

A related synthesis is the reduction[18,19] of an aromatic nitro compound to a nitrene in the presence of diethylamine, when a 2-diethylamino-3*H*-azepine is formed.

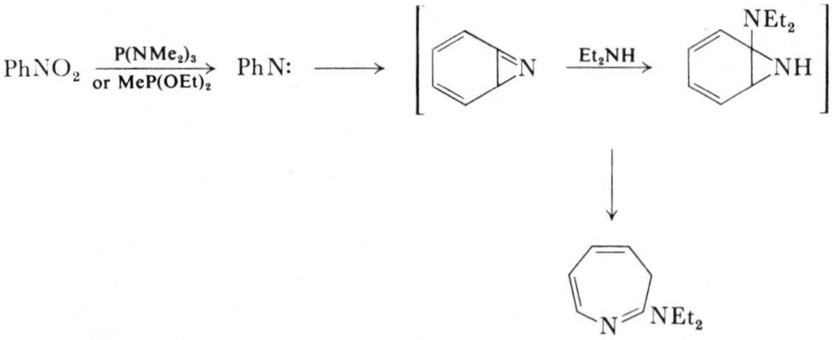

$PhNO_2$ $\xrightarrow[\text{or MeP(OEt)}_2]{\text{P(NMe}_2)_3}$ PhN: → [N $\xrightarrow{Et_2NH}$ NEt_2 NH] → N NEt_2

(2) A much more specific approach, suitable for *N*- and *C*-substituted azepines, starts with a 1,4-cyclohexadiene.[20] The aziridine **31** can be hydrolysed and decarboxylated by aqueous potassium hydroxide to **33**, a new substituent is then placed on the nitrogen atom, and the product is converted to an azepine of general structure **32**.

(a) INCO (b) MeOH

I
NHCO$_2$Me
NaOMe
NCO$_2$Me
31
Br_2
Br
Br
NCO$_2$Me
NaOMe, THF
N
CO$_2$Me
NH
33
(a) Substitution
(b) bromination
(c) dehydrobromination
N
R
32

(3) 4,5-Dimethoxycarbonylazepines can be synthesised[21] from pyrroles possessing a strongly electron-attracting substituent at position 1. A Diels–Alder addition with dimethyl acetylenedicarboxylate yields the adduct **34**, which on photolysis gives the 3-azaquadricyclane **35**. Thermal decomposition of **35** cannot give back **34** by a concerted process, but in fact leads to the azepine **36**.

N
R
$MeO_2CC{\equiv}CCO_2Me$
$AlCl_3$
CO$_2$Me
CO$_2$Me
N
R
34
$h\nu$
CO$_2$Me
CO$_2$Me
N
R
35
Δ
MeO$_2$C
CO$_2$Me
N
R
36

R = CO$_2$Me, COMe, SO$_2$Aryl

(4) The photolysis of anthranils yields 3*H*-azepines.[22]

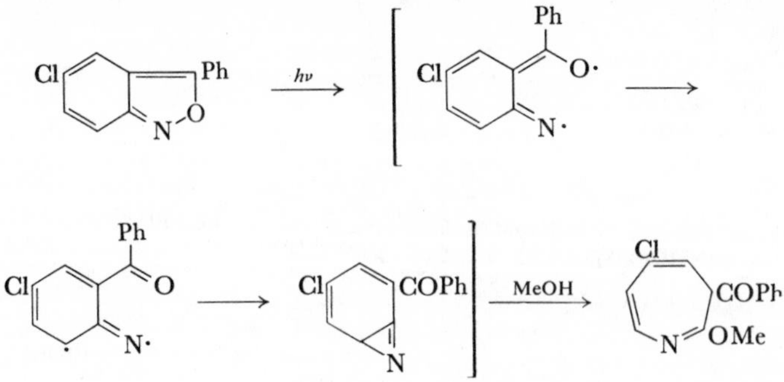

(5) Azirines can undergo Diels–Alder reactions leading to azepines.[23]

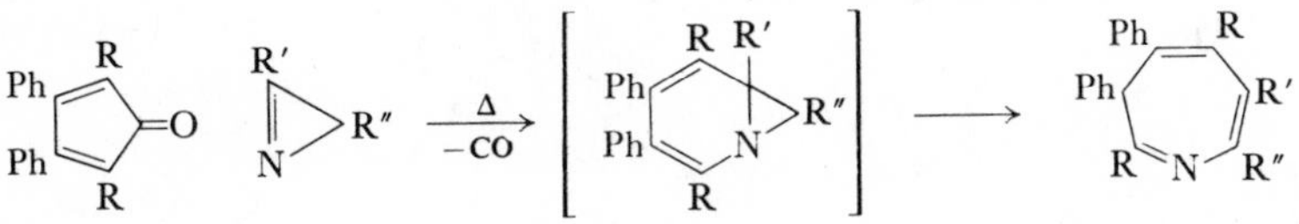

(6) Suitable dinitriles can be cyclised to azepines in high yield.[24]

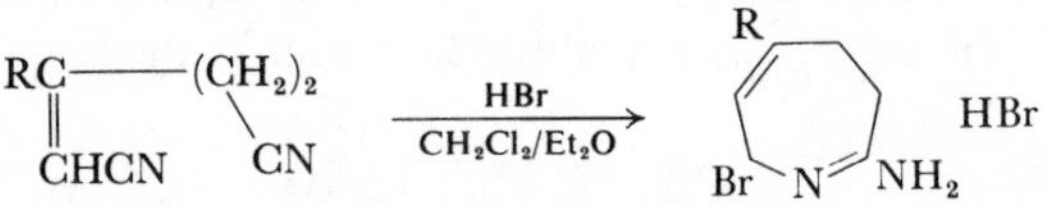

(7) 1-Methylindole, and various derivatives, with dimethyl acetylenedicarboxylate in pure acetonitrile yield[25] benzo[*b*]azepines. The indole behaves as an enamine and suffers electrophilic attack at position 3. This is followed by cyclisation to the cyclobutene (**37**) and ring opening.

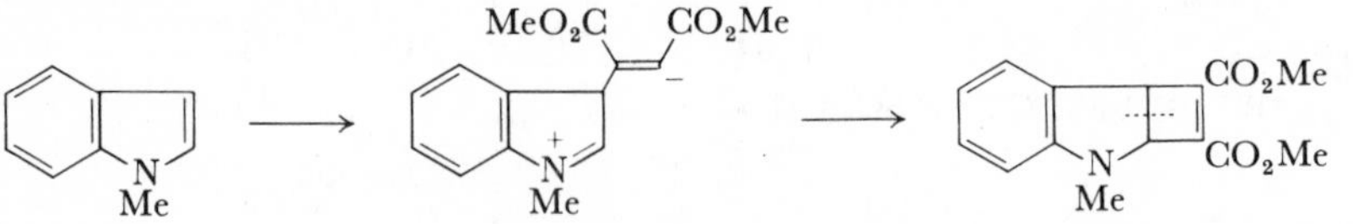

37

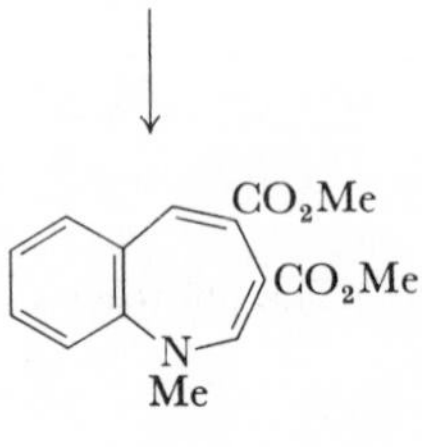

3. DIAZEPINES

A very great deal of work has been done in the last few years on various diazepines, some of which undergo fascinating rearrangements. Others possess valuable pharmacological properties. Examples of the rearrangements are

$$PhCH{=}CMeCO_2H \xrightarrow[\text{(b) } CH_2N_2]{\text{(a) } SOCl_2} PhCH{=}CMeCOCHN_2 \xrightarrow{CH_2N_2} \mathbf{1}$$

1 → 4 → (H^+, cold) → 3 → (H^+, or HO^-) → 2; 1 ⇸ 5 (Ph, Me, $COCH_2OAc$)

provided by 6,7-dihydro-5-methyl-4-phenyldiazepin-6-one (**2**), which was obtained in 80% yield, instead of the expected acetoxy ketone (**5**), from **1** by

2: $h\nu$; RCOCl + pyridine (→ COR); base + H_2O_2 (→ CH_2CO_2H); base (→ OH, NH_2 + H_2N, OH); strong acid; $NaBH_4$ (→ Me, OH, NH); NH_2OH (→ N–OH, O); $NH_2NHCONH_2$ → Normal derivative; Ac_2O (→ NH, N Ac); H^+ R = H; R = Ac (→ CH_2NHNHR); → NH_2

treatment with warm acetic acid. The intermediate compounds, **4** and **3**, are isolable when the correct conditions are employed. Some of the rearrangements and reactions of this diazepine are shown in the flow sheet (p. 449).[26] Mechanisms accounting for all these transformations have been put forward.

The valuable tranquilizing agent, Librium, is the 1,4-diazepine 4-oxide (**7**) and has been synthesized as outlined. It also undergoes a number of rearrangements. The mould product cyclopenin (**6**) is a 1,3-diazepine derivative and has been synthesised.[27]

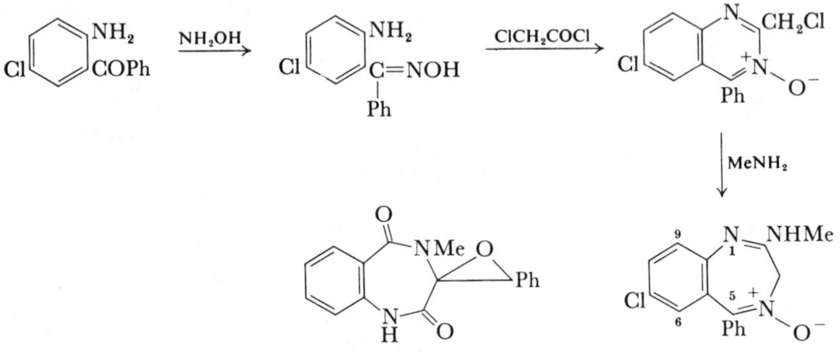

NH_2OH $ClCH_2COCl$ $MeNH_2$

6 **7**

4. OXEPINE

The synthesis of oxepine was first achieved in 1964 by a method which has also been used for the preparation of some derivatives. Oxepine is an orange liquid, b.p. 38°C at 30 mm. The nuclear magnetic resonance spectrum of the liquid is temperature dependent and is interpreted to show the presence of comparable

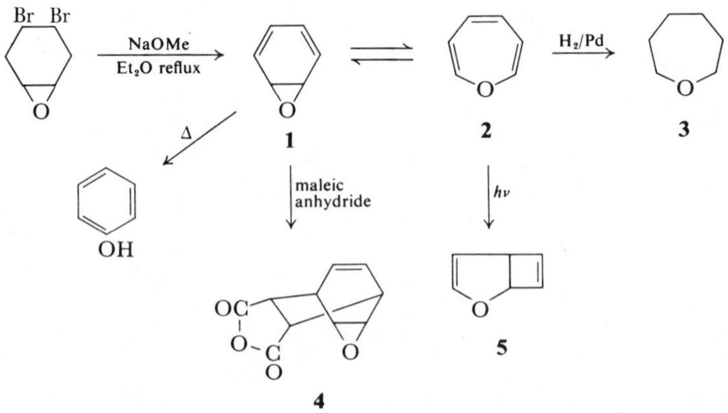

NaOMe, Et_2O reflux; Δ; H_2/Pd; maleic anhydride; $h\nu$

1 **2** **3** **4** **5**

amounts of the oxirane (**1**) and the oxepine (**2**).[28] 1*H*-Azepines do not undergo a comparable easy disrotatory cyclisation to the corresponding aziridines; the reasons for this difference are not known. Hydrogenation of the oxepine gives the known cyclic ether (**3**), the seven-membered ring structure thereby being established, while heating to 70°C isomerizes oxepine to phenol. Oxepine undergoes the Diels–Alder reaction with maleic anhydride yielding (**4**), and all cycloaddition reactions appear to take place with the arene–oxide tautomer (**1**). Photolysis causes quantitative cyclisation to **5**.[29] The nuclear magnetic resonance spectrum of the 3-chloro-6-oxooxepine **6** gives no indication of the presence of any enol, and the compound is very sensitive to both acids and bases.[30] There is therefore no doubt that there is little, if any, aromatic character present in simple oxepines.

AcO — $:CCl_2$ → AcO — Cl_2 — (a) hydrolysis + NaOMe (b) MnO_2 → O — Cl_2 — heat → O — Cl

6

Oxepines can also be synthesised[31] from furans via oxaquadricyclanes (**7**) by a route similar to that used for azepines (p. 447).

$MeO_2CC{\equiv}CCO_2Me$ → CO_2Me, CO_2Me 75% — $h\nu$ → CO_2Me, CO_2Me 60% **7** — 113 °C → MeO_2C CO_2Me 75%

Arene oxides (cf. **1**) are intermediates in the hydroxylation of many aromatic compounds in plants and animals, but as yet there is no evidence which indicates the corresponding oxepines (cf. **2**) are of biochemical significance.[32]

2,3-Dihydrooxepine (**9**) can be obtained[33] from 4*H*-pyran (**8**) by ring expansion, but attempts to carry out similar experiments with pyrylium salts in

CH_2Cl_2 / BuLi; *endo* and *exo*; *exo* only / hot quinoline; H_2/Pt; $h\nu$; H^+; $(CH_2)_4$ CHO CHO; CHO

8 9 10 11 12 13

the hope of obtaining oxepine itself have failed. This dihydrooxepine undergoes a photocyclization to **13**.[34] 4,5-Dihydrooxepine (**10**)[35] behaves as a vinyl ether towards acids. The product is cyclopent-1-enecarbaldehyde (**12**), which is presumably formed via the dialdehyde (**11**).

Benzo[*b*]oxepine (**14**), obtained as shown,[36,37] is a yellow-green oil of b.p. 50°C at 0·5 mm. Its structure has been established[36] by hydrogenation to the known homochroman **16** and is consistent with its nuclear magnetic resonance spectrum,[36] which also suggests that the heterocyclic ring is not planar. It cannot cyclise to an arene oxide (cf. **1**) without negating the aromaticity of the benzene ring, and so this compound on irradiation in the presence of methylene blue and oxygen, a source of singlet oxygen, gives the adduct **15**[38]; under comparable conditions oxepine adds the oxygen across the double bond system

Br Br Br Br; KOH / EtOH; 20%; 50%; chromatography on silica gel; $^1O_2$ 14 H_2 catalyst

14 15 16 17

of structure **1**. Compound **17**, the structure of which seems established, is clearly an oxepine of an unusual type; the nitrogen analogue is also known.[39]

Benzo[*d*]oxepine (**18**) has been obtained[40] from the bis-Wittig reagent derived from **19**, and is much less stable to acid than the diester obtained from

20, with diazomethane.[41] This behaviour is parallel to that of furan and its 2,5-dimethoxycarbonyl derivative. The structure of the diester of **20** is consistent with its nuclear magnetic resonance spectrum.[42]

6 5 7 4 8 3 9 2 O 1 **18** ← $O(CH_2\overset{+}{P}Ph_3)_2$ $2Br^-$ **19** OHC CHO $\xrightarrow[\text{NaOMe}]{O(CH_2CO_2Me)_2}$ HO_2C O CO_2H **20**

5. THIEPIN

Few compounds definitely possessing the thiepin ring are known. Recent[43] attempts to prepare thiepin (**2**) by the action of bases on 1 and 3 have given only benzene and sulphur. However, the 1,1-dioxide (**4**) is a stable compound, m.p.

Br Br S **3** base S Cl Cl **1** base [S ?] → C_6H_6 + S ← [S] 5 4 6 3 7 2 S 1 **2**

117–118°C, which decomposes slowly on melting to benzene and sulphur dioxide.[44] An X-ray structure determination[45] has shown that the molecule exists in the boat form, and n.m.r. studies show that the same shape persists in solution.[46] The C–C bond lengths are very similar to the corresponding ones of cycloheptatriene and show no evidence for delocalisation of the aromatic type, although the lengths of the C–S bonds suggest that these possess some double-bond character.

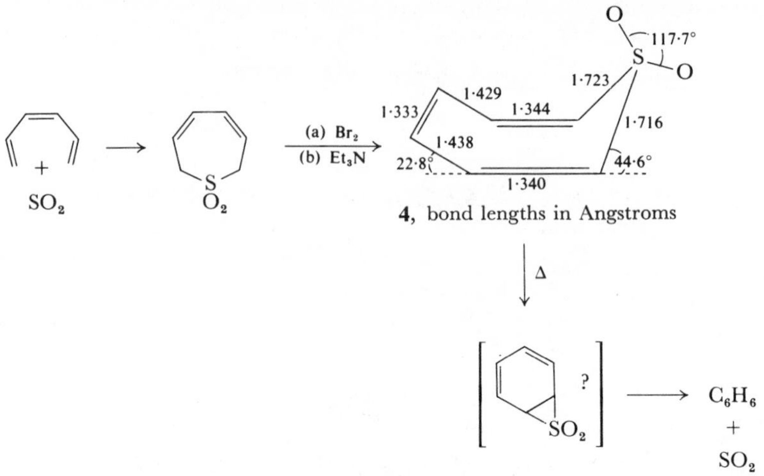

$+ SO_2 \longrightarrow$ (S, O_2) $\xrightarrow{(a)\ Br_2,\ (b)\ Et_3N}$ 4

1·723, 117·7°, 1·429, 1·344, 1·333, 1·716, 1·438, 22·8°, 44·6°, 1·340

4, bond lengths in Angstroms

$\downarrow \Delta$

[SO_2 ?] $\longrightarrow C_6H_6 + SO_2$

The first monocyclic thiepin to be detected is **7**, obtained[47] in a remarkable reaction from the thiophene **5**. This combines, like an enamine, with the acetylenic ester to give the bicyclic compound **6**, slowly yielding the thiepin **7**, which decomposes (via **10**?) to sulphur and the benzene derivative **9**. All these compounds have been detected from their nuclear magnetic resonance spectra. The benzo[*b*]-derivative of **7**, obtained by the same route, is stable at room temperature but undergoes a corresponding loss of sulphur at 100°C in dioxane. The very sterically hindered thiepin **8** has been prepared.[48] It is a yellow solid, stable in hot chloroform or dimethylformamide but which is desulphurised by triphenylphosphine to the corresponding benzene derivative.

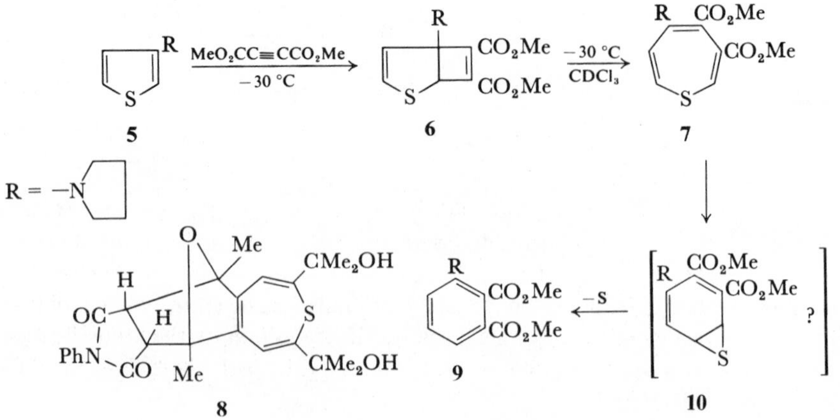

5 $\xrightarrow[-30\ ^\circ C]{MeO_2CC{\equiv}CCO_2Me}$ **6** $\xrightarrow[CDCl_3]{-30\ ^\circ C}$ **7** $\longrightarrow$ **10** $\xrightarrow{-S}$ **9**

R = —N(pyrrolidinyl)

8 (O, Me, H, OC, PhN, CO, Me, S, CMe_2OH, CMe_2OH)

9 (R, CO_2Me, CO_2Me)

10 (R, CO_2Me, CO_2Me, S, ?)

Benzo[*b*]thiepin (**11**) has been obtained[49] as an unstable yellow liquid, m.p. 15–20°C, yielding sulphur and naphthalene at room temperature. With 3-chloroperbenzoic acid it gives the 1,1-dioxide (**12**), which is stable at 200°C.

NCS; DMSO, *t*-BuOK; [O]

NCS = *N*-Chlorosuccinimide

11 **12**

The dioxide nitrates at position 8, and bromine adds to the heterocyclic ring only under the influence of light. Nucleophilic addition, which occurs easily with benzo[*b*]thiophene 1,1-dioxide (p. 221), does not take place.

The benzo[*d*]thiepin (**13**) is easily obtained[50] and readily loses sulphur to form naphthalene-2,3-dicarboxylic acid (**14**); this desulphurization may be compared with a similar one in the benzo[*c*]thiophene series (p. 224). The acid with a limited amount of diazomethane gives the corresponding diester, the nuclear magnetic resonance of which gives no evidence for the presence of a valency tautomer. Excess diazomethane adds onto two of the double bonds.

CHO, CHO; $S(CH_2CO_2Me)_2$, NaOMe; CO_2H, CO_2H; EtOH, boil; CO_2H, CO_2H

13 **14**

Benzo[*d*]thiepin 3,3-dioxide (**16**) has been obtained by a long route from **15**, involving cyclization of the acid chloride, oxidation to the 3,3-dioxide, and several more stages.[51] It is easily reduced by Raney nickel and hydrogen to the 1,2,4,5-tetrahydro derivative and on pyrolysis gives a good yield of naphthalene.

$(CH_2)_2SCH_2CO_2H$; SO_2; heat; SO_2 +

15 **16**

Simple thiepins show a great tendency to lose sulphur, presumably via the arene sulphide (cf. **10**) unless there is a great deal of steric hindrance (e.g., **8**) or additional conjugation present. This indicates high reactivity and suggests that little aromatic stabilisation is present. The *S*-dioxides also possess little aromatic character, but they are very much more stable; the reasons for this have yet to be ascertained.

6. LARGE-RING HETEROCYCLES

Many heterocycles possessing rings with more than seven atoms, including one or more oxygen, nitrogen, sulphur or other heteroatoms, are known. Some are natural products and most contain saturated, or partially saturated, rings. Very many possibilities for isomerism exist, and transannular reactions or interactions can be important. In the conjugated unsaturated series, investigations are only just beginning and great developments may be expected. One antibiotic, for example, can give a dioxocin ring (p. 38). The following brief discussion is restricted to a few eight- and nine-membered ring systems. The endings -ocine and -onine are used for the unsaturated eight- and nine-membered rings, respectively.

A. Azocines

Although cyclooctatetraene has been known for many years, the first aza derivative was obtained[52] in 1968 as a yellow oil, b.p. about 82°C at 14 torr, from 1,4-cyclohexadiene. Addition of chlorosulphonyl isocyanate gave the β-lactam **1**. Alkali then removed the 1-substituent, and alkylation with trimethyloxonium tetrafluoroborate gave the very base-sensitive 1-azetine **3**. Monobromination with *N*-bromosuccinimide and dehydrobromination yielded a mixture of benzonitrile (**4**) and 2-methoxyazocine (**6**). Presumably **5** is an intermediate, and through loss of the 2a proton and the methoxyl group as indicated could give the nitrile **4**.

ClSO₂NCO, 70 °C, PhH → **1** (NSO₂Cl) → 4M NaOH → **2** (NH) → Me_3O^+ BF_4^- → **3** (N, OMe)

(a) NBS, CCl_4 reflux
(b) *t*-BuOK, THF

4 (CN) ← −MeOH ← **5** (N, OMe, H) → **6** (N, OMe)

6 → H_2/Pd → **9**

7 (OC, PhN, CO, N, OMe)

8 (NH, O) → Me_3O^+ BF_4^- → **9** (N, OMe)

The monocyclic structure of 2-methoxyazocine was shown by the reduction to the imino ether **9**, which was synthesised by the alternative route from the amide **8**. Treatment of **6** with potassium *t*-butoxide in tetrahydrofuran gave a large amount of benzonitrile (**4**), thereby suggesting that **6** is in equilibrium with its valency tautomer **5**. Support for this can be adduced from the reaction of **6** with *N*-phenylmaleimide, as the adduct **7** is formed. The ultraviolet and infrared spectra of **6** are very similar to those of cyclooctatetraene, as is expected for a straightforward replacement of a CH group by N. The nuclear magnetic resonance spectrum for the azocine, which is unchanged down to –75°C, is consistent only with the presence of the monocyclic structure. Bicyclic valency tautomers, for which structure **5** is the most likely but not the only possibility, could have been present to the maximum extent of 5% and remained undetected.

No X-ray crystal structure studies have yet been done on azocines, but it is very likely that the ring system will adopt the boat form shown (**6**), like that of cyclooctatetraene.

Derivatives of **6** with methylene bridges across the 3,8-positions have been prepared.[52] With six methylene groups (**10**, $n = 6$) the compound possessed spectral properties almost identical to those of **6**. However, the compound with five methylene groups showed a temperature dependent nuclear magnetic resonance spectrum which indicated a reversible equilibrium between **10** ($n = 5$) and **11** ($n = 5$), the steric restriction clearly increasing the tendency towards the bicyclic structure. With three and four methylene groups in the bridge (**10**, $n = 3$ or 4) only the bicyclic forms (**11**) could be detected and the n.m.r. spectra were temperature independent.

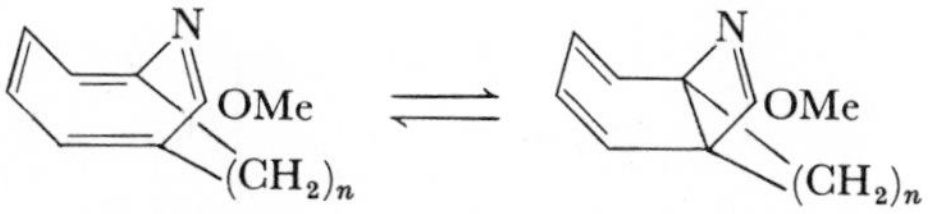

N OMe $(CH_2)_n$ ⇌ N OMe $(CH_2)_n$

10 **11**

In contrast to **6**, the n.m.r. spectrum of its hydrolysis product, the amide (**12**–**13**), showed that the bicyclic tautomer (**13**) was predominant and[53] that the proportion of **12** in tetrachloroethylene solution increased from 2·4 to 15·3% as the temperature was raised from 60 to 115°C.

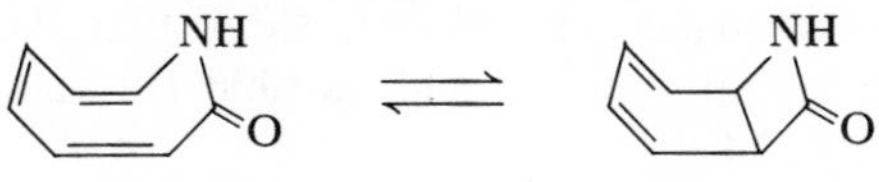

NH O ⇌ NH O

12 **13**

2-Methoxyazocine is converted to the dianion (**14**) by metallic potassium in liquid ammonia, and the n.m.r. spectrum of this salt suggested that some delocalisation of the 10 π-electron system takes place.[54] Quenching the dianion with a proton source gave a mixture of the 3,4- 3,6- (**15**), and 7,8-dihydro derivatives.

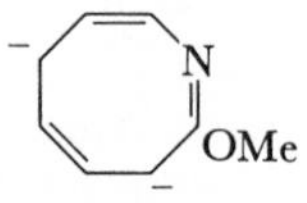

OMe OMe

14 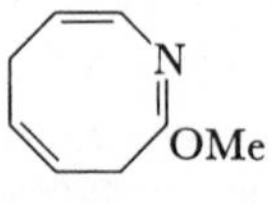 **15**

Two further syntheses leading to azocine derivatives have been discovered. The diester **17** with methyl 5,6-diphenyl-1,2,4-triazine-3-carboxylate (**16**) undergoes reaction leading to **18**. This is in equilibrium with the dihydroazocine **20**, which undergoes a retro Diels–Alder reaction giving the azocine **19** and dimethyl phthalate.[55]

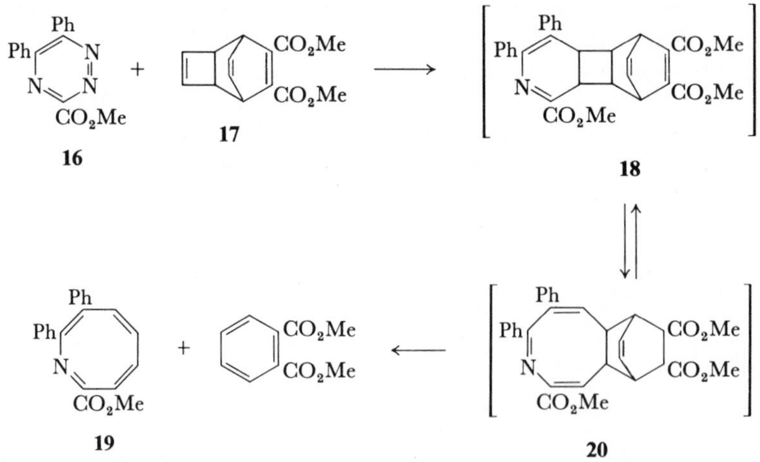

Ph Ph N N N CO_2Me **16** + CO_2Me CO_2Me **17** → [Ph Ph N CO_2Me CO_2Me CO_2Me] **18**

Ph Ph N CO_2Me **19** + CO_2Me CO_2Me ← [Ph Ph N CO_2Me CO_2Me CO_2Me] **20**

In the second synthesis[56] the 1,2-dihydropyridine **21** reacts, as an enamine, by a nonconcerted route, to give the bicyclo compound **22**, which has been detected by its nuclear magnetic resonance spectrum. The central bond opens, presumably by a concerted disrotatory process, to give the dihydroazocine **23**. Nuclear magnetic resonance experiments show that the ring is flexible but becomes rigid at about −40°C, and no bicyclic tautomer is detectable. Photolysis of **23** gives **24**.

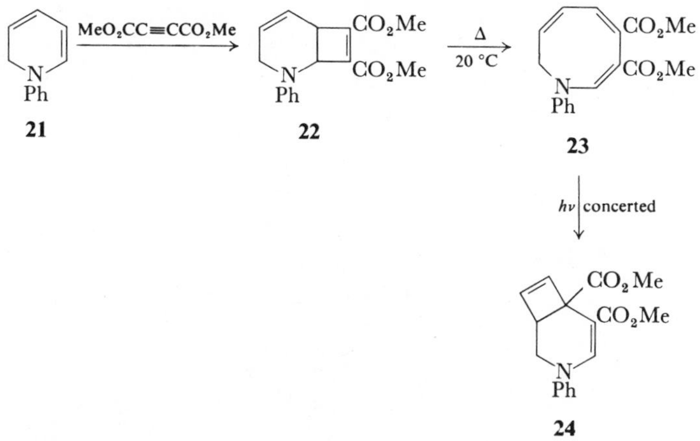

The first synthesis of 1,2-diazacyclooctatetraene (**25**) has been reported.[57] The compound is stable in solution below room temperature, but on warming it decomposes rapidly to tars. On photolysis it gives benzene quantitatively.

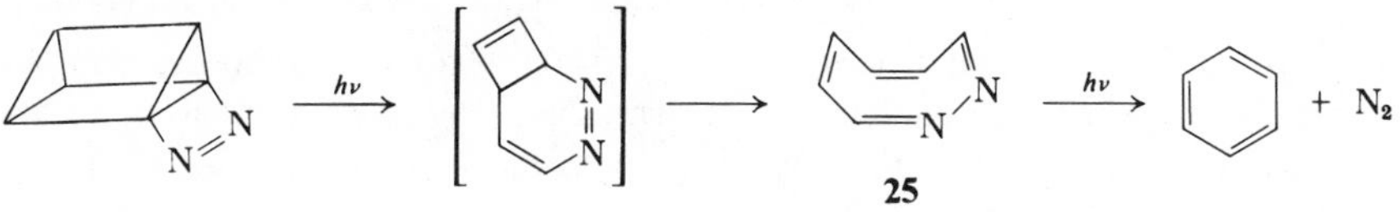

B. Azonines

The azonine system (**1**) has attracted attention recently,[58] and if the lone pair of electrons of the heteroatom is included the ring could contain 10 π-electrons, one requirement for aromaticity.

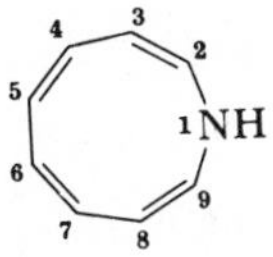

Cyclooctatetraene with ethoxycarbonylnitrene, generated from **2** with triethylamine, gives[59] the aziridine **3**, which on photolysis undergoes a disrotatory ring opening to form the azonine **4**. This compound does not show significant aromatic properties, and on heating undergoes cyclisation to **5**.

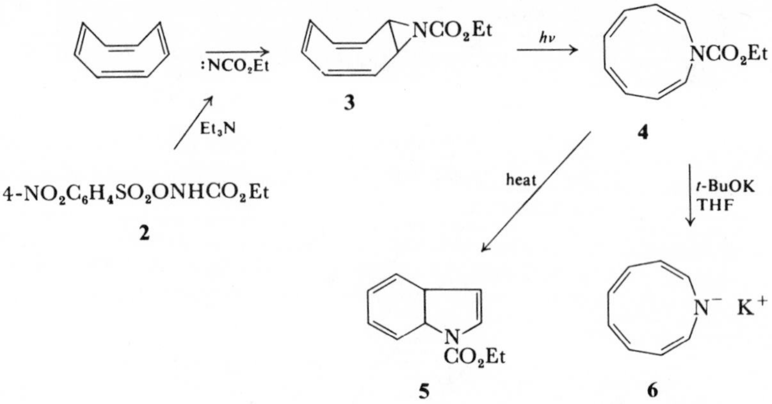

$4\text{-}NO_2C_6H_4SO_2ONHCO_2Et$

2

The azonine with potassium *t*-butoxide in tetrahydrofuran at −20°C is converted[60] to the stable potassium derivative **6**, which possesses some of the spectral properties expected of a cyclic aromatic 10 π-electron system. Protonation appears to give 1*H*-azonine (**1**).[61] The potassium derivative on alkylation with methyl iodide yields the *N*-methyl derivative, which does not appear to be aromatic,[62] and with dimethylaminocarbonyl chloride, **7** is formed. An X-ray crystal structure determination[63] for this compound (**7**) has given the molecular parameters shown. The 2-, 3-, 6-, and 7-carbon atoms are effectively coplanar, while the other carbon atoms, and the nitrogen atom, are between 0·83 and 1·35 Å away from this plane. The bonds leading from the ring nitrogen atom are not coplanar. There is no doubt both from the shape and bond lengths that in this compound there can be no significant electron delocalisation of the aromatic type, as is the case with the seven-membered ring

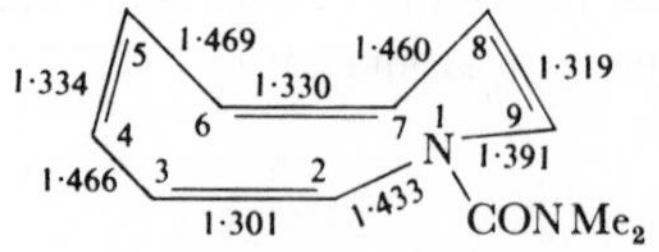

7, bond lengths in Angstroms

of the azepine system. All the double bonds of **7** have the *cis* configuration, but there is some evidence that 1-acetyl-*cis-cis-trans-cis*-azonine can be a reaction intermediate.[64]

GENERAL BIBLIOGRAPHY

A. Rosowsky (ed.), 'Seven-Membered Heterocyclic Compounds Containing Oxygen and Sulphur,' in *Chemistry of Heterocyclic Compounds,* Vol. 26, Wiley-Interscience, New York, 1972.

L. A. Paquette, *Azepines, Oxepines and Thiepins,* in *Non Benzenoid Aromatics,* Vol. I, ed. J. P. Snyder, (1969).

G. A. Archer and L. H. Sternback, 'Benzodiazepines,' *Chem. Rev.,* **68**, 747 (1968).

F. D. Popp and A. C. Noble, 'Chemistry of Diazepines,' *Adv. Heterocyclic Chem.,* **8**, 21 (1967).

J. A. Moore and E. Mitchell, 'Seven-membered Rings containing one or more Nitrogen Atoms and Related Compounds,' in *Heterocyclic Compounds,* Vol. 9, ed. R. C. Elderfield, Wiley, New York, 1967.

REFERENCES

1. G. Brieger, *Diss. Abstr.,* **22**, 1824 (1964).
2. E. van Tamelen, G. Brieger, and K. G. Untch, *Tetrahedron Lett.,* **8**, 14 (1960).
3. K. Hafner and C. König, *Angew. Chem.,* **75**, 89 (1963); K. Hafner, *Angew. Chem. Int. Ed.,* **3**, 165 (1964).
4. W. Lwowski, T. J. Mericich, and T. W. Mattingly, *J. Amer. Chem. Soc.,* **85**, 1200 (1963).
5. I. C. Paul, S. M. Johnson, L. A. Paquette, J. H. Barrett, and R. J. Haluska, *J. Amer. Chem. Soc.,* **90**, 5023 (1968).
6. F. D. Marsh and H. E. Simmons, *J. Amer. Chem. Soc.,* **87**, 3529 (1965).
7. J. R. Wiseman and B. P. Chong, *Tetrahedron Lett.*, 1619 (1969).
8. L. A. Paquette, J. H. Barrett, and D. E. Kuhla, *J. Amer. Chem. Soc.,* **91**, 3616 (1969).
9. G. Habermehl and S. Göttlicher, *Angew. Chem. Int. Ed.,* **6**, 805 (1967).
10. L. A. Paquette and J. H. Barrett, *J. Amer. Chem. Soc.,* **88**, 1718 (1966).
11. M. Anderson and A. W. Johnson, *J. Chem. Soc.*, 2411 (1965).
12. R. F. Childs and A. W. Johnson, *Chem. Commun.*, 95 (1965).
13. L. A. Paquette, *J. Amer. Chem. Soc.,* **85**, 3288 (1963).
14. L. A. Paquette, *J. Amer. Chem. Soc.,* **86**, 4096 (1964), and earlier papers.
15. D. M. James and A. H. Rees, *J. Med. Pharm. Chem.,* **5**, 1234 (1962).
16. R. J. Cotler and W. F. Beach, *J. Org. Chem.,* **29**, 751 (1964); L. E. Chapman and R. F. Robbins, *Chem. Ind.* (*Lond.*), 1266 (1966).
17. W. Lwowski and R. L. Johnson, *Tetrahedron Lett.*, 891 (1967).
18. J. I. G. Cadogan and M. J. Todd, *J. Chem. Soc.* (*C*), 2809 (1969).
19. F. R. Atherton and R. W. Lambert, *J.C.S. Perkin I*, 1079 (1973).
20. L. A. Paquette, D. E. Kuhla, J. H. Barrett, and R. J. Haluska, *J. Org. Chem.,* **34**, 2866 (1969).
21. R. C. Bansal, A. W. McCulloch, and A. G. McInnes, *Can. J. Chem.,* **47**, 2391 (1969); H. Prinzbach, R. Fuchs, R. Kitzing, and H. Achenbach, *Angew. Chem. Int. Ed.,* **7**, 727 (1968), and earlier papers.
22. M. Ogata, H. Kano, and H. Matsumoto, *Chem. Comm.*, 397 (1968).
23. A. Hassner and D. J. Anderson, *J. Org. Chem.,* **39**, 3070 (1974).
24. W. A. Nasutavicus, S. W. Tobey, and F. Johnson, *J. Org. Chem.,* **32**, 3325 (1967); F. Johnson and R. Madroñero, *Adv. Heterocycl. Chem.,* **6**, 95 (1966).
25. R. M. Acheson, J. N. Bridson, and T. S. Cameron, *J.C.S. Perkin I*, 968 (1972), and papers therein quoted.
26. S. M. Rosen and J. A. Moore, *J. Org. Chem.,* **37**, 3770 (1972).
27. H. Smith, P. Wegfahrt, and H. Rapoport, *J. Amer. Chem. Soc.,* **90**, 1668 (1968).
28. E. Vogel, W. A. Böll, and H. Günther, *Tetrahedron Lett.*, 609 (1965).
29. J. M. Holovka and P. D. Gardner, *J. Amer. Chem. Soc.,* **89**, 6390 (1967).

30. S. Masamune and N. T. Castelluci, *Chem. Ind.* (*Lond.*), 184 (1965).
31. H. Prinzbach, P. Vogel, and W. Auge, *Chimia,* **21**, 469 (1967); H. Prinzbach, M. Arguelles, and E. Drickrey, *Angew. Chem.,* **78**, 1057 (1966).
32. J. W. Daly, J. M. Jerina, and B. Witkop, *Experientia,* **28**, 1129 (1972).
33. E. E. Schweizer and W. E. Parham, *J. Amer. Chem. Soc.,* **82**, 4085 (1960).
34. L. A. Paquette, J. H. Barrett, R. P. Spitz, and R. Pitcher, *J. Amer. Chem. Soc.,* **87**, 3417 (1965).
35. R. A. Braun, *J. Org. Chem.,* **28**, 1383 (1963).
36. F. Sondheimer and A. Shani, *J. Amer. Chem. Soc.,* **86**, 3168 (1964).
37. E. Vogel, M. Biskup, W. Pretzer, and W. A. Böll, *Angew. Chem. Int. Ed.,* **3**, 642 (1964).
38. J. E. Baldwin and O. W. Lever, *Chem. Comm.*, 344 (1973).
39. E. Vogel, W. Pretzer, and W. A. Böll, *Tetrahedron Lett.*, 3613 (1965).
40. K. Dimroth and G. Pohl, *Angew. Chem.,* **73**, 436 (1961).
41. K. Dimroth and H. Freyschlag, *Chem. Ber.,* **90**, 1623 (1957).
42. M. J. Jorgenson, *J. Org. Chem.,* **27**, 3224 (1962).
43. T. J. Barton, M. D. Martz, and R. G. Zika, *J. Org. Chem.,* **37**, 552 (1972).
44. W. L. Mock, *J. Amer. Chem. Soc.,* **89**, 1281 (1967).
45. H. L. Ammon, P. H. Watts, and D. M. Stewart, *Acta Crystallogr.*, Sect. B., **26**, 1079 (1970).
46. M. P. Williamson and M. C. Mock, *J. Magn. Resonance,* **2**, 50 (1970).
47. D. N. Reinhoudt and C. G. Kouwenhofen, *Chem. Comm.*, 1232, 1233 (1972).
48. J. M. Hoffman and R. H. Schlessinger, *J. Amer. Chem. Soc.,* **92**, 5263 (1970).
49. V. J. Traynelis, Y. Yoshikama, J. C. Sih, and L. J. Miller, *J. Org. Chem.,* **38**, 3978 (1973).
50. K. Dimroth and G. Lenke, *Chem. Ber.,* **89**, 2608 (1956).
51. W. E. Truce and F. J. Lotspeich, *J. Amer. Chem. Soc.,* **78**, 848 (1956).
52. L. A. Paquette and T. Kakihara, *J. Amer. Chem. Soc.,* **90**, 3897 (1968); L. A. Paquette, T. Kakihara, J. F. Hansen, and J. C. Phillips, *J. Amer. Chem. Soc.,* **93**, 152 (1971).
53. L. A. Paquette, T. Kakihara, J. F. Kelly and J. R. Malpass, *Tetrahedron Lett.*, 1455 (1969).
54. L. A. Paquette, J. F. Hansen, and T. Kakihara, *J. Amer. Chem. Soc.,* **93**, 168 (1971).
55. J. A. Elix, W. S. Wilson, and R. N. Warrener, *Tetrahedron Lett.*, 1837 (1970).
56. R. M. Acheson, G. Paglietti, and P. A. Tasker, *J.C.S. Perkin I*, 2496 (1974).
57. B. M. Trost and R. M. Corey, *J. Amer. Chem. Soc.,* **93**, 5572, 5573 (1971).
58. A. G. Anastassiou, *Acc. Chem. Res.,* **5**, 281 (1972).
59. S. Masamune and N. T. Castellucci, *Angew. Chem. Int. Ed.,* **3**, 582 (1964).
60. R. T. Seidner and S. Masamune, *Chem. Comm.*, 149 (1972).
61. A. G. Anastassiou and J. H. Gebrian, *Tetrahedron Lett.*, 825 (1970).
62. A. G. Anastassiou and H. Yamato, *Chem. Comm.*, 286 (1972).
63. C. C. Chiang, I. C. Paul, A. G. Anastassiou, and S. W. Eachus, *J. Amer. Chem. Soc.,* **96**, 1636 (1974).
64. A. G. Anastassiou, R. L. Elliott, H. W. Wright, and J. Clardy, *J. Org. Chem.,* **38**, 1959 (1973).

COMPOUND INDEX

This index is subdivided into compound and subject indexes. Compounds are listed under the parent heterocyclic system, where possible. Page numbers in boldface type refer to major discussions of the subject or compound, and page numbers in *italics* refer to synthesis.

SUBJECT INDEX